Lignocellulose Bioconversion Through White Biotechnology

Lignocellulose Bioconversion Through White Biotechnology

Edited by

Anuj Kumar Chandel
University of São Paulo, Lorena
Brazil

WILEY

This edition first published 2023

Registered Offices
John Wiley & Sons, Inc., 111 River Street, Hoboken, NJ 07030, USA
John Wiley & Sons Ltd, The Atrium, Southern Gate, Chichester, West Sussex, PO19 8SQ, UK

Editorial Office
The Atrium, Southern Gate, Chichester, West Sussex, PO19 8SQ, UK

For details of our global editorial offices, customer services, and more information about Wiley products visit us at www.wiley.com.

Wiley also publishes its books in a variety of electronic formats and by print-on-demand. Some content that appears in standard print versions of this book may not be available in other formats.

Library of Congress Cataloging-in-Publication Data
Names: Chandel, Anuj Kumar, editor.
Title: Lignocellulose bioconversion through white biotechnology / edited by Anuj K. Chandel.
Description: Hoboken, NJ : Wiley, 2023. | Includes index.
Identifiers: LCCN 2022029465 (print) | LCCN 2022029466 (ebook) | ISBN 9781119735953 (cloth) | ISBN 9781119735960 (adobe pdf) | ISBN 9781119735977 (epub)
Subjects: LCSH: Lignocellulose–Biotechnology. | Renewable natural resources.
Classification: LCC TP248.65.L54 L5349 2022 (print) | LCC TP248.65.L54 (ebook) | DDC 333.79/4–dc23/eng/20220723
LC record available at https://lccn.loc.gov/2022029465
LC ebook record available at https://lccn.loc.gov/2022029466

Cover Design: Wiley
Cover Image: © BanksPhotos/Getty Images

Set in 9.5/12.5pt STIXTwoText by Straive, Pondicherry, India

Contents

List of Contributors

João Ricardo M. Almeida
Laboratory of Genetics and Biotechnology
Embrapa Agroenergia
Brasília
Brazil

Graduate Program of Microbial Biology
Department of Cell Biology
Institute of Biology
University of Brasília
Brasília
Brazil

Ediane S. Alves
Chemical Engineering Graduate Program
Universidade Federal de São Carlos
(UFSCar)
São Carlos, SP
Brazil

A Alzamendi
Consejo Nacional de Investigaciones
Científicas y Técnicas (CONICET)
Buenos Aires
Argentina

Instituto Multidisciplinario de Biología
Celular (IMBICE CONICET-CICPBA-
UNLP), La Plata, Argentina

Jesús J. Ascencio
Department of Biotechnology
Engineering School of Lorena (EEL)
University of São Paulo (USP)
Lorena, São Paulo
Brazil

Patrícia F. Ávila
Bioprocess and Metabolic Engineering
Laboratory
School of Food Engineering,
University of Campinas (UNICAMP)
Campinas, SP
Brazil

Henrique M. Baudel
America Biomass Technologies
Piracicaba
Brazil

Department of Chemical Engineering
Federal University of Pernambuco
Recife
Brazil

Bhima Bhukya
Centre for Microbial and Fermentation
Technology
Department of Microbiology
University College of Science
Osmania University
Hyderabad, Telangana
India

Mohit Bibra
Prairie AquaTech LLC
South Dakota
Brookings, SD
USA

David N. Bryant
Institute of Biological Environmental and Rural Sciences (IBERS)
Aberystwyth University
Aberystwyth, UK

Clara Vida G. C. Carneiro
Laboratory of Genetics and Biotechnology
Embrapa Agroenergia
Brasília, Brazil

Graduate Program of Microbial Biology
Department of Cell Biology
Institute of Biology
University of Brasília
Brasília, Brazil

Anuj Kumar Chandel
Department of Biotechnology
Engineering School of Lorena (EEL)
University of São Paulo (USP)
Lorena, São Paulo
Brazil

Ana F. M. Costa
Centre Nacional de la Recherche Scientifique (CNRS)
Laboratory of Chemical Engineering (LGC) and Laboratory of Fundamental and Applied Heterochemistry (LHFA)
Université Paul Sabatier Toulouse
Toulouse, France

BS Dhanya
Department of Biotechnology
Udaya School of Engineering
Kanyakumari, Tamil Nadu
India

Eduardo Diebold
America Biomass Technologies
Piracicaba
Brazil

Kelly J. Dussán
Department of Engineering, Physics and Mathematics
Institute of Chemistry, São Paulo State University (UNESP)
Araraquara, SP
Brazil

Bioenergy Research Institute (IPBEN)
São Paulo State University (UNESP)
Araraquara, SP
Brazil

Center for Monitoring and Research of the Quality of Fuels, Biofuels, Crude Oil, and Derivatives
Institute of Chemistry (CEMPEQC)
São Paulo State University (UNESP)
Araraquara, SP
Brazil

Andrew M. Elias
Chemical Engineering Graduate Program
Universidade Federal de São Carlos (UFSCar)
São Carlos, SP
Brazil

EBRAPA Instrumentation
Brazilian Agricultural Research Corporation—EMBRAPA
São Carlos, SP
Brazil

Marcos Fellipe da Silva
Bioprocess and Metabolic Engineering Laboratory
School of Food Engineering
University of Campinas (UNICAMP)
Campinas, SP
Brazil

Paula F. Franco
Laboratory of Genetics and Biotechnology
Embrapa Agroenergia
Brasília, Brazil

Felipe F. Furlan
Chemical Engineering Graduate Program
Universidade Federal de São Carlos (UFSCar)
São Carlos, SP
Brazil

Department of Chemical Engineering
Universidade Federal de São Carlos (UFSCar)
São Carlos, SP
Brazil

Joe Gallagher
Institute of Biological Environmental and Rural Sciences (IBERS)
Aberystwyth University
Aberystwyth, UK

J Gamboa-Santos
Consejo Nacional de Investigaciones Científicas y Técnicas (CONICET)
Buenos Aires
Argentina

Centro de Investigación y Desarrollo en Criotecnología de Alimentos (CIDCA-CONICET UNLP)
La Plata, Argentina

Ellen C. Giese
Service of Extractive Metallurgy and Bioprocesses
Centre for Mineral Technology (CETEM)
Rio de Janeiro, RJ
Brazil

Roberto C. Giordano
Chemical Engineering Graduate Program
Universidade Federal de São Carlos (UFSCar)
São Carlos, SP
Brazil

Department of Chemical Engineering
Universidade Federal de São Carlos (UFSCar)
São Carlos, SP
Brazil

Rosana Goldbeck
Bioprocess and Metabolic Engineering Laboratory
School of Food Engineering
University of Campinas (UNICAMP)
Campinas, SP
Brazil

Departamento de Engenharia de Alimentos – DEA
Laboratório de Engenharia Metabólica e Bioprocessos – LEMEB
Universidade Estadual de Campinas – UNICAMP
Cidade Universitária Zeferino Vaz
Campinas-SP
Brazil

Luana C. Grangeiro
Department of Engineering, Physics and Mathematics
Institute of Chemistry, São Paulo State University (UNESP)
Araraquara, SP
Brazil

V. Guadalupe Bustos
Department of Biotechnology
University Autonomous of Tamaulipas
Unidad Académica Multidisciplinaria Mante
Ciudad Mante, Tamaulipas
Mexico

Rishi Gupta
Anton Paar India Pvt Ltd
Gurgaon
Haryana
India

Ruly T. Hilares
Laboratório de Materials
Universidad Católica de Santa Maria—UCSM
Yanahuara, Arequipa
Perú

Phil J. Hobbs
Anaerobic Analytics
Okehampton, UK

Samuel Jacob
Department of Biotechnology
School of Bioengineering, College of Engineering and Technology
Faculty of Engineering and Technology
SRM Institute of Science and Technology
Tamil Nadu, India

Sridevi Jagavati
Department of Microbiology
Indira Priyadarshini Government Degree College for Women
Hyderabad
Telangana, India

Praveen K. Keshav
Centre for Microbial and Fermentation Technology
Department of Microbiology
University College of Science
Osmania University
Hyderabad, Telangana
India

Pratham Khaitan
Department of Biotechnology
School of Bioengineering
College of Engineering and Technology
Faculty of Engineering and Technology
SRM Institute of Science and Technology
Chengalpattu, Tamil Nadu
India

Deepak Kumar
Department of Chemical Engineering
SUNY College of Environmental Science and Forestry
Syracuse, NY
USA

Andreza A. Longati
Department of Materials and Bioprocess Engineering
School of Chemical Engineering
University of Campinas
Campinas, SP
Brazil

Letícia M. Mallmann Ferreira
Laboratory of Genetics and Biotechnology
Embrapa Agroenergia
Brasília, Brazil

Graduate Program of Microbial Biology
Department of Cell Biology
Institute of Biology
University of Brasília
Brasília, Brazil

C. Linda M. Martínez
Department of Biotechnology
University Autonomous of Tamaulipas
Unidad Académica Multidisciplinaria
Mante, Ciudad Mante, Tamaulipas
Mexico

Department of Food Technology,
University Autonomous of Tamaulipas
Unidad Académica Multidisciplinaria
Reynosa-Aztlán,
Reynosa, Tamaulipas, México

Manoela Martins
Bioprocess and Metabolic Engineering Laboratory
School of Food Engineering, University of Campinas (UNICAMP)
Campinas, SP
Brazil

Danielle Matias Rodrigues
America Biomass Technologies
Piracicaba,
Brazil

Post-graduation Program, Bioenergy
University of Campinas (UNICAMP)
Campinas, Brazil

Allan Henrique Felix de Melo
Bioprocess and Metabolic Engineering Laboratory
School of Food Engineering
University of Campinas (UNICAMP)
Campinas, SP
Brazil

Alain E. M. Mera
Department of Biotechnology
Engineering School of Lorena (EEL)
University of São Paulo
Lorena/SP
Brazil

Katarina R. Mihajlovski
Faculty of Technology and Metallurgy
Belgrade
Serbia

Marija D. Milić
Faculty of Technology and Metallurgy
Belgrade, Serbia

Everson A. Miranda
Department of Materials and Bioprocess Engineering
School of Chemical Engineering
University of Campinas
Campinas, SP
Brazil

Alberto M. Moura Lopes
Bioprocess and Metabolic Engineering Laboratory
School of Food Engineering
University of Campinas (UNICAMP)
Campinas, SP
Brazil

Simone C. Myoshi
Chemical Engineering Graduate Program
Universidade Federal de São Carlos (UFSCar)
São Carlos, SP
Brazil

Lucas Ramos
Laboratório de Materials
Universidad Católica de Santa Maria—UCSM
Yanahuara, Arequipa
Perú

Sreenivas R. Ravella
Institute of Biological Environmental and Rural Sciences (IBERS)
Aberystwyth University
Aberystwyth, UK

Eduardo A. Ribeiro
Instituto Federal de Santa Catarina
IFSC, Criciúma/SC
Brazil

Nayanika Sarkar
Department of Biotechnology
School of Bioengineering
College of Engineering and Technology
Faculty of Engineering and Technology
SRM Institute of Science and Technology
Tamil Nadu, India

Department of Biotechnology
Bhupat and Jyoti Mehta School of Biosciences
Indian Institute of Technology Madras
Chennai, Tamil Nadu
India

Luana Assis Serra
Laboratory of Genetics and Biotechnology
Embrapa Agroenergia
Brasília, Brazil

Graduate Program of Microbial Biology
Department of Cell Biology
Institute of Biology
University of Brasília
Brasília, Brazil

Débora D. V. Silva
Department of Biochemistry and Organic Chemistry
Institute of Chemistry
São Paulo State University (UNESP)
Araraquara, SP
Brazil

Marcos H. L. Silveira
Chemnnova Mhl Silveira
Consultoria—Inovacao
Para Novos Negocios LTDA
Centro Santa Rosa de Lima/SC
Brazil

Akhilesh K. Singh
Department of Biotechnology
School of Life Sciences
Mahatma Gandhi Central University
Motihari, Bihar
India

Lakhveer Singh
Department of Environmental Science
SRM University-AP
Amaravati, Andhra Pradesh
India

Yeruva Thirupathaiah
Central Sericultural Research and Training Institute
Mysuru, Karnataka
India

S. Rodolfo Torres
Department of Biotechnology
University Autonomous of Tamaulipas
Unidad Académica Multidisciplinaria
Mante, Ciudad Mante
Tamaulipas, Mexico

Débora Trichez
Laboratory of Genetics and Biotechnology
Embrapa Agroenergia
Brasília, Brazil

R. Daniel Trujillo
Department of Biotechnology
University Autonomous of Tamaulipas
Unidad Académica
Multidisciplinaria Mante
Ciudad Mante
Tamaulipas, Mexico

Nilesh Kumar Sharma
Department of Chemical Engineering
Worcester Polytechnic Institute
Worcester, MA
USA

Adhinarayan Vamsidhar
Department of Biotechnology
School of Bioengineering
College of Engineering and Technology
Faculty of Engineering and Technology
SRM Institute of Science and Technology
Tamil Nadu, India

David J. Warren-Walker
Institute of Biological Environmental and Rural Sciences (IBERS)
Aberystwyth University
Aberystwyth, UK

Ana Winters
Institute of Biological Environmental and Rural Sciences (IBERS)
Aberystwyth University
Aberystwyth, UK

Preface

White biotechnology is a multidisciplinary branch of industrial biotechnology dealing with the sustainable production of renewable fuels, chemicals, food and fibers, enzymes, and value-added products at the industrial level. Bio (chem) transformation of lignocellulosic biomass (LCB) delivers renewable fuels and low carbon bulk and specialty chemicals as well as value-added products such as lignocellulosic-derived oligosaccharides, pre-biotics, additives, etc. These green products are the integral component of white biotechnology. White biotechnology can cater to the growing demand for renewable fuels, chemicals, and materials in a sustainable manner harnessing the potential of LCB leveraging the carbon-neutral circular economy. Essentially, green fuels and chemicals produced from LCB have the potential to reduce the dependence on petroleum-derived products/fuels.

During the recent meeting of COP (Conference of the Parties) 26, which is also known as the 2021 United Nations Climate Change Conference, held in Glasgow, Scotland, between 31 October and 12 November 2021, some important measurements such as the regulation of carbon market mechanisms, zero deforestation and a 30% net reduction in methane emissions were gauged. The primary objective of COP 26 is compliance with the goals established in the Paris Agreement to keep the planet's temperature below 2 °C throughout this century. This goal can be accomplished through the development of low carbon chemicals and renewable fuels from LCB eventually strengthening the renewable economy.

LCB conversion into bioproducts usually has four steps: (i) unlocking the biomass integrity through a robust pretreatment, (ii) depolymerization of complex carbohydrates into free fermentable sugars, (iii) fermentation or chemical conversion of a mix of sugars (pentose and hexoses) into fuels/chemicals, and (iv) product recovery. Process integration employing industrially viable parameters seems an interesting approach for the economic transformation of LCB in chemicals and fuels. Moreover, metabolic engineering tools can play a decisive role in developing superior strains and enzymes which can make the bioprocess economic and environmentally friendly. In a similar vein, many bulk and specialty chemicals manufacturers have come forward to unleash the potential of LCB. This book presents a global overview on the critical aspects of white biotechnology, technical challenges of biomass conversion and biorefinery, the wide range of bio-based products, industrialization aspects, and sustainability.

Efforts have been made to compile the latest and landmark information on white biotechnology in this book. There are 16 specialized book chapters written by the experts of the research area. **Chandel et al.** in **Chapter 1** present a critical review on white biotechnology and its role in developing the sustainable bioeconomy. The chapter summarizes the fundamental aspects of white biotechnology, major routes of biomass conversion, the broad range of plant-biomass-derived products, and a critical overview on sustainability. **Chapter 2** by **Vázquez et al.** describes the importance of three major lignocellulosic residues, that is sugarcane bagasse, and residues of the Tequilana agave, in high-valued products. Biomass processing routes and their economic impacts are also discussed. The circular economy approach aims to utilize resources throughout their shelf-life to recover and regenerate the materials at the end of each life cycle product. The biorefinery is a strategic platform to develop the economically viable bioconversion processes to utilize each fraction of LCB with lesser generation of carbon footprints. In **Chapter 3**, **Gamboa-Santos and Alzamendi** explore the available pretreatment methods and the new developments made in bioconversion strategy focusing on the pretreatment stage.

Pretreatment is the first but a decisive step in lignocellulosic biorefinery. It is the gateway of the success of biorefining technologies. While a number of pretreatment methods are available, steam explosion is recognized as one of the most widely applied physicochemical pretreatment methods. It specifically acts on the hemicellulose fractionation and thus ameliorates the digestibility of LCB into renewable sugars with a mild environmental impact with high energy efficiency. **Chapter 4** by **Bhukya and Keshav** evaluates the steam explosion pretreatment to solubilize the hemicellulose, in turn enhancing the enzymatic digestibility of LCB into second-generation sugars.

For second-generation sugar production, an effective and complete cellulosic enzyme cocktail is necessary for the complete hydrolysis of LCB. Renewable sugar production with desired production cost has an important role in the overall development of LCB refinery. Commercial enzyme suppliers, for example Novozymes, AB Enzymes, DSM, Genencor-DuPont, have developed a highly efficient cellulose-digesting enzyme cocktail for cellulosic sugar production from a variety of lignocellulosic residues. In this context, **Mihajlovski and Milić** in **Chapter 5** describe the role of a complete and effective cellulase cocktail in the overall development of biorefinery.

The production of green bulk chemicals like aldehydes, diols, organic acids, polyols, and others in addition to fuels from LCB is the cornerstone of successful deployment of lignocellulosic refining technologies. **Chapter 6** by **Serra et al.** critically appraises the microbial production of biobased chemicals and the improvements and challenges in synthetic biology to develop designer microbial strains. Beside the production scenario of fuels and chemicals, this chapter presents an overview on the applications, markets, and challenges for the production of each of these biobased chemicals.

Lignocellulose hydrolysates usually have a variety of plant cell wall derived inhibitors in the form of furans, weak acids, lignin-derived phenolics, among others. Therefore, the microorganism must grow in the hydrolysate even in the presence of these inhibitors. Also, the microorganism should have the ability to assimilate different sugars present in the lignocellulose hydrolysate and deliver the products with higher yields and productivities. Different molecular biological approaches have been developed aiming at the heterologous expression, metabolic evolution, and the protein engineering to increase substrate uptake, lower by-product formation, redox balance, increase the resistance against cell wall inhibitors, and multiple sugar utilization for fermentative ethanol production from waste biomass. **Sharma and Bibra** summarize in **Chapter 7** various techniques, for example random mutagenesis, site-specific engineering on conventional and non-conventional yeasts together with biomass pretreatment, for sustainable biorefinery concepts.

Synthetic biology is a powerful branch of systematic biology aiming to fabricate biological constituents such as cells, enzymes, and genetic circuits. High throughput pathways, multigene constructs, and high precision gene editing can be performed with excellent precision using synthetic biology techniques to develop novel biological systems in a cost-efficient manner. **Sarkar et al.** evaluate the recent developments in synthetic biology and their role in developing a lignocellulose bioeconomy in **Chapter 8**.

The effective pretreatment of LCB followed by a precise enzymatic hydrolysis to obtain a higher amount of fermentable sugars are the important criteria for the development of biochemical/biofuel production at cost competitive prices. **Martins et al.** in **Chapter 9** have provided an update on the biochemical methods and strategies for LCB conversion into bioethanol and other platform chemical products in biorefinery.

Beside the carbohydrates, lignin is another plant cell wall component which can be converted into drop-in or bulk chemicals and fuels. The chemical transformation of lignin has been studied more compared with biological transformation. However, recently biological transformation of lignin has gained significant attention. In **Chapter 10, Ribeiro et al.** compare lignin conversion through biological and chemical routes into potential platform chemicals.

Like any other bioprocessing systems, downstream processing is equally important in lignocellulose biorefinery. Biomass conversion at industrial level needs to be environmentally sustainable. Biorefinery fits well in the circular economy agenda and this allows the reuse of by-products in the industrial network, and Industry 4.0 will become smarter in economical, ecological, and social issues. In **Chapter 11**, **Dussán et al.** provide all the key aspects of circular bioeconomy associated with biorefinery and its techno-economical aspects advocating the linkage of biorefinery operation with Industry 4.0.

A smooth scaling up of the biological process with fewer hurdles is necessary for the overall success of lignocellulose biorefinery at industrial scale. Because of the inherent recalcitrant nature of LCB, it is obvious that a number of technological hurdles, for example biomass transfer from one reactor to another, filtration of pretreated slurries and hydrolysates, concentration of liquid samples, mixing of heterogeneous materials, product recovery, among others, appear at the shop floor. **Chapter 12** by **Baudel et al.** discusses the scale-up process challenges in LCB processing and the possible outcomes. This chapter is particularly focused on the challenges, problem-solving issues, and opportunities during the scale-up from bench to pilot (single and integrated) and flagship plants related to operating over longer hours for the production of 2G bioethanol and chemicals from lignocellulosic feedstocks.

For successful LCB refining at commercial scale, techno-economic viability analysis of the process is crucial for the success of the plant. Biomass is relatively of lower economic value but the cost of the biomass still plays a decisive role in biomass refineries. The material and energy balances of all unit processes included in the bioconversion pathway are required to determine the product and by-product yields, capital investments, raw material amounts, and their costs, etc., and to translate these parameters into overall technical and economic feasibility. In this aspect, **Kumar et al.** in **Chapter 13** comprehensively review the critical parameters for the techno-economic analysis of woody biomass transformation into ethanol. In particular, they have addressed the hot-spots (energy or cost-intensive unit processes or process inputs). Eventually, the authors have provided a direction for future research to determine the most sustainable technology approach for woody biomass conversion into ethanol.

A fully integrated biorefinery may have the potential to replace the complex oil refinery petrochemical industry. This integrated biorefinery converts biomass into a myriad of primary and secondary products (either biofuels or biochemicals). However, the most important criteria are to analyze the integrated biorefinery critically through a reliable life cycle assessment in order to meet the futuristic demands for energy, chemicals, and materials. In this context, **Longati et al.** in **Chapter 14** comprehensively review various environmental indicators, life cycle analysis, and ecological perspective on biomass conversion. Life cycle assessment can provide the key information on various stages of the process which effectively contribute to environmental impacts in turn helping to develop a sustainable process for biomass conversion into renewable fuels and chemicals.

The sustainable development of renewable fuels, drop-in chemicals or specialty chemicals is effective if consumers contribute in green consumption. For this, robust policy initiatives are required to be incorporated to combat the challenges. **Chapter 15** by **Dhanya** is focused on the integrated policy initiatives for exploration of the lignocellulose bioeconomy and its development through green consumption.

The commercial success of biorefineries is directly dependent upon the available or possible technology in the future and the economics of the process. However, most of the biorefinery technologies are new and yet to get the final approval from industries in the future. Among the most critical parameters, regular biomass supply chains and technologies utilizing all the waste streams to generate value-added products are essential for the success of biorefineries in the long run. Taking these parameters into consideration, **Ravella et al.** in **Chapter 16** cover the important technological considerations, for example studies of the fermentation microbial communities; application of molecular techniques is crucial to understanding microbial communities, effective biomass pretreatment technology like steam explosion, and computational tools for the better control of reactors in the long run at the shop floor.

Lignocellulose Bioconversion Through White Biotechnology is a unique collection of 16 chapters describing the recent cutting-edge research and technological advancements in lignocellulose biomass conversion into fuels, bulk and specialty chemicals, and materials. Each chapter has clear illustrative text supported with relevant figures and tables. The pertinent information presented in each chapter is well supported with the relevant references from peer reviewed published papers, books, chapters, and conference proceedings.

The book chapters presented have been written by one or more authors who are expert in their research area and have a great amount of teaching and research experience. We hope that this book will be a great source of information on white biotechnology and lignocellulose transformation to commodity chemicals and low carbon fuels. The book will be a reliable source of information for researchers from academia and industry, undergraduate, postgraduate, and doctoral students of various disciplines, for example industrial biotechnology, chemistry, biochemistry, biochemical engineering, environmental science, food science, biofuel and bioenergy industries, and related subjects. Also, the policy makers, regulatory bodies, and biofuel promoting agencies will find this book a pivotal source of information.

The Editor is very grateful to all the contributors for their outstanding efforts to provide state-of-the-art information on the subject matter of their respective chapters. Their efforts will certainly enhance and update the knowledge of readers about lignocellulosic biorefining technologies. We express our sincere thanks to Kerry Powell, Rebecca Ralf, Mahalakshmi Pitchai, Ganesh Kumar Bhavani and the publishing staff at John Wiley and Sons Ltd for their kind cooperation and generous efforts in the development of this book.

The Editor is very thankful to the Research Council for the State of São Paulo (FAPESP), Brazil, for providing financial assistance through the *jovem pesqvisador* grant (Process No. 2020/12559–6) and Brazilian National Council for Scientific and Technological Development (CNPq), Brazil for the scientific productivity program (Process No. 309214/2021-1). We hope that the book will be a pertinent source of information for all readers looking for information on the latest research and advances in the field of white biotechnology.

Prof. Dr Anuj K. Chandel

1

White Biotechnology

Impeccable Role in Sustainable Bioeconomy

Anuj Kumar Chandel[1], *Jesús J. Ascencio*[1], *Akhilesh K. Singh*[2], *Ruly T. Hilares*[3], *Lucas Ramos*[3], *Rishi Gupta*[4], *Yeruva Thirupathaiah*[5], *and Sridevi Jagavati*[6]

[1] *Department of Biotechnology, Engineering School of Lorena (EEL), University of São Paulo (USP), Lorena, São Paulo, Brazil*
[2] *Department of Biotechnology, School of Life Sciences, Mahatma Gandhi Central University, Motihari, 845401, Bihar, India*
[3] *Laboratório de Materials, Universidad Católica de Santa Maria–UCSM, Yanahuara, Arequipa, Perú*
[4] *Anton Paar India Pvt Ltd, Gurgaon, 122016, Haryana, India*
[5] *Central Sericultural Research and Training Institute, Mysuru, Karnataka, 570008, India*
[5] *Department of Microbiology, Indira Priyadarshini Government Degree College for Women, Hyderabad, Telangana, India*

1.1 Introduction

White biotechnology or industrial biotechnology broadly covers the utilization of microorganisms and living cells and/or their metabolites in compound creation or development of new sustainable materials and the advancement of new feasible vitality sources, for instance biofuels and biochemicals (Frazzetto 2003). White biotechnology considers structuring low assets, expanding procedures and items, making them more vitality proficient and less contaminating than customary ones, and in turn developing biodegradable industrial products. This multidisciplinary branch of science is considered as a potential technological platform for producing base chemicals/fuels, food nutrients, and biomaterials in an eco-friendly manner conforming to the sustainability norms. White biotechnology encompasses the recent scientific and technological advancements in functional genomics and proteomics, metabolic engineering, biocatalysis, synthetic biology, and fermentation (Heux et al. 2015).

White biotechnology exploits living cells—from yeast, molds, bacteria, and plants—and other required ingredients to create products that are effectively degradable, require fewer chemicals, and produce less waste during their creation (Frazzetto 2003). For a considerable length of time, bacterial catalysts have been utilized broadly in food production and as dynamic fixings in washing powders to diminish the measure of fake surfactants. The development of hybrid genes and their subsequent transfer to living organisms has become the main focus of biotechnology (Gomes et al. 2020). Transgenic *Escherichia coli, Pichia pastoris, Hansenula polymorpha, Yarrowia lypolitica*, among others, are considered ideal hosts to produce therapeutic molecules, designer enzymes, organic solvents, etc. in large

Lignocellulose Bioconversion Through White Biotechnology, First Edition. Edited by Anuj Kumar Chandel.

scale bioreactors. The advantages of exploring normal procedures and items are complex: they do not depend on fossil assets, are more vitality proficient, and their substrates and waste are naturally degradable, which all assist with diminishing their ecological effect (Dhanya et al. 2020). White biotechnology offers numerous advancements in the production of sustainable fuels, food and feed, energy, and materials. In last two decades, there has been a surge in white biotechnology based industries inspired by the sustainable development of biobased products and solutions empowering the socio-economic outcome (Frazzetto 2003).

Biomass conversion into valuable and diversified chemicals/biofuels in white biotechnology eventually strengthened biorefinery (Chandel et al. 2019). In essence, biorefineries have to explore the hybrid technological development of various areas of knowledge (mainly bioengineering, agriculture, and polymer chemistry) (Fernando et al. 2006; O'Hara et al. 2013). In first-generation (1G) biorefinery, sugarcane juice, molasses, corn grains, sugarbeet juice, oil from oil-seed crops, and so on are transformed into both valuable chemicals and biofuels (ethanol, biodiesel, organic acids, and others). In second-generation (2G) biorefinery, the lignocellulosic biomass (LCB) is further fractionated into valuable components through pretreatments, saccharification reactions, fermentation processes, and pyrolysis under established conditions, including some conventional approaches (Chandel et al. 2020). Moreover, the chemicals can be further transformed into higher-value products, for instance oligosaccharides derived from lignocellulose, xylitol, 2G ethanol, succinic acid, furans, as can be seen in Figure 1.1 (Clark 2007; Bhatia et al. 2020a).

This chapter provides an important overview on the industrial developments in white biotechnology. Critical aspects of biomass conversion like biomass composition, renewable sugar production (pretreatment and enzymatic hydrolysis), conversion of these sugars into a wide range of products, and their commercialization are also discussed.

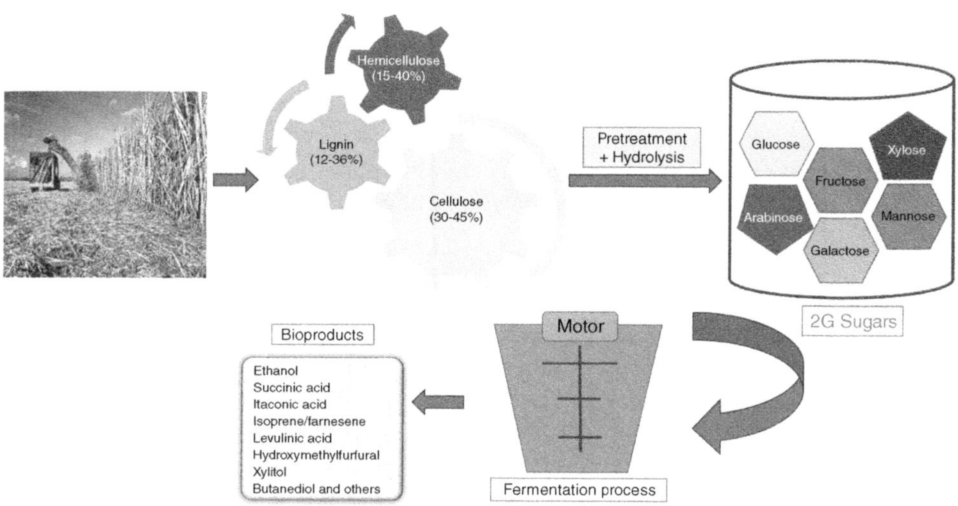

Figure 1.1 Lignocellulosic materials: the backbone of white biotechnology. Adapted from Clark, 2007; Bhatia et al., 2020a.

1.2 Biomass Feedstock: Types and Composition

Biomass is defined as the entire cellular dry weight of organic materials synthesized by an organism usually from sunlight and carbon dioxide (Sharma and Singh 2015). More precisely, biomass comprises any organic material/substance accessible in an ecological, economic, and competitive way from edible and inedible crops, agricultural residues, wood and waste, aquatic plants, animal manure, and many others (Kamm et al. 2016). Photosynthesis is the pathway used by nature to generate 170 billion metric tonnes of biomass each year. Of that amount, 75% is equivalent to carbohydrates. Surprisingly, humans only use 3%–4% of the biomass generated to produce food and non-food materials (Röper 2002). Carbohydrates from biomass pose the most plentiful renewable resource available as a promising raw material toward the green chemistry of the future (Lichtenthaler and Peters 2004). In addition, many industrial sectors including the cattle feed/food industry have been switched over or are switching over to biomass as feedstock, in order to enhance their sustainability. This is supported by the limited availability of green fodder for animal consumption remaining (Sharma and Arora 2015). Further, the worldwide requirement for abundant as well as consistent supplies of meat as well as milk is likely to be double in 2050 (Alexandratos and Bruinsma 2012). For ensuring food security, the upcoming feed constituents must not compete with human food. In order to overcome the issue of cattle/animal feed, the cost-effective biomass that involves non-fossilized as well as decomposable organic substances derived from biobased resources like LCB can act as potential cattle feed (Sharma and Arora 2015). The most promising natural and renewable LCB composed of plant cell walls involving polymeric cellulose, hemicellulose, and lignin contents is one of the most plentiful waste materials available in the biosphere. Various cereal crop remains such as wheat straw, paddy straw, corn stover, etc., which are composed of LCB, are frequently exploited as cattle fodder as well as a feed supplement (Parimala et al. 2007). The overall annual generation of LCB is expected to be almost 10-50 billion metric tonnes, where almost 4 billion tonnes are made up of annual crop remains (Sanchez and Cardona 2008).

Usually, LCB is classified into three types, that is, crop residues from agriculture, residual trees, as well as non-woody stems (Ullah et al. 2015). The main structural constituents of these residual trees and crops are cellulose, hemicellulose, and lignin. Cellulose is the most plentiful constituent of LCB that accounts for up to 40%–60% by dry weight (Sharma et al. 2019). The cellulose constitution, in essence, is β-D-glucopyranose units linked by β-(1,4) bonds alongside cellobiose as a repeating basic unit. The cellulosic chains are composed of 500–1400 D-glucose monomers in which the neighboring cellulosic chains interact through hydrogen bonds as well as hydrophobic linkages. These cellulosic chains are organized collectively to produce microfibrils that act as precursors toward the formation of cellulosic fibrils by undergoing collective organization. In contrast to cellulose, hemicellulose is made up of heterogeneous kinds of sugars, that is, glucose, xylose, as well as mannose, and account for 20%–35% by dry weight (McKendry 2002; Chandel et al. 2018). Monomeric unit numbers of the hemicellulose vary from 100 to 200 which are considerably lower than those of cellulose (500–1400) (Mota et al. 2018). Hemicellulose is amorphous in nature having a lower physical strength and is easily disrupted with dilute acids and bases including hemicellulase (Isikgor and Becer 2015). On the other hand, lignin is a complex polymer of aromatic alcohols. It is the second most plentiful polymeric material of LCB after cellulosic material and contributes 15%–40% by dry weight. It seals the spaces of cell walls present in cellulosic as well as hemicellulosic materials. As a result, lignin hampers the entrance of microorganisms/enzymes to these polymeric substances (Zoghlami and Paës 2019).

1.3 Biomass Pretreatment: An Overview and State-of-the-Art

LCB is recalcitrant in nature owing to its resistance to chemical as well as biological hydrolysis. Some features like cellulosic crystallinity, degree of lignin content, as well as structural heterogeneity together with complex cell wall components are accountable for the biomass recalcitrance, which should be overcome for beneficial use of LCB feedstock like cattle/animal feed and so on. The LCB like agricultural crop residues/grasses can be used prudently as cattle/animal feed as this supplies adequate energy together with nutrients to animals like cattle/ruminants. Cattle/ruminants are exceptional in their capability to transform such LCB into high-quality meat/milk protein for human beings. Nevertheless, a higher lignin level, little digestibility, lesser protein content as well as a considerably reduced flavor of agricultural waste/non-woody stems prevents its use as feed for livestock and animals. Lignin is a recalcitrant material in the plant wall structure that limits the complete availability of sugars for the microbes present within the ruminant's gut (Shrivastava et al. 2014). It is responsible for maintaining cell wall strength, defending against pathogenic organisms, enhancing water conduction together with inhibiting the disintegration of LCB (Grabber 2005). Lignin can be hydrolyzed under aerobic conditions and not anaerobically inside the rumen. It is not only accountable for hydrophobic characteristics as well as structural inflexibility, but also permanently adsorb cellulase enzymes together with other enzymes in the course of the enzymatic breakdown of LCB owing to its hydrophobicity, hydrogen bonding, methoxy groups as well as polyaromatic structure (Zeng et al. 2014). On the other hand, cellulose is available in either crystalline or amorphous form, and the crystalline form is found to be more difficult to hydrolyze with enzymes. Also, the cellulosic fibrils that are entrenched within the matrix of LCB are extremely resistant to enzymatic action. It is believed that long cellulosic chains involve surplus hydrogen bonds and therefore such cellulosic chains are not easy to disrupt. On the other hand, smaller cellulosic chains involve a feebler hydrogen bonding that ultimately aids in enzymatic approachability (Meng et al. 2017). Furthermore, the interactions between microfibers from cellulosic as well as hemicellulosic materials together with the lignin–carbohydrate complex association considerably influence its enzymatic breakdown by decreasing the cellulosic accessible region toward enzymes (Du et al. 2011). LCB is limited in its application as cattle/animal feed owing to crystalline cellulose as well as a high content of lignin. Therefore, elimination of lignin content with a decrease of crystalline cellulose in LCB results in loosening of the cellulosic structure that enhances the cellulosic contact region with useful microbes or enzymatic accessibility (Kejun et al. 2011). The production of sugars like glucose, xylose, arabinose, rhamnose, and galactose from such heterogeneous LCB feedstock needs pretreatment not only for the alteration of the structure of crystalline cellulosic materials but also for removal of lignin content, thereby enabling them to be more accessible toward the attack of enzymes. This enzymatic hydrolysis results in the transformation of the carbohydrate polymeric materials into fermentable sugars (Hsu 1996). The yield of cellulose hydrolysis as a result of pretreatment often surpasses 90% of the theoretical over 20% when pretreatment is not performed (Lynd 1996). The pretreatment process transforms the natural LCB into a crystalline cellulose material without recalcitrant for efficient hydrolysis attained. LCB pretreatment can significantly enhance the approachability and redistribution of the cellulosic surface area and lignin solubilization (Lynd et al. 2002). In this manner, the characteristics of an efficient pretreatment approach imply breaking the LBC complex structure; reducing the cellulose crystallinity; preventing the degradation of hemicellulose sugars; restricting the production of inhibitory compounds that affect hydrolysis and fermentation; decreasing the energy input and the use of toxic chemical products; needing a simple configuration; and the production of lignin as a high-value co-product limiting the formation of dangerous and unsafe residues (Banerjee et al. 2010).

Countless physical, chemical, physicochemical, and even biological pretreatments have been developed to treat LCB. The purpose is to get rid of its lignin together with a reduction in crystallinity in the cellulose region, ultimately enhancing the accessibility and digestibility of LCB by hydrolytic enzyme action. The main hurdle for large-scale sugar production as the LCB backbone is the failed approach to a profitable pretreatment for overcoming the recalcitrance of these materials. Based upon investigation, it was found that these conventional techniques of pretreatment are price intensive as excess chemicals or energy are required (Lim and Wang 2013). At least 20% of the overall production price is contributed as a result of the pretreatment stage in all these various techniques, thus making it a very costly process stage (Yang and Wyman 2008). Therefore, intensified investigation is required for exploring approaches toward reducing the price of the transformation process. A fundamental understanding of each step in a bioprocess is imperative; consequently, its techno-economic viability can guarantee the commercial success of biomass conversion into sugars and other value-added chemicals (Amin et al. 2017).

Physical pretreatment techniques regularly involve mechanical comminution (particle size reduction), ultrasound, and hydrothermolysis. Comminution leads the milling of biomass using dry or wet biomass to facilitate size reduction for its use in consecutive processing steps. Usually, physical pretreatment depicts relatively poorer performance apart from higher prices (Yang and Wyman 2008). However, the physicochemical pretreatment approaches imply acid or base that provokes LCB hydrolysis and improves glucose recovery from the cellulosic material separated from the hemicellulose and the lignin in the course of pretreatment. Sulfuric acid and sodium hydroxide represent the most cited acid and base in physicochemical pretreatments. Amongst the various strategies, the following stand out as economically viable pretreatments: steam explosion; liquid hot water; dilute acid; lime; ammonia; among others (Mosier et al. 2005; Chandel et al. 2019).

1.4 Lignocellulosic Sugar Production

Naturally, LCB is the most abundant and available carbon-based green feedstock in the world. This renewable resource is revealed as an attractive opportunity for sustainable and economic biobased development, seeking to replace energy, chemicals, and materials from conventional fossil resources (Zhu et al. 2020). To meet the expectations of a sustainable world, it is necessary to take full advantage of LCB, such as agricultural residues (corn straw, wheat straw, sugarcane straw, and bagasse), forest residues (residues from wood harvest, such as branches, foliage, roots), and herbaceous and woody biomass (particularly perennial grasses such as miscanthus, grasses, grasslands, and short-rotation forest species, e.g. eucalyptus, poplar, and robinia) (Ullah et al. 2015).

LCB production throughout the world, equivalent to 200×10^9 tons, attracts the interest of many countries for its exploitation under the concept of biorefineries as a complementary alternative to fossil resources to produce biofuels and chemical products of commercial value (Kumar et al. 2020). The polymeric matrix of LCB is mainly composed of 40%–80% cellulose, 15%–30% hemicellulose, and 10%–25% lignin (Halder et al. 2019), and the structural composition is a consequence of various features like origin, age, type, and altitude of the crop, soil, climatic conditions, harvesting, gathering and storage processes, among others (Canilha et al. 2012).

Undoubtedly, the complex characteristics of LCB provide heterogeneity and rigidity to the polymeric matrix, which causes resistance to its deconstruction (Kumar et al. 2020). The recalcitrant nature of biomass protects and prevents the degradation of its carbohydrates by the action of microorganisms or enzymes (Bhatia et al. 2020b). The fermentable sugars obtained from their carbohydrates are made up of glucose units, in the case of cellulose; while hemicellulose has pentoses (xylose and arabinose), hexoses (mannose, glucose, galactose), and residual sugars such as rhamnose and fucose (Halder et al. 2019). The reasons for its recalcitrance are justified by the epidermal tissue and chemical substances, chemical and physical cell wall composition, cellulosic structural surface, and mainly by the presence of lignin (Zhao et al. 2012).

A short time ago, industries and research centers intensified technological development for biomass deconstruction and exploitation as a biotechnological platform, as shown in Figure 1.2. This has generated technical and economic challenges to release fermentable sugars from LCB, essential for their biological conversion into value-added products (Yoo et al. 2020). Pretreatment helps break down the natural resistance to lignin disorder and the breakdown of its carbohydrates, cellulose, and hemicellulose. Therefore, the selective use of these technologies can help to improve the accessibility of hydrolytic enzymes to carbohydrates and subsequently to generate fermentable sugars during the enzymatic hydrolysis stage (Hassan et al. 2018). On the other hand, lignin can be converted into value-added chemicals of great economic importance, such as phenolic resins, epoxy, polyesters, polyurethanes, binders in mortars, sequestration of metals in solutions, hydrogels, additives, and others (Ahmad and Pant 2018; Garlapati et al. 2020).

Most studies conclude that lignin acts as a physical barrier that represents an obstacle between the enzyme and the substrate; it is considered a source of inhibitory compounds for some enzymes like hydrolases and microorganisms during the fermentation process (Zoghlami and Paës 2019). Zhao et al. (2012) point out that the presence of lignin is a factor that limits the speed and extension of the enzymes for the release of sugars, so it is necessary to eliminate lignin to increase the digestibility of cellulose. However, despite the negative reputation in the literature of the lignin effect on enzymatic hydrolysis, some results have shown an opposite profile. Thus, the presence of lignosulfonates with lower molecular weight are more effective during the enzymatic hydrolysis of LCB, due to the hydrophilic surfaces of the lignosulfonates that act as surfactants to block lignin bound to LCB that is hydrophobic and has a high affinity for cellulases, resulting in an unproductive cellulase–lignin binding

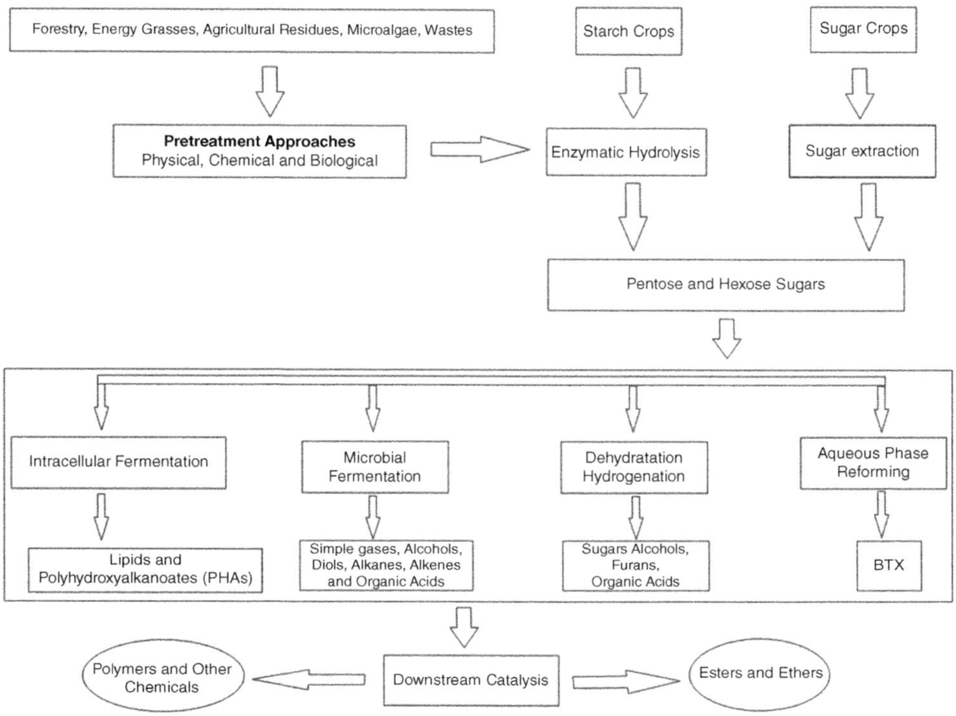

Figure 1.2 Representation of routes through the sugar platform.

and better enzymatic conversion of LCBs (Zhou et al. 2013). The previous result shows that there will not always be a detrimental effect, since the addition of lignin can help the release of sugars.

In addition to recalcitrance due to the chemical composition of the lignocellulosic feedstock (hemicellulose, lignin, acetyl group), other causes could affect the enzymatic hydrolysis process and are closely related to the physical structure (structural proteins in the cell wall, crystallinity in the cellulosic fraction, polymerization degree, surface area, particle and pore size less than 5.1 nm) (Bhatia et al. 2020b). The selective use of hydrolytic enzymes to depolymerize the LCB components has been used extensively to obtain different biofuels and bioproducts. For instance, de Assis Castro et al. (2017) used alkaline deacetylation pretreatment to improve the recovery of the sugar during enzymatic conversion and substantially increased ethanol production from the hemicellulose and cellulose of rice straw. Similarly, other studies show that the production of fermentable sugars is dependent on the type of pretreatment used in the biomass, as happened with the sequential acid–base pretreatment of sugarcane bagasse where hemicellulose and lignin were efficiently removed, allowing higher accessibility of cellulases to treated biomass (Ascencio et al. 2020).

The optimal reaction conditions for the enzymatic conversion of carbohydrates to sugars can vary over a range of 4.8–5.0 pH as well as 45–50 °C temperature. Nonetheless, the saccharification reaction requires a specific group of enzymes for cellulose and hemicellulose degradation: endo-β-(1→4)-glucanases attack the regions of low cellulose crystallinity;

cellobiohydrolases or exoglucanases are responsible for removing the cellobiose units at the ends of the chain, increasing the molecule degradation; β-glucosidases hydrolyze cellobiose to obtain glucose monomers; xylanase, β-xylosidase, glucuronidase, acetyl ester, galactomannanase, and glucomannanase will be necessary to break down hemicellulose into its monomeric sugars (Canilha et al. 2012; Lugani et al. 2020). It is important to note that an economically viable lignocellulosic conversion process is possible when higher loads of solid content are used, generally ≥15%, in such a way that it presents little or no volume of free water (Modenbach and Nokes 2013).

Currently, the use of LCB has aroused great interest in fermentative processes for the synthesis of biofuels (bio-oil, bioethanol, biohydrogen, biogas, syngas, etc.), and added value products such as furfural, 5-hydroxymethylfurfural, sorbitol, xylitol, glycerol, lactic acid, succinic acid, phenolic compounds, biopolymers, pigments, as shown in Table 1.1. Furthermore, LCB has been used for the development of advanced technology products, such as energy storage, transportation, medical applications, biosensors, environmental remediation, among others (Kumar and Sharma 2017).

1.5 Production of Ethanol and Biodiesel

Hydrocarbons, used extensively throughout the world, represent our primary source of fuels. The backbone of the energy sector, hydrocarbons are responsible for 75% of total greenhouse gas emissions in developed countries, mainly CO_2, CH_4, and N_2O (Intergovernmental Panel on Climate Change 2006). Biofuels are a feasible opportunity to produce clean and renewable energy from edible and non-edible crops by biological and physicochemical processes using diversified microbial metabolism (Rodionova et al. 2017).

Many materials can be used as renewable and available feedstocks for the production of biofuels, for example canola, coconut, corn, sugarcane, barley, oat, wheat, sorghum, olive, palm, peanut, rice, safflower, soybean, sunflower, miscanthus, switchgraw, rice or wheat straw, wood chips, paper pulp, *Cyanophyceae*, *Bacilleriophyceae*, *Xanthophyceae*, *Raoultella ornithinolytica*, *Clostridium thermocellum*, etc. (Lin and Lu 2021). In this way, biofuels can typically be classified as first generation, which includes edible feedstocks (starches, sugars, and vegetable oils); second generation, lignocellulosic materials (agricultural and forestry waste) and other inedible lipids (used cooking oils and municipal solid waste); third generation, microorganisms (cyanobacteria, algae, and bacteria); and fourth generation from genetically modified microorganisms (Mat Aron et al. 2020).

Despite the first-generation biofuel industry competing with edible crops (i.e. canola, corn, sugarcane, wheat, sorghum, and rice) for food safety, ethanol and biodiesel production have consolidated within the energy sector of various countries. In addition, technological advances in research on lignocellulosic industrial waste and manipulation of genetically modified microorganisms have opened the doors to take advantage and innovate in the production of biofuels.

The largest ethanol producers are the United States and Brazil, and it is expected that by 2029 they will use 14% of the corn and 25% of the sugarcane grown in the world, while used cooking oils will continue to be essential feedstocks for biodiesel production in countries within the European Union, Canada, and the United States (OECD/FAO 2020). Likewise,

Table 1.1 Lignocellulosic feedstocks for fuel and chemical production through process fermentation.

Lignocellulosic feedstock	Pretreatment type	Microorganism	Fuel and chemical production	Reference
Aspen	Ionic liquid	*Saccharomyces cerevisiae* CCUG 53310	Bioethanol	Goshadrou et al. (2013)
Banana waste	Size reduction	*Pleurotus ostreatus* and *P. sajor-caju*	Lignolytic and cellulolytic enzymes	Reddy et al. (2003)
Barley husk	Size reduction	*Bacillus subtilis* sp.	Xylanase	Ho and Chinonso (2016)
Barley straw	Diluted sulfuric acid	*Clostridium beijerinckii* P260	Butanol	Qureshi et al. (2010)
Coconut husk	Subcritical water	*Enterobacter aerogenes* NBRC 13534	Biohydrogen	Muharja et al. (2018)
Coffee grounds	Diluted sulfuric acid	*Cupriavidus necator* H16 and *Sporobolomyces roseus*	Polyhydroxyalkanoates and carotenoids	Obruca et al. (2015)
Corn cob	Diluted sulfuric acid	*Rhizopus oryzae* HZS6	Lactic acid	Bai et al. (2008)
Corn stover	Dilute sulfuric acid	*Coniochaeta ligniaria* NRRL 30616	Furfuryl alcohol and furoic acid	Nichols et al. (2008)
Cotton stalk	Diluted sulfuric acid	*Candida tropicalis*	Xylitol	Zhang et al. (2014)
Eucalyptus	Steam explosion	*C. saccharoperbutylacetonicum* N1–4	Acetone, butanol and ethanol	Zheng et al. (2015)
Groundnut shell	Diluted sulfuric acid	*Cryptococcus psychrotolerans* IITRFD	Biodiesel	Deeba et al. (2017)
Hybrid poplar	Hot-water extraction	*C. acetobutylicum* ATCC 824	Acetone, butanol and ethanol	Guan et al. (2018)
Maple	Diluted sulfuric acid	*C. acetobutylicum* ATCC824	Butanol	Sun and Liu (2012)
Paper residues	Diluted sodium hydroxide	*Penicillium chrysogenum* and *S. cerevisiae*	Bioethanol	Geetha and Gopalakrishnan (2011)
Pine	Steam explosion (SO_2)	*S. cerevisiae* T_2	Bioethanol	Zhong et al. (2019)
Poplar	Diluted sulfuric acid	*Candida tropicalis* UFMG BX 12-a	Xylitol	Dalli et al. (2017)
Oak	Steam explosion	*Mannheimia succiniciproducens* MBEL55E	Succinic acid	Kim et al. (2004)

(Continued)

Table 1.1 (Continued)

Lignocellulosic feedstock	Pretreatment type	Microorganism	Fuel and chemical production	Reference
Rice straw	Diluted sodium hydroxide	*Trichoderma reesei* ZM4-F3	Biosurfactant	Zhang et al. (2009)
Sorghum stalks	Organosolv	Anaerobic consortia from wastewater treatment	Biogas	Ostovareh et al. (2015)
Soy straw	Diluted hydrochloric acid	*Agaricus brasiliensis*	Pullulan	Lima et al. (2008)
Sunflower	Alkali potassium hydroxide	*Aspergillus oryzae*	Polyhydroxyalkanoates	Kachrimanidou et al. (2013)
Sugarcane bagasse	Hydrothermic	*Rhodosporidium toruloides*	Single-cell oil	Bonturi et al. (2017)
Sugarcane straw	Sequential acid–base	*Lasiodiplodia theobromae* CCT 3966	β-glucan biopolymer	Abdeshahian et al. (2020)
Switchgrass	Ammonia fiber expansion (AFEX)	*Zymomona mobilisstrain* 2032	Bioethanol	Zhang et al. (2020)
Wheat bran	Alkali (CH_3CO_2K)	*Streptomyces* S10	Ferulic acid	Mukherjee et al. (2007)

the European Union has endeavored to create a circular economy based on biofuels produced through LCB, which consequently has led to a search for the technological development and implementation of bioenergy (Hassan et al. 2019). At present, the industrial market for ethanol and biodiesel production from biomass is dominated by the United States, the European Union, and Brazil, in decreasing order.

The energy gain concentrated in biofuels depends on the biomass used, the conversion process, and co-product production (Mat Aron et al. 2020). So, then, the energy efficiency of corn varies between 26% and 68% and sugarcane from 78% to 100% for ethanol production, while the production of biodiesel from palm oil has caused a significant difference of up to nine times compared to rapeseed oil (Garlapati et al. 2019).

Unlike other energy sources, ethanol and biodiesel are already used successfully in flex-fuel vehicles and offer a formidable advantage by being produced from many types of biomass. The most notable differences between them are the origin of their feedstocks and their final application. The chemical production of ethanol can be achieved by ethylene hydration, whereas the biological production of ethanol involves plant-based sugars and saccharolytic activities of certain microorganisms like *Saccharomyces cerevisiae* and other organisms (Lopes et al. 2016; Chandel et al. 2021). As for the production of biodiesel, it is through chemical reactions of esterification and transesterification, where the raw material is vegetable oils, animal fat, or lipids derived from algae (Almeida et al. 2018).

In some countries, flex-fuel cars use blends of fossil fuel/biofuel and some engines use pure biofuels. However, the engine type (diesel or gasoline) will determine the biofuels use as additives. Ethanol and gasoline blends are the most common today; for instance, in the United States the ethanol–gasoline blend E85 (85% ethanol, 15% fuel) is used in flex-fuel cars (Coelho and Goldemberg 2004). Thus, in Brazil, hydrated ethanol (92% ethanol) is used in E100 vehicles and anhydrous ethanol E20-25 (20%–25% ethanol, 80%–75% fuel) in flexible fuel vehicles (Nogueira et al. 2020). Biodiesel can be used as an additive in diesel in different proportions B10, B20, B30 (10%, 20%, 30% biodiesel; 90%, 80%, 70% diesel oil). Biodiesel can even be used as 100% (B100%) offering the advantages of low carbon and sulfur emissions (Kohse-Höinghaus et al. 2010). This particularity in flexible combustion engines will allow the use of alternative fuels to be regulated in a sustainable way with our environment.

1.6 Drop-in Renewable Biofuels: Green Hydrocarbons

Drop-in renewable biofuels are liquid hydrocarbons produced from biomass by oleochemical, thermochemical, biochemical, and hybrid routes; in addition, they are as functional as fossil fuels and compatible with the available petroleum industry (International Energy Agency 2019). Indeed, it is possible to convert the integral components (lignin, cellulose, hemicellulose) of LCB into aromatic hydrocarbons, a liquid biofuel type (Han et al. 2019).

In this sense, LCB is an economically viable option for the production of drop-in biofuels. For the production of this biofuel type, the LCB transformation includes catalytic depolymerization, the last step to obtaining biofuels with specific properties (van Dyk et al. 2019). Recently, Zhang et al. (2020) proposed a new catalytic reaction (SO_4^{2-}/TiO_2–ZrO_2) in inedible oils for drop-in biofuel production, a C_6–C_{26} hydrocarbon (17% alkanes, 56% olefins, and 22%

monocyclic aromatics). Moreover, approaches for the production of long-chain hydrocarbons and aromatic additives through LCB-derived furfural (2,5-dimethylfuran, 2-methylfuran, 5-hydroxymethylfurfural, γ-valerolactone) have been summarized (Bohre et al. 2015).

Although drop-in biofuels are not yet technologically consolidated, they are attracting more and more attention within the energy industry. The feedstocks and catalytic processes used in their manufacture, in general, are inexpensive and represent an environmental strategy for the mitigation of greenhouse gases through different transport options (aviation, maritime, rail, and long-distance road).

1.7 Global Scenario of the Biofuel Industry

This section provides brief details on industries based on lignocellulosic feedstocks from different parts of the world, while a detailed list is given in Table 1.2. As previously described, industrial biofuel production was triggered by ethanol from corn and sugarcane, mainly produced and consumed in America. Despite ethanol being overproduced on a commercial scale, biodiesel also has a consolidated market in the European Union. In this context, the Brazilian company Raízen stands out for maximizing LCB utilization by diversifying biofuel

Table 1.2 List of major white biotechnology companies worldwide.

Country	Company	Location	Country	Company	Location
USA	Abengoa Bioenergy	Hugoton, KS	*European countries*	Abengoa	Spain
		York NE		chemre	Pitea, Sweden
	AE Biofuels	Butte, MT		Sekab	Ornskldsvik, Sweden
	Alltech	Springfield, KY		Stora Enso	Varkaus, Finalnd
	American Energy Enterprises	New Milford, CT		Elson	Denmark
	Bluefire	Corona, CA		UPM Kymmene	Finland
		Lancaster, CA		Choren	Freiburg, Germany
	California Ethanol	Brawley, CA		Butalco	Germany
	Coskata	Madison, PA		Dong energy	Denmark
	DuPont Danisco Cellulosic Ethanol LLC	Vonore, TN		Mossi & Ghisolfi group	Crescentino, Italy
	Ecofin, LLC	Washington County, KY		Biogasohol	Bornholm, Denmark
	Flambeau River Biofuels LLC	Park Falls, WI		TMO Renewables	Guildford, UK

Table 1.2 (Continued)

Country	Company	Location	Country	Company	Location
	Fulcrum Bioenergy	Reno, NV		Inbicon	Kalundborg, Denmark
	ICM Inc.	Shelley, ID		Ensted-Vaerket	Aabenraa, Denmark
	KioR Inc	Columbus, MS		Mbbjerg Energy Concept	Mabjerg
	KL Process	Upton, WY		Clariant	Germany
	Lignol Innovations	Grand Junction, CO	*Canada*	Enerkem	Westbury, Quebec
	Mascoma	Rome, NY		Iogen	Ottawa
		Chippewa County, MI		Lignol	Vancouver, BC
	Mississipi Biofuels LLC	Pantotoc, MS		KL Energy Crop.	South Dakota
	NewPage Corp.	Wisconsin Rapids, WI		Prairie Green Renewable Energy	Hudson Bay, Saskatchewen
	New Planet Energy	Vero Beach, FL	*India*	Mission NewEnergy	Maharashtra, India
	Pacific Ethanol	Boardman, OR		Praj Industries	Maharashtra, India
	POET	Scotland, SD	*China*	China Resources Alcohol Corporation	Zhao Dong, China
		Emmetsburg, IA		Lanzatech	Sanghai, China
	Range Fuels Inc.	Soperton, GA			Caofeidian, China
	Terrabon	Bryan, TX	*Japan*	Bioethanol Japan	Osaka Prefecture
	Vercipia	Highlands County, FL		Honda-Rite	Tokyo, Japan
	Verenium	Jennings, LA	*New Zealand*	Scion	Rotorua, New Zealand
		Highlands County, FL		Lanzatech	Parnell, New Zealand
	ZeaChem	Boardman, OR	*Thailand*	Marubeni	Saraburi, Thailand
South America	Graal Bio	Brazil	*Australia*	Ethtec	Maclean New South Wales
	Dedini	Sao Paulo, Brazil	*Singapore*	Pure Power	Mosque Street, Singapore
	Bioenercal	Chile	*Norway*	Weyland	Bergen, Norway
	Raizen	Sao Paulo, Brazil	*Russia*	Kirov Biochemicals	Kirov, Russia

production (ethanol, cellulosic ethanol, bioelectricity, biogas), as well as value-added bioproducts. Its principal biofuel, ethanol, sustains an annual production of 2500 million liters, while the commercial manufacture of 2G ethanol places the company as the only one in the world using by-products of its sugarcane crops. Comparably, POET-DSM Advanced Biofuels in the United States inaugurated a 2G ethanol plant using corn cobs and/or corn stubble as biomass. However, the plant halted 2G ethanol production due to pressure from the petrochemical industry on the Environmental Protection Agency, reducing biofuel sales and hampering the advancement of cellulosic ethanol. Besides, it has a production of 1.8 billion gallons of ethanol per year from corn crops. Likewise, the Sekab industry (Sweden) takes advantage of the biomass sugars (softwood, straw, and sugarcane bagasse) in ethanol, whereas lignin dehydrated at 50% humidity is used as solid biofuel.

In Europe, rapeseed continues to be the principal feedstock used in the production of biodiesel. Thus, the GBF German Biofuels GmbH company uses rapeseed oil to produce more than 130 000 tons of biodiesel every year, whose properties are similar to conventional diesel. Moreover, its oil plant is also capable of processing palm oil and used cooking oil. Nevertheless, the Spanish company Repsol is working to produce advanced biofuels; for this, it is investing €188 million in a second-generation biodiesel plant for the production of hydrotreated vegetable oil. Its expected capacity is 250 000 tons per year between hydrobiodiesel, biojet, bionaphtha, and biopropane, reducing 900 000 tons of CO_2 emissions each year. Notably, another path toward the sustainable generation of biodiesel from sugar LCB is the phototrophic production of algae. This technology has not yet been commercialized and needs much research for its development.

Unquestionably, fluctuations in the price of oil damage the hydrocarbon market. Therefore, drop-in biofuels are the perfect opportunity to avoid this problem, considering their biological origin from a wide variety of feedstocks. In this sense, Global Bioenergies, a French company, has a pilot plant using wheat straw for bioisobutene production. Nonetheless, There are some catalytic processes for the production of drop-in biofuels and chemicals using firstgeneration and second-generation biomass (Hirani et al. 2018).

The American company Virent Inc., Madison, uses beetsugar, sugarcane, corn starch, bagasse, corn stover, grasses, sorghum, and wood for drop-in fuels and low carbon chemicals production.

1.8 Economic Outcomes

The use of biobased products is increasing globally owing to environment-specific strategies that impose more and more use of renewable resources. Besides, these white biotechnology products also offer several other advantages compared to fossil fuels and/or other renewable energy sources. Biomass can be exploited to generate all sorts of power including heat, electricity, liquid transportation fuels, or combustive gases on a decentralized basis. These bioenergies have shown enormous potential to replace existing fossil fuels, which will not only build the nation's security by reducing the dependence on imported oils but also boost the economy by saving foreign currency. Further, the use and addition of value to its natural resources in the industrial sector will lead to regional and territorial development. Since the growth of biomass-based industries is directly linked to the availability of biomass, the demand for woody biomass will increase continuously and will subsequently increase the price of available biomass. This value addition of biomass will develop more business options for landowners and will promote them to keep their land forested. An important aspect to consider within biomass facilities is to use raw material generated in the region to create local jobs.

Furthermore, these technologies are quite labor intensive and hence will be generating employment, especially for low-income populations. Energy demand from biomass is not only creating direct local jobs but also commercial sectors throughout the entire biomass fuel supply chain. Since biomass serves as an abundant and cheap fuel feedstock, the white biotechnology processes will add value to the biological resources leading to reduced rural poverty. Moreover, these renewable resources will create a sustainable supply chain of feedstock material over the long term for efficient resource utilization and subsequently the development of biomaterials and bioenergy. Moreover, the rural population will also get an advantage from feedstock production and rural entrepreneurs will benefit by being the supplier and provider for agricultural inputs, for instance farm tools, marketers, seeds, fertilizer, pesticides, etc. Another impact of white biotechnology will be the development of a new segment of the market. Replacement of existing products with biobased products will in turn develop new markets, which are based on renewable material and are equally efficient. Besides, scientists and researchers will be involved in further innovations and technological development for improved productivity of feedstock materials; thus, white biotechnology will foster technological empowerment. However, there are several critical factors that need to be fixed with a master plan to scale up biochemicals and biofuel production on a large scale, as described in Figure 1.3.

Also, the cultivation of these energy crops will reduce the soil loss which otherwise would be carried away by water streams into various water bodies and reservoirs leading to altered water flow and storage capacity. Besides, it also has several other environmental

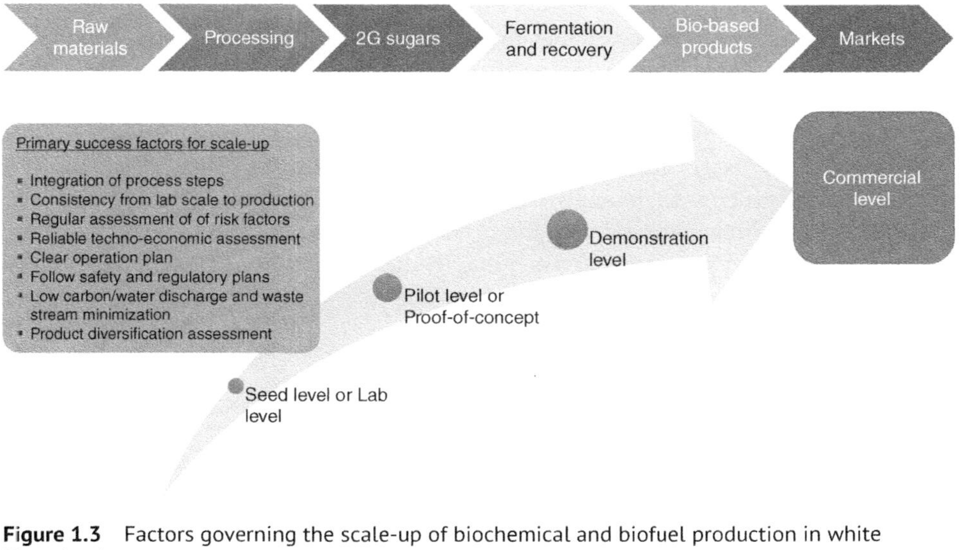

Figure 1.3 Factors governing the scale-up of biochemical and biofuel production in white biotechnology.

benefits including reduced greenhouse gases, lower carbon footprint, and decline of atmospheric pollutants. The health damage (disease and accidents) caused by atmospheric pollutants, such as SO_2, NO_x, CO, NH_3, ozone, and particulate matter, radionuclide emissions, and occupational health issues are a significant economic burden, which will be levied off with the use of the products and processes of white biotechnology.

1.9 Sustainability and Biorefinery

Over the past few decades, serious attempts have been made to shift from fossil-fuel-based systems to biobased systems for the benefit of the environment, society, and the economy. Biorefineries are currently an emerging area of research for future societal needs. The sustainability of biorefineries relies on the conversion of CO_2 and biomass into highly valuable products. Therefore, the production of energy-yielding products for industrial and daily life applications as well as the generation of high-value bioactive products for potential applications in health or pharmaceutics and nutraceuticals need to be considered for the sustainability of a biorefinery. However, the sustainability of biorefineries are mainly dependent on the integration of biomass conversion processes and the right engineering equipment to produce biofuels and biochemicals (Cherubini and Jungmeier 2010; Sacramento-Rivero 2012). The sustainability criteria for biorefineries are the availability of biomass and its renewability, economic feasibility, as well as preservation of the ecosystem and biodiversity, with social acceptance (Dale and Ong 2014). A great variety of renewable sources ranging from microbial biomass, residual biomass, edible crops, and non-edible crops has been attempted seriously in recent years by applying biochemical, chemical, and physical processes. The major products expected to be generated from these renewable sources are biomolecules or biomaterials, biofuels or bioenergy, and food/feedstocks. It is noticed that the sustainability of biorefineries is possible by overcoming the environmental issues connected with green gas emission, acidification, and eutrophication (Gnansounou 2011). Additional areas to focus on for biorefinery sustainability are employment generation by creating operational jobs within the plant and recycling of raw materials to fall farther away from critical limits. For sustainable production of biofuels or biochemicals, fossil fuel displacement potentiality, the replacement of non-renewable energy sources, bypassing or reducing the need for freshwater usage, and increasing the raw materials cost to benefit ratio are the important considerations. Additionally, the designed process would also consider biodiversity preservation, sustainable land use, and benefits to the public from the biorefinery by providing employment and monetary benefit to the local community. However, currently, several attempts have been made to achieve sustainable development goals for the benefit of the economy and society.

1.10 Conclusion

Sustainable development of lignocellulose biorefineries is the cornerstone of the success of white biotechnology. Biofuel and biochemical production from renewable feedstock is an important factor to replace fossil fuels and petroleum-derived chemicals to control global temperature rise and pollution control. Taking into consideration geopolitical issues and reducing the dependence on crude petroleum, white biotechnology is gaining significant attention. Many industries in European countries are developing biobased chemicals, polymers, food, and fuels. Similarly, Brazil has recently launched the "RenovaBio policy" to cut down carbon emissions and give a major push to renewable ethanol and chemicals. India is also now aiming to blend 20% ethanol in petrol by 2023 according to the Indian Government's new policy. However, there is still a long way to go to compete with conventional chemicals and gasoline. Lacking of industrially mature technology and the high production cost of biofuels and biochemicals from LCB are important concerns. These two factors are impeding the commercialization of biorefineries. In order to unleash the full potential of white biotechnology, modern bioprocess engineering techniques and "omics" technologies can develop designer strains and biocatalysts and optimized processes that can revolutionize the overall success of biorefineries.

Acknowledgement

AKC gratefully acknowledges the financial support from FAPESP (Project number: 2020/12559-6) and CNPq for scientific productivity program (Process number: 309214/2021-1).

References

Abdeshahian, P., Ascencio, J.J., Philippini, R.R. et al. (2020). Utilization of sugarcane straw for production of β-glucan biopolymer by *Lasiodiplodia theobromae* CCT 3966 in batch fermentation process. *Bioresour. Technol.* 314: 123716.

Ahmad, E. and Pant, K.K. (2018). Lignin conversion: a key to the concept of lignocellulosic biomass-based integrated biorefinery. In: *Waste Biorefinery* (ed. T. Bhaskar, A. Pandey, S.V. Mohan, et al.), 409–444. Amsterdam: Elsevier.

Alexandratos, N., Bruinsma, J. (2012). World agriculture towards 2030/2050: the 2012 revision. Food and Agriculture Organization, United Nations, Rome.

Almeida, E.L., Andrade, C.M.G., and dos Santos, O.A. (2018). Production of biodiesel via catalytic processes: a brief review. *Int. J. Chem. React. Eng.* 16 (5): 1–13.

Amin, F.R., Khalid, H., Zhang, H. et al. (2017). Pretreatment methods of lignocellulosic biomass for anaerobic digestion. *AMB Express* 7 (1): 1–12.

Ascencio, J.J., Chandel, A.K., Philippini, R.R., and da Silva, S.S. (2020). Comparative study of cellulosic sugars production from sugarcane bagasse after dilute nitric acid, dilute sodium hydroxide and sequential nitric acid-sodium hydroxide pretreatment. *Biomass Convers. Biorefin.* 10 (4): 813–822.

de Assis Castro, R.C., Fonseca, B.G., dos Santos, H.T.L. et al. (2017). Alkaline deacetylation as a strategy to improve sugars recovery and ethanol production from rice straw hemicellulose and cellulose. *Ind. Crop. Prod.* 106: 65–73.

Bai, D.M., Li, S.Z., Liu, Z.L., and Cui, Z.F. (2008). Enhanced L-(+)-lactic acid production by an adapted strain of *Rhizopus oryzae* using corncob hydrolysate. *Appl. Biochem. Biotechnol.* 144 (1): 79–85.

Banerjee, S., Mudliar, S., Sen, R. et al. (2010). Commercializing lignocellulosic bioethanol: technology bottlenecks and possible remedies. *Biofuels Bioprod. Biorefin.* 4 (1): 77–93.

Bhatia, L., Bachheti, R.K., Garlapati, V.K., and Chandel, A.K. (2020a). Third-generation biorefineries: a sustainable platform for food, clean energy, and nutraceuticals production. *Biomass Convers. Biorefin.* 1–16.

Bhatia, S.K., Jagtap, S.S., Bedekar, A.A. et al. (2020b). Recent developments in pretreatment technologies on lignocellulosic biomass: effect of key parameters, technological improvements, and challenges. *Bioresour. Technol.* 300: 122724.

Bohre, A., Dutta, S., Saha, B., and Abu-Omar, M.M. (2015). Upgrading furfurals to drop-in biofuels: an overview. *ACS Sustain. Chem. Eng.* 3 (7): 1263–1277.

Bonturi, N., Crucello, A., Viana, A.J.C., and Miranda, E.A. (2017). Microbial oil production in sugarcane bagasse hemicellulosic hydrolysate without nutrient supplementation by a *Rhodosporidium toruloides* adapted strain. *Process Biochem.* 57: 16–25.

Canilha, L., Chandel, A.K., Milessi, S.S.T. et al. (2012). Bioconversion of sugarcane biomass into ethanol: an overview about composition, pretreatment methods, detoxification of hydrolysates, enzymatic saccharification, and ethanol fermentation. *J. Biomed. Biotechnol.* 2012: 1–15.

Chandel, A.K., Garlapati, V.K., Singh, A.K. et al. (2018). The path forward for lignocellulose biorefineries: bottlenecks, solutions, and perspective on commercialization. *Bioresour. Technol.* 264: 370–381.

Chandel, A.K., Albarelli, J.Q., Santos, D.T. et al. (2019). Comparative analysis of key technologies for cellulosic ethanol production from Brazilian sugarcane bagasse at a commercial scale. *Biofuels Bioprod. Biorefin.* 13 (4): 994–1014.

Chandel, A.K., Garlapati, V.K., Jeevan Kumar, S.P. et al. (2020). The role of renewable chemicals and biofuels in building a bioeconomy. *Biofuels Bioprod. Biorefin.* 14 (4): 830–844.

Chandel, A.K., Forte, M.B., Gonçalves, I.S. et al. (2021). Brazilian biorefineries from second generation biomass: critical insights from industry and future perspectives. *Biofuels Bioprod. Biorefin.* 15: 1190–1208.

Cherubini, F. and Jungmeier, G. (2010). LCA of a biorefinery concept producing bioethanol, bioenergy, and chemicals from switchgrass. *Int. J. Life Cycle Assess.* 15 (1): 53–66.

Clark, J.H. (2007). Green chemistry for the second generation biorefinery—sustainable chemical manufacturing based on biomass. *J. Chem. Technol. Biotechnol.* 82 (7): 603–609.

Coelho, S.T. and Goldemberg, J. (2004). Alternative transportation fuels: contemporary case studies. In: *Encyclopedia of Energy* (ed. C.J. Cleveland), 67–80. São Paulo: University of São Paulo.

Dale, B.E. and Ong, R.G. (2014). Design, implementation, and evaluation of sustainable bioenergy production systems. *Biofuels Bioprod. Biorefin.* 8 (4): 487–503.

Dalli, S.S., da Silva, S.S., Uprety, B.K., and Rakshit, S.K. (2017). Enhanced production of xylitol from poplar wood hydrolysates through a sustainable process using immobilized new strain *Candida tropicalis* UFMG BX 12-a. *Appl. Biochem. Biotechnol.* 182 (3): 1053–1064.

Deeba, F., Pruthi, V., and Negi, Y.S. (2017). Fostering triacylglycerol accumulation in novel oleaginous yeast *Cryptococcus psychrotolerans* IITRFD utilizing groundnut shell for improved biodiesel production. *Bioresour. Technol.* 242: 113–120.

Dhanya, B.S., Mishra, A., Chandel, A.K., and Verma, M.L. (2020). Development of sustainable approaches for converting the organic waste to bioenergy. *Sci. Total Environ.* 723: 138109.

Du, W., Yu, H., Song, L. et al. (2011). The promoting effect of byproducts from *Irpex lacteus* on subsequent enzymatic hydrolysis of bio-pretreated cornstalks. *Biotechnol. Biofuels* 4 (1): 1–8.

van Dyk, S., Su, J., Mcmillan, J.D., and Saddler, J. (2019). Potential synergies of drop-in biofuel production with further co-processing at oil refineries. *Biofuels Bioprod. Biorefin.* 13 (3): 760–775.

Fernando, S., Adhikari, S., Chandrapal, C., and Murali, N. (2006). Biorefineries: current status, challenges, and future direction. *Energy Fuel* 20 (4): 1727–1737.

Frazzetto, G. (2003). White biotechnology: the application of biotechnology to industrial production holds many promises for sustainable development, but many products still have to pass the test of economic viability. *EMBO Rep.* 4 (9): 835–837.

Garlapati, V.K., Tewari, S., and Ganguly, R. (2019). Life cycle assessment of first-, second-generation, and microalgae biofuels. In: *Advances in Feedstock Conversion Technologies for Alternative Fuels and Bioproducts* (ed. M. Hosseini), 355–371. Amsterdam: Woodhead Publishing.

Garlapati, V.K., Chandel, A.K., Kumar, S.J. et al. (2020). Circular economy aspects of lignin: towards a lignocellulose biorefinery. *Renew. Sust. Energ. Rev.* 130: 109977.

Geetha, G.S. and Gopalakrishnan, A.N. (2011). Bioethanol production from paper fibre residue using diluted alkali hydrolysis and the fermentation process. *J. Chem.* 8 (4): 1951–1957.

Gnansounou, E. (2011). Assessing the sustainability of biofuels: a logic-based model. *Energy* 36 (4): 2089–2096.

Gomes, C., Silva, A.C., Marques, A.C. et al. (2020). Biotechnology applied to cosmetics and aesthetic medicines. *Cosmetics* 7 (2): 33.

Goshadrou, A., Karimi, K., and Lefsrud, M. (2013). Characterization of ionic liquid pretreated aspen wood using semi-quantitative methods for ethanol production. *Carbohydr. Polym.* 96 (2): 440–449.

Grabber, J.H. (2005). How do lignin composition, structure, and cross-linking affect degradability? A review of cell wall model studies. *Crop Sci.* 45 (3): 820–831.

Guan, W., Xu, G., Duan, J., and Shi, S. (2018). Acetone–butanol–ethanol production from fermentation of hot-water-extracted hemicellulose hydrolysate of pulping woods. *Ind. Eng. Chem. Res.* 57 (2): 775–783.

Halder, P., Kundu, S., Patel, S. et al. (2019). Progress on the pre-treatment of lignocellulosic biomass employing ionic liquids. *Renew. Sust. Energ. Rev.* 105: 268–292.

Han, X., Guo, Y., Liu, X. et al. (2019). Catalytic conversion of lignocellulosic biomass into hydrocarbons: a mini review. *Catal. Today* 319: 2–13.

Hassan, S.S., Williams, G.A., and Jaiswal, A.K. (2018). Emerging technologies for the pretreatment of lignocellulosic biomass. *Bioresour. Technol.* 262: 310–318.

Hassan, S.S., Williams, G.A., and Jaiswal, A.K. (2019). Lignocellulosic biorefineries in Europe: current state and prospects. *Trends Biotechnol.* 37 (3): 231–234.

Heux, S., Meynial-Salles, I., O'Donohue, M.J., and Dumon, C. (2015). White biotechnology: state of the art strategies for the development of biocatalysts for biorefining. *Biotechnol. Adv.* 33 (8): 1653–1670.

Hirani, A.H., Javed, N., Asif, M. et al. (2018). A review on first- and second-generation biofuel productions. In: *Biofuels: Greenhouse Gas Mitigation and Global Warming* (ed. A. Kumar, S. Ogita and Y.-Y. Yau), 141–154. New Delhi: Springer.

Ho, H.L. and Chinonso, A.M. (2016). Overproduction of xylanase from mutants of *Bacillus subtilis* with barley husk as the prime carbon source under submerged fermentation after random mutagenesis using ethyl methane sulfonate (EMS) and acridine orange (AO). *Microbiol. Res. J. Int.* 14 (1): 1–17.

Hsu, T.A. (1996). Pretreatment of biomass. In: *Handbook on Bioethanol, Production and Utilization*, Applied Energy Technology Series (ed. C. Wyman, J.G. Speight and S. Lee), 179–212. Washington, DC: Taylor & Francis.

Intergovernmental Panel on Climate Change (IPCC) (2006). *2006 IPCC Guidelines for National Greenhouse Gas Inventories* (ed. H.S. Eggleston, L. Buendia, K. Miwa, et al.). Institute for Global Environmental Strategies, Kanagawa.

International Energy Agency (IEA) (2019). *Drop-in Biofuels: The Key Role that Co-processing Will Play in its Production*. USA: IEA Bioenergy.

Isikgor, F.H. and Becer, C.R. (2015). Lignocellulosic biomass: a sustainable platform for the production of bio-based chemicals and polymers. *Polym. Chem.* 6 (25): 4497–4559.

Kachrimanidou, V., Kopsahelis, N., Chatzifragkou, A. et al. (2013). Utilisation of by-products from sunflower-based biodiesel production processes for the production of fermentation feedstock. *Waste Biomass Valorization* 4 (3): 529–537.

Kamm, B., Gruber, P.R., and Kamm, M. (2016). Biorefineries—Industrial Processes and Products. In: *Ullmann's Encyclopedia of Industrial Chemistry* (ed. B. Elvers, S. Hawklins and G. Schulz), 1–38. Hoboken, NJ: Wiley-VCH.

Kejun, S., Juntao, Z., Ying, C. et al. (2011). Accelerating the degradation of green plant waste with chemical decomposition agents. *J. Environ. Manag.* 92 (10): 2708–2713.

Kim, D.Y., Yim, S.C., Lee, P.C. et al. (2004). Batch and continuous fermentation of succinic acid from wood hydrolysate by *Mannheimia succiniciproducens* MBEL55E. *Enzym. Microb. Technol.* 35 (6–7): 648–653.

Kohse-Höinghaus, K., Oßwald, P., Cool, T.A. et al. (2010). Biofuel combustion chemistry: from ethanol to biodiesel. *Angew. Chem. Int. Ed.* 49 (21): 3572–3597.

Kumar, A.K. and Sharma, S. (2017). Recent updates on different methods of pretreatment of lignocellulosic feedstocks: a review. *Bioresour. Bioprocess.* 4 (1): 1–19.

Kumar, B., Bhardwaj, N., Agrawal, K. et al. (2020). Current perspective on pretreatment technologies using lignocellulosic biomass: an emerging biorefinery concept. *Fuel Process. Technol.* 199: 106244.

Lichtenthaler, F.W. and Peters, S. (2004). Carbohydrates as green raw materials for the chemical industry. *C. R. Chim.* 7 (2): 65–90.

Lim, J.W. and Wang, J.Y. (2013). Enhanced hydrolysis and methane yield by applying microaeration pretreatment to the anaerobic co-digestion of brown water and food waste. *Waste Manag.* 33 (4): 813–819.

Lima, L.F.O., Habu, S., Gern, J.C. et al. (2008). Production and characterization of the exopolysaccharides produced by *Agaricus brasiliensis* in submerged fermentation. *Appl. Biochem. Biotechnol.* 151 (2): 283–294.

Lin, C.Y. and Lu, C. (2021). Development perspectives of promising lignocellulose feedstocks for production of advanced generation biofuels: a review. *Renew. Sust. Energ. Rev.* 136: 110445.

Lopes, M.L., de Lima, P.S.C., Godoy, A. et al. (2016). Ethanol production in Brazil: a bridge between science and industry. *Braz. J. Microbiol.* 47: 64–76.

Lugani, Y., Rai, R., Prabhu, A.A. et al. (2020). Recent advances in bioethanol production from lignocelluloses: a comprehensive review with a focus on enzyme engineering and designer biocatalysts. *Biofuel Res. J.* 7 (4): 1267–1295.

Lynd, L.R. (1996). Overview and evaluation of fuel ethanol from cellulosic biomass: technology, economics, the environment, and policy. *Annu. Rev. Energy Environ.* 21 (1): 403–465.

Lynd, L.R., Weimer, P.J., Van, W.H.Z., and Pretorius, I.S. (2002). Microbial cellulose utilization: fundamentals and biotechnology. *Microbiol. Mol. Biol. Rev.* 66 (3): 506–577.

Mat Aron, N.S., Khoo, K.S., Chew, K.W. et al. (2020). Sustainability of the four generations of biofuels—a review. *Int. J. Energy Res.* 44 (12): 9266–9282.

McKendry, P. (2002). Energy production from biomass (part 1): overview of biomass. *Bioresour. Technol.* 83 (1): 37–46.

Meng, X., Pu, Y., Yoo, C.G. et al. (2017). An in-depth understanding of biomass recalcitrance using natural poplar variants as the feedstock. *ChemSusChem* 10 (1): 139–150.

Modenbach, A.A. and Nokes, S.E. (2013). Enzymatic hydrolysis of biomass at high-solids loadings—a review. *Biomass Bioenergy* 56: 526–544.

Mosier, N., Wyman, C., Dale, B. et al. (2005). Features of promising technologies for pretreatment of lignocellulosic biomass. *Bioresour. Technol.* 96 (6): 673–686.

Mota, T.R., de Oliveira, D.M., Marchiosi, R. et al. (2018). Plant cell wall composition and enzymatic deconstruction. *AIMS Bioeng.* 5 (1): 63–77.

Muharja, M., Junianti, F., Ranggina, D. et al. (2018). An integrated green process: subcritical water, enzymatic hydrolysis, and fermentation, for biohydrogen production from coconut husk. *Bioresour. Technol.* 249: 268–275.

Mukherjee, G., Singh, R.K., Mitra, A., and Sen, S.K. (2007). Ferulic acid esterase production by *Streptomyces* sp. *Bioresour. Technol.* 98 (1): 211–213.

Nichols, N.N., Sharma, L.N., Mowery, R.A. et al. (2008). Fungal metabolism of fermentation inhibitors presents in corn stover dilute acid hydrolysate. *Enzym. Microb. Technol.* 42 (7): 624–630.

Nogueira, L.A.H., Souza, G.M., Cortez, L.A.B., and de Brito Cruz, C.H. (2020). Biofuels for transport. In: *Future Energy* (ed. T. Letcher), 173–197. Amsterdam: Elsevier.

Obruca, S., Benesova, P., Kucera, D. et al. (2015). Biotechnological conversion of spent coffee grounds into polyhydroxyalkanoates and carotenoids. *New Biotechnol.* 32 (6): 569–574.

OECD/FAO (Organisation for Economic Co-operation and Development/Food and Agriculture Organization) (2020). *OECD-FAO Agricultural Outlook 2020–2029*. Paris: FAO, Rome/OECD Publishing.

O'Hara, I.M., Zhang, Z., Rackemann, D.W. et al. (2013). Prospects for the development of sugarcane biorefineries. In: *Proceedings of the 28th International Society of Sugar Cane Technologists Conference* (ed. D.M. Hogarth), 54–65. Brazil: Sociedade dos Tecnicos Acucareiros e Alcooleiros do Brasil (STAB).

Ostovareh, S., Karimi, K., and Zamani, A. (2015). Efficient conversion of sweet sorghum stalks to biogas and ethanol using organosolv pretreatment. *Ind. Crop. Prod.* 66: 170–177.

Parimala, V., Krishnani, K.K., Gupta, B.P. et al. (2007). Removal of ammonia and nitrite from coastal water using low-cost agricultural waste. *Bull. Environ. Contam. Toxicol.* 78 (3–4): 288–293.

Qureshi, N., Saha, B.C., Dien, B. et al. (2010). Production of butanol (a biofuel) from agricultural residues: Part I—Use of barley straw hydrolysate. *Biomass Bioenergy* 34 (4): 559–565.

Reddy, G.V., Babu, P.R., Komaraiah, P. et al. (2003). Utilization of banana waste for the production of lignolytic and cellulolytic enzymes by solid substrate fermentation using two *Pleurotus* species (*P. ostreatus* and *P. sajor-caju*). *Process Biochem.* 38 (10): 1457–1462.

Rodionova, M.V., Poudyal, R.S., Tiwari, I. et al. (2017). Biofuel production: challenges and opportunities. *Int. J. Hydrog. Energy* 42 (12): 8450–8461.

Röper, H. (2002). Renewable raw materials in Europe—industrial utilisation of starch and sugar. *Starch-Stärke* 54 (3–4): 89–99.

Sacramento-Rivero, J.C. (2012). A methodology for evaluating the sustainability of biorefineries: framework and indicators. *Biofuels Bioprod. Biorefin.* 6 (1): 32–44.

Sanchez, O.J. and Cardona, C.A. (2008). Trends in biotechnological production of fuel ethanol from different feedstocks. *Bioresour. Technol.* 99 (13): 5270–5295.

Sharma, R.K. and Arora, D.S. (2015). Fungal degradation of lignocellulosic residues: an aspect of improved nutritive quality. *Crit. Rev. Microbiol.* 41 (1): 52–60.

Sharma, L. and Singh, A.K. (2015). Algal biofuel: a sustainable green energy. In: *Biotechnology in India: initiatives and accomplishments*, 205–222. New Delhi: New India Publishing Agency.

Sharma, H.K., X, C., and Qin, W. (2019). Biological pretreatment of lignocellulosic biomass for biofuels and bioproducts: an overview. *Waste Biomass Valorization* 10 (2): 235–251.

Shrivastava, B., Jain, K.K., Kalra, A., and Kuhad, R.C. (2014). Bioprocessing of wheat straw into nutritionally rich and digested cattle feed. *Sci. Rep.* 4 (1): 1–9.

Sun, Z. and Liu, S. (2012). Production of *n*-butanol from concentrated sugar maple hemicellulosic hydrolysate by *Clostridia acetobutylicum* ATCC824. *Biomass Bioenergy* 39: 39–47.

Ullah, K., Sharma, V.K., Dhingra, S. et al. (2015). Assessing the lignocellulosic biomass resources potential in developing countries: a critical review. *Renew. Sust. Energ. Rev.* 51: 682–698.

Yang, B. and Wyman, C.E. (2008). Pretreatment: the key to unlocking low-cost cellulosic ethanol. *Biofuels Bioprod. Biorefin.* 2 (1): 26–40.

Yoo, C.G., Meng, X., Pu, Y., and Ragauskas, A.J. (2020). The critical role of lignin in lignocellulosic biomass conversion and recent pretreatment strategies: a comprehensive review. *Bioresour. Technol.* 301: 122784.

Zeng, Y., Zhao, S., Yang, S., and Ding, S.Y. (2014). Lignin plays a negative role in the biochemical process for producing lignocellulosic biofuels. *Curr. Opin. Biotechnol.* 27: 38–45.

Zhang, Q., He, G., Wang, J. et al. (2009). Mechanisms of the stimulatory effects of rhamnolipid biosurfactant on rice straw hydrolysis. *Appl. Energy* 86: S233–S237.

Zhang, Q., Li, Y., Xia, L., and Liu, Z. (2014). Enhanced xylitol production from statistically optimized fermentation of cotton stalk hydrolysate by immobilized *Candida tropicalis*. *Chem. Biochem. Eng. Q.* 28 (1): 87–93.

Zhang, Y., Serate, J., Xie, D. et al. (2020). Production of hydrolysates from unmilled AFEX-pretreated switchgrass and comparative fermentation with *Zymomonas mobilis*. *Bioresour. Technol. Rep.* 11: 100517.

Zhao, X., Zhang, L., and Liu, D. (2012). Biomass recalcitrance. Part I: The chemical compositions and physical structures affecting the enzymatic hydrolysis of lignocellulose. *Biofuels Bioprod. Biorefin.* 6 (4): 465–482.

Zheng, J., Tashiro, Y., Wang, Q. et al. (2015). Feasibility of acetone–butanol–ethanol fermentation from eucalyptus hydrolysate without nutrients supplementation. *Appl. Energy* 140: 113–119.

Zhong, N., Chandra, R., and Saddler, J.J.N. (2019). Sulfite post-treatment to simultaneously detoxify and improve the enzymatic hydrolysis and fermentation of a steam-pretreated softwood lodgepole pine whole slurry. *ACS Sust. Chem. Eng.* 7 (5): 5192–5199.

Zhou, H., Lou, H., Yang, D. et al. (2013). Lignosulfonate to enhance enzymatic saccharification of lignocelluloses: role of molecular weight and substrate lignin. *Ind. Eng. Chem. Res.* 52 (25): 8464–8470.

Zhu, P., Abdelaziz, O.Y., Hulteberg, C.P., and Riisager, A. (2020). New synthetic approaches to biofuels from lignocellulosic biomass. *Curr. Opin. Green Sust. Chem.* 21: 16–21.

Zoghlami, A. and Paës, G. (2019). Lignocellulosic biomass: understanding recalcitrance and predicting hydrolysis. *Front. Chem.* 7: 874.

2

Lignocellulose Feedstock Availability, Types of Feedstocks, and New Designer Crops

V. Guadalupe Bustos[1], R. Daniel Trujillo[1], C. Linda M. Martínez[1,2], and S. Rodolfo Torres[1]

[1] *Department of Biotechnology, University Autonomous of Tamaulipas, Unidad Académica Multidisciplinaria Mante, Ciudad Mante, Tamaulipas, Mexico*
[2] *Department of Food Technology, University Autonomous of Tamaulipas Unidad Académica Multidisciplinaria Reynosa-Aztlán, Reynosa, Tamaulipas, México*

2.1 Introduction

In recent years, the growing interest of researchers to give value to lignocellulosic biomass of biological origin is due to global problems of environmental pollution (Asgher et al. 2013). Lignocellulosic biomass is the most abundant and cheapest renewable raw material in the world (Qian 2014; Ge et al. 2018), with potential use in biofuel production, cogeneration of electrical and thermal energy, and the generation of chemical fuels, among other applications (Martínez et al. 2015). This annual production has been estimated at 1×10^{10} million tons around the world (Sánchez and Cardona 2008). Lignocellulosic biomass is considered to be the residues from the agricultural, forestry, and urban sectors, as well as agro-industrial residues from the food industry and other industries (Ge et al. 2016).

Lignocellulosic waste from agro-industrial origin is a type of biomass with attractive characteristics of being renewable, abundant, and economical; it is used as a unique natural resource for the collection of bioenergy on a large scale through its transformation by different processes and when brought into contact with microorganisms is capable of transforming them into metabolites of commercial interest (Field et al. 2015). Until 2017, the most important crops available in considerable quantities were barley, corn, rice, soybeans, sugarcane, and wheat in 227 countries and territories in the world—within North and South America, the geographical regions with a production of more than 500 million tons per year, followed by East and South Asia with a waste production of more than 200 million tons per year (Boufi 2017). These lignocellulosic materials (LCMs) represent one of the largest sources of carbon and renewable energy on earth and include various agricultural materials and residues such as bamboo, palms, bagasse, husks, corn stubble, straw, wheat stubble, algae, and others that are accumulated annually as an important renewable biomass raw material (Zhou et al. 2014; Luo et al. 2019).

Lignocellulose Bioconversion Through White Biotechnology, First Edition. Edited by Anuj Kumar Chandel.

Mexico has a great diversity of ecosystems, being a country with potential for the availability of raw materials that can be used as a source of compounds of commercial interest in the food, pharmaceutical, cosmetic, and chemical industries. The interest in the use of these materials is based on the idea of rationally taking advantage of natural and renewable raw material sources in order to obtain a wide range of products adapted to market requirements. An effective use of the constituents of these materials and the reduction of the environmental impact and economy of the process are key factors in the search for alternative materials.

The aim of this chapter focuses mainly on the potential use of lignocellulosic biomass as a renewable source and the different types of raw material to obtain products with high added value, which could represent a solution to the large number of environmental problems in the world.

2.2 Lignocellulosic Biomass

Lignocellulosic biomass is one of the raw materials potentially used for the production of various second-generation cellulosic and biochemical products such as ethanol (Zhou et al. 2018). Lignocellulose is the main and most abundant component of the biomass produced from photosynthesis. In 2012, the total amount of available raw material in the United States was from 341 million metric tonnes; about 70% of this came from agricultural residues and 30% was from forest residues (Balan 2014).

The main components of lignocellulosic biomass are cellulose (35%–50%), hemicelluloses (25%–30%), and (25%–30%) lignin (Anwar et al. 2014), as well as minor components such as oils, resins, tannins, sugars, gums, dyes, and starches (Romero-Uscanga et al. 2014). These components are called structural components and are polymeric in nature since they constitute the cell wall of the LCMs, so they are interpenetrated and linked through physical and chemical interactions (Isikgor and Becer 2015). Table 2.1 shows the content of the different structural components in different substrates.

Table 2.1 Content of cellulose, hemicellulose, and lignin of different lignocellulosic materials (wt% on total dry material).

	Cellulose (%)	Hemicellulose (%)	Lignin (%)	Ash (%)	Extracts (%)	References
Wheat	52.11	20.09	10.68	10.64	–	Cueva and Monzón (2014)
straw	38.7	30.0	12.4	1.6	35.4	Sevilla (2017)
Blueberry bushes	51.9	11.9	26.6	1.8	6.9	Pacheco et al. (2018)
Corn	42.48	26.73	7.02	5.98	–	Cueva and Monzón (2014)
bagasse	28.3	25.0	10.1	8.1	36.7	Sevilla (2017)
Barley	33.38	10.62	10.58	–	–	Rojas-León et al. (2014)
husk	13.2	25–3	19–6	4.1	–	Kemppainen et al. (2016)
Barley	47.14	21.35	10.53	9.75	–	Cueva and Monzón (2014)
straw	18.16	5.37	13.94	–		Rojas-León et al. (2014)

(Continued)

Table 2.1 (Continued)

	Cellulose (%)	Hemicellulose (%)	Lignin (%)	Ash (%)	Extracts (%)	References
Goliath grass	33.79	22.54	16.1	2.46	10.77	Mateus et al. (2012)
Rice husk	36.53	18.41	22.53	13.59	–	Cueva and Monzón (2014)
						Matos et al. (2020)
Sugarcane bagasse	44.87	16.99	14.40	2.57	–	Cueva and Monzón (2014)
	42.94	23.54	24.97	3.03	–	González-Leos et al. (2017)
	42.22	33.95	13.44	–	–	Matos et al. (2020)
Agave	17.72	17.15	7.32	12.45	45.34	Carmona et al. (2017)
Coconut shells	32–43	0.15–0.25	40–45	2	–	Ayrilmis et al. (2011)
	36.0	25.0	28.0	–	–	Dhyani and Bhaskar (2018)
Orange peel	9.21	10.50	0.84	3.50	–	Ángel Siles López et al. (2010)
	9.93	26.45	2.81	2.28	–	Sánchez-Orozco and Vázquez-Velázquez (2017)
Agave bagasse	41.90	12.10	7.20	8.80	29.00	Iñiguez et al. (2005)
	42.0	20.0	15.0	–	–	Salcedo Luna et al. (2010)
Lemon peel	23.1	8.09	7.6	2.50	–	Mamma and Christakopoulos (2014)
Lentil stubble	40.32	18.27	16.12	5.44	–	Cueva and Monzón (2014)
Banana peel	12.17	10.19	2.88	9.81	–	Oberoi et al. (2012)
Citrus peel	20.8	17.2	8.9	3.0	–	Mamma and Christakopoulos (2014)
Wheat bran	9–12	55.0	3–5	5–6	–	Babu et al. (2018)
	4.31	33.95	1.62	–	–	Matos et al. (2020)
Corn stubble	41.41	40.02	4.24	–		Matos et al. (2020)
Oat bran	6.5–9.9	20.8–33.0	2.2–9.0	5.7–6.5	–	Chaquilla-Quilca et al. (2018)
Vine trimming waste	34.1	19.0	27.1	–	7.1	Bustos et al. (2004)
Rye straw	37.6–44.1	22.2–30.5	19.0–22.9	–	–	Gullón et al. (2010)

–, unreported data.

The content of the different components varies relatively from one material to another due to the differences that exist within the same species, which may be due to different factors such as environmental factors, growth (climate, soil), physical location, origin, and age, among others (Wodzicki 2001; Hafızoğlu and Usta 2005).

2.2.1 Plant Cell Wall

The cell wall is located externally in the plasma membrane. The most characteristic component found throughout the cell wall of plants is cellulose, in addition to other polysaccharides and glycoproteins. In some cells, three layers can be distinguished in the secondary cell wall: S1, S2, and S3, for outer, middle, and inner layer, respectively (Sticklen 2008). That is, the different types of plant cells can be identified by the characteristics of their cell wall, as shown in Figure 2.1.

The S1 layer is the thinnest of the S layers, being only 0.1–0.35 μm thick and representing just 5%–10% of the total thickness of the cell wall. The S2 layer is the thickest layer in the secondary cell wall and is the most important with regard to mechanical support, measuring between 1 and 10 μm; it accounts for 75%–85% of the total thickness of the secondary cell wall. Finally, the S3 layer is relatively thin, being only 0.5–1.10 μm thick. The S1 layer contains helices of microfibrils and is highly birefringent, while S2 and S3 do not have helices (Figure 2.1a) (Plomion et al. 2001; Evert 2006; Sticklen 2008).

The secondary cell walls are primarily made up of the polymers cellulose, hemicellulose, and lignin, as shown in Figure 2.1(b) (Loix et al. 2017).

Cellulose is the most abundant organic molecule, susceptible to hydrolysis treatments (Balan et al. 2009); it is a polymer of D-glucose linked by β-1,4 glucosidic bonds structured in long linear chains (microfibrils) linked by hydrogen bonds and intramolecular van der Waals forces forming a crystalline structure resistant to hydrolysis and amorphous regions susceptible to enzymatic degradation (quoted by Cuervo et al. (2009) from Ovando-Chacón and Waliszewski (2005)). Cellulose is synthesized, to a lesser extent, by bacteria of the *Acetobacter* genus (Sasakura et al. 2005; Czaja et al. 2007).

On the other hand, hemicellulose as a complex polymer of heteropolysaccharides formed by pentoses (D-xylose and L-arabinose) and hexoses (D-glucose, D-mannose, and D-galactose) forms branched chains and 4-O-methylglucuronic acid, D-galacturonic acid and D-glucuronic acid. Sugars are linked by β-1,4 bonds and occasionally by β-1,3 bonds (quoted by Cuervo et al. (2009) from Pérez et al. (2002)). This polymer binds to the surface of the cellulose microfibrils and can be segregated into four groups: xyloglycans (xylans), mannoglycans (mannans), xyloglucans, and mixed linkage β-glucans (Ebringerová 2006). Meanwhile, pectins consist of four polysaccharide domains: homogalacturonan (HGA), xylogalacturonan (XGA), rhamnogalacturonan I (RGI), and rhamnogalacturonan II (RGII) (Figure 2.1c) (Trujillo-Ramírez et al. 2018), which provide resistance and support to plants by maintaining cell consistency (mechanical resistance), besides influencing various properties of the cell wall such as porosity, surface charge, pH, and ion balance and transport (Voragen et al. 2009). Generally, pectin is most abundant in the middle lamella, followed by the primary cell walls of the plant cell (Suárez et al. 2013).

Lignin is an amorphous, three-dimensional, and branched heteropolymer formed by aromatic alcohols that gives structural support, rigidity, impermeability, and protection to structural polysaccharides (cellulose and hemicellulose) and is highly resistant to chemical and biological degradation (Aro et al. 2005). Lignified tissues resist the attack of microorganisms, preventing the penetration of destructive enzymes into the cell wall. Lignin is optimally degraded in aerobic conditions through various dioxygenase-type enzymes, although some anaerobic bacteria can degrade it slowly (Zhou et al. 2014); it is soluble in

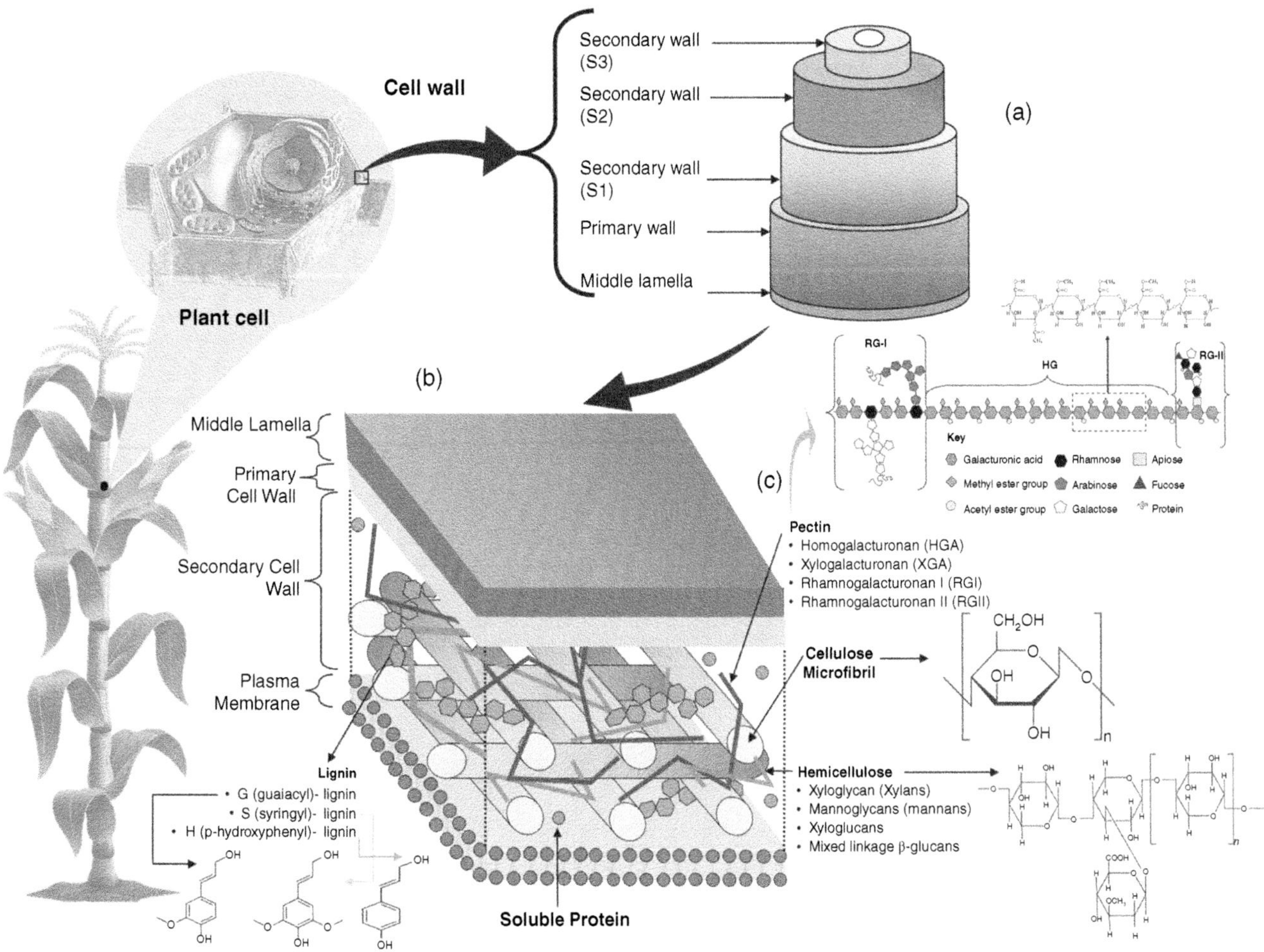

Figure 2.1 Schematic representation of (a) the composition of the cell wall of plants; (b) the structure and composition of the primary and secondary cell walls of plants; (c) a hypothetical model of the structure and main functional groups of pectin.

water and has a density of 1.3 g mL^{-1} at 25 °C; its structural characteristics are not well defined. Chávez-Sifontes and Domine (2013) mention that both its structure and composition depend on factors such as the source material and its extraction form.

Lignin originates from the oxidative polymerization of monolignol subunits. Its biosynthesis starts with the conversion of the amino acid phenylalanine to cinnamic acid by the key enzyme phenylalanine ammonia lyase and subsequently to *p*-coumaric acid (Figure 2.1b). After conversion of *p*-coumaric acid to *p*-coumaroyl-CoA by 4-coumarate-CoA ligase, *p*-coumaroyl-CoA can follow one of three pathways that are specific but still share common enzymes, leading to the formation of three possible alcohols. These alcohols are further converted to specific monolignol subunits: guaiacyl (G), *p*-hydroxyphenyl (H), and syringyl (S) lignin. The last step of monolignol synthesis is not fully elucidated yet (Loix et al. 2017).

2.3 Biomass Conversion Pathways

The use of lignocellulosic biomass to obtain products of commercial interest dates back to 1930. In this decade William Jay developed the idea of taking advantage of carbohydrates of agro-industrial origin as raw material to obtain various commercial products where agriculture would be considered as a source with potential to obtain a high range of chemical compounds (Sifontes 2019). The major challenge encountered in the utilization of biomass for bioconversion in a wide range of applications is the complex structure and characteristics of the cell wall components of the biomass which need to be pretreated prior to starting the bioconversion process.

2.3.1 Lignocellulosic Biomass Pretreatment

In general, the fractionation of LCM is conditioned by the fact that cellulose, hemicelluloses, and lignin are polymeric in nature and are interpenetrated with each other, with physical and chemical bonds between them, in such a way that the classic fractionation (crystallization, precipitation, extraction) are not satisfactory for their individual separation since the isolation of one of the polymers implies the degradation of the structure of the others (Nimz and Casten 1986). The literature reports that there are characteristics of great importance that must be considered in a biomass pretreatment process, such as (i) low-cost chemical reagents; (ii) low waste generation; (iii) that rapid reactions are generated and they must be non-reactive, non-corrosive to minimize maintenance costs in reactors; (iv) that the yields in acid hydrolysis must be good to decrease the costs of enzymatic hydrolysis; (v) that lignin recovery must be facilitated (Wyman et al. 2005a, 2005b, 2005c).

Therefore, to increase the access and biodegradability of cellulose and hemicellulose for enzymatic or chemical action, Barakat et al. (2013) and Alonso et al. (2013) proposed the importance of carrying out a pretreatment of lignocellulosic biomass before any other treatment. According to its composition, the degree of humidity, and the product to be obtained, there are four main groups for the processes or technologies of biomass transformation (Figure 2.2).

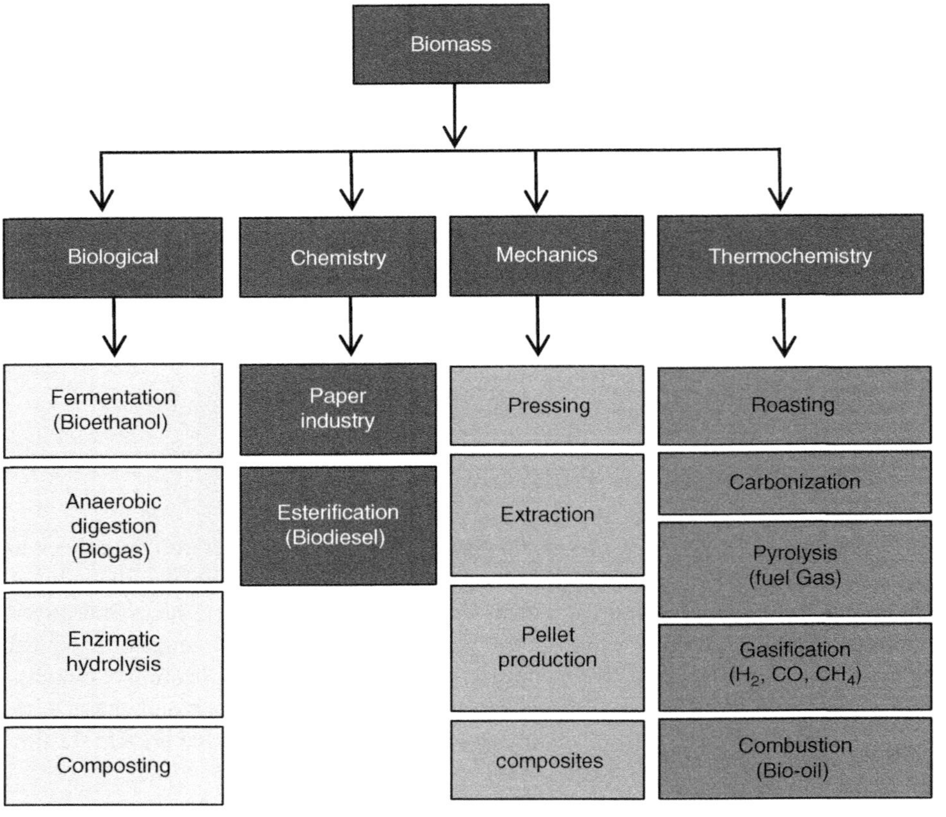

Figure 2.2 Biomass transformatión technologies.

The objective of using these treatments is in order to reduce the size of the biomass and open its physical structure: there are advantages and disadvantages to each one (cited by Isikgor and Becer (2015), taken from Taherzadeh and Karimi (2008) and Saha (2005)). Table 2.2 shows some most used pretreatment options in the fractionation of biomass as well as the advantages and disadvantages that each one presents (Menon and Rao 2012).

Table 2.2 Lignocellulosic biomass treatments.

Treatments	Advantages	Disadvantages	References
Biological			
Microorganisms	• Lignin and hemicellulose degradation • Requires little energy	• Slow hydrolysis • Requires long residence times	Putro et al. (2016), Singh et al. (2016), Ishola and Taherzadeh (2014), and Sharma et al. (2019)

Table 2.2 (Continued)

Treatments	Advantages	Disadvantages	References
Physical			
Mechanical fragmentation and pyrolysis	• Reduces cellulose crystallinity	• Major energy consumption	Behera et al. (2014) and Sun et al. (2016)
Steam explosion	• Hemicellulose degradation and lignin transformation • Cost efficiency	• Destruction of xylan fractions • Lignin incomplete breaking off • Forms inhibitors	Sun et al. (2016), Kemppainen et al. (2016), and Maniet et al. (2017)
Explosion with CO_2	• Reduction of lignin content • Does not produce toxic wastes	• Biomass deficiency with high content of lignin	Behera et al. (2014), Putro et al. (2016), and Sun et al. (2016)
Ammonia fiber explosion	• Effectivity in costs • Without formation of inhibitors	• Deficiency in biomass with high lignin content	Behera et al. (2014) and Sun et al. (2016)
Pyrolysis by pulsed magnetic field	• Obtaining gaseous and liquid products under ambient conditions • Alter plant cells • Depolymerization and fragmentation of the biopolymers (cellulose, hemicellulose, and lignin)	• Requires high temperatures • Ash production	Huang et al. (2016), Bridgwater (2012), and Nanduri et al. (2021)
Chemical			
Acid hydrolysis	• Hydrolyzes hemicellulose to xilose and other sugars • Alters lignin structure	• High costs • Formation of toxic substances • Reduces sugar recovery • Equipment corrosion	Behera et al. (2014) and Singh et al. (2016)
Alkaline hydrolysis	• Removes lignin and hemicellulose • Reduction of crystallinity	• Requires long residence times • Salt formation • Low temperature and pressure	Bali et al. (2015), Sun et al. (2016), and Singh et al. (2016)
Organsolvents	• Lignin and hemicellulose hydrolysis • Recovery of the extracted lignin with high purity	• High costs • Requires draining reactor solvents	Rabemanolontsoa and Saka (2012), Li et al. (2016), Zhang et al. (2016), and Behera et al. (2014)
Ozonolysis	• Reduction of lignin content • Does not produce toxic remains	• High production costs • It requires a great amount of ozone	Chaturvedi and Verma (2013), Singh et al. (2016), and Travaini et al. (2016)

To select the technology or the most appropriate method for a better biomass conversion Álvarez Alonso et al. (2008) recommend multicriteria decision analysis through a method called the analytical hierarchy process, which consists in the selection of alternatives based on well-defined variables or criteria; this method is currently widely used in research.

2.3.2 Enzymatic Hydrolysis

Factors such as crystallinity, degree of polymerization, moisture content, lignin content, and available contact surface make the enzymatic hydrolysis of LCMs very limited (Puri 1984; Koullas et al. 1992; Chang and Holtzapple 2000). Therefore, pretreatment of lignocellulosic substrates is considered as a necessary step for optimal enzymatic hydrolysis. In order to achieve the hydrolysis of lignocellulosic biomass, different types of physical pretreatments have been proposed, including mechanical and thermal pretreatments and chemical pretreatments which use acids, bases, or a combination of these (Ortiz 2016).

Due to the non-fermentable nature of lignin, the bioconversion of lignocellulosic biomass requires (i) a method to release cellulose and hemicellulose from the lignin complex, (ii) the hydrolysis of carbohydrate polymers to produce free sugars and (iii) the fermentation of mixed sugars of hexose and pentose to produce marketable biologically based products (biomaterials) and bioenergy (biofuels) (Kumar et al. 2009; Nanda et al. 2014; Galbe and Wallberg 2019; Sankaran et al. 2020).

The enzymatic hydrolysis of cellulose present in native LCMs requires pretreatments to increase the susceptibility of the substrates. Griggs et al. (2011, 2012a) mention that this reaction requires an "enzyme complex," where three types of activities act synergistically: (i) endoglucanases—β-(1,4) glucan glucanhydrolase (EC 3.2.1.4) that attacks the internal bonds of low crystallinity cellulose, producing a rapid reduction in the average polymerization index (Nidetzky and Steiner 1993); (ii) exoglucanases—β-(1,4) cellobiohydrolase (EC 3.2.1.91) which catalyzes breakage of the extreme bonds present in amorphous cellulose, in crystalline celluloses, and in glucose oligosaccharides, releasing cellobiose, and (iii) cellobiase—β-(1,4) glucosidase (EC 3.2.1.21) which catalyzes the hydrolysis of cellobiose and glucose units of the non-reducing ends of short chain oligosaccharides to give glucose (Carrillo Navarrete 2002; Griggs et al. 2012b).

The catalytic activity of cellulases depends on their ability to adsorb from the substrate and form an active substrate enzyme complex. This last aspect is influenced by factors that affect the substrate, such as physical configuration, morphology, crystallinity, and chemical structure (Focher et al. 1991), owing to the characteristics of the enzyme protein (Klyosov et al. 1986; Gaviria-Acosta et al. 2015), by the relative concentrations of substrate and enzyme, by temperature (Ooshima et al. 1983), and by factors that affect the transfer of matter in the reaction medium (Wang et al. 2013).

The use of cellulases in the hydrolysis of cellulose is made very difficult because of the presence of lignin, which constitutes a physical barrier that prevents the penetration of enzymes (Kumar et al. 2012). Chang and Holtzapple (2000) consider that a factor which limits hydrolytic action is the degree of crystallinity. The advantage of the enzymatic processes of cellulose compared to other hydrolysis methods such as acid hydrolysis are the low costs of utility and elimination since the reactions are carried out with specificity, at moderate temperatures, and in non-corrosive media (Sun and Cheng 2002).

2.3.3 Conversion of Lignocellulosic Components

2.3.3.1 Biofuels

One of the most important stages for the conversion of lignocellulosic biomass into products of industrial interest such as biofuels is the treatment of biomass. According to the dictionary of the Royal Academy of the Spanish Language (RAE because its name is Spanish), biomass is defined as "organic matter originating in a biological process, spontaneous or provoked, usable as a source of energy." Although there is an infinity of interpretations where its use includes the production of a large number of products, the transformation of biomass is relevant because it is the only renewable source available for the production of materials, fuels, and carbon-based products (Arauzo et al. 2014).

In recent years, researchers have focused their studies on the use of lignocellulosic biomass, for which they have launched several industrial processes aimed at the production of first-, second-, third-, and fourth-generation biofuels depending on the complexity of the processes for its production (Dutta et al. 2014).

In first-generation biofuels, processes such as (i) the fermentation of sugars and/or carbohydrates is used to obtain ethanol, methanol, and *n*-butanol; (ii) the *trans*-esterification of oils and fats is used to obtain biodiesel; and (iii) anaerobic digestion in the case of organic waste is used to obtain biogas. The great disadvantage of these first-generation biofuels is that crops such as sugarcane or corn are used for the production of energy vectors (Demirbas 2007; Naik et al. 2010).

Agricultural and forest residues such as sugarcane bagasse, sawdust maize, wheat straw, leaves, and dry tree branches composed mainly of cellulose, hemicellulose, and lignin have been used to develop processes for obtaining biofuels from second generation where more complex processes such as saccharification and simultaneous fermentation processes are used to obtain ethanol and *n*-butanol or the Fischer–Tropsch process is used to synthesize liquid fuels such as ethanol through catalytic processes (Dry 2002; Van Steen and Claeys 2008). The great advantage of second-generation fuels is that there are no deviations from food crops.

For the production of third-generation biofuels, study focuses on non-food crops such as perennial grasses, algae, trees, and fast-growing plants with a stored energy density so that they are called "energy crops," although these processes are still under development. The advantage is that some species of energy crops are from low nutritional requirements (Ho et al. 2014).

On the other hand, fourth-generation biofuels exist in which genetically modified bacteria are used that fully carry out the entire production stage and that use carbon dioxide as raw material. For this process it is important to develop optimization technologies for the production of high-quality fuels, so it is considered that this biofuel generation is still in the investigation phase (Aro 2016).

The production of biofuels from lignocellulosic feedstocks can be achieved through three different processing pathways (Nanda et al. 2014): (i) biochemical, where enzymes and microorganisms are used to convert the cellulosic and hemicellulosic sugars to alcohols; (ii) hydrothermal, where supercritical water acts as a medium in the biomass conversion to fermentable sugars and hydrogen-rich syngas; and (iii) thermochemical, where pyrolysis

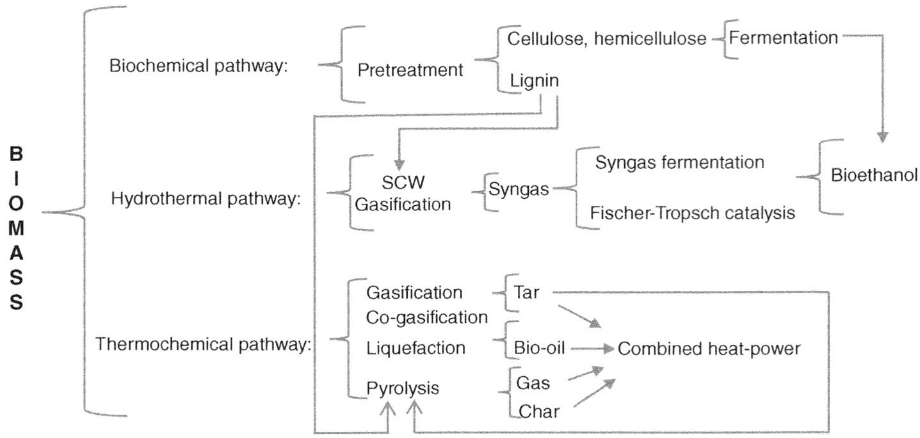

Figure 2.3 Pathways for lignocellulosic biomass conversion to biofuels using biochemical, hydrothermal, and thermochemical routes.

and gasification technologies produce a synthesis gas ($CO + H_2$) from which a wide range of long carbon chain biofuels such as synthetic gasoline and bio-oils can be reformed (Figure 2.3). In the future, biofuels such as ethanol are expected to be used worldwide, due to the fact that ethanol has low toxicity, is easily biodegradable, and its use produces less air pollutants than petroleum fuel.

2.3.3.2 Pectins

Pectin is the most complex polysaccharide according to its structure and the function of plant cell walls; it serves as a gelling and stabilizing gel in various foods and specialized products and has positive effects on human health and in multiple biomedical uses (Mohnen 2008).

With the popularization of "green chemistry," new technologies have been developed and used as well as some innovative approaches in the extraction of pectins, such as the extraction assisted by enzymes, microwaves, ultrasound, and sub-critical water. These eco-efficient methods and technologies have been developed to reduce or eliminate the use of toxic solvents, provide higher performance, and improve the quality of pectin (Wang et al. 2016; Marić et al. 2018). In addition, the ethanol precipitation stage has been shown to have a considerable impact on the performance and physicochemical properties of pectin and therefore could be used to modulate its structure and functional properties (Garna et al. 2007; Xu et al. 2014).

Trujillo-Ramírez et al. (2018) extracted pectin from cacao pod husks by acid hydrolysis using citric acid (1:25 w/v) put in a water bath with constant stirring. The extraction conditions utilized were pH 3.0, temperature 85 °C, and 90 min extraction time. The precipitation of pectin is carried out by adding 96% ethanol in a 2:1 volume ratio and keeping it for 24 h at 4 ± 1 °C. Then pectin is separated by filtration and washed twice with 70% ethanol in a 1:1 volume ratio for removal of free sugars, pigments, and other impurities; afterwards the pectin is dried at 55 °C in an oven with air circulation until constant weight is achieved. Finally, it is milled until a fine powder (Figure 2.4).

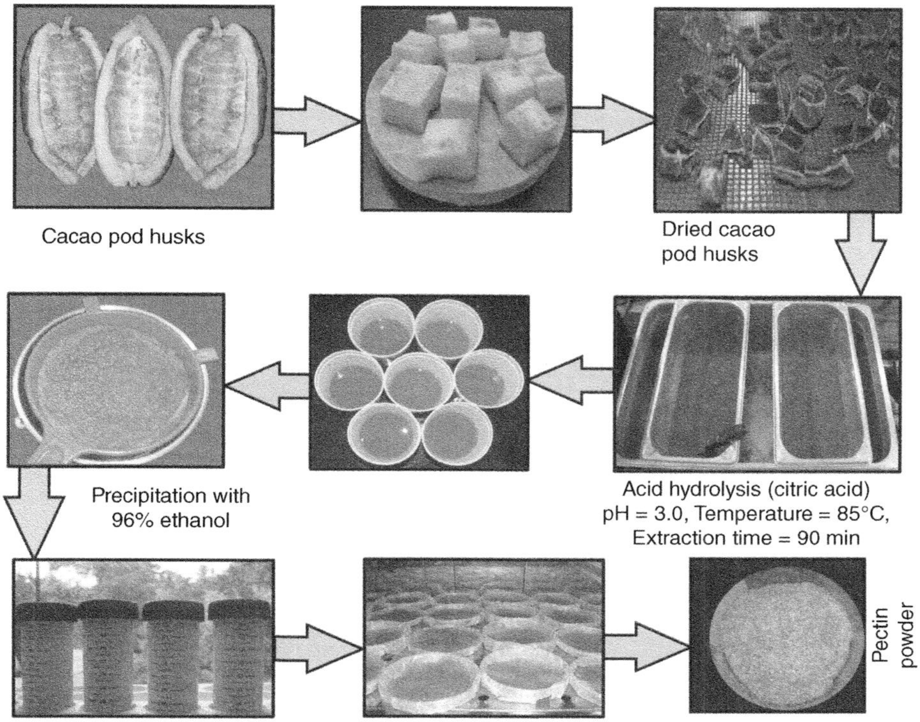

Figure 2.4 A schematic representation of a pectin extraction process by an ecofriendly method (Trujillo-Ramírez et al. 2018).

2.3.3.3 Cellulose Nanofibers

Nanocrystals, microfibrillated cellulose, microfibril aggregates, or nanofibers are single fibers of cellulose of nanometer size and can be obtained from various sources of lignocellulosic biomass available anywhere in the world. Recently, these nanocomposites have attracted considerable interest in many applications such as paper making, coating additives, security papers, food packaging, dispersions, foams, gas barriers, and in many other areas (Khalil et al. 2014).

The individualization process of cellulose nanofibers consists of chemical pretreatment and high-intensity ultrasonication (Figure 2.5) (Chen et al. 2011). For the chemical pretreatment, first lignin is removed from the samples, using acidified sodium chlorite solution at 75 °C for an hour, and this process is repeated until the product becomes white. Next, the samples are treated in 3% potassium hydroxide at 80 °C for 2 h and then in 6% potassium hydroxide at 80 °C for 2 h in order to leach hemicellulose, residual starch, and pectin. Then the samples are filtered and rinsed with distilled water until the residues are neutralized.

For high-intensity ultrasonication, after chemical pretreatment the solution containing chemically purified cellulose fibers is placed in a common ultrasonic generator for 30 min to isolate the nanofibers. The ultrasonic treatment is carried out in a water bath, keeping the temperature constant throughout the entire ultrasonication process.

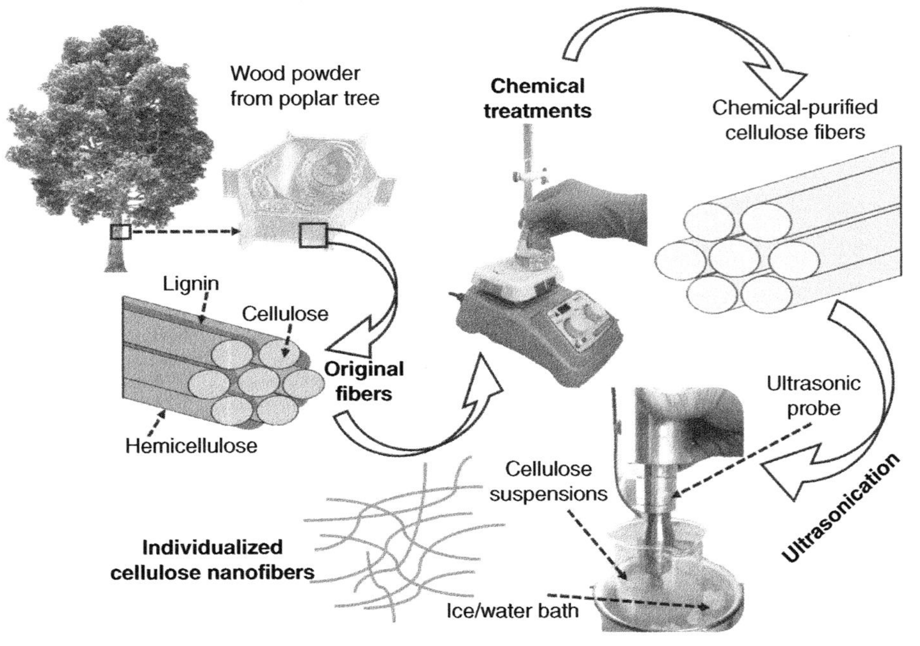

Figure 2.5 Procedure for individualizing cellulose nanofibers.

2.4 Different Types of Biomass Available in Mexico

The literature demonstrates that the great amount of lignocellulosic biomass existing in the world is used as a renewable alternative energy which can solve existing pollution problems (Fatma et al. 2018).

In Mexico, around 76 million tons of biomass waste are produced per year (Valdez-Vazquez et al. 2010; Garcia-Peña et al. 2011), which could be used to help solve the country's energy needs through digestion technologies (Wang et al. 2009; Weiland 2010; Koupaie et al. 2014). Lignocellulosic biomass is currently considered an energy alternative to the oil crisis and for its energy use it undergoes various extraction processes such as (i) combustion through exposure to high temperatures, (ii) thermochemical processes through oxidation of the biomass, (iii) biochemical processes through its degradation by microorganisms, or (iv) esterification from oil seeds. Figure 2.6 shows the different types of biomass available in Mexico, which can be used as lignocellulose raw material.

2.4.1 Coconut Shells

The coconut tree (*Cocos nucifera*) is of the Arecaceae family, a tree whose origin is not well defined because, while some say that it is native from Asia, others consider that it comes from South America (Da Silva Rodríguez et al. 2017). Coconut is a fruit of great importance which has driven the development of various industries due to the large number of nutritional, medicinal, and artisanal properties it possesses. Due to its high saturated fat

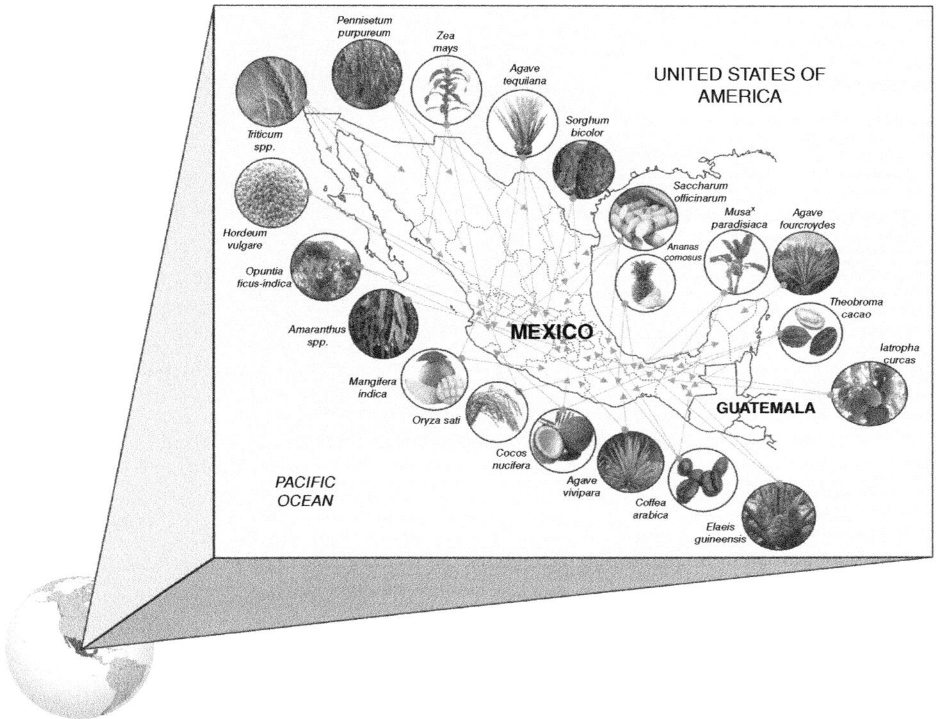

Figure 2.6 Types of biomasses available in Mexico, which can be used as lignocellulose feedstock. The crop is indicated around the map and the main states producing each crop are indicated with arrows.

content, it is used to obtain cooking oil and vinegar, molasses. Coconut sugar is obtained from the tuba or nectar and is used as an isotonic drink consumed by athletes as a rehydrating and natural diuretic (Cortazar and Carrillo 2008; Uzcanga Pérez et al. 2015; Da Silva Rodríguez et al. 2017). Likewise, this derivative helps the proper functioning of the body since it contains sugars, vitamins B and C, phosphorus, sodium, potassium, chlorides, copper, sulfa, minerals, amino acids, and essential vitamins (Yong et al. 2009; Cortazar et al. 2010; Zavaleta et al. 2010). Its varieties are used in its genetic improvement (SAGARPA 2005).

Coconut shells constitute an agro-industrial waste, considered as a natural source of fiber, representing about 85% of the weight of this fruit (Ayrilmis et al. 2011) with an average composition of 36% cellulose, 25% hemicellulose, and 28% lignin (Dhyani and Bhaskar 2018). The world area planted with coconut is around 12 million hectares, with the main production regions in the coastal and subtropical islands of Asia and Oceania, representing 86% of the total planted area (Prades et al. 2016). Its industrial applications are very varied such as in the interior coatings of cars, due to its mechanical properties (such as resistance to tensile stress, compression, bending and torsion, hardness, elasticity, etc.,) that are as good, if not better, than those from polyester fibers (Martinez 2009). It is also used in the treatment and removal of pollutants from wastewater. Furthermore, Talbot et al. (2007) scientifically demonstrated its effectiveness as a biofiltration material in a wastewater treatment system. Other investigations focus on its use as activated carbon, used as a separating element in hardness treatment of phenols, fluorides, iron, lead, and chromium in water samples (Karri et al. 2017; Talat et al. 2018; Chandana et al. 2020).

The literature shows that coconut shell is a waste with great potential to be used in industry in different forms, for example in the form of fiber or powder, for the production of semicrystalline magnetite nanoparticles (Sebastian et al. 2018); biosorbents for the sorption of Ag ions (Staron et al. 2017); thermal insulation (Araujo et al. 2015); agglomerates (Fiorelli et al. 2012); panels (Ayrilmis et al. 2011); substrates for planting vegetables and fruits (Ramos et al. 2008; Moreira et al. 2010; Dias et al. 2011); covers for soil protection (Babu and Vasudevan 2008); organic filters (Igwe et al. 2008; Monaco et al. 2009); as reinforcement in polymer matrices (Das and Biswas 2016); and in civil construction in the aggregation of cement or concrete compounds (Ali et al. 2012).

2.4.2 Sugarcane Bagasse

Across the world the accumulation of solid residues, mainly of vegetable origin, has generated an environmental impact due to growing development, especially in the agro-industrial sector such as the sugar industry. The diversification in the use of sugar cane, in addition to representing an alternative to the crisis in the sugar market, constitutes a logical and economically advantageous development strategy with the generation of value-added products which increase the process economy (Bustos et al. 2003; Suárez and Morín 2005).

The sugarcane agro-industry has a particularity by which, by diversifying to obtain energy and derivatives, it produces secondary waste that must be treated or used appropriately to avoid environmental contamination. Within these residues (by-products) are sugarcane bagasse (Bustos-Vázquez et al. 2011); other by-products rich in structural carbohydrates and low nitrogen content are the tips or buds of sugarcane and cachaça, although residual cane or harvest residues are also included (Rabelo et al. 2015; Palma et al. 2016).

Sugarcane bagasse is a lignocellulosic natural resource that has very interesting characteristics, such as its abundance, renewable energy, and low costs. Therefore, it has an increasing potential as a raw material for the chemical and biotechnology industry (Bustos-Vázquez et al. 2011) due to its limited use, as it is considered as garbage and not as a by-product (SAGARPA 2013; Rabelo et al. 2015).

Sugarcane bagasse is considered a fibrous residue of sugarcane obtained after crushing and juice extraction in the last mill. It constitutes a heterogeneous set of particles of different sizes. The bagasse has a uniform size varying in length from 0.5 to 5 cm and does not exceed 2 cm in width. On average, bagasse has 50% humidity, 46% fiber, 2% soluble solids, and 2% mineral impurities (Garcés and Martínez 2007). Likewise, the composition of the fiber present in bagasse is 48% cellulose, 25% hemicelluloses, and 27% lignin, the main natural polymers (Muñoz 2014).

In Mexico, sugarcane bagasse is the largest source of fiber for the pulp and paper industry (Aguilar-Rivera 2011), although its use is not limited to this industry and extends to obtaining ethanol, considered of great application as fuel (Agüero-Rodríguez et al. 2015), and the obtaining of various chemical products (Aguilar-Rivera 2007, 2011, 2013; Rivera 2013).

2.4.2.1 Applications of Sugarcane Bagasse

Sugarcane bagasse has been extensively studied for different uses; some studies on the application of sugarcane bagasse are described below.

- Sugarcane bagasse is used in the co-generation system of electrical energy in the sugar agro-industry (Gongora and Villafranco 2018). In the processing of 1 ton of cane, the availability of bagasse (with 50% humidity) is 0.13 and 0.17 ton of dry bagasse (Zumalacárregui and Zumalacárregui 2008). Dried bagasse has 50% cellulose, 30% hemicelluloses, and 20% lignin (Mesa-Orama and González-Penichet 2003).
- It has also been used to obtain sugars derived from its lignocellulose structure to be used as raw material for the production of lactic acid, a precursor of renewable bioplastics (Van der Pol et al. 2016) or food additives using lactic acid bacteria (Bustos-Vázquez et al. 2011).
- Vera et al. (2016) and Wang et al. (2017) have taken advantage of the absorption power that sugarcane bagasse presents to remove heavy metal ions from aqueous solutions and wastewater.
- From hydrothermally pretreated sugarcane bagasse, Soares et al. (2017) have designed and optimized the production of hydrogen.
- Likewise, different fermentation strategies have been implemented for the production of ethanol using yeast isolation technologies (Rivera 2013; González-Leos et al. 2017), using different pretreatments for hydrolysis by *Spathaspora passalidarum* and *Scheffersomyces stipitis* (Neves et al. 2016; Nakanishi et al. 2017).
- Sugarcane bagasse is also used in bioremediation processes in contaminated soils (Ordaz et al. 2011).

2.4.3 Tequilana Agave

One of the widely distributed crops in the country is species of the genus Agave (belonging to the Agavaceae family which is native to America), with around 75% of the 295 existing species of this genus, including *Agave tequilana*, *Agave salmiana*, *Agave angustifolia*, *Agave atrovirens*, *Agave americana* and *Agave fourcroydes* (Iñiguez-Covarrubias et al. 2001a; García-Mendoza 2002). These plants are highly efficient in the use of water and require minimal supplementation with water or fertilizers (Holtum et al. 2011). The morphological characteristics of these plants are that they form a large rosette of thick fleshy leaves, each ending generally in a sharp needle at the apex, arranged in a spiral around a short stem, whose edges may or may not have marginal spines (Moreno-Vilet et al. 2016); in addition to possessing the photosynthetic route of crassulacean acid metabolism, they can be grown in arid and semi-arid regions that are not suitable for conventional agricultural crops (such as corn or wheat) or lignocellulosic raw materials (such as pasture or sugarcane) (Garcia-Moya et al. 2011; Núñez et al. 2011; Rijal et al. 2016; Yan et al. 2020). Industrialization of this type of plant is generally in order to produce distillates of alcoholic beverages (Mandujano Bueno et al. 2018).

2.4.3.1 Heart or Piña

Undoubtedly, one of the emblematic drinks and a symbol of Mexicanness that has gained popularity inside and outside the country is "tequila"; this alcoholic beverage is a distillate from *Agave tequilana* Weber var. *azul*, and the species is the best known and economically important in Mexico because it allows the production of the drink due to its original designation (CRT 2020). 351.7 million liters of tequila (Statista 2020) were produced in 2019. The production of this alcoholic beverage uses the stem of the agave, commonly known as "piña" (when the plants are between 8 and 10 years old) (Figure 2.1), and the method consists in removing the inflorescence in order to concentrate sugars on the stem; then the piña are cooked or hydrolyzed and subjected to alcoholic fermentation with yeasts; cultivated or not, the musts are susceptible to being enriched and mixed in the formulation with other sugars, up to a proportion of no more than 49% of total reducing sugars (NOM-006-SCFI-2005 n.d.; Flores-Gómez et al. 2018; Ramírez-Guzmán et al. 2019). After cooking and crushing the piña for the extraction of the must, a residual fiber called "bagasse" (composed of the ring and fibrovascular bundles dispersed throughout the interior of the *Agave* head) is obtained. This residual represents approximately 40% of the total weight of the milled *Agave* on a wet weight and is the main waste of the tequila industry (Iñiguez-Covarrubias et al. 2001a; Avila-Gaxiola et al. 2017); it has been used as a fibrous supplement for animal feeding in ruminants (Iñiguez-Covarrubias et al. 2001a; Ramírez-Cortina et al. 2012; Iñiguez et al. 2014), for the production of fiber boards (Iñiguez-Covarrubias et al. 2001a), as a substrate for tomato production (Iñiguez et al. 2011, 2014) for the production of bioplastics, as an alternative to synthetic petroleum materials (Alva Munoz and Riley 2008), and as a sustainable lignocellulosic feedstock for biofuel production (Pérez-Pimienta et al. 2016; Flores-Gómez et al. 2018). Likewise, Bustos-Vázquez et al. (2017) mention that the roots are still used in the preparation of brushes, brooms, and handcrafted baskets. This activity is particularly practiced in the municipalities of Jaumave, Palmillas, Miquihuana, Bustamante, and Tula in the state of Tamaulipas, Mexico (Bustos-Vázquez et al. 2017).

2.4.3.2 Leaves

The leaves of *Agave tequilana* Weber var. *azul* are some of the lignocellulosic feedstocks that have received little attention because they are left in the field during the harvest process (Corbin et al. 2015), despite the fact that they make up 32%–46% of the total weight of the plant (Iñiguez-Covarrubias et al. 2001b; Montañez-Soto et al. 2011; Alemán-Nava et al. 2018). The leaves are 90–120 cm long × 8–12 cm wide; they are lanceolate, acuminate, firm fiber, rigidly extended, concave, ascending to horizontal, widest through the middle, narrowed and thickened toward the base, generally bluish-green to grayish (Figure 2.7) (Iñiguez-Covarrubias et al. 2001b). The freshly cut leaves are rich in moisture, soluble sugars, and pectin, while the fiber of the dried leaf is composed of crystalline cellulose (31%–50%), non-cellulosic polysaccharides (15%–22%), and lignin (12%–17%) (USDE 2013; Corbin et al. 2015).

2.4.3.3 Conditioning of *Agave tequilana* Leaves to Obtain Fermentable Sugars

Agave tequilana leaves represent a viable, economical, and renewable source of potential lignocellulosic biomass for the production of second-generation biofuels (Avila-Gaxiola et al. 2017, 2018). However, in order for this lignocellulosic feedstock to be used to produce ethanol, physical, chemical, or biological pretreatments must first be carried out to release fermentable sugars; that is, to break the lignin structure down and disrupt the crystalline structure of cellulose so that acids or enzymes can easily access and hydrolyze the cellulose and thus

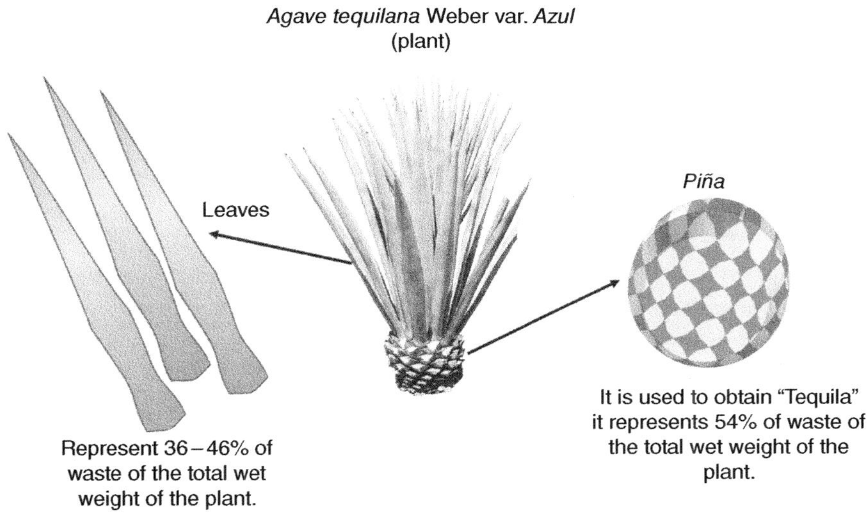

Figure 2.7 *Agave tequilana* Weber var. *azul* components.

decrease the recalcitrance and increase biofuel yield (Mosier et al. 2005; Alvira et al. 2010; Agbor et al. 2011; Avila-Gaxiola et al. 2017; Contreras-Hernández et al. 2018).

Some investigations that have been carried out with regard to the leaves of *Agave tequilana* Weber var. *azul* include those by Close et al. (2017), where they used the leaves from 2-year-old plants without the need to apply any treatment and it was determined that they can be used as a microbial fermentation biomass substrate, that is, a sole carbon source that could support yeast growth. This may be due to the fact that, in young plants, the majority of the total reducing sugars are retained in the leaves (Iñiguez-Covarrubias et al. 2001b; Arrizon et al. 2010), while Agave leaves powdered (dried at 100 °C and with particle size ≤0.3 mm) represent a good source of carbohydrates (52%) and crude fiber (32%) (Avila-Gaxiola et al. 2017). Fructans (37%), followed by reducing sugars (15%), are the main non-structural carbohydrates in Agave leaves, while structural carbohydrates are composed of holocellulose (16%) and lignin (15%). The chemical composition of the non-structural carbohydrates present in these leaves make it an attractive material compared to other biomass of agro-industrial waste that does not contain monosaccharides and thus by-products of industrial interest (Avila-Gaxiola et al. 2018).

On the other hand, the effect of the addition of H_2SO_4 (0.5% v/v) and enzymatic hydrolysis using commercial cellulases for 18 h allowed the aqueous extraction of *A. tequilana* leaf powder to yield a better conversion of carbohydrates from LCM and fructans to monosaccharides than those in which HCl was used; these results are desirable because fructans, cellulose, and hemicellulose are polymers that can easily be hydrolyzed for the production of biofuels (Avila-Gaxiola et al. 2018).

Rijal et al. (2016) evaluated five different pretreatment conditions (time, temperature, and acid concentrations) in terms of cellulose digestibility, hemicellulose solubilization, and lignin content in leaves of *A. tequilana* plants. They determined the H_2SO_4 pretreatment at 2%, 60 min at 120 °C, and enzymatic saccharification (enzymatic mixture of cellulase) at 6%. Under these conditions with *Saccharomyces cerevisiae*, a glucose to ethanol conversion rate of 68% and 61% was achieved for the raw bagasse and for the hydrolyzates of the *A. tequilana* juice, respectively.

2.5 Conclusion

The use of lignocellulosic biomass in a rational, ecological, and sustainable way represents a possible alternative to face the various world problems existing in the field of energy. The great diversity of biomass in the world allows it to be used rationally through various treatments applied to each raw material.

This chapter describes the importance of three residues (coconut, sugarcane bagasse, and residues of the blue variety *Tequilana agave*) considered to be important for the purpose of being lignocellulosic biomass with high potential to be transformed into products with high added value. However, for these materials to be used as lignocellulosic biomass, a series of physical, chemical, or biological pretreatments is required to release fermentable sugars. Each pretreatment has a specific effect on the cellulose, hemicellulose, and lignin fraction. Therefore, different pretreatment methods and conditions should be chosen to make the production of the selected product more efficient or the biomass more digestible when it is required as a fibrous supplement for the feeding of ruminants.

In addition, with the use of these residues the environmental impact generated by the agro-industries can be reduced.

References

Agbor, V.B., Cicek, N., Sparling, R. et al. (2011). Biomass pretreatment: fundamentals toward application. *Biotechnol. Adv.* 29 (6): 675–685.

Agüero-Rodríguez, J.C., Tepetla-Montes, J., and Torres-Beristaín, B. (2015). Producción de biocombustibles a partir de la caña en Veracruz, México: perspectivas y riesgos socio-ambientales. *CienciaUAT.* 9 (2): 74–84.

Aguilar-Rivera, N. (2007). Bioetanol de la caña de azúcar. *Av. Investig. Agropecu.* 11 (3): 25–39.

Aguilar-Rivera, N. (2011). Efecto del almacenamiento de bagazo de caña en las propiedades físicas de celulosa grado papel. *Ing. Investig Tecnol.* 12 (2): 189–197.

Aguilar-Rivera, N. (2013). Análisis de la productividad de etanol de caña de azúcar en ingenios azucareros de México. Universidad Autónoma del Estado de México, Toluca, México. *Cienc. Ergo Sum.* 20 (1): 17–28.

Alemán-Nava, G.S., Gatti, I.A., Parra-Saldivar, R. et al. (2018). Biotechnological revalorization of tequila waste and by-product streams for cleaner production—a review from bio-refinery perspective. *J. Clean. Prod.* 172: 3713–3720.

Ali, M., Liu, A., Sou, H., and Chouw, N. (2012). Mechanical and dynamic properties of coconut fibre reinforced concrete. *Constr. Build. Mater.* 30: 814–825.

Alonso, D.M., Wettstein, S.G., and Dumesic, J.A. (2013). Gamma-valerolactone, a sustainable platform molecule derived from lignocellulosic biomass. *Green Chem.* 15 (3): 584–595.

Alva Munoz, L.E. and Riley, M.R. (2008). Utilization of cellulosic waste from tequila bagasse and production of polyhydroxyalkanoate (PHA) bioplastics by *Saccharophagus degradans*. *Biotechnol. Bioeng.* 100 (5): 882–888.

Álvarez Alonso, M., Arquero Hidalgo, A., and Martínez Izquierdo, E. (2008). Empleo del AHP (Proceso Analítico Jerárquico) incorporado en SIG para definir el emplazamiento óptimo de equipamientos universitarios. Aplicación a una biblioteca. *Cuadernos Geográficos* 39: 579–595.

Alvira, P., Tomás-Pejó, E., Ballesteros, M., and Negro, M.J. (2010). Pretreatment technologies for an efficient bioethanol production process based on enzymatic hydrolysis: a review. *Bioresour. Technol.* 101 (13): 4851–4861.

Ángel Siles López, J., Li, Q., and Thompson, I.P. (2010). Biorefinery of waste orange peel. *Crit. Rev. Biotechnol.* 30 (1): 63–69.

Anwar, Z., Gulfraz, M., and Irshad, M. (2014). Agro-industrial lignocellulosic biomass a key to unlock the future bio-energy: a brief review. *J. Radiat. Res. Appl. Sci.* 7 (2): 163–173.

Araujo, P.M.M., Felix Filho, L.F., and Barbosa, J.J. (2015). Estudo das propriedades termofísicas da fibra de coco minimamente processada visando aplicação como isolante térmico. *Revista Interdisciplinar de Pesquisa e Inovação* 1 (1): 134–142.

Arauzo, J., Bimbela, F., Ábrego, J. et al. (2014). Introducción a las tecnologías de aprovechamiento de biomasa. Boletín del Grupo Español del Carbón., 33, 2–6.

Aro, E.M. (2016). From first generation biofuels to advanced solar biofuels. *Ambio* 45 (1): 24–31.

Aro, N., Pakula, T., and Penttilä, M. (2005). Transcriptional regulation of plant cell wall degradation by filamentous fungi. *FEMS Microbiol. Rev.* 29 (4): 719–739.

Arrizon, J., Morel, S., Gschaedler, A., and Monsan, P. (2010). Comparison of the water-soluble carbohydrate composition and fructan structures of *Agave tequilana* plants of different ages. *Food Chem.* 122 (1): 123–130.

Asgher, M., Ahmad, Z., and Iqbal, H.M.N. (2013). Alkali and enzymatic delignification of sugarcane bagasse to expose cellulose polymers for saccharification and bio-ethanol production. *Ind. Crop. Prod.* 44: 488–495.

Avila-Gaxiola, E., Avila-Gaxiola, J., Velarde-Escobar, O. et al. (2017). Effect of drying temperature on *Agave tequilana* leaves: a pretreatment for releasing reducing sugars for biofuel production. *J. Food Process Eng.* 40 (3): e12455.

Avila-Gaxiola, J., Velarde-Escobar, O.J., Millan-Almaraz, J.R. et al. (2018). Treatments to improve obtention of reducing sugars from agave leaves powder. *Ind. Crop. Prod.* 112: 577–583.

Ayrilmis, N., Jarusombuti, S., Fueangvivat, V. et al. (2011). Coir fiber reinforced polypropylene composite panel for automotive interior applications. *Fibers Polym.* 12 (7): 919.

Babu, G.L.S. and Vasudevan, A.K. (2008). Strength and stiffness response of coir fiber-reinforced tropical soil. *J. Mater. Civ. Eng.* 20 (9): 571–577.

Babu, C.R., Ketanapalli, H., Beebi, S.K., and Kolluru, V.C. (2018). Wheat bran-composition and nutritional quality: a review. *Adv. Biotechnol. Microbiol.* 9 (1): 1–7.

Balan V. (2014). Current Challenges in Commercially Producing Biofuels from Lignocellulosic Biomass. Hindawi Publishing Corporation ISRN Biotechnology Volume, Article ID 463074, 31 pp.

Balan, V., Bals, B., Chundawat, S.P. et al. (2009). Lignocellulosic biomass pretreatment using AFEX. In: *Biofuels: Methods and Protocols, vol. 581 of Methods in Molecular Biology* (ed. J.R. Mielenz), 61–77. Springer Science + Business Media: Humana Press.

Bali, G., Meng, X., Deneff, J.I. et al. (2015). The effect of alkaline pretreatment methods on cellulose structure and accessibility. *ChemSusChem* 8 (2): 275–279.

Barakat, A., de Vries, H., and Rouau, X. (2013). Dry fractionation process as an important step in current and future lignocellulose biorefineries: a review. *Bioresour. Technol.* 134: 362–373.

Behera, S., Arora, R., Nandhagopal, N., and Kumar, S. (2014). Importance of chemical pretreatment for bioconversion of lignocellulosic biomass. *Renew. Sust. Energ. Rev.* 36: 91–106.

Boufi, S. (2017). Agricultural crop residue as a source for the production of cellulose nanofibrils. In: *Cellulose-Reinforced Nanofibre Composites: Production, Properties and Applications* (ed. M. Jawaid, S. Boufi and H.P.S. Abdul Khalil), 129–152. Woodhead Publishing Series in Composites Science and Engineering.

Bridgwater, A.V. (2012). Review of fast pyrolysis of biomass and product upgrading. *Biomass Bioenergy* 38: 68–94.

Bustos, G., Ramírez, J.A., Garrote, G., and Vázquez, M. (2003). Modeling of the hydrolysis of sugar cane bagasse with hydrochloric acid. *Appl. Biochem. Biotechnol.* 104 (1): 51–68.

Bustos, G., Moldes, A.B., Cruz, J.M., and Domínguez, J.M. (2004). Production of fermentable media from vine-trimming wastes and bioconversion into lactic acid by *Lactobacillus pentosus*. *J. Sci. Food Agric.* 84 (15): 2105–2112.

Bustos-Vázquez, M.G., Cervantes-Martínez, J.E., García-Delgado, M. et al. (2011). Residuos de la industria azucarera: una alternativa para la obtención de ácido láctico. *CienciaUAT* 5 (4): 61–66.

Bustos-Vázquez, M.G., Rodríguez-Durán, N.A., del Ángel-del Ángel A. et al. (2017). El agave tequilero: una perspectiva de aprovechamiento biotecnológico. In Bustos Vázquez, M.G., del Ángel del Ángel, J.A. (eds), Tecnología y desarrollo sustentable: avances en el aprovechamiento de recursos agroindustriales, Ciudad de México, Universidad Autónoma de Tamaulipas y Colofón. ISBN 978-607-8513-40-6, pp. 377–385.

Carmona, J.E., Morales-Martínez, T.K., Mussatto, S.I. et al. (2017). Propiedades químicas, estructurales y funcionales de la lechuguilla (*Agave lechuguilla* Torr.). *Rev. Mex. Cienc. Forestales* 8 (42): 100–122.

Carrillo Navarrete, F. (2002). *Caracterización estructural de fibras lyocell y su comportamiento frente a procesos de degradación*. Tesis doctoral, Universitat Politècnica de Catalunya Departament d'Enginyeria Química https://upcommons.upc.edu/handle/2117/93739?show=full.

Nimz, H.H. and Casten, R. (1986). Chemical processing of lignocellulosics. *Holz Roh Werk.* 44: 207–212.

Chandana, L., Krushnamurty, K., Suryakala, D., and Subrahmanyam, C. (2020). Low-cost adsorbent derived from the coconut shell for the removal of hexavalent chromium from aqueous medium. *Mater. Today Proc.* 26 (1): 44–51.

Chang, V.S. and Holtzapple, M.T. (2000). Fundamental factors affecting biomass enzymatic reactivity. In: *Twenty-First Symposium on Biotechnology for Fuels and Chemicals*. Appl. Biochem. Biotechnol., (ed. M. Finkelstein and B.H. Davison). Totowa, NJ: Humana Press https://doi.org/10.1007/978-1-4612-1392-5_1.

Chaquilla-Quilca, G., Balandrán-Quintana, R.R., Mendoza-Wilson, A.M., and Mercado-Ruiz, J.N. (2018). Propiedades y posibles aplicaciones de las proteínas de salvado de trigo. *CienciaUAT* 12 (2): 137–147.

Chaturvedi, V. and Verma, P. (2013). An overview of key pretreatment processes employed for bioconversion of lignocellulosic biomass into biofuels and value added products. *Biotech* 3 (5): 415–431.

Chávez-Sifontes, M. and Domine, M.E. (2013). Lignina, estructura y aplicaciones: métodos de despolimerización para la obtención de derivados aromáticos de interés industrial. *Av. Cienc. Ing.* 4 (4): 15–46.

Chen, W., Yu, H., Liu, Y. et al. (2011). Individualization of cellulose nanofibers from wood using high-intensity ultrasonication combined with chemical pretreatments. *Carbohydr. Polym.* 83 (4): 1804–1811.

Close, D., Rodriguez, M. Jr., Hu, R., and Yang, X. (2017). Disposition and bioavailability of inulin and free sugar in untreated and dilute acid pretreated *Agave tequilana* leaves. *Biomass Bioenergy* 106: 176–181.

Contreras-Hernández, M.G., Ochoa-Martínez, L.A., Rutiaga-Quiñones, J.G. et al. (2018). Effect of ultrasound pre-treatment on the physicochemical composition of *Agave durangensis* leaves and potential enzyme production. *Bioresour. Technol.* 249: 439–446.

Corbin, K.R., Byrt, C.S., Bauer, S. et al. (2015). Prospecting for energy-rich renewable raw materials: agave leaf case study. *PLoS One* 10 (8): e0135382.

Cortazar, M., Carrillo, H. (2008). Manual para el aprovechamiento integral de la palma de coco. Comité Editorial del CIRSE. Primera Edición. Campo Experimental Chetumal, Quintana Roo. Folleto Técnico Serie INIFAP. 52 pp.

Cortazar, M., Flores, R., Sánchez, B., Góngora, R. 2010. La tuba de coco: consumo potencial en Quintana Roo. Comité Editorial del CIRSE. Campo Experimental Chetumal, Quintana Roo. Desplegable para productores Núm. 4. Serie INIFAP.

CRT (2020). Consejo Regulador del Tequila. Available from: https://www.crt.org.mx/index.php/en/pages-2/geografia-de-la-dot (Accessed May 25, 2020).

Cuervo, L., Folch, J.L., and Quiroz, R.E. (2009). Lignocelulosa como fuente de azúcares para la producción de etanol. *BioTecnología* 13 (3): 11–25.

Cueva, M.B.R. and Monzón, A.H. (2014). Evaluación de residuos agrícolas para la producción del hongo Pleurotus ostreatus. *ICIDCA. Sobre Deriv. Caña Azúcar* 48 (1): 54–59.

Czaja, W.K., Young, D.J., Kawecki, M., and Brown, R.M. (2007). The future prospects of microbial cellulose in biomedical applications. *Biomacromolecules* 8 (1): 1–12.

Da Silva Rodríguez, C.P., Valencia, F.I.R., Montalvo, I.A.P. et al. (2017). Extracción de aceite de Coco (*Cocos nuciferas*) como estrategias de aprovechamiento de los productos locales de Mitú. *Revista Vaupés Innova* 1: 82–89.

Das, G. and Biswas, S. (2016). Effect of fiber parameters on physical, mechanical and water absorption behaviour of coir fiber–epoxy composites. *J. Reinf. Plast. Compos.* 35 (8): 644–653.

Demirbas, A. (2007). Progress and recent trends in biofuels. *Prog. Energy Combust. Sci.* 33 (1): 1–18.

Dhyani, V. and Bhaskar, T. (2018). A comprehensive review on the pyrolysis of lignocellulosic biomass. *Renew. Energy* 129: 695–716.

Dias, N.D.S., Oliveira, A.M.D., Sousa Neto, O.N.D. et al. (2011). Concentração salina e fases de exposição à salinidade do meloeiro cultivado em substrato de fibra de coco. *Rev. Bras. Frutic.* 33 (3): 915–921.

Dry, M.E. (2002). The Fischer–Tropsch process: 1950–2000. *Catal. Today* 71 (3–4): 227–241.

Dutta, K., Daverey, A., and Lin, J.G. (2014). Evolution retrospective for alternative fuels: first to fourth generation. *Renew. Energy* 69: 114–122.

Ebringerová, A. (2006). Structural diversity and application potential of hemicelluloses. *Macromol. Symp.* 232: 1–12.

Evert, R.F. (2006). *Esau's Plant Anatomy: Meristems, Cells, and Tissues of the Plant Body: Their Structure, Function, and Development*, 3e. Wiley. ISBN: 9780471738435.

Fatma, S., Hameed, A., Noman, M. et al. (2018). Lignocellulosic biomass: a sustainable bioenergy source for the future. *Protein Pept. Lett.* 25 (2): 148–163.

Field, S.J., Ryden, P., Wilson, D. et al. (2015). Identification of furfural resistant strains of *Saccharomyces cerevisiae* and *Saccharomyces paradoxus* from a collection of environmental and industrial isolates. *Biotechnol. Biofuels* 8 (1): 33.

Fiorelli, J., Curtolo, D.D., Barrero, N.G. et al. (2012). Particulate composite based on coconut fiber and castor oil polyurethane adhesive: an eco-efficient product. *Ind. Crop. Prod.* 40: 69–75.

Flores-Gómez, C.A., Silva, E.M.E., Zhong, C. et al. (2018). Conversion of lignocellulosic agave residues into liquid biofuels using an AFEX™-based biorefinery. *Biotechnol. Biofuels* 11 (1): 1–18.

Focher, B., Marzetti, A., Beltrame, P.L., and Carniti, P. (1991). Structural features of cellulose and cellulose derivatives and their effects on enzymatic hydrolysis. In: (ed. C.H. Haigler and P.J. Weimer) Biosynthesis and Biodegradation of Cellulose, 293–310. New York, NY: Marcel Dekker.

Galbe, M. and Wallberg, O. (2019). Pretreatment for biorefineries: a review of common methods for efficient utilisation of lignocellulosic materials. *Biotechnol. Biofuels* 12: 294.

Garcés, R.V., Martínez, S.V. (2007). Estudio del poder calorífico del bagazo de caña de azúcar en la industria azucarera de la zona de Risaralda. Tesis, Colección Tecnología Química, https://hdl.handle.net/11059/825

García-Mendoza, A. (2002). Distribution of agave (Agavaceae) in México. *Cactus Succul. J.* 74 (4): 177–188.

Garcia-Moya, E., Romero-Manzanares, A., and Nobel, P.S. (2011). Highlights for Agave productivity. *GCB Bioenergy* 3 (1): 4–14.

Garcia-Peña, E.I., Parameswaran, P., Kang, D.W. et al. (2011). Anaerobic digestion and co-digestion processes of vegetable and fruit residues: process and microbial ecology. *Bioresour. Technol.* 102 (20): 9447–9455.

Garna, H., Mabon, N., Robert, C. et al. (2007). Effect of extraction conditions on the yield and purity of apple pomace pectin precipitated but not washed by alcohol. *J. Food Sci.* 72 (1): C001–C009.

Gaviria-Acosta, E.D.I.E.R., Benítez-Benítez, R.I.C.A.R.D.O., Lenis, L., and Hoyos-Concha, J.L. (2015). Optimization of enzymatic hydrolysis of proteins present in seeds of Guandul (*Cajanus-cajan*). *Biotecnol. Sector Agropecuario Agroind.* 13 (2): 114–122.

Ge, X., Xu, F., and Li, Y. (2016). Solid-state anaerobic digestion of lignocellulosic biomass: recent progress and perspectives. *Bioresour. Technol.* 205: 239–249.

Ge, X., Chang, C., Zhang, L. et al. (2018). Chapter five: Conversion of lignocellulosic biomass into platform chemicals for biobased polyurethane application. *Adv. Bioenergy* (3): 161–213.

Gongora, A. and Villafranco, D. (2018). Sugarcane bagasse cogeneration in Belize: a review. *Renew. Sust. Energ. Rev.* 96: 58–63.

González-Leos, A., Del Angel-Del Ángel, J.A., González-Castillo, J.L. et al. (2017). Evaluation of producing ethanol native yeasts present in sugar cane bagasse. *CienciaUAT* 11 (2): 80–92.

Griggs, A.J., Stickel, J.J., Lischeske, J.J. (2011). Mechanistic modeling of enzymatic saccharification of cellulose using continuous distribution kinetics. In *Abstracts of Papers of the American Chemical Society* (Vol. 242). Washington, DC, American Chemical Society.

Griggs, A.J., Stickel, J.J., and Lischeske, J.J. (2012a). A mechanistic model for enzymatic saccharification of cellulose using continuous distribution kinetics I: Depolymerization by EGI and CBHI. *Biotechnol. Bioeng.* 109 (3): 665–675.

Griggs, A.J., Stickel, J.J., and Lischeske, J.J. (2012b). A mechanistic model for enzymatic saccharification of cellulose using continuous distribution kinetics II: Cooperative enzyme action, solution kinetics, and product inhibition. *Biotechnol. Bioeng.* 109 (3): 676–685.

Gullón, B., Yáñez, R., Alonso, J.L., and Parajó, J.C. (2010). Producción de oligosacáridos y azúcares a partir de paja de centeno: un enfoque cinético. *Tecnol. Bioresour.* 101 (17): 6676–6684.

Hafızoğlu, H. and Usta, M. (2005). Chemical composition of coniferous wood species occurring in Turkey. *Holz als Roh-und Werkstoff* 63 (1): 83–85.

Ho, S.H., Ye, X., Hasunuma, T. et al. (2014). Perspectives on engineering strategies for improving biofuel production from microalgae—a critical review. *Biotechnol. Adv.* 32 (8): 1448–1459.

Holtum, J.A., Chambers, D.O.N., Morgan, T., and Tan, D.K. (2011). Agave as a biofuel feedstock in Australia. *GCB Bioenergy* 3 (1): 58–67.

Huang, Y.F., Chiueh, P.T., and Lo, S.L. (2016). A review on microwave pyrolysis of lignocellulosic biomass. *Sustain. Environ. Res.* 26 (3): 103–109.

Igwe, J.C., Abia, A.A., and Ibeh, C.A. (2008). Adsorption kinetics and intraparticulate diffusivities of Hg, As and Pb ions on unmodified and thiolated coconut fiber. *Int. J. Environ. Sci. Technol.* 5 (1): 83–92.

Iñiguez, G., Acosta, N., Martínez, L. et al. (2005). Utilización de subproductos de la industria tequilera. Parte 7. Compostaje de bagazo de agave y vinazas tequileras. *Rev. Int. Contam. Ambient.* 21 (1): 37–50.

Íñiguez, G., Martínez, G.A., Flores, P.A., and Virgen, G. (2011). Utilización de subproductos de la industria tequilera: Parte 9. Monitoreo de la evolución del compostaje de dos fuentes distintas de bagazo de agave para la obtención de un substrato para jitomate. *Rev. Int. Contam. Ambient.* 27 (1): 47–59.

Iñiguez, C.G., Bernal, C.J.J., Ramírez, M.W., and Villalvazo, N.J. (2014). Recycling agave bagasse of the tequila industry. *Adv. Chem. Eng. Sci.* 4 (2): 135–142.

Iñiguez-Covarrubias, G., Lange, S.E., and Rowell, R.M. (2001a). Utilization of byproducts from the tequila industry: part 1: Agave bagasse as a raw material for animal feeding and fiberboard production. *Bioresour. Technol.* 77 (1): 25–32.

Iñiguez-Covarrubias, G., Díaz-Teres, R., Sanjuan-Dueñas, R. et al. (2001b). Utilization of by-products from the tequila industry. Part 2: Potential value of *Agave tequilana* Weber azul leaves. *Bioresour. Technol.* 77: 101–108.

Ishola, M.M. and Taherzadeh, M.J. (2014). Effect of fungal and phosphoric acid pretreatment on ethanol production from oil palm empty fruit bunches (OPEFB). *Bioresour. Technol.* 165: 9–12.

Isikgor, F.H. and Becer, C.R. (2015). Lignocellulosic biomass: a sustainable platform for the production of bio-based chemicals and polymers. *Polym. Chem.* 6 (25): 4497–4559.

Karri, R.R., Jayakumar, N.S., and Sahu, J.N. (2017). Modelling of fluidised-bed reactor by differential evolution optimization for phenol removal using coconut shells based activated carbon. *J. Mol. Liq.* 231: 249–262.

Kemppainen, K., Rommi, K., Holopainen, U., and Kruus, K. (2016). Steam explosion of Brewer's spent grain improves enzymatic digestibility of carbohydrates and affects solubility and stability of proteins. *Appl. Biochem. Biotechnol.* 180 (1): 94–108.

Khalil, H.A., Davoudpour, Y., Islam, M.N. et al. (2014). Production and modification of nanofibrillated cellulose using various mechanical processes: a review. *Carbohydr. Polym.* 99: 649–665.

Klyosov, A.A., Mitkevich, O.V., and Sinitsyn, A.P. (1986). Role of the activity and adsorption of cellulases in the efficiency of the enzymic hydrolysis of amorphous and crystalline cellulose. *Biochemistry* 25 (3): 540–542.

Koullas, D.P., Christakopoulos, P., Kekos, D. et al. (1992). Correlating the effect of pretreatment on the enzymatic hydrolysis of straw. *Biotechnol. Bioeng.* 39 (1): 113–116.

Koupaie, E.H., Leiva, M.B., Eskicioglu, C., and Dutil, C. (2014). Mesophilic batch anaerobic co-digestion of fruit-juice industrial waste and municipal waste sludge: process and cost–benefit analysis. *Bioresour. Technol.* 152: 66–73.

Kumar, P., Barrett, D.M., Delwiche, M.J., and Stroeve, P. (2009). Methods for pretreatment of lignocellulosic biomass for efficient hydrolysis and biofuel production. *Ind. Eng. Chem. Res.* 48: 3713–3729.

Kumar, L., Arantes, V., Chandra, R., and Saddler, J. (2012). The lignin present in steam pretreated softwood binds enzymes and limits cellulose accessibility. *Bioresour. Technol.* 103 (1): 201–208.

Li, M.F., Yang, S., and Sun, R.C. (2016). Recent advances in alcohol and organic acid fractionation of lignocellulosic biomass. *Bioresour. Technol.* 200: 971–980.

Loix, C., Huybrechts, M., Vangronsveld, J. et al. (2017). Reciprocal interactions between cadmium-induced cell wall responses and oxidative stress in plants. *Front. Plant Sci.* 8: 1867.

Luo, C., Li, Y., Chen, Y. et al. (2019). Degradation of bamboo lignocellulose by bamboo snout beetle *Cyrtotrachelus buqueti* in vivo and vitro: efficiency and mechanism. *Biotechnol. Biofuels* 12 (1): 75.

Mamma, D. and Christakopoulos, P. (2014). Biotransformation of citrus by-products into value added products. *Waste Biomass Valorization* 5 (4): 529–549.

Mandujano Bueno, A., Pons Hernández, J.L., Paredes Melesio, R., and García Meza, P. (2018). Diversidad genética de maguey (Agave spp.) en las sierras y llanuras del norte de Guanajuato. *Rev. Mexicana Cienc. Agric.* 9 (3): 511–523.

Maniet, G., Schmetz, Q., Jacquet, N. et al. (2017). Effect of steam explosion treatment on chemical composition and characteristic of organosolv fescue lignin. *Ind. Crop. Prod.* 99: 79–85.

Marić, M., Grassino, A.N., Zhu, Z. et al. (2018). An overview of the traditional and innovative approaches for pectin extraction from plant food wastes and by-products: ultrasound-, microwaves-, and enzyme-assisted extraction. *Trends Food Sci. Technol.* 76: 28–37.

Martinez, Y. (2009). Tendencias de la ingeniería, "Utilizan cáscara de coco para fabricar materiales de automóviles". Available from https://www.tendencias21.net/Utilizan-cascaras-de-coco-para-fabricar-materiales-de-automoviles_a2884.htmlf

Martínez, J.M.R., Mejía, R.G., Samperio, J.M.I., and de Lizarde, J. (2015). Capítulo 1. Potencial de la biomasa y perspectivas en diferentes países. In: (ed. J.M. Rincón Martínez and E. Silva Lora) Red Iberamericana de Residuos Orgánicos en Prducción de Energía, 17–55. Bioenergía: Fuentes, conversión y sustentabilidad, Charlie's Impresores Ltda. (CYTED).

Mateus, L., Hernández, O., Velásquez, M., and de Jesús Díaz, J. (2012). Evaluación del pretratamiento con ácido sulfúrico diluido del pasto maralfalfa (*Pennisetum glaucum* × *Pennisetum purpureum*) para la producción de etanol. *Rev. Colomb. Biotecnol.* 14 (1): 146–156.

Matos, M., Hernández, Y.P., Valdivia, A. et al. (2020). Use of agroindustrial residues for producing enzymes by *Bacillus subtilis* E. *Cuba. J. Agric. Sci.* 54 (1).

Menon, V. and Rao, M. (2012). Trends in bioconversion of lignocellulose: biofuels, platform chemicals and biorefinery concept. *Prog. Energy Combust. Sci.* 38 (4): 522–550.

Mesa-Orama, J. González-Penichet, L. (2003). Balance de CO_2 en la obtención de energía renovable a partir de caña de azúcar. Available from http://www.ilustrados.com/documentos/balanceco2.doc

Mohnen, D. (2008). Pectin structure and biosynthesis. *Curr. Opin. Plant Biol.* 11 (3): 266–277.

Monaco, P.A.V.L., de Matos, A.T., Sarmento, A.P. et al. (2009). Desempenho de filtros constituídos por fibras de coco no tratamento de _aguas residu_arias de suino- cultura. *Rev. Eng. Agricul.-Reveng* 17 (6): 473–480.

Montañez-Soto, J., Venegas-González, J., Vivar-Vera, M., and Ramos-Ramírez, E. (2011). Extracción, caracterización y cuantificación de los fructanos contenidos en la cabeza y en las hojas del *Agave tequilana* Weber azul. *Bioagro* 23 (3): 199–206.

Moreira, M.A., Dantas, F.M., Bianchini, F.G., and Viégas, P.R.A. (2010). Produção de mudas de berinjela com uso de pó de coco. *Rev. Bras. Prod. Agroind.* 12 (2): 163–170.

Moreno-Vilet, L., Camacho-Ruiz, R.M., and Portales-Pérez, D.P. (2016). Chapter 11. Prebiotic agave fructans and immune aspects. In: *Probiotics, Prebiotics, and Synbiotics. Bioactive Foods in Health Promotion* (ed. R.R. Watson and V.R. Preedy), 165–179. USA, Academic Press Elsevier Inc.

Mosier, N., Wyman, C., Dale, B. et al. (2005). Features of promising technologies for pretreatment of lignocellulosic biomass. *Bioresour. Technol.* 96 (6): 673–686.

Muñoz, M. (2014). XIV. Perspectivas de los coproductos de la caña de azúcar. In: *El Cultivo de la Caña de Azúcar en Guatemala* (ed. M. Melgar, A. Meneses, H. Orozco, et al.), 420–446. Guatemala, Artemis Edinter.

Naik, S.N., Goud, V.V., Rout, P.K., and Dalai, A.K. (2010). Production of first and second generation biofuels: a comprehensive review. *Renew. Sust. Energ. Rev.* 14 (2): 578–597.

Nakanishi, S.C., Soares, L.B., Biazi, L.E. et al. (2017). Fermentation strategy for second generation ethanol production from sugarcane bagasse hydrolyzate by *Spathaspora passalidarum* and *Scheffersomyces stipitis*. *Biotechnol. Bioeng.* 114 (10): 2211–2221.

Nanda, S., Mohammad, J., Reddy, S.N. et al. (2014). Pathways of lignocellulosic biomass conversion to renewable fuels. *Biomass Convers. Bioref.* 4: 157–191.

Nanduri, A., Kulkarni, S.S., and Mills, P.L. (2021). Experimental techniques to gain mechanistic insight into fast pyrolysis of lignocellulosic biomass: a state-of-the-art review. *Renew. Sust. Energ. Rev.* 148: 111262.

Neves, P.V., Pitarelo, A.P., and Ramos, L.P. (2016). Production of cellulosic ethanol from sugarcane bagasse by steam explosion: effect of extractives content, acid catalysis and different fermentation technologies. *Bioresour. Technol.* 208: 184–194.

Nidetzky, B. and Steiner, W. (1993). A new approach for modeling cellulase–cellulose adsorption and the kinetics of the enzymatic hydrolysis of microcrystalline cellulose. *Biotechnol. Bioeng.* 42 (4): 469–479.

NOM-006-SCFI-2005. (n.d.) Norma Oficial Mexicana NOM-006-SCFI-2005, Bebidas Alcohólicas-Tequila-Especificaciones.

Núñez, H.M., Rodriguez, L.F., and Khanna, M. (2011). Agave for tequila and biofuels: an economic assessment and potential opportunities. *GCB Bioenergy* 3 (1): 43–57.

Oberoi, H.S., Sandhu, S.K., and Vadlani, P.V. (2012). Statistical optimization of hydrolysis process for banana peels using cellulolytic and pectinolytic enzymes. *Food Bioprod. Process.* 90 (2): 257–265.

Ooshima, H., Sakata, M., and Harano, Y. (1983). Adsorption of cellulase from Trichoderma viride on cellulose. *Biotechnol. Bioeng.* 25 (12): 3103–3114.

Ordaz, J.A., Toledo, Á.M., Morales, F.R.R. et al. (2011). Biorremediación de un suelo contaminado con petróleo mediante el empleo de bagazo de caña con diferentes tamaños de partícula. *Multiciencias* 11 (2): 136–145.

Ortiz, W.G.C. (2016). Tratamientos aplicables a materiales lignocelulósicos para la obtención de etanol y productos químicos. *Rev. Tecnol.* 13 (1).

Ovando-Chacón, S.L. and Waliszewski, K.N. (2005). Preparativos de celulasas comerciales y aplicaciones en procesos extractivos. *Univ. Cienc.* 21 (42): 113–122.

Pacheco, C.M., Bustos, C., Reyes, G. et al. (2018). Characterization of residues from Chilean blueberry bushes: a potential source of cellulose. *Bioresources* 13 (4): 7345–7359.

Palma, D., Zavala, J., Cámara, J. et al. (2016). Uso de residuos de la agroindustria de la caña de azúcar (*Saccharum* spp.) para elaborar abonos orgánicos. *Agroproductividad* 9 (7): 29–34.

Pérez, J., Munoz-Dorado, J., De la Rubia, T.D.L.R., and Martinez, J. (2002). Biodegradation and biological treatments of cellulose, hemicellulose and lignin: an overview. *Int. Microbiol.* 5 (2): 53–63.

Pérez-Pimienta, J.A., Flores-Gómez, C.A., Ruiz, H.A. et al. (2016). Evaluation of agave bagasse recalcitrance using AFEX™, autohydrolysis, and ionic liquid pretreatments. *Bioresour. Technol.* 211: 216–223.

Plomion, C., Leprovost, G., and Stokes, A. (2001). Wood formation in trees. *Plant Physiol.* 127: 1513–1523.

Prades, A., Salum, U.N., and Pioch, D. (2016). New era for the coconut sector. What prospects for research? *OCL* 23 (6): D607.

Puri, V.P. (1984). Effect of crystallinity and degree of polymerization of cellulose on enzymatic saccharification. *Biotechnol. Bioeng.* 26 (10): 1219–1222.

Putro, J.N., Soetaredjo, F.E., Lin, S.Y. et al. (2016). Pretreatment and conversion of lignocellulose biomass into valuable chemicals. *RSC Adv.* 6 (52): 46834–46852.

Qian, E.W. (2014). Chapter 7: Pretreatment and saccharification of lignocellulosic biomass. In: *Research Approaches to Sustainable Biomass Systems* (ed. S. Tojo and T. Hirasawa), 181–204. Elsevier.

Rabelo, S.C., da Costa, A.C., and Rossel, C.E.V. (2015). Chapter 17: Industrial waste recovery. In: *Sugarcane: Agricultural Production, Bioenergy and Ethanol* (ed. F. Santos, A. Borém and C. Caldas), 365–381. Academic Press.

Rabemanolontsoa, H. and Saka, S. (2012). Characterization of Lake Biwa macrophytes in their chemical composition. *J. Jpn. Inst. Energy* 91 (7): 621–628.

Ramírez-Cortina, C.R., Alonso-Gutiérrez, M.S., and Rigal, L. (2012). Valorization of tequila agroindustrial residues for ruminants feeding. *Rev. Chapingo Ser. Cienc. For. Ambiente* 18 (3): 449–457.

Ramírez-Guzmán, K.N., Torres-León, C., Martinez-Medina, G.A. et al. (2019). 15—Traditional fermented beverages in Mexico. In: *Fermented Beverages*, vol. 5 (ed. A.M. Grumezescu and A.L. Holban), 605–635. The Science of Beverages. Woodhead Publishing.

Ramos, S.J., Guilherme, D., Junior, C.F.C. et al. (2008). Produção de mudas de tomate em substrato contendo fibra de coco e resíduo de cultivo de cogumelo. *Rev. Bras. Ciênc. Avic.* 3: 7–24.

Rijal, D., Vancov, T., McIntosh, S. et al. (2016). Process options for conversion of *Agave tequilana* leaves into bioethanol. *Ind. Crop. Prod.* 84: 263–272.

Rivera, N.A. (2013). Análisis de la productividad de etanol de caña de azúcar en ingenios azucareros de México. *CIENCIA Ergo-Sum* 20 (1): 17–28.

Rojas-León, A., Otazo-Sánchez, E.M., Bolarín-Miró, A.M. et al. (2014). Residuos agrícolas: Caracterización y estrategias sustentables para su aprovechamiento. *Rev. Iberoam. Cienc.* 1: 254–262.

Romero-Uscanga, E., Montero-Alpírez, G., Toscano-Palomar, L. et al. (2014, October). Determinación de los principales componentes de la biomasa lignocelulósica; celulosa, hemicelulosa y lignina de la paja de trigo para su posterior pretratamiento biológico. In *Conference: XVII Congreso Internacional en Ciencias Agrícolas, Agricultura sustentable: Uso eficiente del agua, suelo y fertilizantes. Mexicali, Baja California*, pp. 114–118.

SAGARPA (Secretaría de Agricultura Ganadería Desarrollo Rural Pesca y Alimentación) (2005). Plan Rector Sistema Producto Nacional Palma de Coco. México D. F. Comité Sistema Producto Palma de Coco. 54 pp.

SAGARPA (Secretaría de Agricultura Ganadería Desarrollo Rural Pesca y alimentación) (2013). Plan Nacional de Desarrollo 2013-2018, Programa Nacional de la agroindustria de la caña de azúcar, pp. 1–45.

Saha, B.C. (2005). Enzymes as biocatalysts for conversion of lignocellulosic biomass to fermentable sugars. In: *Handbook of Industrial Biocatalysis* (ed. C.T. Hou), 1–12. CRC Press.

Salcedo Luna, J., Castro Montoya, A.J., Rico, J.L., and Campos García, J. (2010). Optimization of acid hydrolysis of bagasse from *Agave tequilana* Weber. *Rev. Mex. Ing. Quim.* 9 (1): 91–97.

Sánchez, O.J. and Cardona, C.A. (2008). Trends in biotechnological production of fuel ethanol from different feedstocks. *Bioresour. Technol.* 99 (13): 5270–5295.

Sánchez-Orozco, R. and Vázquez-Velázquez, J. (2017). Hidrólisis de residuos de naranja para la obtención de carbohidratos fermentables. *Rev. Investig. Desarro.* 3 (9): 63–70.

Sankaran, R., Parra Cruz, R.A., Pakalapati, H. et al. (2020). Recent advances in the pretreatment of microalgal and lignocellulosic biomass: a comprehensive review. *Bioresour. Technol.* 298: 122476.

Sasakura, Y., Nakashima, K., Awazu, S. et al. (2005). Transposon-mediated insertional mutagenesis revealed the functions of animal cellulose synthase in the ascidian *Ciona intestinalis*. *Proc. Natl. Acad. Sci.* 102 (42): 15134–15139.

Sebastian, A., Nangia, A., and Prasad, M.N.V. (2018). A green synthetic route to phenolics fabricated magnetite nanoparticles from coconut husk extract: implications to treat metal contaminated water and heavy metal stress in Oryza sativa L. *J. Clean. Prod.* 174: 355–366.

Sevilla, I.R. (2017). Caracterización química de tres residuos lignocelulósicos generados en la región del Cantón Alausí. *Revista del Instituto de Investigación de la Facultad de Ingeniería Geológica, Minera, Metalúrgica y Geográfica* 20 (40): 80–85.

Sharma, H.K., Xu, C., and Qin, W. (2019). Biological pretreatment of lignocellulosic biomass for biofuels and bioproducts: an overview. *Waste Biomass Valorization* 10 (2): 235–251.

Sifontes, M.C. (2019). La biomasa: fuente alternativa de combustibles y compuestos químicos. *Anales de Química de la RSEQ* 115 (5): 399–407.

Singh, R., Krishna, B.B., Kumar, J., and Bhaskar, T. (2016). Opportunities for utilization of non-conventional energy sources for biomass pretreatment. *Bioresour. Technol.* 199: 398–407.

Soares, L.A., Braga, J.K., Motteran, F. et al. (2017). Design and optimization of hydrogen production from hydrothermally pretreated sugarcane bagasse using response surface methodology. *Water Sci. Technol.* 76 (1): 95–105.

Staron, P., Chwastowski, J., and Banach, M. (2017). Sorption and desorption studies on silver ions from aqueous solution by coconut fiber. *J. Clean. Prod.* 149: 290–301.

Statista (2020). Tequila production in Mexico from 1995 to 2019. Available from: https://www.statista.com/statistics/311696/mexico-s-tequila-production (Accessed May 25, 2020).

Sticklen, M.B. (2008). Plant genetic engineering for biofuel production: towards affordable cellulosic ethanol. *Nat. Rev. Genet.* 9: 433–443.

Suárez, R. and Morín, R. (2005). Caña de azúcar y sostenibilidad: enfoques y experiencias cubanas. *Desarrollo Alternativo AC Desal* 19 (41): 26.

Suárez, C., Zienkiewicz, A., Castro, A.J. et al. (2013). Cellular localization and levels of pectins and arabinogalactan proteins in olive (*Olea europaea* L.) pistil tissues during development: implications for pollen–pistil interaction. *Planta* 237 (1): 305–319.

Sun, Y. and Cheng, J. (2002). Hydrolysis of lignocellulosic materials for ethanol production: a review. *Bioresour. Technol.* 83 (1): 1–11.

Sun, S., Sun, S., Cao, X., and Sun, R. (2016). The role of pretreatment in improving the enzymatic hydrolysis of lignocellulosic materials. *Bioresour. Technol.* 199: 49–58.

Taherzadeh, M.J. and Karimi, K. (2008). Pretreatment of lignocellulosic wastes to improve ethanol and biogas production: a review. *Int. J. Mol. Sci.* 9 (9): 1621–1651.

Talat, M., Mohan, S., Dixit, V. et al. (2018). Effective removal of fluoride from water by coconut husk activated carbon in fixed bed column: experimental and breakthrough curves analysis. *Groundw. Sustain. Dev.* 7: 48–55.

Talbot, P., Pettigrew, D., Lacasse, R. et al. (2007). Uso de un material biofiltrante basado en mesocarpio de coco en un sistema de tratamiento de aguas residuales, Patente de utilidad ES 2 285 173 T3, November 16, 2007.

Travaini, R., Martín-Juárez, J., Lorenzo-Hernando, A., and Bolado-Rodríguez, S. (2016). Ozonolysis: an advantageous pretreatment for lignocellulosic biomass revisited. *Bioresour. Technol.* 199: 2–12.

Trujillo-Ramírez, D., Lobato-Calleros, C., Román-Guerrero, A. et al. (2018). Complexation with whey protein hydrolysate improves cacao pods husk pectin surface active and emulsifying properties. *React. Funct. Polym.* 123: 61–69.

USDE (2013). United States Department of Energy: Energy Efficiency and Renewable Energy. Biomass feedstock composition and property database. Available from http://www.afdc.energy.gov/biomass/progs/search1.cgi.

Uzcanga Pérez, N.G., Camarena Gómez, D.M.J., Cortazar Ríos, M., and Góngora Pérez, R.D. (2015). Preferencias de consumo por productos derivados del cocotero en la Península de Yucatán, México. *Rev. Mex. Cienc. Agrícolas* 6 (1): 45–57.

Valdez-Vazquez, I., Acevedo-Benítez, J.A., and Hernández-Santiago, C. (2010). Distribution and potential of bioenergy resources from agricultural activities in Mexico. *Renew. Sust. Energ. Rev.* 14 (7): 2147–2153.

Van der Pol, E.C., Eggink, G., and Weusthuis, R.A. (2016). Production of L(+)-lactic acid from acid pretreated sugarcane bagasse using *Bacillus coagulans* DSM2314 in a simultaneous saccharification and fermentation strategy. *Biotechnol. Biofuels* 9 (1): 248.

Van Steen, E. and Claeys, M. (2008). Fischer–Tropsch catalysts for the biomass-to-liquid (BTL) process. *Chem. Eng. Technol.: Industrial Chemistry-Plant Equipment-Process Engineering-Biotechnology* 31 (5): 655–666.

Vera, L., Uguña, M.F., García, N. et al. (2016). Eliminación de los metales pesados de las aguas residuales mineras utilizando el bagazo de caña como biosorbente. *Afinidad* 73 (573): 43–49.

Voragen, A.J., Coenen, G.-J., Verhoef, R., and Schols, H. (2009). Pectin, a versatile polysaccharide present in plant cell walls. *Struct. Chem.* 20 (2): 263–275.

Wang, Y., Zhang, Y., Wang, J., and Meng, L. (2009). Effects of volatile fatty acid concentrations on methane yield and methanogenic bacteria. *Biomass Bioenergy* 33 (5): 848–853.

Wang, Y., Nie, X., and Liu, Z. (2013). Study on preparation of levulinic acid from biomass and its prospects. *J. For. Prod. Ind.* 2 (3): 5–7.

Wang, W., Ma, X., Jiang, P. et al. (2016). Characterization of pectin from grapefruit peel: a comparison of ultrasound-assisted and conventional heating extractions. *Food Hydrocoll.* 61: 730–739.

Wang, F., Pan, Y., Cai, P. et al. (2017). Single and binary adsorption of heavy metal ions from aqueous solutions using sugarcane cellulose-based adsorbent. *Bioresour. Technol.* 241: 482–490.

Weiland, P. (2010). Biogas production: current state and perspectives. *Appl. Microbiol. Biotechnol.* 85 (4): 849–860.

Wodzicki, T.J. (2001). Natural factors affecting wood structure. *Wood Sci. Technol.* 35 (1–2): 5–26.

Wyman, C.E., Dale, B.E., Elander, R.T. et al. (2005a). Comparative sugar recovery data from laboratory scale application of leading pretreatment technologies to corn stover. *Bioresour. Technol.* 96 (8): 2026–2032.

Wyman, C.E., Dale, B.E., Elander, R.T. et al. (2005b). Coordinated development of leading biomass pretreatment technologies. *Bioresour. Technol.* 96 (18): 1959–1966.

Wyman, C.E., Decker, S.R., Himmel, M.E. et al. (2005c). Hydrolysis of cellulose and hemicellulose. *Renew. Energy* 160 (1): 995–1033.

Xu, D., Yuan, F., Gao, Y. et al. (2014). Influence of whey protein–beet pectin conjugate on the properties and digestibility of β-carotene emulsion during in vitro digestion. *Food Chem.* 156: 374–379.

Yan, X., Corbin, K.R., Burton, R.A., and Tan, D.K. (2020). Agave: a promising feedstock for biofuels in the water–energy–food–environment (WEFE) nexus. *J. Clean. Prod.* 261: 121283, 1–11.

Yong, J.W., Ge, L., Ng, Y.F., and Tan, S.N. (2009). The chemical composition and biological properties of coconut (*Cocos nucifera L.*) water. *Molecules* 14 (12): 5144–5164.

Zavaleta, C., Oropeza, C. Cortazar, M. (2010). Agua de coco una alternativa para la salud. Comité Editorial del CIRSE. Campo Experimental Chetumal, Quintana Roo. Desplegable Informativa Núm. 1. Serie INIFAP.

Zhang, K., Pei, Z., and Wang, D. (2016). Organic solvent pretreatment of lignocellulosic biomass for biofuels and biochemicals: a review. *Bioresour. Technol.* 199: 21–33.

Zhou, M., Yang, J., Wang, H. et al. (2014). Chapter 9 – Bioelectrochemistry of microbial fuel cells and their potential applications in bioenergy. In: *Bioenergy Research: Advances and Application* (ed. V.K. Gupta, M.G. Tuohy, C.P. Kubicek, et al.), 131–152. Elsevier.

Zhou, Z., Lei, F., Li, P., and Jiang, J. (2018). Lignocellulosic biomass to biofuels and biochemicals: a comprehensive review with a focus on ethanol organosolv pretreatment technology. *Biotechnol. Bioeng.* 115 (11): 2683–2702.

Zumalacárregui, L. and Zumalacárregui, B. (2008). Cálculo del beneficio ambiental de la caña de azúcar para la producción de etanol combustible. *Ing. Compet.* 10 (1): 65–71.

3

Lignocellulose Bioconversion

Technical Aspects and New Developments

J Gamboa-Santos[1,2] *and A Alzamendi*[1,3]

[1] *Consejo Nacional de Investigaciones Científicas y Técnicas (CONICET), Buenos Aires, Argentina*
[2] *Centro de Investigación y Desarrollo en Criotecnología de Alimentos (CIDCA-CONICET UNLP), La Plata, Argentina*
[3] *Instituto Multidisciplinario de Biología Celular (IMBICE-CONICET-CICPBA-UNLP), La Plata, Argentina*

3.1 Introduction

A huge global effort to reduce the consumption of crude oil with renewable sources is currently urgent. Since the total world oil reserves are only 239.4 thousand million tonnes, our global production will only be sufficient to sustain our lifestyle for another 50 years (British Petroleum 2020). As is evident, an effective and comprehensive renewable energy is needed. Lignocellulosic (LC) biomass is considered the most abundant sustainable carbon resource on the planet (Liu et al. 2020; Ong and Wu 2020). About 933 million tons of LC biomass are generated every year by agricultural practices and various agro-based industries; only 2% is reused by humans (De Bhowmick et al. 2018). In this context, the development of fuels and chemicals from LC wastes is attracting significant attention from governments worldwide (Liu et al. 2020).

Industries that process any kind of plant biomass generate residues and by-products with specific characteristics, varying in volume, physical state, and composition (Lorenci Woiciechowski et al. 2020). LC biomass can even come from organic wastes, contributing to the ecological cycle that is being demanded around the globe (Ravindran and Jaiswal 2016; Chrysikou et al. 2018; Sewsynker-Sukai et al. 2020). Even though LC biomass is residues, these materials are still biodegradable; however, they usually need to be treated before final disposal, in order to avoid environmental damage. The recycling of these residues, for either economic or environmental reasons, can also valorize them as value-added products (Lorenci Woiciechowski et al. 2020). In LC biomass, cellulose and hemicellulose are sugar-rich resources that are suitable for biofuel production such as ethanol, whereas lignin is a useful feedstock for synthesizing aromatic-based chemicals (Soltanian et al. 2020). LC biomass appears to be the perfect substitute for petroleum for the production of fuels and fine chemicals with net zero carbon emission (Isikgor and Becer 2015; Dhyani and Bhaskar 2018). Pretreatment has been considered as an expensive stage in LC biomass transformation

Lignocellulose Bioconversion Through White Biotechnology, First Edition. Edited by Anuj Kumar Chandel.

processes and therefore the development of an efficient and cost-effective pretreatment method is imperative for the economical production of fuels and chemicals. The aim of this chapter is to explore the available pretreatment methods to address the best bioconversion performance during LC biomass processing. After a short description of the LC biomass composition, a detailed overview of the pretreatments commonly used and those that are seen as promising technological developments is given. This chapter could help to identify a suitable LC biomass pretreatment in order to promote a biobased circular economy.

3.2 Lignocellulosic (LC) Biomass Composition

LC materials consist mainly of cellulose, hemicellulose, and lignin along with trace amounts of extractive fractions such as tannins, resins, fatty acids, and inorganic salts (Soltanian et al. 2020). The cellulose, hemicellulose, and lignin contents of LC are reported to be in the range 30%–55%, 15%–40%, and 10%–35% (weight basis), respectively. The rigid structure of LC is largely due to the crosslinked lignin and hemicellulose surrounding the crystalline cellulose nanofibrils (Ferrand et al. 2020).

3.2.1 Cellulose

Cellulose is a semicrystalline macromolecule comprising both highly ordered crystalline domains and amorphous regions. Basically, it is a linear polymer of D-anhydroglucopyranose molecules joined by β-1,4-glycosidic bonds (Alexandridis et al. 2018). Their linearity and regular structure promote association of cellulose molecules over large regions, forming bundles of polycrystalline fibers. Due to the huge amount of hydroxyl groups, cellulose presents an intramolecular and intermolecular hydrogen bonding network, which makes the crystalline structure robust. To solubilize cellulose polymer, the majority of hydroxyl bonds should break simultaneously. Thus, at mild temperatures, cellulose is insoluble not only in water but also in most typical solvents. The degree of polymerization of cellulose largely depends on the origin and pretreatments applied to the raw LC biomass. Cellulose is one of the most important biomass resources due to its abundance in nature, with the peculiarity of not being included in the human food chain (Ferrand et al. 2020).

3.2.2 Hemicellulose

Hemicellulose is a highly branched heteropolymer including pentoses (mainly xylose and arabinose) and hexoses (glucose, galactose, mannose) along with some other compounds (glucuronic acid and acetyl groups). In hardwoods, the hemicellulose (15%–30%) structure is composed of xylan that is a polysaccharide consisting of D-xylose units linked together through β-1,4-glycosidic bonds. In softwoods, the hemicellulose (15%–20%) is formed by galactoglucomannan which is a polysaccharide comprising D-galactose and D-glucose units connected by β-1,4-glycosidic linkages (Bugg et al. 2011). Hemicellulose usually presents a lower degree of polymerization compared with cellulose. Besides, hemicellulose is highly hydrophilic and amenable to hydrolysis (Saha et al. 2019). Among the three main components in biomass, hemicellulose is a promising material to produce value-added chemicals (Ferrand et al. 2020).

3.2.3 Lignin

Lignin is a highly complex phenolic structure formed by three different hydroxycinnamyl alcohol monomers, namely monolignols including *p*-coumaryl, coniferyl, and sinapyl alcohols (Upton and Kasko 2016). These units are linked mainly by ether linkage (e.g. α-O-4, β-O-4, and 4-O-5) and C–C linkage (e.g. β-β, β-5, and 5-5) (Wang et al. 2012). The predominant type of interconnection in the structure of lignin is β-O-4 aryl ether that is responsible for the highly branched aromatic structure. In wood lignin these linkages are above 45% of the total bonds (Higuchi et al. 2019). At a molecular and cellular level, lignin controls permeability, improves the mechanical properties of cells, and protects them from microorganism and chemical degradation, being the primary factor in biomass recalcitrance (Lorenci Woiciechowski et al. 2020). For the sustainable production of biofuels and bulk chemicals, lignin is considered the second most abundant renewable resource, after cellulose (Agarwal et al. 2018).

3.3 Biorefinery Concept in the Era of Sustainable Circular Economy

The circular economy approach utilizes the resources throughout their shelf-life toward maximally profitable usage, whereas in a linear economy the resources are created, utilized, and disposed of. Besides, the circular economy focuses on the recovery and regeneration of materials and resources at the end of each product's life cycle. Thus, a circular economy replaces the "end of life" idea toward the utilization of sustainable power sources, avoiding the utilization of poisonous synthetic compounds and focusing on proper materials handling (Yaashikaa et al. 2020). LC valorization aims to integrate the whole process that fits the circular bioeconomy principle (Garlapati et al. 2020).

Technologies that use biomass as a feedstock for producing biobased products are referred to as biorefineries (Yaashikaa et al. 2020). Biorefineries carry out the transformation of carbon held from rural biomass and natural wastes into broadened final results and materials, including feed, food, biobased synthetic chemicals, bioenergy, powers, and bio-polymers (Zeller et al. 2019). LC biomass feedstock can come from three categories: agricultural products (e.g. corn), waste from the agro-food or forestry industry (e.g. oil palm fronds), and specialized energy crops (e.g. willow) (Ong and Wu 2020). The adoption of the biorefinery concept (circular economy approach), particularly bolstering lignin biorefinery, is a strategic point not only to make the cost of the process economically viable but also as a major push toward less greenhouse gas emissions (Garlapati et al. 2020). Additionally, biomass contains oxygen functionality for developing high value-added products. The latter fit the biorefinery concept that highlights the valorization of lignin with other biomass components (cellulose and hemicellulose) (Garlapati et al. 2020). Unlike conventional petroleum refineries, LC biorefinery-based biomass is locally available, inexpensive, and easy to scale up. LC feedstock provides critical advantages to sustain current and future production of biofuels, biomolecules, and biomaterials, while boosting rural economies in order to drive the transition to a biobased economy (Govil et al. 2020).

3.4 Biorefinery Treatments

To achieve biomass utilization for fuels and chemicals production several biorefinery models have been proposed. The biorefinery model based on the sugar platform is favored over other models due to its milder processing conditions and overall sustainability (Zhao et al. 2020). The hydrolysis of LC biomass has led to the release of several reducing sugars which are greatly esteemed for the production of biogas, phenols, bioethanol, and aldehydes (Ferrand et al. 2020). However, efficient biomass deconstruction into sugars is still a challenge and has limited large-scale commercialization of cellulosic biorefineries (Wang et al. 2017). The bioconversion of different biomass sources requires the selection of a suitable biorefinery strategy since feedstock characteristics could considerably differ from each other, and consequently the treatment will not be as effective as would be expected. In addition, pretreatment effectiveness depends on the expected value chain outputs (Bichot et al. 2020). Bioethanol production depends on three continuous stages, namely pretreatment, hydrolysis, and fermentation. The cost of bioethanol production could be reduced by creating productive pretreatment technologies, hydrolysis, and fermentation processes focusing also on integration of the processing stages (Yaashikaa et al. 2020).

3.4.1 Pretreatments

The robust and recalcitrant structure of LC biomass requires a pretreatment process to disintegrate crosslinked fractions of LC biomass for enhancing the accessibility and biodegradability of cellulose microfibrils and hemicellulose matrices in order to facilitate the subsequent bioconversion steps such as enzymatic hydrolysis, fermentation, or anaerobic digestion (Soltanian et al. 2020). Several pretreatments (Figure 3.1) such as thermal (high temperature, hydrothermal), biological (consortium, bacteria, enzymatic, fungal), chemical (acids/alkalis, organosolvents, etc.) and physical (ball milling, extrusion, irradiation) approaches have been analyzed as effective methods for the successful breakdown of recalcitrant moieties from LC waste (Rebello et al. 2020; Sankaran et al. 2020; Sewsynker-Sukai et al. 2020). Since the pretreatment stage is considered to take up 20%–48% of the total operational cost, a pretreatment process can remarkably affect the overall sustainability of LC bioconversion (Ofori-Boateng and Lee 2013; Rebello et al. 2020; Soltanian et al. 2020). In this sense, it is critical to analyze the pretreatment methods based on energy, environmental impact, and economic indicators to arrive at a single method/combination of methods with optimal product extraction (Rebello et al. 2020). Independently of the final product target, the following factors need to be considered: the pretreatment efficiency, energy demands associated with processing, cost effectiveness, scale-up possibilities, operational and maintenance pretreatment costs, effect of pretreatment method on product quality, environmental impact, fermenting bacteria, and saccharification choice (Nagarajan et al. 2020). The selection of a suitable pretreatment strategy also depends on the process configurations employed in the downstream hydrolysis and fermentation (Soltanian

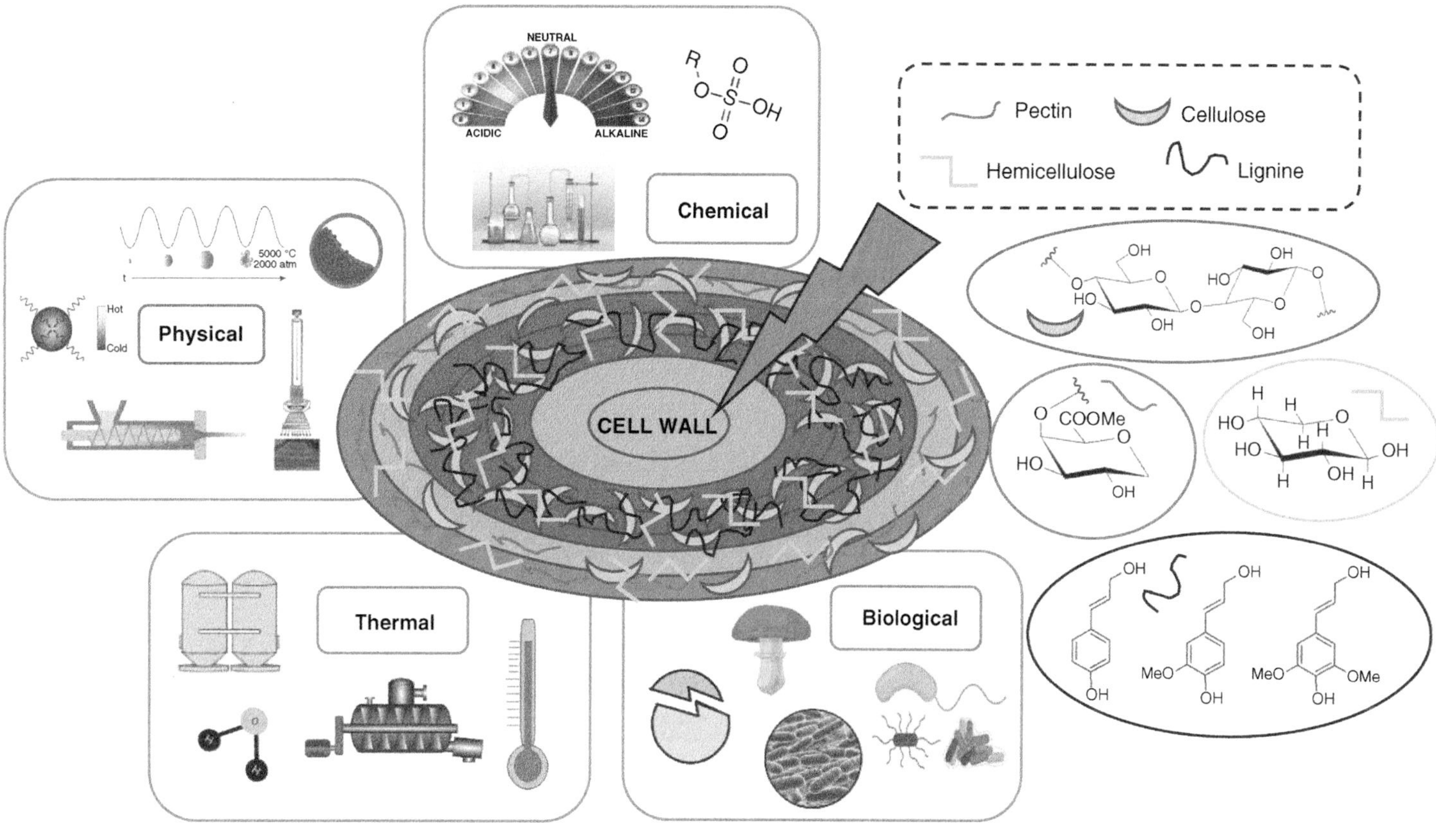

Figure 3.1 Several traditional pretreatment methods applied during lignocellulosic biomass processing and building blocks of the basic constituents of lignocellulose.

et al. 2020). Besides, the formation of inhibitory compounds is unavoidable in the pretreatment of LC feedstock, which is an additional challenge to be faced (Jönsson and Martín 2016). Among the inhibitors formed during the pretreatment step, the following are relevant: furan compounds (i.e. furfural, hydroxymethylfurfural), weak acids (i.e. formic acid, acetic acid, levulinic acid), and phenolic derivatives, since they could hamper the activity of microorganisms and consequently delay the hydrolysate fermentation (Almeida et al. 2007). In the following sections, several pretreatment methods grouped by their physical, thermal, chemical, and biological mechanisms are discussed in order to reveal the main advantages and drawbacks when selecting a suitable processing strategy for bioconversion of LC into value-added products.

3.4.1.1 Physical Methods

The physical technique involves the utilization of either mechanical or irradiation techniques and avoids the utilization of external elements, for example chemicals and microorganisms, in the pretreatment process. Normally, the physical method is applied to alter the surface area, size of particles, crystallinity index, and degree of polymerization of the biomass (Jędrzejczyk et al. 2019). Common techniques involving mechanical pretreatment, such as milling, grinding, and chipping, are applied to reduce the size along with modification of the biomass structure, breaking the crystalline structure of cellulose, which subsequently increases the surface area of the biomass for further processing. Size reduction is utilized in most of the hydrolysis research and it is known to be the most efficient technique in enhancing the enzymatic reaction of LC biomass. However, this method entails high power and energy consumption that results in high production cost (Sankaran et al. 2020).

Ball Milling

Ball milling advantages include the reduction of particle size, degree of polymerization, and cellulose crystallinity of biomass feedstock. Also remarkable is the milling ability to work at high solids loadings along with no formation of inhibitory compounds (Soltanian et al. 2020). Ball milling has been effectively used to reduce wood fiber (Gu et al. 2018), rice straw (Zhang et al. 2018), wheat straw (Liu et al. 2018), and corn stover (Bai et al. 2018) at times of 5–240 min and at temperatures ranging from 30 to 180 °C, respectively. The mechanical agitation favors cell disruption by means of a simple and effective technique. However, bead beating generates a high amount of heat which increases the energy consumption in large-scale processing (Nagarajan et al. 2020). Optimization of cell disruption by bead milling could be performed by considering the cell wall structure and integrity, the biomass concentration, the morphological properties of the beads such as size and shape, and the operative parameters such as time, agitation speed, and design (Nagarajan et al. 2020).

Extrusion

Since the 1980s, extrusion technology has been widely used in the processing of LC mass, such as the papermaking industry (De Choudens et al. 1989), components extraction (N'Diaye et al. 1996), and pretreatment for biofuel production (Zheng and Rehmann 2014). Extrusion involves the combination of thermal and mechanical techniques. The biomass is altered during its route through the extruder, causing an increase in the specific area of the biomass for enhanced enzymatic reaction (Sankaran et al. 2020). The pretreatment using the extrusion method can be improved by combining with other treatments to enhance sugar recovery. Extrusion pretreatment has been popular as one of the continuous and extra feasible techniques for bioethanol production, with high sugar recoveries from various types of biomass (Sankaran et al. 2020). Furthermore, the extrusion technology can be used alone to treat LC biomass and can also be used in combination with other pretreatment methods such as chemical treatment with either alkali or acid, steam explosion (SE), ammonia fiber explosion, vacuum extrusion, and bio-extrusion for more efficient biomass component fractionation (Duque et al. 2017).

Irradiation

By using the irradiation method, high energy radiation is used to disrupt the chemical bonds of the biomass. This radiation may result in characteristic changes in cellulose crystallinity, depolymerization of lignin, and hydrolysis of hemicelluloses (Chaturvedi and Verma 2013). Four different radiation methods with varying mechanisms affecting LC biomass have been generally used; they are microwaves, gamma rays, electron beams, and ultrasonication (Sankaran et al. 2020). Despite the efficiency and benefits of the irradiation technique, the high operative cost associated with it is still a barrier for its large-scale industrial application.

Microwaves Microwaves are electromagnetic waves in the frequency range 0.3–300 GHz. Microwave irradiation causes cell disruption by interacting with the dielectric and polar molecules in the solution causing local heating and increased pressure. As a consequence, cell disruption occurs (Nagarajan et al. 2020). Microwaves pretreatment has earned interest for biorefinery applications because of its advantages, such as non-contact and volumetric heating, reduced reaction time, low solvent consumption, reduced side reactions, and ease of parameter control (Bundhoo 2018), but industrial application could be impaired by the high maintenance costs and additional downstream methods necessary for the desired product extraction (Nagarajan et al. 2020). Many studies working with microwaves focus on microwave effects on LC biomass composition and structure to valorize the biomass in the form of sugars for biofuel production (Tsegaye et al. 2019), for biomethane production (Kan et al. 2018), or for improving enzymatic hydrolysis (Moodley and Gueguim Kana 2017).

Gamma Rays or Electron Beam Irradiation The high energy oxidative irradiation pretreatment technique uses γ-ray or electron beam irradiation to penetrate the fiber lattice, to activate LC molecules, and to generate free radicals, which eventually lead to depolymerization of LC (Wu et al. 2020). Electron beam irradiation treatment has been demonstrated to be a promising pretreatment method applied in wood, agricultural residues, reed, bagasse, and other biomasses (Kapoor et al. 2017; Liu et al. 2017; Masyutin et al. 2017). During electron beam irradiation treatment the mechanical strength in biomass decreases, due to disruption in the biomass structure by the breakage of hydrogen bonds of cellulose, promoting the conversion of cellulose into fermented sugars (Sankaran et al. 2020; Wu et al. 2020). Simultaneously, the treatment has the advantages of being environmentally friendly while obtaining a high yield of reducing sugar, by applying a simple operation involving low energy consumption and low temperature. Despite being one of the less studied pretreatments compared to other methods, electron beam irradiation is able to decompose cellulose and hemicellulose to water-soluble oligosaccharides, damage the apparent morphology of agricultural residues, rupture the hydrogen bonds of LC molecules, break the glycosidic bonds of cellulose and hemicellulose, destroy the crystal structure of cellulose, and decrease the thermal stability of agricultural residues (Wu et al. 2020).

Ultrasonication Ultrasonication pretreatment comprises the use of ultrasonic wave energy for cell disruption at frequencies ranging between 10 kHz and 20 MHz. The uses of ultrasonication irradiation technology have garnered the attention of researchers due to physical and chemical effects, namely mechanoacoustic and sonochemical, respectively (Ong and Wu 2020). During ultrasonication treatment, the compression and rarefaction of acoustic pressure in a liquid will cause a decrease in local pressure. When it falls below the saturated vapor pressure, the formation of microbubbles occurs (Ong and Wu 2020). The microbubbles present in the liquid will undergo a growth phase by absorbing ultrasonic energy and growing in size and instability (Shirkavand et al. 2016). The physical effect over biomass feedstock results in a vigorous hydromechanical shear force from shockwaves and microjets. The cavitation of microbubbles could generate turbulence and agitation within the liquid as a secondary effect (Loow et al. 2016). The shockwave propagation could reach pressures above 100 MPa resulting in effective cell disruption (Nagarajan et al. 2020). Ultrasonication also produces a chemical effect by oxidizing radicals that attack the LC matrix and disrupt chemical linkages of lignin (Lari et al. 2019; Sankaran et al. 2020). Physical and chemical effects of microbubble cavitation increase the extent of delignification, assist in breakdown of the crystalline nature of cellulose, enhance the heat and mass transfer through vigorous agitation, and improve the solubilization of organic matters (Bundhoo and Mohee 2018; Muthuvelu et al. 2019; Zhang et al. 2020). The wave frequency also influences the intensity of cellular disruption since higher frequencies promote a cavitation effect (Kurokawa et al. 2016). For thermolabile compound extraction, cooling is required since temperature could rise significantly (Nagarajan et al. 2020).

3.4.1.2 Thermal Methods

Thermal pretreatment consists in the use of high temperatures to promote biomass degradation, the higher the pretreatment temperatures the higher the degradation of organic substances. Hemicellulose is partially hydrolyzed through the supplied heat. Besides, concomitant release of acetic acid in the reaction medium causes partial autocatalytic hydrolysis of acid-labile hemicellulose (Huang et al. 2018; Espirito Santo et al. 2020). In the following sections high temperature/hydrothermal and SE pretreatments are described as the main exponents of thermal methods found in the literature.

High Temperature/Hydrothermal Pretreatment

High temperature treatments operate on the basic principle of the denaturation of cell walls at high temperatures (80–160 °C). Liquid hot water pretreatment has several merits including relatively low cost in terms of capital investment, energy, and chemical inputs, as well as minimal generation of inhibitory products and waste, among others (Wells et al. 2020). The recovery of structural sugars by liquid hot water pretreatment followed by enzymatic hydrolysis of LC biomass is dependent on the hemicellulose and lignin content and structure (Wells et al. 2015). Hydrothermal pretreatment is most often used in combination with chemical methods to enhance the product yield (Nagarajan et al. 2020).

Steam Explosion

SE has been studied for decades, and it is considered an ideal and cost-effective thermal pretreatment method for breaking the LC bonds in biomass (Ong and Wu 2020). SE treatment solubilizes the hemicellulose to increase the accessibility of the cellulose for enzymatic hydrolysis and to avoid the formation of inhibitors such as furfural, hydroxymethylfurfural, and lignin products (Sorek et al. 2014). SE is considered a mature technology that can be used as an industrial pretreatment for LC bioethanol production (Duval and Lawoko 2014). Due to the high temperatures (180–240 °C) and pressures (1–3.45 MPa) applied during SE, degradation of hemicellulose and lignin occurs, which enhances the cellulose hydrolysis and increases the biogas yield. Moreover, an increase in the pore volume and a decrease in the particle size of the biomass particle structure is produced (Agarwal et al. 2018). However, owing to its operative processing condition requirements, SE treatment could substantially increase the overall processing cost (Nagarajan et al. 2020).

3.4.1.3 Chemical Methods

Chemical pretreatments are often conducted by using strong chemicals like acids and alkalis (Nagarajan et al. 2020). Chemicals alter the degree of polymerization, reducing the crystallinity of cellulose while increasing its surface area (Sewsynker-Sukai et al. 2020). Chemical pretreatments have shown favorable results with regard to fermentable sugar recovery of approximately 80% compared with pretreatments employed without chemicals (Sewsynker-Sukai et al. 2020). In the following sections, the main chemical methods, namely acids, alkalis, and organosolv pretreatment applied during LG bioconversion, are addressed.

Acid Pretreatment

From LC biomass sources, acids mainly solubilize the hemicellulosic biomass fraction, but they also solubilize part of the lignin making cellulose more accessible to further enzymatic attack (Lorenci Woiciechowski et al. 2020). Inorganic acids, such as phosphoric acid, nitric acid, hydrochloric acid, and sulfuric acid, have been evaluated for biomass pretreatment, allowing the process to be scaled up to an industrial level (Solarte-Toro et al. 2019). Among them, sulfuric acid is the most commonly used acid as it is cheaper than nitric and phosphoric acid (Cao et al. 2018). Acid hydrolysis of the cell wall of polymers results in cell disruption due to the cleavage of acetyl groups from xylan polymers (Kumari and Singh 2018). Acid hydrolysis performance is improved by higher temperature (Nagarajan et al. 2020). However, high concentrations of toxic inhibitor compounds, such as furfural and 5-hydroxymethyl furfural, are produced during harsh operating conditions and from substrates with high lignin content (Hendriks and Zeeman 2009). The effects of furfural and 5-hydroxymethyl furfural on microorganisms include reduction of the biological and enzymatic activities, DNA breakdown, inhibition of protein, and RNA synthesis (Lorenci Woiciechowski et al. 2020). Therefore, after acidic pretreatments an additional detoxification unit operation is absolutely necessary prior to saccharification and/or fermentation (Zhao et al. 2020).

Alkaline Pretreatment

Alkaline pretreatment is very efficient at solubilizing lignin and part of the hemicellulose, and in reducing cellulose crystallinity (Lorenci Woiciechowski et al. 2020). Bases saponify membrane cell wall lipids leading to cellular structure disintegration, and consequently the removal of lignin compounds takes place (Sewsynker-Sukai et al. 2020). It is the most traditional biomass pretreatment method used in pulp and paper processing with lignin and hemicelluloses considered as by-products (Lorenci Woiciechowski et al. 2020). Alkaline chemicals have demonstrated various advantages over acid treatment, since they result in high sugar yields with minimal damage to reactors due to its non-corrosive characteristics (Sewsynker-Sukai et al. 2020). Compared with acid pretreatments, the reaction conditions by using alkalis are generally milder with little to almost no major inhibitors generated during the process. To conduct alkaline pretreatments, ammonia, sodium hydroxide, and calcium hydroxide are preferred as treatment reagents (Zhao et al. 2020). Nevertheless, alkaline pretreatments surface a major concern for industrial scale processes that is mainly attributed to the high cost, thus making it infeasible (Nagarajan et al. 2020; Sewsynker-Sukai et al. 2020).

Organosolv

Organosolv pretreatment employs organic solvent(s) for the removal of lignin and hemicellulose. As a consequence, after pretreatment and separation, a high purity solid-glucan-rich fraction can be obtained (Ferreira and Taherzadeh 2020). Organosolv pretreatment is often chosen due to its capacity to selectively separate the LC polymers into high quality streams (Nitsos et al. 2017). The method applies combinations of different organic solvents (organosolv) and conditions for dissolution of hemicellulose and lignin (Ferreira and Taherzadeh 2020). For carrying out organosolv pretreatments, different solvents such as short-chain aliphatic alcohols (e.g. methanol, ethanol), polyhydric alcohols (e.g. glycerol, ethylene glycol, triethylene glycol), alkylene carbonates, organic acids, among others (e.g. acetone, ionic liquids [ILs]) are commonly utilized (Ferreira and Taherzadeh 2020). As expected, the selected organosolv system and the operative conditions applied during LC biomass fractionation, including solvent type and concentration, catalyst addition, temperature, retention time, lignin precipitation methods, etc., could notably influence both polymer purity and recovery (Ferreira and Taherzadeh 2020).

3.4.1.4 Biological Methods

Biological pretreatment methods include the use of cellulase and hemicellulase enzymes derived from fungi, bacteria, and actinomycetes in order to achieve lignin and hemicellulose decomposition (Maurya et al. 2015; Garlapati et al. 2020). Enzymatic treatments mainly involve enzymatic hydrolysis of the cell wall and intracellular components to alter the cell wall permeability so that the intracellular nutritious components are accessible to fermenting bacteria. Enzymatic hydrolysis is carried out under mild conditions without leaving behind secondary pollution. In some cases, substrate enzymatic reaction is intensified by using ozone affecting the lignin solubility (Garlapati et al. 2020). Biological pretreatments show promise of decreased inhibitor compounds and good energy efficiency and maintain mild operations that are eco-friendly (Sindhu et al. 2016). The main limitations in the large volume industrial applications of enzymes are their slow rate of reaction along with their high cost (Nduko et al. 2012; Zabed et al. 2019).

Compared to acid hydrolysis, the enzymatic method offers a prospective for almost complete hydrolysis of cellulose content by applying an environmentally innocuous, mild operational conditions treatment, with no need for corrosion resilient processing equipment and lower release of inhibitors (Govil et al. 2020). In general, a mixture of enzymes is necessary for effective cell disruption and hydrolysis; thus, costs are expected to further increase along with the treatment efficiency. In addition, enzymatic treatment has the disadvantage of requiring long hydrolysis times to achieve a satisfactory level of sugar yield (Sun et al. 2013). On the other hand, the biomass carbohydrate components are hydrolyzed to simple sugars; thus cell disruption and carbohydrate liquefaction are reduced to a single step (Nagarajan et al. 2020). The main effects on LC biomass of the pretreatments described above along with those of the new developments addressed in the next sections are summarized in Table 3.1.

Table 3.1 Main mechanisms and effects of some of the traditional and innovative pretreatment methods used in LC biomass bioconversion.

Pretreatment	Effect on LC biomass	Literature
Physical methods		
Ball milling	Cell wall disruption, depolymerization and decrystallization of cellulose	Zheng and Rehmann (2014), Gu et al. (2018), Soltanian et al. (2020)
Extrusion	Particle size reduction and increase in the specific surface area	N'Diaye et al. (1996), Duque et al. (2017), Sankaran et al. (2020)
Irradiation		
Microwave	Disruption of crystalline structure of cellulose and lignin. Oscillating electromagnetic waves interact with polar molecules which leads to dissipation of heat via molecular friction	Bundhoo (2018), Kan et al. (2018), Liu et al. (2018), Tsegaye et al. (2019), Bichot et al. (2020), Nagarajan et al. (2020)
Gamma ray	Generation of high activity radicals through rapid localization of the absorbed energy within LC biomass	Kapoor et al. (2017), Wu et al. (2020)
Electron beam	High energy electrons degrade internal structure. Breakage of hydrogen bonds of cellulose, promoting conversion to fermented sugars	Kumar and Tumu (2019), Fei et al. (2020)
Ultrasonication	Mechanoacoustic effect: cavitation bubbles generate a stream of liquid that physically disrupts the biomass surface Sonochemical effect: extreme temperature and pressure foam cavitation leading to highly reactive radical formation	Kurokawa et al. (2016), Loow et al. (2016), Shirkavand et al. (2016), Ong and Wu (2020)

(Continued)

Table 3.1 (Continued)

Pretreatment	Effect on LC biomass	Literature
Thermal methods		
High temperature/ hydrothermal	Denaturation of cell walls at high temperatures (>180 °C)	Ruiz et al. (2020), Wells et al. (2020)
Steam explosion	Solubilization of hemicellulose to increase cellulose accessibility for enzymatic hydrolysis. Avoid inhibitor formation such as furfural, hydroxymethylfurfural and lignin products	Sorek et al. (2014), Shi et al. (2019), Espirito Santo et al. (2020)
Chemical methods		
Dilute acids	Solubilization of hemicellulose and fragmentation into simple sugars. Disruption of hydrogen, covalent and van der Waals forces reducing the requirement of enzymatic hydrolysis	Loow et al. (2016), Lorenci Woiciechowski et al. (2020)
Alkaline	Selectively lignin and hemicellulose removal without degrading cellulose, generating an enriched cellulose biomass	Solarte-Toro et al. (2019), Sewsynker-Sukai et al. (2020)
Organoslov	Cleavage of functional groups such as α-O-4 and β-O-4 found in lignin	Wang et al. (2012), Nitsos et al. (2017), Ferreira and Taherzadeh (2020)
New technologies		
Ionic liquids	Degradation of the complex interactions between cellulose, hemicellulose and lignin with reduced degradation products. Removal of lignin and disruption in the crystalline structure of cellulose	Hou et al. (2012), Sun et al. (2013), Capolupo and Faraco (2016), Financie et al. (2016), Alayoubi et al. (2020), Usmani et al. (2020)
Supercritical fluids		
SC water	Hydrolysis of cellulose and hydrolysis and dealkylation of lignin	Gillet et al. (2017)
SC CO_2	Penetration of SC CO_2 into small pores of LC decreasing the degree of crystallinity and increasing permeability due to β-O-4 bond cleavage, increasing accessibility and surface area of cellulose, lowering reducing sugar yields without causing sugar degradation	Daza Serna et al. (2016)
SC ammonia	Physical breaking down of the lignin-polysaccharide crosslinks, hemicellulose degradation to oligomeric sugars, successively deacetylated, and partial decrystallization of cellulose while lignin remains unaffected	Zhao et al. (2020)
Deep eutectic solvents	Lignin removal and hemicellulose fractionation by solubilization of lignin and hemicellulose.	Martins et al. (2019), Ali et al. (2020), Xu et al. (2020)

Table 3.1 (Continued)

Pretreatment	Effect on LC biomass	Literature
Microbial bioconversion	Degradation of lignin with lignocellulytic enzymes producing commercially interesting value-added products under mild conditions. Hemicellulose degradation increase porosity.	Ravindran and Jaiswal (2016), Shirkavand et al. (2016), Saldarriaga-Hernández et al. (2020), Govil et al. (2020)
Combined methods		
Dilute acids + ultrasonication	To maximize efficiency of dilute acids	Ong and Wu (2020)
Alkali + ultrasonication	Ultrasonication intensifies alkaline treatment due to cavitation bubbles that disrupt hydrophobic LC layer and weaken covalent bonds between lignin and hemicelluloses, facilitating penetration of alkaline solution	Muthuvelu et al. (2019)

3.5 New Innovative Technologies and Developments

As previously mentioned, the major difficulty encountered in using LC biomass is their natural recalcitrance; this means the intrinsic resistance of cell walls to deconstruction, mainly due to hydrolysis resistance and high pentose sugar content (Carpita and McCann 2020; Ko et al. 2020; Usmani et al. 2020). Biomass recalcitrance is a bottleneck in their usage, considering the mainly toxic by-products generated due to pretreatment choice and the lowering in product yield. Future perspectives tend to overcome recalcitrance and highlight LC biomass potential to be used in the emerging bioeconomy. To deal with this issue four domains should be considered: (i) molecular scale interactions among structurally complex polymers; (ii) nanoscale assembly of individual and bundled cellulose microfibrils with hemicellulose and lignin; (iii) mesoscale with wall domains of distinct composition within and between cells within a tissue; and (iv) macroscale organization of cells and tissues that determine the chemo-physical properties of biomass (Carpita and McCann 2020).

On the other hand, sustainability, energy consumption, capital cost investment, and overall efficacy of the methods must be considered prior to an industrial application of pretreatment in an LC biorefinery on a commercial basis (Haldar and Purkait 2021). Capital cost investment is an important issue considered prior to industrial application of LC biomass pretreatment, because it accounts for nearly 40% of total process cost. To reduce such cost, more insight is needed in the pretreatment used to develop an integrated techno-economic assessment of the process, considering new technologies, economic cost, benefits, and sustainability of the whole process. However, because of several existing bottlenecks, commercial adaptation to new technologies is still waiting to convert the process to a green sustainable alternative, with rapid pretreatment and high processing facilities (Korányi et al. 2020). Therefore, more research is needed to overcome the high capital cost and unavailability of comparative analysis on the use of green technologies during LC biomass bioconversion treatments. Development technologies should deal with the efficient use of raw feedstock, pointing to a reduction of waste products and negligible use of toxic chemicals. Moreover, the idea of using alternative substrates decreasing the price of the final bioproduct is gaining importance. In this sense, the use of LC substrates, mainly agro-industrial residues, to produce cellulases and other lignocellulolytic enzymes is a viable approach for reducing process material costs (Siqueira et al. 2020). On the other hand, the combination of pretreatment methods, physical, chemical, biological, or physicochemical, has been shown to reduce the time necessary for biomass conversion, as well as being more cost effective. However, more studies are needed to investigate the environmental impact and energy balance in combined processes (Saldarriaga-Hernández et al. 2020). Figure 3.2 schematizes the new developments and technologies recently applied during the pretreatment stage in LC biomass bioconversion, which are widely described in the following sections.

3.5.1 Development of Green/Environmentally Friendly Methods

In the recent past several technologies have emerged as green and sustainable alternatives due to rapid pretreatment with high processing facilities. Among these techniques irradiation-based pretreatment is mostly considered for its high effectivity on a variety of LC biomass (Haldar and Purkait 2021). Nevertheless, green technologies are in continuous development due to the fact that methods involved in pretreatment techniques in use

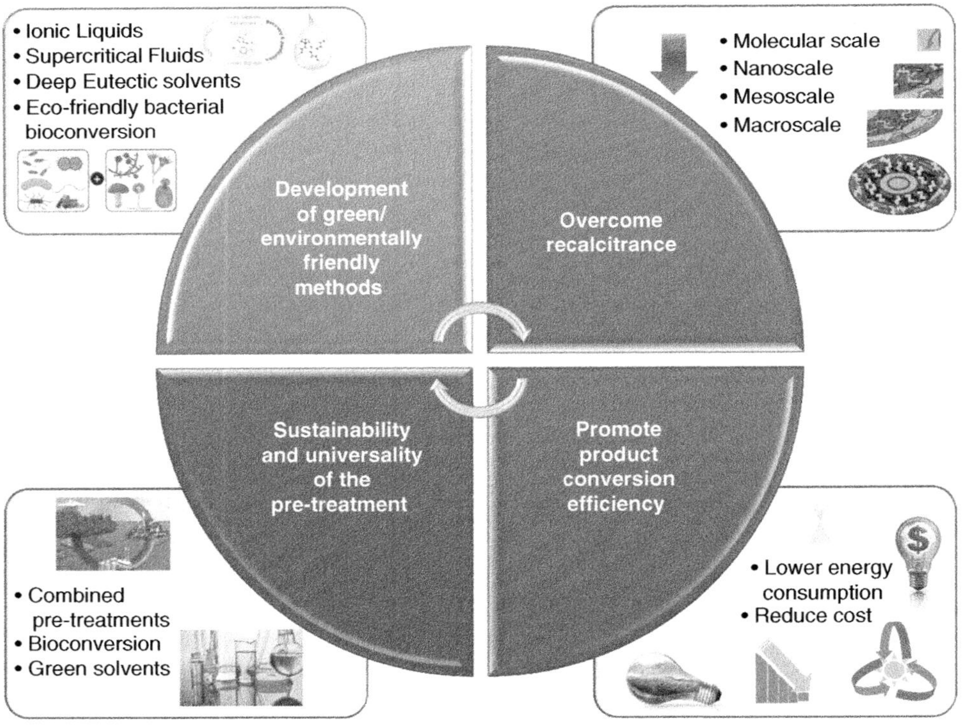

Figure 3.2 Main challenges of lignocellulosic biomass bioconversion and the new developments applied to deal with them.

generally have elevated costs, long reaction times, high energy consumption, low sugar degradation, and environmental pollution. Green technologies aim to achieve high efficiency and economic viability of pretreatments considering (i) a lower concentration of fermentation step inhibitors; (ii) high yields of sugar from fermentation; and (iii) hemicellulose and lignin recovery for production of valuable by-products, all of which with minimal energy, water, and hazard chemicals.

3.5.1.1 Green Solvents

Pretreatment of LC biomass at temperatures lower than 100 °C can save up to 50% energy consumption compared to pretreatments operating at temperatures ranging between 160 and 180 °C. While water is the most benign, environmentally friendly, and importantly cheap solvent, it provides limited impact on biomass deconstruction under mild pretreatment temperatures of ≤100 °C unless it is used in conjunction with concentrated mineral acids. Sustainability of the pretreatment process can be achieved by the use of green solvents such as ILs, supercritical (SC) fluids, and deep eutectic solvents (DESs), all of which are environmentally friendly and recyclable (Nagarajan et al. 2020).

Ionic Liquids

ILs have gained significant attraction in the pretreatment process due to their unique and green properties combined with their potential to completely break down and dissolve LC materials (Usmani et al. 2020). They are considered green alternatives to organic solvents, since no hazardous chemicals are formed by their application. These organic salts in nature can produce both anionic and cationic species and possess a melting point lower than 100 °C (Alayoubi et al. 2020). ILs are composed entirely by ions: cations are generally organic (e.g. imidazolium, ammonium-based, pyrrolidinium-based, pyridinium, sulfonium-based, phosphonium-based) while anions can be both inorganic and organic (Usmani et al. 2020). At room temperature most ILs stay in the liquid state, thus making these solvents easy to work with. The mechanism through which ILs interact with biomass is based on the specific combination of anions and cations along with specific constituents of the biomass (hemicelluloses, lignin, and cellulose) of interest. Cellulose can be efficiently dissolved in anionic ILs, such as chloride, acetate, or amino acids, because these form strong hydrogen bonds with sugars' hydroxyl protons. On the other hand, the π–π interactions of the cations with lignin help lignin solubilization. As a result, a degradation of the complex network of interactions between cellulose, hemicellulose, and lignin is observed, with reduced degradation products (Capolupo and Faraco 2016).

The ability to recycle and reuse ILs further improves the economic efficiency of their application, due to their interesting characteristics such as non-flammability, low vapor pressure, high viscosity, low conductivity, and thermal and chemical stability (Paul et al. 2020; Usmani et al. 2020). These ILs are non-corrosive, non-toxic, and economical with also good biodegradability properties.

The role of IL pretreatment is to increase the effective surface area of biomass accessible to enzymes, leading to better enzymatic delignification in terms of lignin degradation and cellulose-rich fiber separation (Financie et al. 2016). However, the performances of pretreatments vary with feedstock types, IL types, pretreatment conditions (e.g. temperature and time), and their impacts on subsequent enzymatic delignification have not been researched well. Recently, recycle and reuse of ILs has been the study focus, due to the fact that they are costly solvents, a problem to the scale application in industry. Hou et al. (2012) conducted a study on the regeneration and reapplication of eight ILs concluding that repeated use of recycled IL (five times) resulted in a marginal reduction in lignin recovery from 58.8% to 52.5% (Hou et al. 2012). Various recycling methods have been developed and studied at laboratory scale but more insight is still needed.

Supercritical Fluids

SC fluids are substances above their critical conditions of temperature and pressure. In these conditions a fluid does not present a vapor–liquid phase transition, existing only in a homogeneous phase condition whose properties such as diffusivity, viscosity, and density lie between those of gases and liquids (Escobar et al. 2020). Among the most common SC fluids, carbon dioxide, ammonia, water, and hydrocarbons such as propane and butane are found. Nevertheless, carbon dioxide is the most used SC fluid due to its non-toxic, recyclable, low-cost, environmentally friendly characteristics (considering it in a closed loop as a solvent), and low critical temperature and pressure (31.1 °C and 7.36 MPa, respectively). When used as pretreatment, SC fluids enhance substrate accessibility to enzymatic hydrolysis. However, they can also be used during a reactive extraction procedure to yield value-added products, dealing with carbohydrate hydrolysis and lignin transformation.

Supercritical Water Above 374 °C and 221 bar, the dielectric constant of water decreases, increasing the solubility of organic compounds (Escobar et al. 2020). The LC biomass physical structure changes when subjected to a treatment involving water at high pressure and high temperature. SC water hydrolyzes cellulose into oligomers and monomers, to form glucose. As temperature increases, hydrolysates can be converted into intermediates (e.g. phenols, dimers, and furfurals) and low molecular weight constituents (e.g. acids, aldehydes, ketones, and alcohols) by a variety of reactions (e.g. dehydration, condensation, isomerization, polymerization). Lignin is first decomposed through hydrolysis and dealkylation into low molecular weight fragments with highly reactive functional groups, for example formaldehyde and phenolic compounds (guaiacols, anisoles, and catechols) (Gillet et al. 2017).

The use of SC water with these characteristics has several benefits, considering the environmentally friendly treatment and post-treatment biomass manipulations. Nevertheless, this technique involves extremely high temperatures, which are cost effective, and reaction time should be under strict control to avoid sugar degradation, limiting its application to ultra-fast reactors with residence times of only 1 s (Cantero et al. 2015).

Supercritical Carbon Dioxide SC carbon dioxide shows an excellent potential for LC biomass pretreatment when dissolved in water or a water/alcohol mixture. The energy demand needed to bring CO_2 to its SC point is lower than those required for other solvents, due to its low critical temperature (31 °C) and pressure (73.8 bar) (Capolupo and Faraco 2016). CO_2 molecules are capable of penetrating into small pores of LC materials, as water and ammonia, decreasing the degree of crystallinity and increasing permeability due to β-O-4 bond cleavage, increasing the accessibility and surface area of cellulose, and lowering reducing sugar yields without causing sugar degradation, due to low temperatures used without inhibitory compound production (Daza Serna et al. 2016). The SC CO_2 economic advantage is that no fermentation inhibitors are produced; this is why, once finished, pretreatment biomass is ready to be hydrolyzed and fermented. Despite all the benefits exposed, the SC CO_2 process seems to be too expensive for the development of an industrial-scale plant because of the high cost associated with the large quantity of CO_2 required (Capolupo and Faraco 2016). However, improvements are necessary to carry out the process on a large scale.

Supercritical Ammonia Biomass exposed to ammonia under high pressure (17–20 bar) and temperatures between 60 and 100 °C for 5–30 min induces a pressure release that expands ammonia causing a physical breaking down of the lignin–polysaccharide crosslinks, hemicellulose degradation to oligomeric sugars, and successive deacetylation and partial decrystallization of cellulose, while lignin remains unaffected (Zhao et al. 2020). The major advantages of this pretreatment include high reducing sugar yields (up to 80%–90%), low sugar degradation, and no generation of inhibitory compounds, even though some phenolic fragments of lignin and other cell wall extractives may remain (Capolupo and Faraco 2016). A great limitation is the high ammonia volumes that increase process costs. However, its recycling during the process can decrease the operating costs and environmental pollution.

Deep Eutectic Solvents

A eutectic solution is the homogeneous mixture of two different constituents at a certain ratio that melt or freeze at a temperature lower than the melting point of the constituents (Martins et al. 2019). During the formation of these solvents, one acts as hydrogen bond donor while the other is a hydrogen bond acceptor. This type of solvent is named DES to differentiate it from ILs. Treatment of biomass using ILs is costly due to the sophisticated synthesis process and the DES is considered a cheaper alternative to ILs.

In the case of LC pretreatment, carbohydrates (such as glucose, fructose, xylitol), acids (such as acetic acid, formic acid, lactic acid), bases (such as amide, amine, imidazole), and phenols (such as hydroxybenzyl alcohol, catechol, vanillin) are investigated as potential hydrogen bond donors, and choline chloride is recognized as the most common form of hydrogen bond acceptor (Haldar and Purkait 2021).

During DES pretreatment, the decomposition of biomass components mainly depends on the physicochemical properties of the DES substance, provoking lignin removal and hemicellulose fractionation (Ali et al. 2020). The most suitable DES treatment type and optimal process parameters are still a matter of debate. The basic knowledge of the important information of structure–property–function–effect relationship of DESs and the detailed inner relationship of pretreatment condition variables could guide the future development of DES utilization on a biorefinery industrial scale (Xu et al. 2020).

3.5.2 Biological New Developments

The valorization of LC waste materials has been a subject of extensive consideration around the globe for environmental sustainability as well as to exploit these wastes in added value applications. In line with this, the utilization of LC wastes has gained incredible interest, for example as smart carriers for enzyme immobilization owing to their biocompatibility, cheapness, and massive availability attributes (Ahmad et al. 2020). Moreover, biological conversion or bioconversion of solid organic wastes by insect larvae has been investigated recently to overcome the setbacks experienced by the currently employed techniques. The outcomes from the studies confirmed the feasibility of a novel bioconversion technique to stabilize waste organics via valorization, while benefitting the environment (Raksasat et al. 2020).

3.5.2.1 Eco-friendly Bacterial Bioconversion

LC biomass heterogeneity led to extensive characterization of the bioconversion process. Biological platforms consider several bacteria and fungi with the ability to degrade lignin producing commercially interesting value-added products (e.g. microbial lipids and polyhydroxyalkanoates, vanillin, and muconic acid) under mild conditions (room temperature and neutral pH). In nature, several types of microorganisms (mainly fungal and bacterial species) have been identified to depolymerize and degrade lignin using specific mechanisms. This is possible because several fungi and bacteria produce lignocellulytic enzymes consisting of (i) ligninolytic enzymes such as laccases, lignin peroxidase, manganese peroxidase, and versatile peroxidase; (ii) cellulolytic enzymes such as endoglucanase, exoglucanase, and β-glucosidase; and (iii) hemicellulolytic enzymes such as endoxylanases, arabinofuranosidase, β-xylosidases, feruloyl esterase, and others, capable of degrading LC biomass in a synergistic system.

Recently, a possible route to lignin valorization was adopted, switching from chemo-catalytic or enzymatic conversion of lignin into depolymerized lignin and subsequent biocatalytic processing using microbial cell factories. Significant improvements were made on alternative chemo-catalytic methods and, in this context, the discovery and engineering of novel oxidative enzymes, such as laccases and peroxidases, play a crucial role (Gonçalves et al. 2020).

A few years ago, cultivation was the preferred method to scan the ligninolytic enzyme activity of each microorganism. Then, the computational tools available made it possible to develop sequence-based methods to screen for biological circuits that enable identification of novel ligninolytic activities from environmental samples. Nowadays, scanning is focused on whole-cell biosensors, which consist of genetically engineered cells that are able to identify targeted compounds and thus they can be applied to detect important aromatic lignin derivatives, while they measure extracellular and intracellular metabolites during bioprocessing. Generally, a biosensor consists of a host cell containing a plasmid, which contains an inducible promoter and an exogenous gene of interest, which in most cases is obtained through environmental samples.

3.5.2.2 Fungal Depolymerization

Fungi, white and brown rot in particular, have been reported to be the most powerful lignin degradation microorganisms, due to their extracellular ligninolytic enzyme systems based on heme peroxidase (lignin peroxidase, manganese peroxidase, and versatile peroxidase) and phenol oxidase (laccase) (Li and Zheng 2020). Lignin peroxidase catalyzes aromatic non-phenolic lignin structure oxidation to produce aryl-cation radicals. Manganese peroxidase in H_2O_2 oxidizes Mn^{2+} to Mn^{3+} that catalyzes alkyl-aryl cleavage and α-carbon oxidation in lignin. Versatile peroxidase exhibits oxidative abilities on both phenolic and non-phenolic aromatic compounds. Lastly, laccase is a copper-containing enzyme and its function does not require the presence of H_2O_2. Laccase can generate reactive radicals with O_2 as a final electron acceptor to catalyze the ring cleavage of aromatic compounds. While white rot fungi progressively break down the three main LC biomass polymers secreting ligninolytic enzymes, brown rot fungi degrade the main polysaccharides, but only when lignin is previously oxidized. Although white rot fungi are highly efficient in lignin degradation, they do not usually produce value-added bioproducts except for CO_2 and H_2O through mineralization. Another drawback is that these fungal enzymes are often challenging to express in high yield, and their fungal hosts are not readily amenable to genetic modification for metabolic engineering (Bugg et al. 2020).

3.5.2.3 Bacterial Depolymerization

Lignin breakdown is achieved by three bacterial groups: actinomycetes, α-proteobacteria, and γ-proteobacteria. Different bacterial strains, such as *Rhodoccocus jostii*, *Pseudomonas putida*, *Bacillus* sp., *Novosphingobium* sp., and *Citrobacter freundii*, have demonstrated lignin degradation abilities (Li and Zheng 2020). Special bacterial peroxidases and dye-decoloring peroxidase were identified as responsible for the oxidative degradation of lignin. In addition, bacteria have other oxidative enzymes capable of decomposing lignin by hydroxylation or demethylation, such as cytochrome P450s, non-heme iron enzymes, or Mn/Cu-containing oxidases. Also, laccase has been identified and studied in several bacterial species (Priyadarshinee et al. 2016). Most bacterial strains mentioned previously are capable of degrading lignin-derived oligomers or aromatic monomers. These strains converted high molecular weight polymers to compounds of lower molecular weight, followed by uptake to provide a source of energy.

Bacterial bioconversion is not limited to a few organisms; an increasing number of species are being identified with lignin-degrading ability, such as *Ochrobactrum tritici*, *Neurospora discrete*, *Bjerkandera adusta*, *Cedecea lapagei*, etc. Despite not being as effective in lignin degradation, bacterial treatment produces commercially interesting value-added bioproducts. Moreover, bacterial LC biomass bioconversion has a great potential advantage at an industrial scale, due to the thermostability, great resistance to denaturing agents, and great yield in enzyme production at lower cost (Saldarriaga-Hernández et al. 2020).

Finally, multiple technologies should be combined to achieve the desired results, because most microorganisms lack all groups of desired specific enzymes for ideal LC biomass degradation. Genetic and metabolic engineering techniques employed for novel enzyme assembly capable of hydrolyzing an extensive variety of polysaccharide linkages combined with a high specific-activity-based selection and combination of multiple enzymes will enable complete degradation of LC biomass to its monomeric forms. New technologies should focus on improvement of LC biomass saccharification by enzymatic hydrolysis, including genetic manipulation to increase the yield in enzyme production in a cost-effective way, as well as predictions of genetic expression and metabolic responses through transcriptomes.

3.5.3 Combined Pretreatment Methods

The generation of a universal process is often perplexing because of the diversity of LC biomass. Nonetheless, a combination of two methods might complement the shortcoming of each single method and could be universally employed. Sustainability of the process can be achieved by the use of green solvents such as ILs, SC fluids, and DESs which are environmentally friendly and recyclable (Nagarajan et al. 2020). In the last few years, irradiation techniques combined with chemical and physical methods have emerged as potential strategies to overcome many economic, environmental, and technological issues of a single pretreatment method, for instance the integration of IL pretreatment with microwave or ultrasonication methods instead of conventional heating, dilute acid or alkali treatments and SE (Halder et al. 2019).

Actual studies also focused on enzymes, trying to improve their activity and stability in extreme conditions of temperature and pH, and combining genetic and metabolic engineering techniques employed for novel enzyme assembly capable of hydrolyzing an extensive variety of polysaccharide linkages. Combining a high-specific-activity-based selection and combination of multiple enzymes will enable complete degradation of the complex structure of LC biomass to its monomeric forms (Saldarriaga-Hernández et al. 2020). Also, interdisciplinary research in plant and yeast engineering combined with the systematic evaluation of new alternatives from biomass to microbial fermentation will yield an economically feasible, sustainable, high performing biofuel (Ko et al. 2020).

3.6 Final Remarks

LC biomass has gained a reputation of being a non-competitive and sustainable resource of organic carbon on Earth. Also, it is the non-edible portion of the plant, and therefore it does not adversely affect food supplies (Govil et al. 2020). Since a pretreatment process is required to remove the recalcitrant structure of lignin (at least partially), decrystallize cellulose, partially depolymerize hemicellulose, and increase the porosity and accessibility for hydrolysis of the LC materials, the present chapter offers a wide description of several pretreatment methods, including their main advantages and drawbacks. Traditional techniques such as chemical, physical, mechanical, and biological methods along with recent advances in the fractionation of LC, including SC fluids, green technologies, and ILs, are reviewed. The latter innovative methods have been highlighted as attractive substitutes, due to the fast reaction rate and the use of green solvents. In addition, selecting less recalcitrant feedstock with concurrent mild pretreatment conditions represents a promising approach to reduce the release of toxic inhibitors during LC biomass utilization.

Compared with the first-generation bioethanol production (from carbohydrates, e.g. starch), second-generation bioethanol production prepared from LC biomass fractions (cellulose, hemicellulose, and lignin) has already begun. Better results are expected to be achieved by depending on methods using solar-based energy sources (Rebello et al. 2020).

In the bioplastic industry, the development of an eco-friendly consolidated bioprocessing system at low capital investment employing microorganisms for the single step conversion of LC to polyhydroxyalkanoate, without the need of pretreatment and enzymatic additional steps, would be an important breakthrough toward the replacement of petroleum-based polymers (Govil et al. 2020).

The future looks promising: an improvement in combined pretreatment methods and fermentation processes with inexpensive forestry and agricultural waste can completely change the shape of the global bio-polymer, biofuel, and bioproduct industries. This goal could be achieved by the combination of existing pretreatments with alternative energy sources, lowering cost and taking advantage of nature, avoiding waste generation and contamination. All efforts should be placed in a circular green economy, to obtain valuable sub-products without endangering the planet.

References

Agarwal, B., Kailasam, K., Singh, S.R. et al. (2018). Traversing the history of solid catalysts for heterogeneous synthesis of 5-hydroxymethylfurfural from carbohydrate sugars: a review. *Renew. Sust. Energ. Rev.* 82, 2408–2425. https://doi.org/10.1016/j.rser.2017.08.088.

Ahmad, B., Yadav, V., Yadav, A. et al. (2020). Integrated biorefinery approach to valorize winery waste: a review from waste to energy perspectives. *Sci. Total Environ.* 719: 137315. https://doi.org/10.1016/j.scitotenv.2020.137315.

Alayoubi, R., Mehmood, N., Husson, E. et al. (2020). Low temperature ionic liquid pretreatment of lignocellulosic biomass to enhance bioethanol yield. *Renew. Energy* 145: 1808–1816. https://doi.org/10.1016/j.renene.2019.07.091.

Alexandridis, P., Ghasemi, M., Furlani, E.P. et al. (2018). Solvent processing of cellulose for effective bioresource utilization. *Curr. Opin. Green Sustain. Chem.* 15: 40–52. https://doi.org/10.1016/j.cogsc.2018.05.008.

Ali, N., Zhang, Q., Liu, Z. et al. (2020). Emerging technologies for the pretreatment of lignocellulosic materials for bio-based products. *Appl. Microbiol. Biotechnol.* 104 (2): 455–473. https://doi.org/10.1007/s00253-019-10158-w.

Almeida, J.R.M., Nodig, T., and Petterson, A. (2007). Increased tolerance and conversion of inhibitors in lignocellulosic hydrolysates by *Saccharomyces cerevisiae*. *J. Chem. Technol. Biotechnol.* 82 (4): 340–349. https://doi.org/10.1002/jctb.1676.

Bai, X., Wang, G., Yu, Y. et al. (2018). Changes in the physicochemical structure and pyrolysis characteristics of wheat straw after rod-milling pretreatment. *Bioresour. Technol* 250: 770–776. https://doi.org/10.1016/j.biortech.2017.11.085.

Bichot, A., Lerosty, M., Radoiu, M. et al. (2020). Decoupling thermal and non-thermal effects of the microwaves for lignocellulosic biomass pretreatment. *Energy Convers. Manag* 203: 112220. https://doi.org/10.1016/j.enconman.2019.112220.

British Petroleum (2020). Renewable energy BP statistical review of world energy. London. Available at: https://www.imemo.ru/files/File/ru/events/2021/BP-2021.pdf (accessed: 15/02/22).

Bugg, T.D.H., Ahmad, M., Hardiman, E.M., and Rahmanpour, R. (2011). Pathways for degradation of lignin in bacteria and fungi. *Nat. Prod. Rep* 28 (12): 1883–1896. https://doi.org/10.1039/c1np00042j.

Bugg, T.D.H., Williamson, J.J., and Rashid, G.M.M. (2020). Bacterial enzymes for lignin depolymerisation: new biocatalysts for generation of renewable chemicals from biomass. *Curr. Opin. Chem. Biol.* 55: 26–33. https://doi.org/10.1016/j.cbpa.2019.11.007.

Bundhoo, Z.M.A. (2018). Microwave-assisted conversion of biomass and waste materials to biofuels. *Renew. Sust. Energ. Rev.* 82 (1): 1149–1177. https://doi.org/10.1016/j.rser.2017.09.066.

Bundhoo, Z.M.A. and Mohee, R. (2018). Ultrasound-assisted biological conversion of biomass and waste materials to biofuels: a review. *Ultrason. Sonochem.* 40 (Part A): 298–313. https://doi.org/10.1016/j.ultsonch.2017.07.025.

Cantero, D.A., Vaquerizo, L., and Mato, F. (2015). Energetic approach of biomass hydrolysis in supercritical water. *Bioresour. Technol* 179: 136–143. https://doi.org/10.1016/j.biortech.2014.12.006.

Cao, L., Yu, I.K.M., Kiu, Y. et al. (2018). Lignin valorization for the production of renewable chemicals: state-of-the-art review and future prospects. *Bioresour. Technol.* 269: 465–475. https://doi.org/10.1016/j.biortech.2018.08.065.

Capolupo, L. and Faraco, V. (2016). Green methods of lignocellulose pretreatment for biorefinery development. *Appl. Microbiol. Biotechnol.* 100 (22): 9451–9467. https://doi.org/10.1007/s00253-016-7884-y.

Carpita, N.C. and McCann, M.C. (2020). Redesigning plant cell walls for the biomass-based bioeconomy. *J. Biol. Chem* 295 (44): 15144–15157. https://doi.org/10.1074/jbc.rev120.014561.

Chaturvedi, V. and Verma, P. (2013). An overview of key pretreatment processes employed for bioconversion of lignocellulosic biomass into biofuels and value added products. *3 Biotech* 3 (5): 415–431. https://doi.org/10.1007/s13205-013-0167-8.

Chrysikou, L.P., Bezergianni, S., and Kiparissides, C. (2018). Environmental analysis of a lignocellulosic-based biorefinery producing bioethanol and high-added value chemicals. *Sustain. Energy Technol. Assess* 28: 103–109. https://doi.org/10.1016/j.seta.2018.06.010.

Daza Serna, L.V., Orrego Alzate, C.E., and Cardona Alzate, C.A. (2016). Supercritical fluids as a green technology for the pretreatment of lignocellulosic biomass. *Bioresour. Technol* 199: 113–120. https://doi.org/10.1016/j.biortech.2015.09.078.

De Bhowmick, G., Sarmah, A.K., and Sen, R. (2018). Lignocellulosic biorefinery as a model for sustainable development of biofuels and value added products. *Bioresour. Technol.* 247: 1144–1154. https://doi.org/10.1016/j.biortech.2017.09.163.

De Choudens, C., Angelier, R., Combette, P.L. (1989). *Procédé de fabrication de pâtes chimicomécaniques ou chimicothermo-mécaniques blanchies (EU Patent No. 0351330).* Office européen des brevets.

Dhyani, V. and Bhaskar, T. (2018). A comprehensive review on the pyrolysis of lignocellulosic biomass. *Renew. Energy* 129: 695–716. https://doi.org/10.1016/j.renene.2017.04.035.

Duque, A., Manzanares, P., and Ballesteros, M. (2017). Extrusion as a pretreatment for lignocellulosic biomass: fundamentals and applications. *Renew. Energy* 114 (Part B): 1427–1441. https://doi.org/10.1016/j.renene.2017.06.050.

Duval, A. and Lawoko, M. (2014). A review on lignin-based polymeric, micro- and nano-structured materials. *React. Funct. Polym* 85: 78–96. https://doi.org/10.1016/j.reactfunctpolym.2014.09.017.

Escobar, E.L.N., Da Siva, T.A., Pirich, C. et al. (2020). Supercritical fluids: a promising technique for biomass pretreatment and fractionation. *Front. Bioeng. Biotechnol.* 8 (252): 1–18. https://doi.org/10.3389/fbioe.2020.00252.

Espirito Santo, M.C., Fockink, D.N., and Pellegrini, V.O.A. (2020). Physical techniques shed light on the differences in sugarcane bagasse structure subjected to steam explosion pretreatments at equivalent combined severity factors. *Ind. Crop. Prod.* 158: 113003: https://doi.org/10.1016/j.indcrop.2020.113003.

Fei, X., Jia, W., Wang, J. et al. (2020). Study on enzymatic hydrolysis efficiency and physicochemical properties of cellulose and lignocellulose after pretreatment with electron beam irradiation. *Int. J. Biol. Macromol.* 145: 733–739. https://doi.org/10.1016/j.ijbiomac.2019.12.232.

Ferrand, L., Vasco, F., and Gamboa-Santos, J. (2020). Bulk and specialty chemicals from plant cell wall chemistry. In: *Lignocellulosic Biorefining Technologies* (ed. A.P. Ingle, A.K. Chandel and S.S. Da Silva), 7–28. Wiley https://doi.org/10.1002/9781119568858.ch2.

Ferreira, J.A. and Taherzadeh, M.J. (2020). Improving the economy of lignocellulose-based biorefineries with organosolv pretreatment. *Bioresour. Technol.* 299: 122695. https://doi.org/10.1016/j.biortech.2019.122695.

Financie, R., Moniruzzaman, M., and Uemura, Y. (2016). Enhanced enzymatic delignification of oil palm biomass with ionic liquid pretreatment. *Biochem. Eng. J* 110: 1–7. https://doi.org/10.1016/j.bej.2016.02.008.

Garlapati, V.K., Chandal, A.K., Kumar, S.P.J. et al. (2020). Circular economy aspects of lignin: towards a lignocellulose biorefinery. *Renew. Sust. Energ. Rev.* 130: 109977. https://doi.org/10.1016/j.rser.2020.109977.

Gillet, S., Aguedo, N., Petitjean, L. et al. (2017). Lignin transformations for high value applications: towards targeted modifications using green chemistry. *Green Chem.* 19: 4200–4233. https://doi.org/10.1039/c7gc01479a.

Gonçalves, C.C., Bruce, T., De Oliveira Gorgulho Silva, C. et al. (2020). Bioprospecting microbial diversity for lignin valorization: dry and wet screening methods. *Front. Microbiol.* 11: 1081. https://doi.org/10.3389/fmicb.2020.01081.

Govil, T., Priya, S., Dipayan, S. et al. (2020). Adaptive enrichment of a thermophilic bacterial isolate for enhanced enzymatic activity. *Microorganisms* 8 (6): 1–16. https://doi.org/10.3390/microorganisms8060871.

Gu, B.J., Wang, J., Wolcott, M.P., and Gonjyal, G.M. (2018). Increased sugar yield from pre-milled Douglas-fir forest residuals with lower energy consumption by using planetary ball milling. *Bioresour. Technol* 251: 93–98. https://doi.org/10.1016/j.biortech.2017.11.103.

Haldar, D. and Purkait, M.K. (2021). A review on the environment-friendly emerging techniques for pretreatment of lignocellulosic biomass: mechanistic insight and advancements. *Chemosphere* 264: 128523. https://doi.org/10.1016/j.chemosphere.2020.128523.

Halder, P., Kundu, S., Patel, S. et al. (2019). Progress on the pre-treatment of lignocellulosic biomass employing ionic liquids. *Renew. Sust. Energ. Rev.* 105: 268–292. https://doi.org/10.1016/j.rser.2019.01.052.

Hendriks, A.T.W.M. and Zeeman, G. (2009). Pretreatments to enhance the digestibility of lignocellulosic biomass. *Bioresour. Technol.* 100: 10–18. https://doi.org/10.1016/j.biortech.2008.05.027.

Higuchi, Y., Kato, R., Tsubota, K. et al. (2019). Discovery of novel enzyme genes involved in the conversion of an arylglycerol-β-aryl ether metabolite and their use in generating a metabolic pathway for lignin valorization. *Metab. Eng.* 55: 258–267. https://doi.org/10.1016/j.ymben.2019.08.002.

Hou, X.D., Smith, T.S., Min-Hua, Z. et al. (2012). Novel renewable ionic liquids as highly effective solvents for pretreatment of rice straw biomass by selective removal of lignin. *Biotechnol. Bioeng* 109 (10): 2484–2493. https://doi.org/10.1002/bit.24522.

Huang, Y.F., Chihueh, P.-T., Kuan, W.-H. et al. (2018). Product distribution and heating performance of lignocellulosic biomass pyrolysis using microwave heating. *Energy Procedia* 152: 910–915. https://doi.org/10.1016/j.egypro.2018.09.092.

Isikgor, F.H. and Becer, C.R. (2015). Lignocellulosic biomass: a sustainable platform for the production of bio-based chemicals and polymers. *Polym. Chem.* 6 (25): 4497–4559. https://doi.org/10.1039/c5py00263j.

Jędrzejczyk, M., Souzka, E., Czapnik, M. et al. (2019). Physical and chemical pretreatment of lignocellulosic biomass. In: *Second and Third Generation of Feedstocks: The Evolution of Biofuels* (ed. A. Basile and F. Dalena), 143–196. Elsevier https://doi.org/10.1016/B978-0-12-815162-4.00006-9.

Jönsson, L.J. and Martín, C. (2016). Pretreatment of lignocellulose: formation of inhibitory by-products and strategies for minimizing their effects. *Bioresour. Technol.* 199: 103–112. https://doi.org/10.1016/j.biortech.2015.10.009.

Kan, X., Zhang, J., Tong, Y.W., and Wang, C.-H. (2018). Overall evaluation of microwave-assisted alkali pretreatment for enhancement of biomethane production from brewers' spent grain. *Energy Convers. Manag.* 158: 315–326. https://doi.org/10.1016/j.enconman.2017.12.088.

Kapoor, K., Garg, N., Diwan, R.K. et al. (2017). Study the effect of gamma radiation pretreatment of sugarcane bagasse on its physcio-chemical morphological and structural properties. *Radiat. Phys. Chem.* 141: 190–195. https://doi.org/10.1016/j.radphyschem.2017.07.010.

Ko, J.K., Lee, J.H., Jung, J.H. et al. (2020). Recent advances and future directions in plant and yeast engineering to improve lignocellulosic biofuel production. *Renew. Sust. Energ. Rev.* 134: 110390. https://doi.org/10.1016/j.rser.2020.110390.

Korányi, T.I., Fridrich, B., Pineda, A. et al. (2020). Development of "lignin-first" approaches for the valorization of lignocellulosic biomass. *Molecules* 25 (12): 2815. https://doi.org/10.3390/molecules25122815.

Kumar, A. and Tumu, V.R. (2019). Physicochemical properties of the electron beam irradiated bamboo powder and its bio-composites with PLA. *Compos. Part B* 175: 107098. https://doi.org/10.1016/j.compositesb.2019.107098.

Kumari, D. and Singh, R. (2018). Pretreatment of lignocellulosic wastes for biofuel production: a critical review. *Renew. Sust. Energ. Rev.* 90: 877–891. https://doi.org/10.1016/j.rser.2018.03.111.

Kurokawa, M., King, P.M., Wu, X. et al. (2016). Effect of sonication frequency on the disruption of algae. *Ultrason. Sonochem.* 31: 157–162. https://doi.org/10.1016/j.ultsonch.2015.12.011.

Lari, Z., Ahmadzadeh, H., and Hosseini, M. (2019). Cell wall disruption: a critical upstream process for biofuel production. In: *Advances in Feedstock Conversion Technologies for Alternative Fuels and Bioproducts: New Technologies, Challenges and Opportunities* (ed. M. Hosseini), 21–35. Elsevier https://doi.org/10.1016/B978-0-12-817937-6.00002-3.

Li, X. and Zheng, Y. (2020). Biotransformation of lignin: mechanisms, applications and future work. *Biotechnol. Prog.* 36 (1): https://doi.org/10.1002/btpr.2922.

Liu, F., Xiao, L., Ye, Y. et al. (2017). Integrated Cherenkov radiation emitter eliminating the electron velocity threshold. *Nat. Photonics* 11 (5): 289–292. https://doi.org/10.1038/nphoton.2017.45.

Liu, H., Zhang, Y.X., Hou, T. et al. (2018). Mechanical deconstruction of corn stover as an entry process to facilitate the microwave-assisted production of ethyl levulinate. *Fuel Process. Technol.* 174: 53–60. https://doi.org/10.1016/j.fuproc.2018.02.011.

Liu, Y.J., Li, B., Feng, Y., and Cui, Q. (2020). Consolidated bio-saccharification: leading lignocellulose bioconversion into the real world. *Biotechnol. Adv.* 40: 107535. https://doi.org/10.1016/j.biotechadv.2020.107535.

Loow, Y.L., Wu, T.Y., Jahim, J. et al. (2016). Typical conversion of lignocellulosic biomass into reducing sugars using dilute acid hydrolysis and alkaline pretreatment. *Cellulose* 23 (3): 1491–1520. https://doi.org/10.1007/s10570-016-0936-8.

Lorenci Woiciechowski, A., Dalmas Neto, C.J., Porto de Souza Vandenberghe, L. et al. (2020). Lignocellulosic biomass: acid and alkaline pretreatments and their effects on biomass recalcitrance—conventional processing and recent advances. *Bioresour. Technol.* 304: 122848. https://doi.org/10.1016/j.biortech.2020.122848.

Martins, M.A.R., Pinho, S.P., and Coutinho, J.A.P. (2019). Insights into the nature of eutectic and deep eutectic mixtures. *J. Solut. Chem* 48 (7): 962–982. https://doi.org/10.1007/s10953-018-0793-1.

Masyutin, Y.A., Gushchina, L.A., Ivanova, L.A. et al. (2017). Oxidative and radiative pretreatment of lignocellulose feedstock for producing biofuel. *Chem. Technol. Fuels Oils* 53 (5): 633–637. https://doi.org/10.1007/s10553-017-0844-0.

Maurya, D.P., Singla, A., and Negi, S. (2015). An overview of key pretreatment processes for biological conversion of lignocellulosic biomass to bioethanol. *3 Biotech* 5: 597–609. https://doi.org/10.1007/s13205-015-0279-4.

Moodley, P. and Gueguim Kana, E.B. (2017). Development of a steam or microwave-assisted sequential salt-alkali pretreatment for lignocellulosic waste: effect on delignification and enzymatic hydrolysis. *Energy Convers. Manag* 148: 801–808. https://doi.org/10.1016/j.enconman.2017.06.056.

Muthuvelu, K.S., Rajarathinam, R., Kanagaraj, L.O. et al. (2019). Evaluation and characterization of novel sources of sustainable lignocellulosic residues for bioethanol production using ultrasound-assisted alkaline pre-treatment. *Waste Manag.* 87: 368–374. https://doi.org/10.1016/j.wasman.2019.02.015.

Nagarajan, D., Dong, C., Chen, C. et al. (2020). Biohydrogen production from microalgae—major bottlenecks and future research perspectives. *Biotechnol. J.* 2000124. https://doi.org/10.1002/biot.202000124.

N'Diaye, S., Rigal, L., Larocque, P., and Vidal, P.F. (1996). Extraction of hemicelluloses from poplar, *Populus tremuloides*, using an extruder-type twin-screw reactor: a feasibility study. *Bioresour. Technol.* 57 (1): 61–67. https://doi.org/10.1016/0960-8524(96)00041-7.

Nduko, J.M., Suzuki, W., Matsumoto, K. et al. (2012). Polyhydroxyalkanoates production from cellulose hydrolysate in *Escherichia coli* LS5218 with superior resistance to 5-hydroxymethylfurfural. *J. Biosci. Bioeng.* 113 (1): 70–72. https://doi.org/10.1016/j.jbiosc.2011.08.021.

Nitsos, C., Rova, U., and Christakopoulos, P. (2017). Organosolv fractionation of softwood biomass for biofuel and biorefinery applications. *Energies* 11 (1): 50. https://doi.org/10.3390/en11010050.

Ofori-Boateng, C. and Lee, K.T. (2013). Sustainable utilization of oil palm wastes for bioactive phytochemicals for the benefit of the oil palm and nutraceutical industries. *Phytochem. Rev.* 12 (1): 173–190. https://doi.org/10.1007/s11101-013-9270-z.

Ong, V.Z. and Wu, T.Y. (2020). An application of ultrasonication in lignocellulosic biomass valorisation into bio-energy and bio-based products. *Renew. Sust. Energ. Rev.* 132: 109924. https://doi.org/10.1016/j.rser.2020.109924.

Paul, A., Muthukumar, S., and Prasad, S. (2020). Review—room-temperature ionic liquids for electrochemical application with special focus on gas sensors. *J. Electrochem. Soc.* 167 (3): 037511. https://doi.org/10.1149/2.0112003jes.

Priyadarshinee, R., Kumar, A., Mandal, T., and Dasguptamandal, D. (2016). Unleashing the potential of ligninolytic bacterial contributions towards pulp and paper industry: key challenges and new insights. *Environ. Sci. Pollut. Res.* 23 (23): 23349–23368. https://doi.org/10.1007/s11356-016-7633-x.

Raksasat, R., Lim, J.W., Kiatkittipong, W. et al. (2020). A review of organic waste enrichment for inducing palatability of black soldier fly larvae: wastes to valuable resources. *Environ. Pollut.* 267: 115488. https://doi.org/10.1016/j.envpol.2020.115488.

Ravindran, R. and Jaiswal, A.K. (2016). Microbial enzyme production using lignocellulosic food industry wastes as feedstock: a review. *Bioengineering* 3: 4, 30. https://doi.org/10.3390/bioengineering3040030.

Rebello, S., Anoopkumar, A.N., Aneesh, E.M. et al. (2020). Sustainability and life cycle assessments of lignocellulosic and algal pretreatments. *Bioresour. Technol.* 301: 122678. https://doi.org/10.1016/j.biortech.2019.122678.

Ruiz, H.A., Conrad, M., Sun, S.N. et al. (2020). Engineering aspects of hydrothermal pretreatment: from batch to continuous operation, scale-up and pilot reactor under biorefinery concept. *Bioresour. Technol.* 299: 122685. https://doi.org/10.1016/j.biortech.2019.122685.

Saha, M., Saynik, P.B., Borah, A. et al. (2019). Dioxane-based extraction process for production of high quality lignin. *Bioresour. Technol. Rep.* 5: 206–211. https://doi.org/10.1016/j.biteb.2019.01.018.

Saldarriaga-Hernández, S., Velasco-Ayala, C., Leal-Isla Flores, P. et al. (2020). Biotransformation of lignocellulosic biomass into industrially relevant products with the aid of fungi-derived lignocellulolytic enzymes. *Int. J. Biol. Macromol.* 161: 1099–1116. https://doi.org/10.1016/j.ijbiomac.2020.06.047.

Sankaran, R., Parra Cruz, R.A., Pakalapati, H. et al. (2020). Recent advances in the pretreatment of microalgal and lignocellulosic biomass: a comprehensive review. *Bioresour. Technol.* 268: 122476. https://doi.org/10.1016/j.biortech.2019.122476.

Sewsynker-Sukai, Y., Naomi David, A., and Gueguim Kana, E.B. (2020). Recent developments in the application of kraft pulping alkaline chemicals for lignocellulosic pretreatment: potential beneficiation of green liquor dregs waste. *Bioresour. Technol.* 306: 123225. https://doi.org/10.1016/j.biortech.2020.123225.

Shi, Q., Li, Y., Li, Y. et al. (2019). Effects of steam explosion on lignocellulosic degradation of, and methane production from, corn stover by a co-cultured anaerobic fungus and methanogen. *Bioresour. Technol.* 290: 121796. https://doi.org/10.1016/j.biortech.2019.121796.

Shirkavand, E., Baroutian, S., Gapes, D.J., and Young, B.R. (2016). Combination of fungal and physicochemical processes for lignocellulosic biomass pretreatment—a review. *Renew. Sust. Energ. Rev.* 54: 217–234. https://doi.org/10.1016/j.rser.2015.10.003.

Sindhu, R., Binod, P., and Pandey, A. (2016). Biological pretreatment of lignocellulosic biomass—an overview. *Bioresour. Technol.* 199: 76–82. https://doi.org/10.1016/j.biortech.2015.08.030.

Siqueira, J.G.W., Rodrigues, C., Vandenberghe, L.P.S. et al. (2020). Current advances in on-site cellulase production and application on lignocellulosic biomass conversion to biofuels: a review. *Biomass Bioenergy* 132: 105419. https://doi.org/10.1016/j.biombioe.2019.105419.

Solarte-Toro, J.C., Romero-García, J.M., Martínez-Patiño, J.C. et al. (2019). Acid pretreatment of lignocellulosic biomass for energy vectors production: a review focused on operational conditions and techno-economic assessment for bioethanol production. *Renew. Sust. Energ. Rev.* 107: 587–601. https://doi.org/10.1016/j.rser.2019.02.024.

Soltanian, S., Aghbashlo, M., Almasi, F. et al. (2020). A critical review of the effects of pretreatment methods on the exergetic aspects of lignocellulosic biofuels. *Energy Convers. Manag.* 212: 112792. https://doi.org/10.1016/j.enconman.2020.112792.

Sorek, N., Yeats, T.H., Szemenyei, H. et al. (2014). The implications of lignocellulosic biomass chemical composition for the production of advanced biofuels. *Bioscience* 64 (3): 192–201. https://doi.org/10.1093/biosci/bit037.

Sun, Y.C., Xu, J.K., Xu, F., and Sun, R.C. (2013). Structural comparison and enhanced enzymatic hydrolysis of eucalyptus cellulose via pretreatment with different ionic liquids and catalysts. *Process Biochem.* 48 (5–6): 844–852. https://doi.org/10.1016/j.procbio.2013.03.023.

Tsegaye, B., Balomajumder, C., and Roy, P. (2019). Optimization of microwave and NaOH pretreatments of wheat straw for enhancing biofuel yield. *Energy Convers. Manag* 186: 82–92. https://doi.org/10.1016/j.enconman.2019.02.049.

Upton, B.M. and Kasko, A.M. (2016). Strategies for the conversion of lignin to high-value polymeric materials: review and perspective. *Chem. Rev.* 116 (4): 2275–2306. https://doi.org/10.1021/acs.chemrev.5b00345.

Usmani, Z., Sharma, M., Gupta, P. et al. (2020). Ionic liquid based pretreatment of lignocellulosic biomass for enhanced bioconversion. *Bioresour. Technol.* 304: 123003. https://doi.org/10.1016/j.biortech.2020.123003.

Wang, K., Bauer, S., and Sun, R.C. (2012). Structural transformation of miscanthus × giganteus lignin fractionated under mild formosolv, basic organosolv, and cellulolytic enzyme conditions. *J. Agric. Food Chem.* 60 (1): 144–152. https://doi.org/10.1021/jf2037399.

Wang, C., Hong, F., Lu, Y. et al. (2017). Improved biogas production and biodegradation of oilseed rape straw by using kitchen waste and duck droppings as co-substrates in two-phase anaerobic digestion. *PLoS One* 12 (8): https://doi.org/10.1371/journal.pone.0182361.

Wells, J.M., Crow, S.E., Ogoshi, R. et al. (2015). Optimizing feedstock selection for biofuel production in Hawaii: CuO oxidative lignin products in C4 grasses. *Biomass and Bioenergy* 83: 511–515. https://doi.org/10.1016/j.biombioe.2015.10.027.

Wells, J.M., Drielak, E., Surendra, K.C., and Kumar Khanal, S. (2020). Hot water pretreatment of lignocellulosic biomass: modeling the effects of temperature, enzyme and biomass loadings on sugar yield. *Bioresour. Technol.* 300: 122593. https://doi.org/10.1016/j.biortech.2019.122593.

Wu, X., Chen, L., He, W. et al. (2020). Characterize the physicochemical structure and enzymatic efficiency of agricultural residues exposed to γ-irradiation pretreatment. *Ind. Crop. Prod.* 150: 112228. https://doi.org/10.1016/j.indcrop.2020.112228.

Xu, H., Peng, J., Kong, Y. et al. (2020). Key process parameters for deep eutectic solvents pretreatment of lignocellulosic biomass materials: a review. *Bioresour. Technol.* 310: 123416. https://doi.org/10.1016/j.biortech.2020.123416.

Yaashikaa, P.R., Kumar, P.S., Saravanan, A. et al. (2020). Bioconversion of municipal solid waste into bio-based products: a review on valorisation and sustainable approach for circular bioeconomy. *Sci. Total Environ.* 748: 141312. https://doi.org/10.1016/j.scitotenv.2020.141312.

Zabed, H.M., Akter, S., Yun, J. et al. (2019). Recent advances in biological pretreatment of microalgae and lignocellulosic biomass for biofuel production. *Renew. Sust. Energ. Rev.* 105: 105–128. https://doi.org/10.1016/j.rser.2019.01.048.

Zeller, V., Towa, E., Degrez, M., and Achten, W.M.J. (2019). Urban waste flows and their potential for a circular economy model at city-region level. *Waste Manag* 83: 83–94. https://doi.org/10.1016/j.wasman.2018.10.034.

Zhang, H., Zhang, P., Ye, J. et al. (2018). Comparison of various pretreatments for ethanol production enhancement from solid residue after rumen fluid digestion of rice straw. *Bioresour. Technol.* 247: 147–156. https://doi.org/10.1016/j.biortech.2017.09.065.

Zhang, Y., Li, T., Shen, Y. et al. (2020). Extrusion followed by ultrasound as a chemical-free pretreatment method to enhance enzymatic hydrolysis of rice hull for fermentable sugars production. *Ind. Crop. Prod.* 149: 112356. https://doi.org/10.1016/j.indcrop.2020.112356.

Zhao, C., Shao, Q., and Chundawat, S.P.S. (2020). Recent advances on ammonia-based pretreatments of lignocellulosic biomass. *Bioresour. Technol.* 298: 122446. https://doi.org/10.1016/j.biortech.2019.122446.

Zheng, J. and Rehmann, L. (2014). Extrusion pretreatment of lignocellulosic biomass: a review. *Int. J. Mol. Sci.* 15 (10): 18967–18984. https://doi.org/10.3390/ijms151018967.

4

An Evaluation of Steam Explosion Pretreatment to Enhance the Digestibility of Lignocellulosic Biomass

Bhima Bhukya and Praveen K. Keshav

Centre for Microbial and Fermentation Technology, Department of Microbiology, University College of Science, Osmania University, Hyderabad, Telangana, India

4.1 Introduction

Various chemicals and materials utilized in everyday life are chemically manufactured from fossil resources via petro-based processes. However, our solid reliance on non-renewable fossil energy sources has been affecting environmental issues and climate change, encouraging us to shift toward the production of bioproducts from sustainable and renewable resources. Furthermore, fermentable monomeric sugars produced from lignocellulosic biomass, including agricultural and forest residues, wood residues, and dedicated crops, have significant possibilities for the production of biochemicals, biomaterials, and biofuels (Chandel and Singh 2011; Keshav et al. 2016; Auxenfans et al. 2017; Chae et al. 2020). The bioconversion of lignocellulosic feedstocks to various chemicals and biofuels offers several advantages but its progress is still hindered by techno-economic barriers. The biorefinery concept is a newly introduced concept for the sustainable utilization of lignocellulosic materials in various biobased products such as biochemicals, biofuels, and products related to cosmetics, food, and feed supplements. It is predicted that the biorefining process will play a significant role in building a new biobased industrial sector that exchanges chemicals and fuels currently derived from petroleum oil and also mitigates greenhouse gas emissions (Chandel et al. 2018a, 2018b; FitzPatrick et al. 2010).

Generally, lignocellulosic biomass is principally composed of polymers such as cellulose, hemicelluloses, and lignin, which are strongly interconnected and chemically bonded by covalent and noncovalent forces. Cellulose is the principal component of plant cell wall, composed of a homopolymer of β-D-glucopyranose units connected by β-(1,4) glycosidic bonds. The cellulose chains are bound together to form microfibrils, which are again organized into cellulose fibers; due to the formation of interchain hydrogen bonds, crystalline and amorphous regions are introduced into the structure of the cellulose. Hemicellulose is a branched heteropolymer of various pentose, hexose sugars and uronic acids with non-crystallized structure. Lignin is an amorphous, phenylpropanoid polymer, consisting of

Lignocellulose Bioconversion Through White Biotechnology, First Edition. Edited by Anuj Kumar Chandel.

three units such as sinapyl, *p*-coumaryl, and coniferyl alcohol (Chandel et al. 2007; Agbor et al. 2011; Brethauer and Studer 2015; Wang et al. 2015). The fractionation of biomass into its core components such as cellulose, hemicellulose, and lignin is an important step in the development of biorefinery processes; however, the implementation of lignocellulose fractionation technologies that are economically viable is still in its naive stages (FitzPatrick et al. 2010; Behera et al. 2014).

Generally, lignocellulose conversion often involves an appropriate pretreatment method which ultimately aims to deconstruct the various biomass components and render the inner lignocellulosic matrix easily susceptible to the action of various enzymes. An ideal pretreatment must effectively solubilize hemicellulose and lignin fractions of biomass, and increase the porosity and surface area of cellulose. An ideal pretreatment should also be cost effective, exhibit lower energy demand, and avoid the formation of inhibitors, which are detrimental to the subsequent fermentation step (Agbor et al. 2011; Chandel and Singh 2011; Ebrahimi et al. 2017; Rastogi and Shrivastava 2017; Liu et al. 2019). Different types of pretreatment methods have been investigated previously and are mainly categorized into physicochemical and biological pretreatment methods. A combination of these methods has also been studied to increase the digestibility of lignocellulosic biomass. As shown in Figure 4.1, in a typical combined pretreatment process it is recommended to separate the hemicellulose fraction of biomass in the first step with suitable alkali, acid, hydrothermal, and steam explosion pretreatment methods, while, in the second step, dilute alkali, organosolv, oxidative, and biological pretreatment methods are known to effectively remove lignin from the biomass (Keshav et al. 2016; Sun et al. 2016; Lemoes et al. 2018).

Among various pretreatment methods, steam explosion (autohydrolysis) is the most widely applied physicochemical pretreatment method used to fractionate different

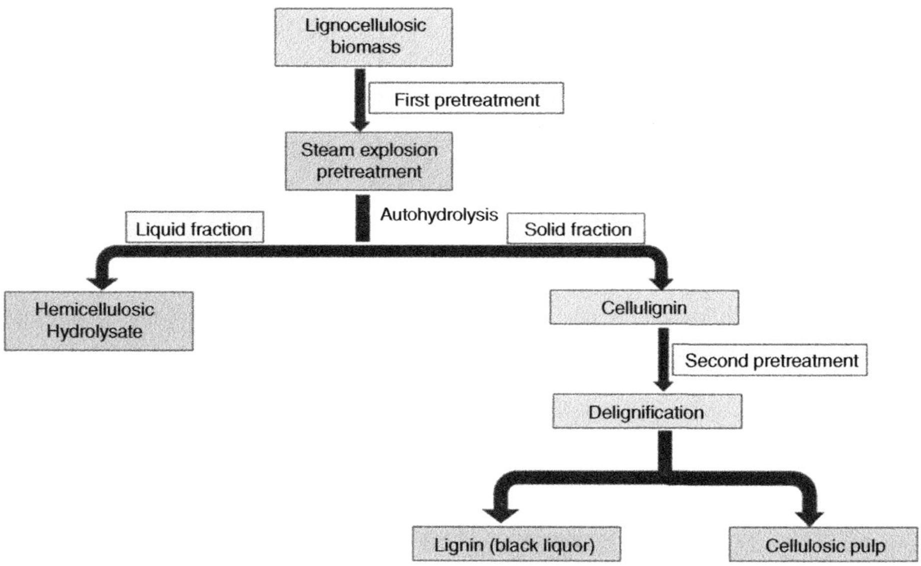

Figure 4.1 Overview of combined steam explosion pretreatment of lignocellulosic biomass.

components of lignocellulosic materials and is also known to improve biomass digestibility in subsequent bioconversion processes. During the pretreatment, the biomass is exposed to highly pressurized saturated steam for a particular time period ranging from seconds to several minutes, and then swiftly depressurized. After steam explosion pretreatment, biomass exhibits considerable disruption in the lignocellulosic matrix with hemicellulose solubilization and partial transformation in the lignin structure (Figure 4.2) (Ruiz et al. 2008; Behera et al. 2014; Singh et al. 2015; Balan et al. 2020). Steam explosion pretreatment has been applied on a wide range of lignocellulosic feedstocks such as rice straw (Chen et al. 2013; Steinbach et al. 2019), wheat straw (García-Aparicio et al. 2011), cotton stalk (Huang et al. 2015; Keshav et al. 2016), corn stover (Li et al. 2011), sugarcane bagasse (Rocha et al. 2015), bamboo biomass (Xing et al. 2013), and *Eucalyptus globulus* (Martin-Sampedro et al. 2014). Steam explosion has been successfully carried out with agricultural or herbaceous residues and hardwoods but the studies with softwoods were not very successful due to the presence of a lower amount of acetyl groups in the hemicellulosic fraction of biomass (Sun and Cheng 2002; Alvira et al. 2010; Balan et al. 2020).

The fractionation of lignocellulosic biomass components using steam explosion is a promising technology because of various advantages such as limited use of chemicals, which avoids recycling of chemicals, low energy input, with moderate costs, and increased accessibility of enzymes to steam exploded biomass which facilitates better sugar recovery

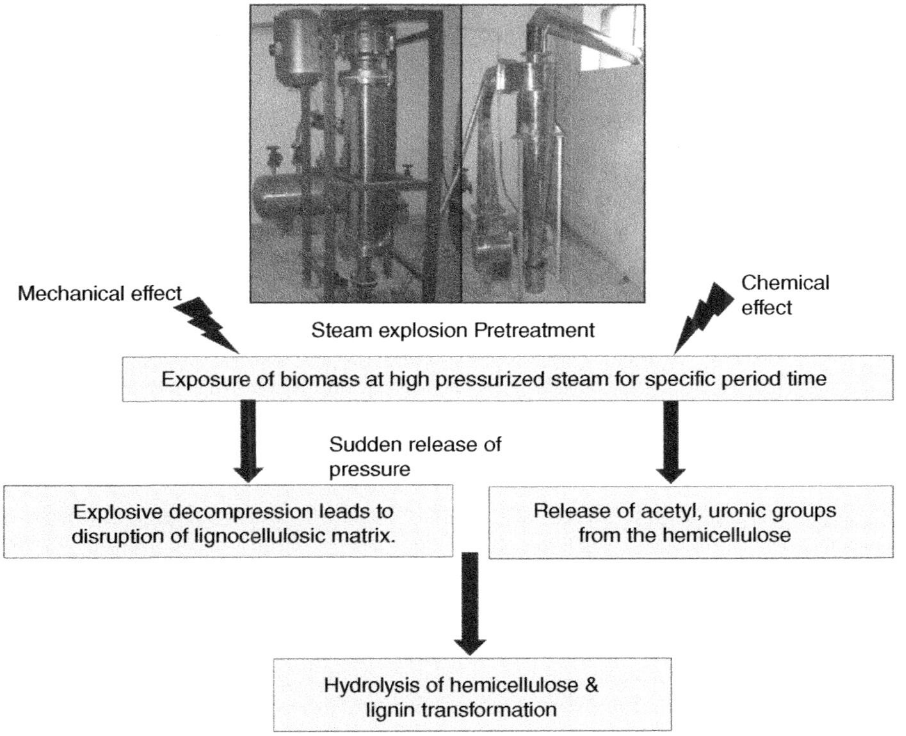

Figure 4.2 Mode of action of steam explosion pretreatment on lignocellulosic biomass.

(Alvira et al. 2010; Agbor et al. 2011; Wang et al. 2015; Ziegler-Devin et al. 2019). The main drawbacks of this method are avoidance of equipment corrosion by addition of acids and less generation of inhibitory compounds from sugar degradation genera hemicellulose and other cell wall polymer degradation (Alvira et al. 2010; Agbor et al. 2011; Maurya et al. 2015; Steinbach et al. 2019). Among biomass saccharification methods, enzymatic hydrolysis is an environmentally friendly method that comprises a cocktail of cellulolytic enzymes, which depolymerizes the polysaccharide fraction of pretreated residue with a higher degree of catalytic activity. The factors that affect the enzymatic saccharification of biomass can be mainly categorized into two types. Primarily, enzyme-linked factors augment enzyme activity by improving thermal tolerance synergism while alleviating the end-product inhibition and non-productive binding of various components of biomass. The second group includes substrate-linked factors, mainly focused on increasing the accessibility of cellulolytic enzymes to cellulose. The digestibility of cellulose can be influenced by many factors, such as the crystallinity of cellulose, the amount and distributions of hemicelluloses, lignin, and the accessible surface area (Alvira et al. 2010; Chandel et al. 2012; Keshav et al. 2016; Sun et al. 2016). Although there are few reviews on steam explosion pretreatment (Jacquet et al. 2015; Singh et al. 2015; Wang et al. 2015), the purpose of this chapter is to review and update the literature on various aspects of steam explosion pretreatment studies with the main focus on improving the digestibility of lignocellulosic biomass. The present review also discusses the mode of action of steam explosion and various pretreatment parameters affecting steam explosion pretreatment. Moreover, the crucial role of steam explosion in various combined pretreatment technologies is also discussed with the focus on improving the enzymatic saccharification of biomass.

4.2 Mode of Action and Types of Steam Explosion Pretreatment

Steam explosion was initially investigated and patented by W.H. Mason for an efficient process to defibrate wood into fibers, and this process was later employed for the pretreatment of lignocellulosic biomass for biofuel production (Jacquet et al. 2015). During steam explosion pretreatment, biomass is exposed to highly pressurized saturated steam for a specific time period and then swiftly depressurized. The main working principle beyond steam explosion pretreatment is autohydrolysis at higher temperature, and various chemical effects are exerted because of the release of acetyl groups from the hemicellulosic fraction of biomass, while water acts as an acid at higher temperatures (Figure 4.2). Furthermore, the mechanical forces exerted by steam explosion are due to the sudden release of pressure with explosive decompression which results in disruption of the lignocellulosic matrix (Alvira et al. 2010; Hu and Ragauskas 2012; Jacquet et al. 2015; Wang et al. 2015; Auxenfans et al. 2017). The mode of action of steam explosion pretreatment is comparable with that of dilute acid hydrolysis except that steam explosion pretreatment renders a higher sugar yield upon the post hydrolysis step. During this pretreatment process, the hemicellulose content of the lignocellulosic biomass is solubilized and further degraded into various sugar inhibitors (furfural and 5-hydroxymethylfurfural) due to application of the harsher conditions of the steam explosion. The major effects of steam explosion pretreatment on lignocellulosic biomass include higher hemicellulose solubilization with partial transformation in the lignin structure, which ultimately improves the digestibility of the lignocellulosic biomass to acid and/or enzymatic hydrolysis. During the steam explosion pretreatment, lignin is partially altered due to the breakage of various bonds in the lignin structure generating various types of cinnamyl alcohol derivatives and condensation by-products. The formation of these strong inhibitors further necessitates a suitable detoxification step to increase the fermentability of sugars present in the lignocellulosic hydrolysates (Singh et al. 2015; Wang et al. 2015).

Based on the conditions of temperature and residence times, the steam explosion of biomass can be performed in two different ways. First, high severity steam explosion pretreatment, which involves the application of a higher temperature with shorter residence times, results in effective hemicellulose and lignin solubilization. The main drawback of using high severity conditions is the formation of various inhibitory compounds from the lignin and hemicellulose fractions of biomass, which necessitates a suitable detoxification step prior to the fermentation of sugar hydrolysates, which ultimately leads to extra costs in the entire process of bioprocessing (Alvira et al. 2010; Agbor et al. 2011; Singh et al. 2015). In an alternative method, low pressure steam explosion pretreatment involves application of lower temperatures with longer residence times (150–190 °C; 5–300 s). The main disadvantage of using low pressure steam explosion pretreatment is the requirement of a post-delignification step for efficient enzymatic saccharification (Xing et al. 2014). In another classification, steam explosion pretreatment can be categorized into two types based on the addition of chemicals as catalyst or an impregnation method to improve the digestibility of biomass. In an uncatalyzed steam explosion, the feedstock is subjected to highly pressurized steam without the addition of any chemical and terminated by an explosive decompression (Balat et al. 2008). In chemically catalyzed steam explosion, the digestibility of biomass can be improved by addition of chemicals as impregnating agents. Various chemicals such as NaOH (Park et al. 2012), H_2SO_4 (Chen et al. 2013), and SO_2 (Carrasco et al. 2010) have been used extensively to improve the enzymatic hydrolysis of lignocellulosic biomass.

4.3 Factors Affecting the Steam Explosion Pretreatment

There are numerous key factors which affect the rate of steam explosion pretreatment of lignocellulose. The major factors that affect steam explosion pretreatment are biomass particle size, moisture content, combined severity factor (SF) of pretreatment (combined effect of both temperature and residence time), and addition of a chemical catalyst prior to steam pretreatment.

4.3.1 Effect on Particle Size of Biomass

The size reduction of biomass prior to any pretreatment method is a labor-intensive and an expensive procedure, but it is essential for bioconversion processes. The maintenance of an appropriate particle size of biomass will significantly increase the pretreatment efficiency owing to the presence of high mass and heat transfer. For biomass with smaller particles, hemicellulose may be degraded into various inhibitory by-products because of the presence of a strong intensity of heat. Hence, modifying the particle size of biomass is prerequisite for attaining high sugar yields with lower production costs (Zhu et al. 2009; Liu et al. 2019; Simangunsong et al. 2020). Various previous pretreatment studies of biomass have given valuable information on the effect of particle size on the digestibility of biomass, as these studies commonly examined biomass particle sizes less than 12 mm. Most studies on the steam explosion of hardwood or woody biomass described how bigger biomass particles showed significantly higher sugar yields in the subsequent enzymatic saccharification compared to biomass particles with a smaller size (Ballesteros et al. 2000; Cullis et al. 2004). These findings are in agreement with earlier studies performed by Ballesteros et al. (2002), who reported that the use of very small size biomass particles in steam explosion would not be necessary to optimize the efficiency of the process and to increase the overall process economy. They also investigated steam explosion of *Brassica carinata* at various particle sizes ranging from 1 to 12 mm and reported higher cellulose recovery at bigger particle sizes (8–12 mm), whereas lower cellulose recovery was achieved at smaller particle sizes (2–5 mm), indicating that the biomass particle size has a major effect on the recovery of cellulose. Liu et al. (2019) studied the effect of different particle sizes of biomass on the efficacy of pretreatment and subsequent enzymatic saccharification. They reported a higher sugar conversion (83% for xylan, 100% for glucan) with larger particle size compared to smaller particle size and concluded that the rate of enzymatic hydrolysis of pretreated residue was increased with increasing biomass particle size. A similar observation was also reported by Alvira et al. (2010), who reported 91.7% glucose yield after enzymatic hydrolysis of wheat straw with particle size of 2–10 mm.

4.3.2 Effect of Moisture Content

The efficiency of steam explosion pretreatment can also be determined by the initial moisture content of biomass which is one of the significant factors influencing the SF of steam explosion pretreatment. The higher amount of moisture could diminish the ability of heat and chemicals to enter into the biomass and then lessen the kinetics, because the cavities in the biomass are occupied with condensate before the predetermined temperature is achieved. An irregular heating of biomass possibly leads to the degradation of various chemical components, obviously being reflected by the discrepancy in the recovery of polysaccharides, lignin, and furans in the liquid and solid fractions (Wang et al. 2015). Oslen et al. (2015) explained that the original moisture content of the biomass has a major effect on the yield of steam pretreated solid fraction with maximum results achieved when biomass showed either a lower or higher moisture content, while lowest yields were obtained with moderate moisture content in the biomass.

4.3.3 Effect of Combined Severity Factor

The combined SF of steam explosion is the effect of temperature and residence time on the overall pretreatment efficiency using a severity parameter (Fang et al. 2011). The severity of pretreatment of steam explosion can be determined by using the SF, which integrates the temperature and residence time of pretreatment in a single factor (Agudelo et al. 2016;

Keshav et al. 2016). The following formula can be applied to calculate the SF of steam explosion pretreatment:

$$SF = \log_{10}\left[t \times \exp\left(\frac{T-100}{14.75}\right)\right]$$

where t and T indicate residence time (minutes) and temperature (°C) of the pretreatment, respectively; 100 is the standard temperature which is presumed to be constant. If the steam explosion pretreatment is carried out under acidic conditions, the pH value could be considered by the combined severity and expressed as (Fang et al. 2011)

$$\text{combined severity} = \log(R_0) - \text{pH}$$

The use of high severity pretreatment usually fractionates lignocellulosic biomass more efficiently and enhances the enzymatic saccharification of pretreated residue. However, higher severe conditions result in greater degradation of cell wall polymers of biomass, forming more inhibitory by-products such as furans and phenolic compounds which inhibit the subsequent fermentation step. The application of less severe pretreatment conditions limited the formation of inhibitory compounds but could not disrupt the lignocellulosic matrix adequately for enzymatic saccharification (Dou et al. 2016). Barbanera et al. (2015) conducted olive tree pruning in pilot scale at different temperatures and reported 28.25 g sugars per 100 g overall glucose yield at 210 °C for 15 min. They also reported that the application of a low temperature with longer residence times avoids the formation of sugar degradation by-products, which generally hamper the subsequent fermentation step. Alvira et al. (2010) studied the steam explosion of wheat straw at five different temperatures (170, 180, 190, 200, and 210 °C). Among them, steam explosion at 200 °C for 10 min resulted in 91.7% higher enzymatic saccharification efficiency with the formation of various inhibitory compounds; in contrast, steam explosion at 190 °C for 10 min resulted in minimal inhibitory sugar degradation compounds. A two-stage steam explosion pretreatment was suggested by some previous studies to maximize the sugar recoveries from steam exploded biomass. The first stage of steam explosion pretreatment is carried out at lower temperatures to solubilize maximum hemicellulose fraction, while in the second stage higher temperatures are applied (>210 °C) to depolymerize the cellulose fraction of biomass. The application of a two-step pretreatment has some benefits including improved utilization of feedstock and higher sugar yields at lower enzyme loadings during enzymatic saccharification. However, an economic appraisal of the whole process is necessary to decide the efficiency of an additional step of steam explosion treatment (Söderström et al. 2003; Monavari et al. 2009; Chen et al. 2011; Galbe and Zacchi 2012).

4.3.4 Effect of Addition of Catalyst

The efficiency of steam explosion pretreatment can be enhanced by the addition of various chemicals as a catalyst or impregnating agent. Prior to steam treatment, different types of chemicals (acid or alkali) can be used to impregnate the biomass material (Wang et al. 2015). For the digestibility of softwood biomass, an addition of acid catalyst is required with steam explosion pretreatment to improve the subsequent enzymatic saccharification of the cellulosic fraction while facilitating the recovery of the hemicellulosic sugars (Olsen et al. 2015). Impregnation with various chemicals, namely H_2SO_4 (Ballesteros et al. 2006; Chen et al. 2011, 2013; McIntosh et al. 2016), SO_2 (Bura et al. 2002; Carrasco et al. 2010; Ouyang et al. 2011; Tooyserkani et al. 2013), NaOH (Park et al. 2012), $Ca(OH)_2$ (Capecchi et al. 2016), and H_3PO_4 (Neves et al. 2016), with various lignocellulosic biomass prior to steam explosion pretreatment has been shown to increase the extraction of hemicellulosic sugars as well as the digestibility of the pretreated residue. Carrasco et al. (2010) carried out SO_2 catalyzed steam explosion (180 and 205 °C, 5–10 min) of sugarcane bagasse and reported an overall sugar yield of 87% (68 g per 100 g dried material) at 2% water insoluble fraction of biomass. Park et al. (2012) conducted NaOH catalyzed steam explosion (210 °C, 9 min) of *Eucalyptus grandis* and observed that the removal of lignin was significantly improved on the addition of NaOH and a maximum 65.5% sugar recovery was obtained, indicating impregnation of NaOH prior to steam explosion pretreatment was an effective approach to enhance the enzymatic digestibility of biomass. McIntosh et al. (2016) studied pilot-scale steam explosion (180 °C, 15 min) of *E. grandis* using 2.4 wt% H_2SO_4 as an impregnating agent and reported a maximum 71.8% glucan digestibility with 63.6% glucose yield. Table 4.1 shows catalyzed and uncatalyzed steam explosion pretreatment key parameters for the extraction of sugars (cellulose and hemicellulose) and lignin from different lignocellulosic feedstocks.

Table 4.1 Summary of steam explosion pretreatment key parameters for extraction of sugars (cellulose and hemicellulose) and lignin from different lignocellulosic feedstocks.

Feedstock	Steam explosion conditions	Catalyst	Bioconversion efficiency	References
Sugarcane bagasse	190 °C, 5 min	0.5% H_2SO_4	40% xylan recovery	Carvalho et al. (2018)
Corncob	180 °C, 10 min and 1.0 MPa	0.5% H_2SO_4	57.3% lignin extraction (99% purity)	Zhang et al. (2019)
Sweet sorghum stems	215 °C, 2 min	Water soaked biomass	37.3 wt% total sugars	Damay et al. (2018)
Elephant grass biomass	161 °C, 11.5 min	0.5% H_2SO_4	Extraction of 52.13% (w/w) cellulose and 31.00% (w/w) lignin; 55% cellulose saccharification	Kataria et al. (2017)
Cotton stalk	200 °C and 5 min	None	71.90% hemicellulose solubilization	Keshav et al. (2016)
Bamboo	2.0 MPa (pressure) and 4 min	None	62.5% cellulose–glucose conversion	Li et al. (2015)
Corn stover	184 °C, 15 min (1.0 MPa)	Choline chloride	84.7% lignin and 78.9% xylan removal	Nasir et al. (2020)
Corn stover	205 °C, 2.0 MPa for 6 min	None	72.5% cellulose conversion	Lu et al. (2010)
Corn stover	200 °C, 5 min	None	100% glucan recovery; 83% xylan recovery	Liu et al. (2013)

Table 4.1 (Continued)

Feedstock	Steam explosion conditions	Catalyst	Bioconversion efficiency	References
Duckweed (*Lemna minor*)	210 °C, 10 min	None	60% pectic polysaccharide hydrolysis	Zhao et al. (2015)
Eucalyptus biomass	210 °C, 9 min	NaOH	65.55% glucose recovery	Park et al. (2012)
Oak wood	230 °C and 28 bar (pressure)	None	79.9% sugar yield	Cotana et al. (2015)
Elephant grass (*Pennisetum purpureum*)	200 °C, 10 min	None	863.42 mg g^{-1} sugar yield	Scholl et al. (2015)
Oil palm empty fruit bunches	195 °C, 6 min	None	68.12% hemicellulose reduction with 34.69% in glucan content	Medina et al. (2016)
Canola straw	180 °C, 4 min	None	29.40% cellulose conversion	Garmakhany et al. (2014)

4.4 Various Post-pretreatment Approaches to Improve Saccharification of Steam Exploded Biomass

The use of a single pretreatment method can exert less effect on the partial factors that restrain the enzymatic saccharification of lignocellulosic biomass; therefore the combination of pretreatment methods has recently been studied to enhance their digestibility and exploit the use of isolated hemicelluloses and lignin from the biomass. The selection of an appropriate combination of pretreatment methods should facilitate the digestibility of lignocellulosic biomass at comparatively low economic cost (Mood et al. 2013; Sun et al. 2016). Sun et al. (2016) suggested the application of steam explosion as the first step in the combined pretreatment of biomass and it should be effective in the solubilization of hemicelluloses. In the second step alkaline, oxidative delignification and biological pretreatments are considered as effective post-delignification methods for improving the enzymatic saccharification of steam exploded biomass (Wang et al. 2015; Sun et al. 2016). Table 4.2 shows a summary of different combinations of steam explosion pretreatment methods and their major changes on the composition of biomass observed for different lignocellulosic feedstocks.

4.5 Summary and Conclusions

Steam explosion pretreatment plays a significant role in lignocellulosic biorefineries for the bioconversion of lignocellulose to various value-added bioproducts, since steam explosion pretreatment can be employed not only as a pretreatment method but also as a process for fractionation of various components of lignocellulosic biomass. In any ideal lignocellulosic biorefinery, fractionation of lignocellulosic material into constituent polysaccharides (cellulose, hemicellulose), lignin and other biomass components can be a first step in biorefinery.

Table 4.2 Summary of different combinations of steam explosion pretreatment methods and their major composition changes observed on different lignocellulosic feedstocks.

Pretreatment strategy	Feedstock	Major effects	References
Steam explosion and ionic liquid pretreatment	Miscanthus	90% (w/w) xylan dissolution; 50% glucan conversion to gluco-oligosacchaides	Bhatia et al. (2020)
Steam explosion and hydrothermal pretreatment	*Bambusa stenostachya* bamboo	72% saccharification efficiency; 39% lignin removal	Dai et al. (2020)
Organosolv and steam explosion pretreatment	Spruce biomass	72% (w/w) cellulose extraction; 79.4% (w/w) delignification	Matsakas et al. (2019)
Sequential acid and steam explosion pretreatment	Olive stone biomass	71% xylose recovery; 83% sugar yield	Padilla-Rascón et al. (2020)
Steam explosion and alkaline extraction	Cotton stalk	85.07% lignin removal; complete hemicellulose solubilization; 82.13% saccharification efficiency	Keshav et al. (2016)
Steam explosion and sequential acid and enzymatic saccharification	Cotton stalk	69.16% holocellulose hydrolysis	Keshav et al. (2018)
Steam explosion and extrusion method	Barley straw	84%, 91%, 87% glucan, hemicellulose, and lignin recoveries respectively	Oliva et al. (2017)
Steam explosion and alkaline treatment	Bamboo	73.8% enzymatic saccharification	Sun et al. (2014)
Steam explosion and alkaline peroxide treatment	Bamboo	63.2% hemicellulose solubilization; 93.1% lignin removal; 90.5% glucose yield	Xing et al. (2013)
Steam explosion and fungal treatment	Corn stalk	36.58% lignin degradation	Li and Chen (2014)
Steam explosion and monobasic alkali treatment	Cotton stalk	73.22% sugar yield	Wang et al. (2019)
Dilute acid pretreatment and steam explosion	Eucalyptus biomass	71.8% glucan digestibility with 63.6% glucose yield	McIntosh et al. (2016)
Steam explosion and alkaline hydrogen peroxide pretreatment	*Eucommia ulmoides* oliver wood	91.69% saccharification efficiency	Zhu et al. (2015)
Steam explosion and alkaline extraction	*Cistus ladanifer* (rockrose)	75% glucose yield	Ferro et al. (2015)
Steam explosion and alkaline extraction	*Cynara cardunculus* (cardoon)	72.8% saccharification efficiency	Fernandes et al. (2015)

Steam explosion pretreatment is selective for the extraction of hemicellulose fraction and can be recovered in larger amounts with a comparatively high degree of purity. Steam explosion pretreatment extracts more hemicellulosic fraction than any other pretreatment method such as dilute acid pretreatments and other steam associated pretreatments. A suitable selection of pretreatment methods and their correct operational sequence can be used to improve the fractionation of biomass components, which is an important characteristic feature of an ideal lignocellulosic biorefinery. If steam explosion pretreatment is employed in the first phase of pretreatment, low hemicellulosic material with higher lignin is attained in the biomass; and the resultant hemicellulose will then be employed in further bioconversion of value-added platform chemicals and biofuels. While the second pretreatment step such as the dilute alkaline pretreatment method is recommended to eliminate maximum lignin from the steam exploded biomass, such an extracted lignin will then be employed in bioconversion of value-added product including fuel additives and biobased polymers. Finally, the solids remaining after combined treatment steam explosion and dilute alkaline pretreatment will be rich in cellulosic residue, which is a promising substrate for various bioconversion applications.

Acknowledgements

This chapter was sponsored by RUSA 2.0 Grant programme, Ministry of Human Resource Development of India.

References

Agbor, V.B., Cicek, N., Sparling, R. et al. (2011). Biomass pretreatment: fundamentals toward application. *Biotechnol. Adv.* 29 (6): 675–685.

Agudelo, R.A., García-Aparicio, M.P., and Görgens, J.F. (2016). Steam explosion pretreatment of triticale (× *Triticosecale Wittmack*) straw for sugar production. *New Biotechnol.* 33 (1): 153–163.

Alvira, P., Tomás-Pejó, E., Ballesteros, M., and Negro, M.J. (2010). Pretreatment technologies for an efficient bioethanol production process based on enzymatic hydrolysis: a review. *Bioresour. Technol.* 101 (13): 4851–4861.

Auxenfans, T., Crônier, D., Chabbert, B., and Paës, G. (2017). Understanding the structural and chemical changes of plant biomass following steam explosion pretreatment. *Biotechnol. Biofuels* 10 (1): 1–16.

Balan, R., Antczak, A., Brethauer, S. et al. (2020). Steam explosion pretreatment of beechwood. Part 1: Comparison of the enzymatic hydrolysis of washed solids and whole pretreatment slurry at different solid loadings. *Energies* 13 (14): 3653.

Balat, M., Balat, H., and Öz, C. (2008). Progress in bioethanol processing. *Prog. Energy Combust. Sci.* 34 (5): 551–573.

Ballesteros, I., Oliva, J.M., Navarro, A.A. et al. (2000). Effect of chip size on steam explosion pretreatment of softwood. In: *Twenty-First Symposium on Biotechnology for Fuels and Chemicals*, 97–110. Totowa, NJ: Humana Press.

Ballesteros, I., Oliva, J.M., Negro, M.J. et al. (2002). Enzymic hydrolysis of steam exploded herbaceous agricultural waste (*Brassica carinata*) at different particule sizes. *Process Biochem.* 38 (2): 187–192.

Ballesteros, I., Negro, M.J., Oliva, J.M. et al. (2006). Ethanol production from steam-explosion pretreated wheat straw. In: *Twenty-Seventh Symposium on Biotechnology for Fuels and Chemicals*, 496–508. Totowa, NJ: Humana Press.

Barbanera, M., Buratti, C., Cotana, F. et al. (2015). Effect of steam explosion pretreatment on sugar production by enzymatic hydrolysis of olive tree pruning. *Energy Procedia* 81: 146–154.

Behera, S., Arora, R., Nandhagopal, N., and Kumar, S. (2014). Importance of chemical pretreatment for bioconversion of lignocellulosic biomass. *Renew. Sust. Energ. Rev.* 36: 91–106.

Bhatia, R., Lad, J.B., Bosch, M. et al. (2020). Production of oligosaccharides and biofuels from Miscanthus using combinatorial steam explosion and ionic liquid pretreatment. *Bioresour. Technol.* 323: 124625.

Brethauer, S. and Studer, M.H. (2015). Biochemical conversion processes of lignocellulosic biomass to fuels and chemicals—a review. *CHIMIA Int. J. Chem.* 69 (10): 572–581.

Bura, R., Mansfield, S.D., Saddler, J.N., and Bothast, R.J. (2002). SO_2-catalyzed steam explosion of corn fiber for ethanol production. *Appl. Biochem. Biotechnol.* 98: 59–72.

Capecchi, L., Galbe, M., Wallberg, O. et al. (2016). Combined ethanol and methane production from switchgrass (*Panicum virgatum L.*) impregnated with lime prior to steam explosion. *Biomass Bioenergy* 90: 22–31.

Carrasco, C., Baudel, H.M., Sendelius, J. et al. (2010). SO_2-catalyzed steam pretreatment and fermentation of enzymatically hydrolyzed sugarcane bagasse. *Enzym. Microb. Technol.* 46 (2): 64–73.

Carvalho, A.F.A., Marcondes, W.F., de Oliva Neto, P. et al. (2018). The potential of tailoring the conditions of steam explosion to produce xylo-oligosaccharides from sugarcane bagasse. *Bioresour. Technol.* 250: 221–229.

Chae, T.U., Ahn, J.H., Ko, Y.S. et al. (2020). Metabolic engineering for the production of dicarboxylic acids and diamines. *Metab. Eng.* 58: 2–16.

Chandel, A.K. and Singh, O.V. (2011). Weedy lignocellulosic feedstock and microbial metabolic engineering: advancing the generation of 'biofuel'. *Appl. Microbiol. Biotechnol.* 89 (5): 1289–1303.

Chandel, A.K., Chan, E.S., Rudravaram, R. et al. (2007). Economics and environmental impact of bioethanol production technologies: an appraisal. *Biotechnol. Mol. Bio. Revs* 2 (1): 14–32.

Chandel, A.K., Chandrasekhar, G., Silva, M.B., and Silvério da Silva, S. (2012). The realm of cellulases in biorefinery development. *Crit. Rev. Biotechnol.* 32 (3): 187–202.

Chandel, A.K., Garlapati, V.K., Singh, A.K. et al. (2018a). The path forward for lignocellulose biorefineries: bottlenecks, solutions, and perspective on commercialization. *Bioresour. Technol.* 264: 370–381.

Chandel, A.K., Antunes, F.A., Terán-Hilares, R. et al. (2018b). Bioconversion of hemicellulose into ethanol and value-added products: commercialization, trends, and future opportunities. In: *Advances in Sugarcane Biorefinery.*, 97–134. Elsevier.

Chen, W.H., Pen, B.L., Yu, C.T., and Hwang, W.S. (2011). Pretreatment efficiency and structural characterization of rice straw by an integrated process of dilute-acid and steam explosion for bioethanol production. *Bioresour. Technol.* 102 (3): 2916–2924.

Chen, W.H., Tsai, C.C., Lin, C.F. et al. (2013). Pilot-scale study on the acid-catalyzed steam explosion of rice straw using a continuous pretreatment system. *Bioresour. Technol.* 128: 297–304.

Cotana, F., Buratti, C., Barbanera, M., and Lascaro, E. (2015). Optimization of the steam explosion and enzymatic hydrolysis for sugars production from oak woods. *Bioresour. Technol.* 198: 470–477.

Cullis, I.F., Saddler, J.N., and Mansfield, S.D. (2004). Effect of initial moisture content and chip size on the bioconversion efficiency of softwood lignocellulosics. *Biotechnol. Bioeng.* 85 (4): 413–421.

Dai, N.H., Huynh, K.T.T., Nguyen, T.A.D. et al. (2020). Hydrothermal and steam explosion pretreatment of *Bambusa stenostachya* bamboo. *Waste Biomass Valor.* 12: 4103–4112.

Damay, J., Duret, X., Ghislain, T. et al. (2018). Steam explosion of sweet sorghum stems: optimisation of the production of sugars by response surface methodology combined with the severity factor. *Ind. Crop. Prod.* 111: 482–493.

Dou, C., Ewanick, S., Bura, R., and Gustafson, R. (2016). Post-treatment mechanical refining as a method to improve overall sugar recovery of steam pretreated hybrid poplar. *Bioresour. Technol.* 207: 157–165.

Ebrahimi, M., Caparanga, A.R., Ordono, E.E. et al. (2017). Effect of ammonium carbonate pretreatment on the enzymatic digestibility, structural characteristics of rice husk and bioethanol production via simultaneous saccharification and fermentation process with *Saccharomyces cerevisiae Hansen* 2055. *Ind. Crop. Prod.* 101: 84–91.

Fang, H., Deng, J., and Zhang, T. (2011). Dilute acid pretreatment of black spruce using continuous steam explosion system. *Appl. Biochem. Biotechnol.* 163 (4): 547–557.

Fernandes, M.C., Ferro, M.D., Paulino, A.F. et al. (2015). Enzymatic saccharification and bioethanol production from *Cynara cardunculus* pretreated by steam explosion. *Bioresour. Technol.* 186: 309–315.

Ferro, M.D., Fernandes, M.C., Paulino, A.F. et al. (2015). Bioethanol production from steam explosion pretreated and alkali extracted *Cistus ladanifer* (rockrose). *Biochem. Eng. J.* 104: 98–105.

FitzPatrick, M., Champagne, P., Cunningham, M.F., and Whitney, R.A. (2010). A biorefinery processing perspective: treatment of lignocellulosic materials for the production of value-added products. *Bioresour. Technol.* 101 (23): 8915–8922.

Galbe, M. and Zacchi, G. (2012). Pretreatment: the key to efficient utilization of lignocellulosic materials. *Biomass Bioenergy* 46: 70–78.

García-Aparicio, M.P., Oliva, J.M., Manzanares, P. et al. (2011). Second-generation ethanol production from steam exploded barley straw by *Kluyveromyces marxianus* CECT 10875. *Fuel* 90 (4): 1624–1630.

Garmakhany, A.D., Kashaninejad, M., Aalami, M. et al. (2014). Enhanced biomass delignification and enzymatic saccharification of canola straw by steam-explosion pretreatment. *J. Sci. Food Agric.* 94 (8): 1607–1613.

Hu, F. and Ragauskas, A. (2012). Pretreatment and lignocellulosic chemistry. *Bioenerg. Res.* 5 (4): 1043–1066.

Huang, Y., Wei, X., Zhou, S. et al. (2015). Steam explosion distinctively enhances biomass enzymatic saccharification of cotton stalks by largely reducing cellulose polymerization degree in *G. barbadense* and *G. hirsutum*. *Bioresour. Technol.* 181: 224–230.

Jacquet, N., Maniet, G., Vanderghem, C. et al. (2015). Application of steam explosion as pretreatment on lignocellulosic material: a review. *Ind. Eng. Chem. Res.* 54 (10): 2593–2598.

Kataria, R., Mol, A., Schulten, E. et al. (2017). Bench scale steam explosion pretreatment of acid impregnated elephant grass biomass and its impacts on biomass composition, structure and hydrolysis. *Ind. Crop. Prod.* 106: 48–58.

Keshav, P.K., Naseeruddin, S., and Rao, L.V. (2016). Improved enzymatic saccharification of steam exploded cotton stalk using alkaline extraction and fermentation of cellulosic sugars into ethanol. *Bioresour. Technol.* 214: 363–370.

Keshav, P.K., Banoth, C., Anthappagudem, A. et al. (2018). Sequential acid and enzymatic saccharification of steam exploded cotton stalk and subsequent ethanol production using *Scheffersomyces stipitis* NCIM 3498. *Ind. Crop. Prod.* 125: 462–467.

Lemoes, J.S., Samios, D., Peralba, M.D.C.R. et al. (2018). Chemical pretreatment of *Arundo donax* L. for second-generation ethanol production. *Electron. J. Biotechnol.* 31: 67–74.

Li, G. and Chen, H. (2014). Synergistic mechanism of steam explosion combined with fungal treatment by *Phellinus baumii* for the pretreatment of corn stalk. *Biomass Bioenergy* 67: 1–7.

Li, Y., Gao, K., Tian, S. et al. (2011). Evaluation of *Saccharomyces cerevisiae* Y5 for ethanol production from enzymatic hydrolysate of non-detoxified steam-exploded corn stover. *Bioresour. Technol.* 102 (22): 10548–10552.

Li, Z., Fei, B., and Jiang, Z. (2015). Effect of steam explosion pretreatment on bamboo for enzymatic hydrolysis and ethanol fermentation. *Bioresources* 10 (1): 1037–1047.

Liu, Z.H., Qin, L., Pang, F. et al. (2013). Effects of biomass particle size on steam explosion pretreatment performance for improving the enzyme digestibility of corn stover. *Ind. Crop. Prod.* 44: 176–184.

Liu, C.G., Xiao, Y., Xia, X.X. et al. (2019). Cellulosic ethanol production: progress, challenges and strategies for solutions. *Biotechnol. Adv.* 37: 491–504.

Lu, Y., Wang, Y., Xu, G. et al. (2010). Influence of high solid concentration on enzymatic hydrolysis and fermentation of steam-exploded corn stover biomass. *Appl. Biochem. Biotechnol.* 60 (2): 360–369.

Martin-Sampedro, R., Revilla, E., Villar, J.C., and Eugenio, M.E. (2014). Enhancement of enzymatic saccharification of *Eucalyptus globulus*: steam explosion versus steam treatment. *Bioresour. Technol.* 167: 186–191.

Matsakas, L., Raghavendran, V., Yakimenko, O. et al. (2019). Lignin-first biomass fractionation using a hybrid organosolv–steam explosion pretreatment technology improves the saccharification and fermentability of spruce biomass. *Bioresour. Technol.* 273: 521–528.

Maurya, D.P., Singla, A., and Negi, S. (2015). An overview of key pretreatment processes for biological conversion of lignocellulosic biomass to bioethanol. *3 Biotech* 5 (5): 597–609.

McIntosh, S., Zhang, Z., Palmer, J. et al. (2016). Pilot-scale cellulosic ethanol production using eucalyptus biomass pre-treated by dilute acid and steam explosion. *Biofuels Bioprod. Biorefin.* 10 (4): 346–358.

Medina, J.D.C., Woiciechowski, A., Zandona Filho, A. et al. (2016). Steam explosion pretreatment of oil palm empty fruit bunches (EFB) using autocatalytic hydrolysis: a biorefinery approach. *Bioresour. Technol.* 199: 173–180.

Monavari, S., Galbe, M., and Zacchi, G. (2009). The influence of solid/liquid separation techniques on the sugar yield in two-step dilute acid hydrolysis of softwood followed by enzymatic hydrolysis. *Biotechnol. Biofuels* 2 (1): 6.

Mood, S.H., Golfeshan, A.H., Tabatabaei, M. et al. (2013). Lignocellulosic biomass to bioethanol, a comprehensive review with a focus on pretreatment. *Renew. Sust. Energ. Rev.* 27: 77–93.

Nasir, A., Chen, H.Z., and Wang, L. (2020). Novel single-step pretreatment of steam explosion and choline chloride to de-lignify corn stover for enhancing enzymatic edibility. *Process Biochem.* 94: 273–281.

Neves, P.V., Pitarelo, A.P., and Ramos, L.P. (2016). Production of cellulosic ethanol from sugarcane bagasse by steam explosion: effect of extractives content, acid catalysis and different fermentation technologies. *Bioresour. Technol.* 208: 184–194.

Oliva, J.M., Negro, M.J., Manzanares, P. et al. (2017). A sequential steam explosion and reactive extrusion pretreatment for lignocellulosic biomass conversion within a fermentation-based biorefinery perspective. *Fermentation* 3 (2): 15.

Olsen, C., Arantes, V., and Saddler, J. (2015). Optimization of chip size and moisture content to obtain high, combined sugar recovery after sulfur dioxide-catalyzed steam pretreatment of softwood and enzymatic hydrolysis of the cellulosic component. *Bioresour. Technol.* 187: 288–298.

Ouyang, J., Ma, R., Huang, W. et al. (2011). Enhanced saccharification of SO_2 catalyzed steam-exploded corn stover by polyethylene glycol addition. *Biomass Bioenergy* 35 (5): 2053–2058.

Padilla-Rascón, C., Ruiz, E., Romero, I. et al. (2020). Valorisation of olive stone by-product for sugar production using a sequential acid/steam explosion pretreatment. *Ind. Crop. Prod.* 148: 112279.

Park, J.Y., Kang, M., Kim, J.S. et al. (2012). Enhancement of enzymatic digestibility of *Eucalyptus grandis* pretreated by NaOH catalyzed steam explosion. *Bioresour. Technol.* 123: 707–712.

Rastogi, M. and Shrivastava, S. (2017). Recent advances in second generation bioethanol production: an insight to pretreatment, saccharification and fermentation processes. *Renew. Sust. Energ. Rev.* 80: 330–340.

Rocha, G.J.M., Gonçalves, A.R., Nakanishi, S.C. et al. (2015). Pilot scale steam explosion and diluted sulfuric acid pretreatments: comparative study aiming the sugarcane bagasse saccharification. *Ind. Crop. Prod.* 74: 810–816.

Ruiz, E., Cara, C., Manzanares, P. et al. (2008). Evaluation of steam explosion pre-treatment for enzymatic hydrolysis of sunflower stalks. *Enzym. Microb. Technol.* 42 (2): 160–166.

Scholl, A.L., Menegol, D., Pitarelo, A.P. et al. (2015). Ethanol production from sugars obtained during enzymatic hydrolysis of elephant grass (*Pennisetum purpureum, Schum.*) pretreated by steam explosion. *Bioresour. Technol.* 192: 228–237.

Simangunsong, E., Ziegler-Devin, I., Chrusciel, L. et al. (2020). Steam explosion of beech wood: effect of the particle size on the xylans recovery. *Waste Biomass Valor.* 11 (2): 625–633.

Singh, J., Suhag, M., and Dhaka, A. (2015). Augmented digestion of lignocellulose by steam explosion, acid and alkaline pretreatment methods: a review. *Carbohydr. Polym.* 117: 624–631.

Söderström, J., Pilcher, L., Galbe, M., and Zacchi, G. (2003). Combined use of H_2SO_4 and SO_2 impregnation for steam pretreatment of spruce in ethanol production. *Appl. Biochem. Biotechnol.* 105 (1–3): 127–140.

Steinbach, D., Wüst, D., Zielonka, S. et al. (2019). Steam explosion conditions highly influence the biogas yield of rice straw. *Molecules* 24 (19): 3492.

Sun, Y. and Cheng, J. (2002). Hydrolysis of lignocellulosic materials for ethanol production: a review. *Bioresour. Technol.* 83 (1): 1–11.

Sun, S.L., Wen, J.L., Ma, M.G., and Sun, R.C. (2014). Enhanced enzymatic digestibility of bamboo by a combined system of multiple steam explosion and alkaline treatments. *Appl. Energy* 136: 519–526.

Sun, S., Sun, S., Cao, X., and Sun, R. (2016). The role of pretreatment in improving the enzymatic hydrolysis of lignocellulosic materials. *Bioresour. Technol.* 199: 49–58.

Tooyserkani, Z., Kumar, L., Sokhansanj, S. et al. (2013). SO_2-catalyzed steam pretreatment enhances the strength and stability of softwood pellets. *Bioresour. Technol.* 130: 59–68.

Wang, K., Chen, J., Sun, S.N., and Sun, R.C. (2015). Steam explosion. In: *Pretreatment of Biomass.*, 75–104. Elsevier https://doi.org/10.1016/B978-0-12-800080-9.00006-2.

Wang, Y., Gong, X., Hu, X., and Zhou, N. (2019). Lignin monomer in steam explosion assist chemical treated cotton stalk affects sugar release. *Bioresour. Technol.* 276: 343–348.

Xing, Y., Yu, H., Zhu, L., and Jiang, J. (2013). Efficient enzymatic hydrolysis of bamboo by pretreatment with steam explosion and alkaline peroxide. *Bioresources* 8 (4): 5392–5408.

Xing, Y., Su, Z., Wang, K. et al. (2014). Combination of low-pressure steam explosion and alkaline peroxide pretreatment for separation of hemicellulose. *Bioresources* 9 (2): 3384–3395.

Zhang, X., Zhu, J., Sun, L. et al. (2019). Extraction and characterization of lignin from corncob residue after acid-catalyzed steam explosion pretreatment. *Ind. Crop. Prod.* 133: 241–249.

Zhao, X., Moates, G.K., Wilson, D.R. et al. (2015). Steam explosion pretreatment and enzymatic saccharification of duckweed (*Lemna minor*) biomass. *Biomass Bioenergy* 72: 206–215.

Zhu, J.J., Yong, Q., Xu, Y. et al. (2009). Adaptation fermentation of *Pichia stipitis* and combination detoxification on steam exploded lignocellulosic prehydrolyzate. *Nat. Sci.* 1 (01): 47.

Zhu, M.Q., Wen, J.L., Wang, Z.W. et al. (2015). Structural changes in lignin during integrated process of steam explosion followed by alkaline hydrogen peroxide of *Eucommia ulmoides* Oliver and its effect on enzymatic hydrolysis. *Appl. Energy* 158: 233–242.

Ziegler-Devin, I., Menana, Z., Chrusciel, L. et al. (2019). Steam explosion pretreatment of willow grown on phytomanaged soils for bioethanol production. *Ind. Crop. Prod.* 140: 111722.

5

The Role of Plant Cell Wall Degrading Enzymes in Biorefinery Development

Katarina R. Mihajlovski and Marija D. Milić

Faculty of Technology and Metallurgy, Belgrade, Serbia

5.1 Introduction

One of the greatest challenges of modern society today is the demand for energy and energy resources that are renewable, environmentally friendly, and cost-effective [1]. A depletion of conventional fossil fuels causes environmental pollution via greenhouse gas emission, so there has been growing global interest for the production of biofuels [2]. According to predictions, it has been estimated that about 10%–50% of fossil fuel will be substituted by biofuel by 2030 [3].

The production of biofuels using lignocellulosic biomass as the substrate is becoming very attractive worldwide, since an enormous amount of biomass is being generated and disposed of as waste. Bioethanol, as a liquid biofuel, is produced using plant biomass and it can be obtained either from renewable sources, like food and feed crops (mainly starch and sugar crops such as corn, wheat, or sugarcane) called the first generation of bioethanol, or from lignocellulosic materials, named the second generation of bioethanol or cellulosic ethanol [4–6]. Although the first generation of bioethanol competes with food production, the second generation of bioethanol is produced from waste and abundant lignocellulosic biomass/feedstock but requires a more complex processing compared to the first generation of ethanol production [5]. This concept, however, has a viable perspective and contributes strongly to biorefinery development.

Cellulases play a crucial role in bioethanol production and demand for them is rapidly growing, due to their potential application in the degradation of lignocellulosic biomass [7–10]. They are capable of degrading the insoluble cellulose polymer present in the lignocellulosic biomass into fermentable sugars [7, 11]. In the last decade, cellulases record an increasing application in almost all industrial sectors (food, feed, pulp and paper, textiles, laundry and detergents, animals, and the brewing industry) [12]. The recent reports are based on very intensive research for cellulase improvement. The commercial cellulases that are used for biofuel production or in other industrial sectors are now developed in the form of cellulosic cocktails (enzyme mixtures) for providing better performance.

Lignocellulose Bioconversion Through White Biotechnology, First Edition. Edited by Anuj Kumar Chandel.

The benefits of tailored cellulase cocktails are reflected in more efficient lignocellulose hydrolysis, a lower dose of enzymes, the possibility of using high substrate loadings, an increased yield of reducing sugars, solving the recalcitrant nature of lignocellulosic biomass and more effective economic feasibility of cellulosic ethanol production.

Considering the previous chapters, this chapter deals with the aspects of the commercial cellulase use for lignocellulosic biomass hydrolysis within the biorefinery development concept. First, the main view of lignocellulose biomass structure is described, followed by an introduction to the plant cell wall degradation enzymes, with emphasis on the cellulase role. Furthermore, the innovation in the form of cellulase preparation, designed by leading companies for enzyme manufacturing, and the characteristics and applications for degrading lignocellulosic biomass are presented. Also, the recent production of cellulases from newly isolated microorganisms as a future perspective for possible application in bioethanol production is briefly discussed.

5.2 Lignocellulosic Biomass—the Plant Cell Wall

Lignocellulosic biomass is a by-product of industrial agricultural processing. It includes agricultural waste material or residues, forest wood residues, industrial and municipal waste, energy crops, and paper waste [13, 14]. Lignocellulose is composed of the plant cell wall that contains three polymer components: cellulose, hemicellulose, and lignin.

Cellulose is the major and the most abundant component of the cell wall [15]. It is a linear homopolysaccharide polymer and consists of D-glucose residues linked by β-1,4-glucosidic bonds. The cellulose polymers are present as microfibrils (2–4 nm) and they are crosslinked by a hemicellulose matrix covered by lignin with a main function to ensure the rigidity and stability of the cell wall [16, 17]. The structure of cellulose is not uniform; there are crystalline and non-crystalline (amorphous) parts of cellulose polymers. The crystallinity of cellulose and the whole matrix structure make the cellulose highly recalcitrant to hydrolysis [18]. In nature, cellulose is always present with two other polymers, hemicellulose and lignin. Cellulose, together with hemicellulose, makes up more than 50% of the total plant biomass.

Hemicelluloses are the second most abundant heteropolymers that are strongly associated and linked to the cellulose polymer [15, 18, 19]. They have a more complexed, branched and amorphous structure than cellulose. They consist of various pentoses (like D-xylose and D-arabinose), hexoses (D-mannose, D-glucose, and D-galactose), and uronic acid attached through various linkages [17, 18]. The major component of hemicellulose in cereals and hardwoods is xylan while another two hemicelluloses in the plant cell wall are galacto(gluco)mannans and xyloglucans [17]. Unlike cellulose and lignin, hemicelluloses can be readily solubilized when treated at different temperatures with different concentrations of alkali, acid, and other chemicals [15].

Hemicellulose is interlinked with lignin and together they form a matrix around cellulose microfibrils and present an impenetrable barrier in the plant cell wall [14, 15].

Lignin is a heteroplymer of complex structure and consists of aromatic rings such as syringyl, guaiacol, and *p*-hydroxyphenyl [15]. Lignin is non-crystalline and forms covalent crosslinks to polysaccharide in cell walls. Cellulose, hemicellulose, and lignin entangle to form a complex matrix that is resistant to enzyme attack to cellulose [15].

The content of cellulose (30%–50% of the total dry matter), hemicellulose (20%–30%) and lignin (10%–25%) in lignocellulose biomass can vary depending on the source of lignocellulose [14]. Lignocellulose is widely used in almost every part of the industrial sector; however, before its use in biorefinery, some kind of pretreatment is necessary. There is an increasing demand to replace classical chemical treatments of lignocellulosic biomass with "white biotechnology" [20]. White or industrial biotechnology uses microorganisms and their enzymes for obtaining products and energy from renewable resources in a more energy-effective manner, with significant reductions in carbon dioxide emissions and water use [21]. With regard to the use of enzymes as biocatalyst for applications in white biotechnology, cellulases and cellulase cocktails are extensively studied for effective cellulose degradation.

5.3 The Cell Wall Degrading Enzymes

Different enzymes are involved in the degradation of polymers present in lignocellulosic materials. The predominant polymer in lignocellulose is cellulose and enzymes involved in its degradation are cellulases. The insoluble and recalcitrant nature of plant cellulose poses great challenges for biochemical degradation of cellulose into monomeric sugars.

Cellulases are a multienzyme complex consisting of three different enzymes: endoglucanase (endo β-1,4-D-glucan glucanohydrolase, EC 3.2.1.4); exoglucanase or cellobiohydrolase (β-1,4-D-glucan cellobiohydrolase, EC 3.2.1.91); and β-glucosidase or cellobiase (β-D-glucoside glucohydrolase, EC3.2.1.21) [22] (Figure 5.1).

The first group of the enzyme cellulolytic complex of endoglucanase are enzymes that catalyze the hydrolytic degradation of β-1,4-glycosidic bonds anywhere in the interior of cellulose molecules at random to form glucose, cellobiose, and cello-oligosaccharides. These enzymes are able to act on amorphous parts of cellulose, as well as on cellulose

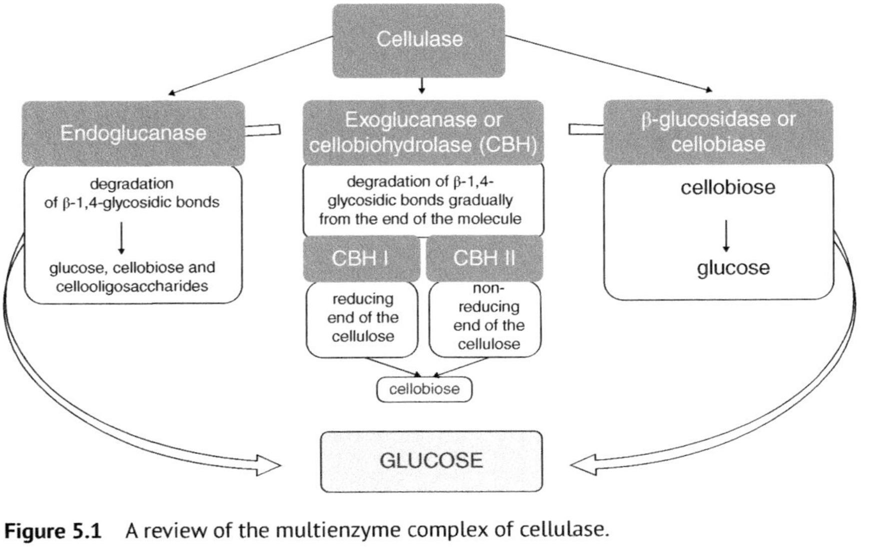

Figure 5.1 A review of the multienzyme complex of cellulase.

derivatives (carboxymethyl cellulose and hydroxyethyl cellulose), while they have little effect on crystalline parts of cellulose. The second group of enzymes is exoglucanases or cellobiohydrolases (CBH) and they catalyze the hydrolytic degradation of β-1,4-glycosidic bonds gradually, starting from the end of the macromolecules. It should be emphasized that they can effectively decompose the crystalline parts of cellulose. The two main types of CBH exist: CBH I and CBH II working from the reducing and from the non-reducing ends of the cellulose chains, respectively, and produce mainly cellobiose. Cellobiase is the third group of enzymes and it can only break down molecules of lower molecular weight, such as the disaccharide cellobiose and other soluble oligosaccharides to glucose. All three enzymes, endoglucanase, exoglucanase, and cellobiase, must work synergistically for the effective hydrolysis of cellulosic biomass to glucose [17, 23].

Due to the complex structure of hemicelluloses, the hydrolysis of hemicelluloses into simple constituents like monomeric and dimeric sugars or oligosaccharides requires the action of a wide spectrum of enzymes. The enzymes can be categorized as main-chain degrading enzymes (xylanase, β-xylosidase, mannanase, β-mannosidase, arabinase) and side-chain cleaving enzymes (α-L-arabinofuranosidase, esterases). Cooperative action of both types of enzymes is required for complete biodegradation of hemicelluloses [17].

Enzymes involved in lignin degradation are classified as phenol oxidase (laccases) and manganese peroxidase, lignin peroxidase, and multifunctional (versatile) peroxidase [24].

Under ideal conditions, cellulose hydrolysis requires a small amount of cellulases. However, in lignocellulose hydrolysis, where above the cellulose layers is a network of lignin, the situation is quite different. The enzymes are irreversibly bound to the lignin and are no longer able to serve as catalysts, so to obtain a glucose the dosage of cellulase must be increased (about ten times) [25]. To overcome this problem, the leading companies in enzyme manufacturing, such as Novozymes (Denmark), Genencor (USA), today DuPont, AB Enzymes (Germany), DSM (Netherlands), etc., developed innovative enzyme solutions, mainly cellulase cocktails, for enhancing cellulolytic activities for efficient hydrolysis of lignocellulosic biomass.

5.4 Cellulases in a Biorefinery Development

A biorefinery is a facility that integrates biomass conversion processes and equipment to produce fuels, power, and value-added chemicals from biomass [25]. Enzymes are essential in the biorefinery concept due to their role in the conversion of renewable lignocellulosic biomass to biofuel and value-added products [26].

In biorefinery, one of the main challenges is to improve the enzymatic hydrolysis, which mostly refers to the use of a lower enzyme dosage and reduction of the enzyme production costs. This may be achieved with the development of tailored and more specific enzyme solutions and extremely high volumes of processed biomass, due to the low consistency. Additionally, the use of a novel substrate pretreatment technology may be a useful tool for enhancing the enzymatic hydrolysis [5, 26–28]. The overall goal of the biorefinery is to make the system cost-competitive [26].

In detail, the first goal is to overcome the recalcitrant nature of lignocellulose. This means separating and removing the lignin from the other parts of biomass; and also

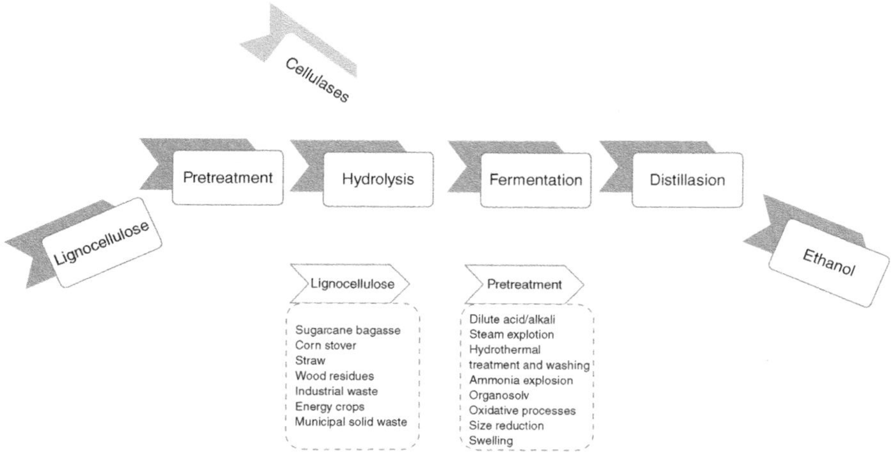

Figure 5.2 The main processing steps for ethanol production from renewable lignocellulosic biomass.

breaking down the lignin structure and furthermore removing and fracturing the hemicellulose to facilitate the enzyme attack. This pretreatment step is essential for liberating the cellulosic fibers from the lignin network, in order to increase water access and enzyme accessibility while minimizing sugar loss and inhibitor generation. It will ensure a high yield and make a substrate suitable for enzymatic hydrolysis and fermentation [29, 30]. However, the pretreatment itself is highly costly and in order to select the most appropriate approach it is necessary to consider feedstock type, economic viability, and environmental impact [26]. Several processing steps for ethanol production from renewable lignocellulosic biomass are presented in Figure 5.2.

Compared to first-generation ethanol, second-generation ethanol is a more complex process and therefore a more expensive process [31]. In second-generation ethanol production, cellulase costs are meaningful and count about 28% of the full costs [32]. The reduction of this cost is another important challenge in lignocellulose biorefining. To date, there is a constant effort to reduce the cellulase production costs in order to produce cost-effective bioethanol fuel.

In the last few decades, the researchers from leading companies for enzyme manufacturing, like Novozymes (Denmark), AB Enzymes (Germany), DSM (Germany), etc., have been intensively studying cellulases for improvement in enhancing cellulolytic activities. To reduce the costs, the National Renewable Energy Laboratory, partnering with two leading enzyme companies, Novozymes and Genencor, made an effort to engineer novel cellulase enzymes for more efficient cellulose hydrolysis [31].

To increase and stimulate development of the biorefinery, the European Union has funded many projects in the utilization of lignocellulosic biomass for second-generation ethanol production. Other companies have helped too. For example, Novozymes helped to establish the first biorefinery in Italy for cellulosic ethanol production. This biorefinery is located near Crescentino (Italy) and uses agricultural residues (rice straw, wheat straw), energy crops (reed, switchgrass, woody crops), and forestry residues/waste as lignocellulose to produce cellulosic ethanol. Innovation in this biorefinery is using Proesa™ Technology for the pretreatment of cellulosic biomass from agricultural residues and energy crops. The process is based on a first phase in which the biomass is subjected to high temperatures and pressures, allowing the cellulose and hemicellulose to be separated from the lignin. Polysaccharides are treated with enzymes, provided by Novozymes, to release simple sugars that can be fermented by yeast into ethanol. The plant in Crescentino has a capacity of 40 000 tons of bioethanol per year [33].

The BIOKOSH project (2016–2022, under Horizon 2020) with Novozymes company, as one of the participants, is designed to establish a breakthrough for 2G biorefineries in Europe, demonstrating a sustainable, both environmentally and economically, second-generation biorefinery within the European Union. Furthermore, for the BIOFOREVER project (2017–2019, Horizon 2020), DSM company provides enzyme technology for lignocellulose hydrolysis (https://bioforever.org). This is a European project for demonstrating the feasibility of conversion of lignocellulosic feedstocks like wood into chemical building blocks and high value-added products (https://bioforever.org).

5.4.1 Commercial Cellulase Cocktails for Lignocellulosic Biomass Degradation

Currently, cellulosic ethanol is on the way to large-scale commercialization in the United States, Europe, and China [3, 34]. For this purpose, a choice of a proper cellulase, with high hydrolysis performance and low cost, is very important due to the high cellulase contribution to the cost in the lignocellulose biorefining process [35].

For the economical production of cellulase, many factors such as high substrate loading, low enzyme loading, and a short hydrolysis period are crucial [7]. The pioneers in enzyme manufacturing, Novozymes (Denmark) and Genencor (USA), and today DuPont, developed the first cellulose-digesting enzyme cellulases with high effectiveness [36]. Over time, enzyme preparations have been improved and novel enzyme innovations in the form of enzyme cocktails or a mixture have been launched on the market for industrial use.

The first generation of cellulases manufactured by Novozymes was Celluclast®. Celluclast is produced by a selected strain of the fungus *Trichoderma reesei*. This cellulase was used for cellulose degradation, and not only for lignocellulose biomass but also in biological washing powders to prevent "pilling" of cotton fabrics. Its use was extended for denim abrasion in the textile industry instead of using pumice stones [37].

This company launched several cellulase solutions under various commercial names and for different purposes. Some most used commercial cellulases and their applications for lignocellulose degradation are presented in Table 5.1.

For example, Celluclast 1.5L is mainly cellobiohydrolase (exoglucanase) with endoglucanase, while Novozymes 188 is β-glucosidase, from *Aspergillus niger*, that contains a mixture of compounds selected to improve enzyme performance and stability [43]. To ensure maximum production of fermentable sugars, Celluclast is often used in combination with another enzyme [61]. Celluclast is also often used for the reduction of the viscosity of the reaction mixture that contains wheat, barley, and sorghum, due to the enzyme action on beta-glucans and other non-starch polysaccharides [61]. Novozymes 188 has the ability to hydrolyze lignocellulosic biomass efficiently in combination with *T. reesei* cellulases,

Table 5.1 The most used commercial cellulases for lignocellulose degradation.

Commercial solutions	Enzyme type	Description and benefits	Producer (company)	References
Cellic	Cellulase complex (endoglucanase)	For hydrolysis of multiple lignocellulosic feedstocks	Novozymes, Denmark	[38]
Cellic CTeC2	Cellulase complex (endoglucanase with high level of β-glucosidases and hemicellulases)	For hydrolysis of multiple lignocellulosic feedstocks; lower enzyme dosage, operating and capital costs for optimization process and lower ethanol cost	Novozymes, Denmark	[39, 40]
Cellic® HTec2	Cellulase complex (cellulase with predominant endoglucanase with high specificity toward soluble hemicellulose)	For hydrolysis of multiple lignocellulosic feedstocks; lower enzyme usage cost, higher ethanol yields and provides increased flexibility for pretreatment technology; in combination with Cellic® CTeC2 for improved cellulose hydrolysis	Novozymes, Denmark	[40]
Cellic® CTeC3	Cellulase complex (cellulase and hemicellulase complex with improved β-glucosidases)	For hydrolysis of multiple lignocellulosic feedstocks; a cost-efficient conversion of pretreated lignocellulosic materials to fermentable sugars	Novozymes, Denmark	[29, 41]
Cellic® HTec3	Cellulase complex (mixture of hemicellulytic enzymes for xylan hydrolysis of pretreated lignocellulosic substrates)	For hydrolysis of multiple lignocellulosic feedstocks; a cost-efficient conversion of pretreated lignocellulosic materials to fermentable sugars; low enzyme loading and high substrate loading	Novozymes, Denmark	[29, 42]
Celluclast 1.5L	Cellobiohydrolase (exoglucanase) with endoglucanase activity	For cellulose degradation	Novozymes, Denmark	[37, 43]
Novozymes 188	β-glucosidase	An additional enzyme for cellulose hydrolysis; mainly used in a combination with Celluclast® 1.5L	Novozymes, Denmark	[43]
MetZyme® SUNO	Enzymes cocktail (cellulase and hemicellulase)	For reduction of the time of hydrolysis, reduction of inhibitors and enabling valorizing process side stream	MetZyme, Finland	[44, 45]

(*Continued*)

Table 5.1 (Continued)

Commercial solutions	Enzyme type	Description and benefits	Producer (company)	References
Accellerase® 1000	Exo-1,4-β-glucanase (cellobiohydrolase), endo-1,4-β-glucanase, hemicellulase, and β-glucosidase	For biomass hydrolysis	Genencor-DuPont	[46]
Accellerase® 1500	Cellulase complex (exoglucanase, endoglucanase, hemicellulase, and high levels of β-glucosidase)	Enhanced saccharification performance on a variety of feedstocks; high β-glucosidase activity to minimize residual cellobiose	Genencor-DuPont	[47]
Accellerase® DUET	Cellulase complex (contains a potent combination of multiple enzyme activities; mainly exoglucanase, endoglucanase, β-glucosidase and hemicellulases including xylanase)	Capable of efficiently and synergistically hydrolyzing lignocellulosic biomass into fermentable monosaccharides such as glucose and xylose; cost effective	Genencor-DuPont	[48]
Accellerase® TRIO	Exoglucanase, endoglucanase, hemicellulases (with xylanases), and β-glucosidase	Dose-efficiency and cost-effectiveness—often a 2-fold dose reduction, compared to Accelerase duet; improved viscosity reduction; broader operating conditions	Genencor-DuPont	[49, 50]
Accellerase® BG	β-glucosidase	An accessory enzyme to supplement whole cellulases deficient in β-glucosidase; work synergistically to achieve high glucan conversion for the lignocellulosic biomass processing industry	Genencor-DuPont	[51]
Viscozyme®	Enzyme mixture (arabanase, cellulase, β-glucanase, hemicellulase, and xylanase)	Reduced viscosity level and peaks due to raw material variation or change in particle size; wheat, triticale, barley, and rye can all be used	Novozymes, Denmark	[52, 53]
Spezyme® CP	Endoglucanase	No available data	Genencor-DuPont	[54, 55]

NS22086	Cellulase complex	Reduce the viscosity and increase the extraction yield of various products of plant origin; tested for synergy with NS22118 and NS22083	Novozymes, Denmark	[56, 57]
NS22083	Xylanase	Purified endoxylanase with a high specificity toward soluble pentosans; able to liberate pentose sugars from biomass hemicellulose fractions	Novozymes, Denmark	[56, 58]
NS22118	β-glucosidase or cellobiase	Can be used to supplement NS22086 in order to increase the yield of fermentable sugars	Novozymes, Denmark	[56, 59]
NS2119	Enzyme complex (arabinase, β-glucanase, cellulase, hemicellulase, pectinase, and xylanase)	Can break cell wall for the extraction of useful components from plant tissue	Novozymes, Denmark	[56]
NS2002	β-glucanase and xylanase (possesses additional side activities including cellulase, hemicellulase, and pentosanase)	Can be used to supplement NS22086 for pretreatment that leaves a significant portion of the hemicellulose intact (neutral pH or alkaline pretreatment methods)	Novozymes, Denmark	[56, 60]

mostly with Celluclast 1.5L [62]. The commercial Novozymes 188 and Celluclast 1.5L have been used in combination for the hydrolysis of steam-pretreated barley straw. The specific aims were to achieve hydrolysis of high substrate levels, low viscosity during hydrolysis, and high glucose concentrations. The highest final glucose concentration, 78 g L^{-1}, after 72 h of hydrolysis, was obtained with an initial, full substrate loading of 15% (w/w of dry matter) [63]. Conversely, the glucose yields were highest at lower substrate concentrations. Rapid cellulose degradation was accompanied by rapid decreases in viscosity before addition of extra substrate, but when extra substrate or substrate plus enzymes were added, the viscosities of the slurries increased and the hydrolytic efficiencies decreased temporarily [63]. High viscosity limits the dry substance level in the process, increasing the energy and water consumption and thus lowering the ethanol yield [64].

In 2005, Novozymes launched a new enzymatic solution, which offered both financial and environmental benefits, aimed especially for bioethanol producers in Europe, where production of wheat, barley, and rye was high. The new enzymes Viscozyme® Wheat, Viscozyme Barley and Viscozyme Rye enable more CO_2-neutral bioethanol to be produced from wheat, barley, and rye than was previously possible. These solutions also saved water and energy in the production process, making the new enzymes environmentally friendly. Viscozyme is a multienzyme product containing a wide range of carbohydrases, including arabanase, cellulase, β-glucanase, hemicellulase, and xylanase. This enzyme mixture is produced from a selected strain of the *Aspergillus* group. Viscozyme enzymes enable higher ethanol production capacity and lower operating costs.

Further, in 2009, Novozymes launched one of the first improved enzymatic solutions, from the cellulase family, called Cellic® ("specially designed cellulase complex") for cellulose and hemicellulose hydrolysis of lignocellulosic biomass, under a wide range of conditions [65]. This enzymatic solution could hydrolyze multiple types of lignocellulosic feedstocks. The main benefit of using Cellic for lignocellulose hydrolysis is lowering the costs of producing cellulosic ethanol. Other benefits are a reduction in total ethanol fuel production costs, great process and substrate versatility, increased sugar conversion from biomass, decrease in enzyme dosing, increase in total solids loading, reduction in hydrolysis time, and reduction in the severity of lignocellulosic pretreatment [38].

Cellic CTeC2 was launched by Novozymes in 2012 and it was an improved version of the previously developed Cellic enzyme for the biofuel industry (Figure 5.3) [39]. The Cellic CTeC2 is a cellulase complex of aggressive cellulases, with a high level of β-glucosidases and hemicellulases, for degradation of cellulose to fermentable sugars. Its features are related to high conversion yield, inhibitor tolerance, effectiveness at a high solids

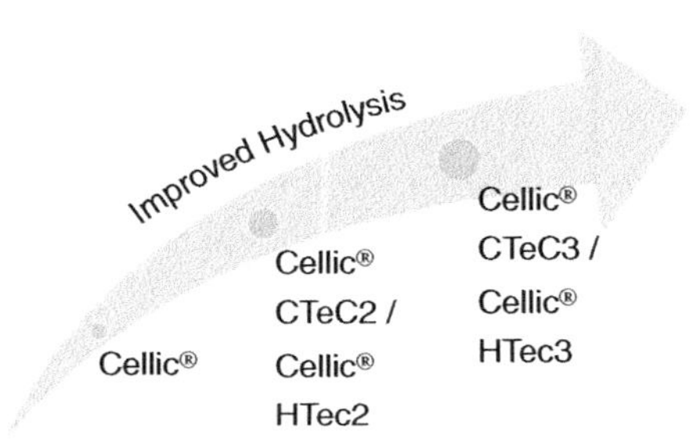

Figure 5.3 The development flow of the Cellic enzyme (Novozymes) for the biofuel industry. Based on [40].

concentration. The benefits of using this enzymatic solution are up to 50% lower enzyme dosage, lower operating and capital costs for the optimization process, and lower ethanol cost [40]. Sun et al. (2016) [66] reported that a new cellulase preparation Cellic CTec2 has a remarkable qualification in its toleration to high solid consistency. They tested different concentrations of sugarcane bagasse (2%–18% solids), previously pretreated with atmospheric glycerol organosolv, with the commercial cellulase preparation Cellic CTec2 at a relatively low enzyme loading, 6 FPU g^{-1} and 10 FPU g^{-1} dried substrate (FPU, filter paper unit). At 10 FPU g^{-1} dried substrate, all substrates presented a fast enzymatic hydrolysis and the 72 h enzymatic hydrolysis of pretreated substrate at less than 15% solid concentration exceeded 90%. At 6 FPU g^{-1} dried substrate, the 72 h enzymatic hydrolysis of substrates (≤12% consistency) was greater than 80%. 90% of enzymatic hydrolysis was achieved after 72 h of enzymatic hydrolysis at 15% solid content with Cellic CTec2 at 10 FPU g^{-1} dried substrate [66].

The hydrolysis of wheat straw was studied using Cellic CTec2 and Celluclast (cellulase from *T. reesei*) mixed with Novozymes 188 (β-glucosidase from *A. niger*). The efficiency of hydrolysis with an enzyme loading of 10 FPU g^{-1} cellulose reached more than 98% using Cellic CTec2 with 12% (w/v) wheat straw, while for Celluclast a conversion of 52% and 81%, was observed without and with β-glucosidase supplementation, respectively [67].

Cellic HTec2 is an improved version of Cellic CTeC2 solution that contains cellulase with predominant endoglucanase with high specificity towards soluble hemicellulose. It can improve cellulose hydrolysis when combined with Cellic CTeC2. This cellulase enables 20% lower enzyme usage cost, higher ethanol yields, and provides increased flexibility for pretreatment technology. Both types of cellulases, Cellic CTeC2 and Cellic HTec2 could be used in combinations if the pretreated lignocellulosic feedstock contains additional hemicellulose for degradation. In this way, Cellic CTeC2 and Cellic HTec2 work in synergy to release maximum reducing sugars for further ethanol production. The optimal temperature for Cellic CTeC2 is 45–50 °C and pH between 5.0 and 5.5. For the Cellic HTec2 the optimal temperature is 45–50 °C and pH 5.0 [40]. Both Cellic cellulases can be effectively used for the hydrolysis of different lignocellulosic biomass such as wood [68], sugarcane bagasse [69], corn cob [70], corn stover [71], grass [72], etc.

Later, in 2012, Novozymes launched the last generation of the Cellic family called Cellic CTeC3/Cellic® CTec3 HS. This is a highly efficient cellulase and hemicellulase complex that ensures a cost-efficient conversion of pretreated lignocellulosic materials to fermentable sugars. Cellic CTec3 contains proficient cellulase components boosted by proprietary enzyme activities, including advanced AA9 molecules, improved β-glucosidases, as well as a new array of hemicellulase activities. Compared to the previously launched Cellic CTeC2, Cellic CTeC3 works 1.5 times better [29]. Cellic CTec3 has proven to work across a variety of feedstocks including corn cobs and stalks, wheat straw, municipal solid waste, sugarcane bagasse, wood chips, and energy crops [73]. It works at high substrate loadings (20% w/v) [74] and on different pretreated substrates like acid, auto-hydrolyzed, and alkaline pretreatments. However, if the pretreated feedstock/lignocellulosc biomass contains an appreciable amount of hemicellulose, it is advisable to combine Cellic CTec3 HS with other lignocellulytic enzymes to further boost overall conversion yields and potentially reduce the severity of pretreatment. The enzyme type and dosage recommendation will depend on product, feedstock type, pretreatment technology, and processing conditions [29].

Novozymes claimed that 50 kg of Cellic CTec3 makes 1 ton of ethanol from biomass and also made the comparative claim that at least 250 kg of a "competing enzyme product" makes the same amount of ethanol. With the introduction of Cellic CTec3 HS, there are more options available for consideration when selecting optimal process layout and targeting cost-efficient cellulose conversion. These include simultaneous saccharification and fermentation (SSF), separate hydrolysis and fermentation (SHF), and a modified version of these referred to as hybrid hydrolysis and fermentation (HHF). The HHF process can also be described as viscosity reduction and pre-saccharification followed by SSF for ethanol production and this setup can be configured in relation to overall process layout and conversion targets. The optimal performance of Cellic CTec3 HS occurs in a temperature range 50–55 °C and at pH 4.75–5.25.

Cellic CTec3 was successfully used for enzymatic hydrolysis of pretreated steam-exploded materials obtained from *Eucalyptus urograndis* chips and sugarcane bagasse at high solids loadings (20 wt%) and relatively low enzyme loadings (13.3 FPU g^{-1} total solid of Cellic CTec3). Enzymatic hydrolysis of steam-treated *E. urograndis* reached 125 g L^{-1} of glucose in 72 h, while steam-treated bagasse gave 25% lower yields [41].

There is an interesting study about comparing three commercial Novozymes cellulase preparations Cellic CTec (CTec 1/2/3) for hydrolysis of steam-pretreated lignocellulosic substrates at an identical filter paper activity loading [75]. With the same enzyme loading, CTec3 resulted in 94% glucan hydrolysis of pretreated poplar at 24 h, which was higher than was reached by CTec1 (86%) and CTec2 (80%). By using steam-pretreated sweet sorghum bagasse, CTec3 reached 66% of glucan hydrolysis (24 h), while for CTec1 it was 47% and 55% for CTec2. With an addition of CTec1 at the same protein level as CTec2 and CTec3, the glucan hydrolysis of pretreated poplar and sweet sorghum bagasse was only up to 38% and 20%, respectively. This result indicated that the older cellulase preparation, CTec1, has a relatively lower hydrolytic efficiency than the new products, CTec2 and CTec3, on pretreated lignocellulosic substrates. In addition, the CTec3 is more efficient than CTec2 at the same specific enzyme loading and enzyme activity. The results also showed that, at a fixed enzyme loading, the three cellulase preparations presented surprisingly different hydrolytic profiles on the different lignocellulosic substrates. Since these Cellic CTec preparations are all derived from Novozymes, it is evident that the product upgrade was achieved not only by improving the protein expression but also by mixing and matching various enzymes as accessory enzymes to strengthen their synergistic cooperation in the lignocellulosic hydrolysis [75].

Finally, in 2014, the newest upgrade or innovation came from Novozymes called Cellic HTeC3. This is a mixture of hemicellulytic enzymes for xylan hydrolysis of pretreated lignocellulosic substrates. This enzyme solution could hydrolyze soluble and insoluble forms of hemicellulose from a variety of pretreated lignocellulosic substrates for fermentable sugar production. Compared to previous Cellic HTec2, a newly Cellic HTec3 provides better conversion [76].

Cellic CTec3 and Cellic HTec3 were tested for the hydrolysis of steam-treated cane bagasse. Cellic CTec3 was able to release more than 100 g L^{-1} glucose equivalent from steam-treated cane bagasse in 72 h at 20% total solids, using 38.6 FPU g L^{-1} glucan. Cellic HTec3 had a boosting effect on the enzymatic hydrolysis of cane bagasse glucans regardless of the low hemicellulose content of the steam-treated material. Cellic HTec3 improved the total glucan conversion by 8%, but only when a relatively high enzyme loading was used for hydrolysis [42]. Commercial cellulases, Cellic CTec3, Cellic HTec3 and Multifect® Pectinase (Genencor-DuPont, composed of diverse hemicellulase activities such as arabinofuranosidase, xylan esterase, pectinase, pectin lyase, alpha galactosidase, mannanase, mannosidase, and other activities) were used for the hydrolysis of agave leaf and bagasse residues from species *Agave tequilana* and *Agave salmiana*, respectively, previously pretreated using the ammonia fiber expansion (AFEX) process. When using Cellic CTec3 enzyme alone, glucose and xylose were released from lignocellulose (6% glucan loading; 17%–19% total solids). The addition of small amounts of accessory enzymes Cellic® HTec3 to cellulose further enhanced the sugar conversion in pretreated agave biomass samples [77].

MetGen, a Finland company for the design, manufacturing, and tailoring of industrial enzymes, developed an enzyme solution for application in biorefinery, as well. An innovative enzyme mix, called MetZyme® SUNO™, was designed to significantly improve the costs of lignocellulosic chemical production and biofuels. It contains cellulase and hemicellulase for breaking down the recalcitrant lignocellulosic biomass for obtaining better hydrolysis. Also, it contains oxidoreductase for eliminating inhibitors that may slow down the process. MetZyme SUNO enzyme cocktail is designed to reduce enzyme dosing and technology cost while obtaining higher reducing sugars yield. The main benefits of applying this enzyme mix is reduced time of hydrolysis, reduced inhibitors, and enabling valorizing process side stream. MetZyme SUNO has wide pH and temperature capabilities allowing it to be readily adapted to many biorefinery processes. Depending on the lignocellulosic biomass, pretreatment with MetZyme LIGNO™ at pH 8 prior to hydrolysis results in higher yields of reducing sugars. MetZyme SUNO could be worked at a temperature around 60 °C, or even higher. Higher operating temperatures result in a higher enzymatic activity providing lower enzyme dosage. MetZyme SUNO will function during the whole hydrolysis time enabling faster and more thorough hydrolysis [44]. It was reported that MetZyme SUNO can improve the valorization of lignocellulosic biomass by reducing the production cost of reducing sugars. Commercial cellulase, MetZyme SUNO, and a combination of cellulase and MetZyme SUNO were used for 72 h hydrolysis of pretreated spruce. The results showed that MetZyme increases sugar yield by 25% with only 50% of cellulase complex used [45].

Genencor, today DuPont, as a world leader in industrial biotechnology and a pioneer in enzyme innovation, has a 25 year history in biofuels research. Genencor launched a first commercial cellulase preparation called Spezyme CP, produced by *T. reesei*. This preparation contains endoglucanase which lacks in adequate β-glucosidase activity to achieve complete hydrolysis to glucose. Therefore, Spezyme CP is usually supplemented with β-glucosidase to promote better hydrolysis. In the literature, the ratios of cellulose and β-glucosidase have ranged widely and have depended upon the nature of the lignocellulosic substrate used [55]. For the hydrolysis of ammonia pretreated sugar cane bagasses a ratio of Spezyme CP/Novozymes 188 (in an activity ratio of 1:2) was used [55]; ratios of 1 FPU:2 cellobiase units [78] and 1 FPU:3.2 cellobiase units was used for the hydrolysis of various softwood substrates and barley straw [79].

Genecor, today DuPont, was the first company that in 2007 launched a commercial-scale enzyme for cellulosic ethanol production, Accelerase® 1000; later this company introduced on the market Accelerase 1500, Accelerase DUET, and Accelerase TRIO. The family Accelerase are cellulases produced with a genetically modified strain of *T. reesei*.

Accelerase 1500 cellulase is an enzyme complex of exoglucanase, endoglucanase, hemicellulase, and high levels of β-glucosidase designed for the lignocellulosic biomass hydrolysis in processing industries, including renewable fuels and chemicals. A potent combination of enzymes in this enzyme mixture effectively modifies and digests non-starch carbohydrates, the structural material of lignocellulosic biomass. Feedstocks, including paper pulp, corn stover and cob, sugar cane bagasse, wheat straw, wood chips, waste paper, and many others, can be hydrolyzed within 24 h using Accelerase 1500. The optimal working temperature is 50–60 °C and pH 4.0–5.0. Accelerase 1500 can work with a variety of ligocellulosic pretreatments including dilute acid, AFEX, and steam expansion.

Accelerase BG β-glucosidase is designed as an accessory product to supplement whole cellulases deficient in beta-glucosidase and work synergistically to achieve high glucan conversion. Accelerase BG enzyme can effectively convert cellobiose into fermentable monosaccharides like glucose. Cellobiose is also an inhibitor of cellulase enzymes, so low levels of enzyme loadings are desirable. Supplementation with Accelerase BG enzyme can result in faster and higher ethanol production. Accelerase BG accessory enzyme, when added to other cellulase products deficient in β-glucosidase, will enhance saccharification of glucan in lignocellulosic biomass and increase the yield of fermentable glucose by alleviating product inhibition to the cellulolytic enzymes. The optimal working temperature is between 30 and 55 °C and optimal pH between 4.0 and 6.0. The newest addition to Accelerase products is Accelerase TRIO. This enzymatic solution contains a potent combination of multiple enzyme activities including exoglucanase, endoglucanase, hemicellulases (with xylanases), and β-glucosidase. Accelerase TRIO is designed for improved lignocellulosic hydrolysis into glucose and xylose and further to enable commercialization for second-generation biorefineries. The key benefits of using this enzyme are dose efficiency and cost effectiveness compared to previously launched Accelerase; increased glucan conversion, and reduced product inhibition regardless of engineered β-glucosidase; improved viscosity reduction at high biomass solids loadings; wide feedstock and pretreatment flexibility; and broader operating conditions for greater flexibility of process configuration including saccharification, SHF, HHF, and SSF. The commercial enzyme mixture Accelerase TRIO was successfully used for (hemi)cellulose hydrolysis of different lignocellulosic substrates, previously pretreated using a mild acetone organosolv process. For enzymatic hydrolysis, enzyme loading of 10 FPU g^{-1} solids of Accelerase TRIO was used at 10% (w/v) solids loading (different lignocellulosic substrates). Full enzymatic cellulose conversion to glucose was achieved using wheat straw and corn stover pulp [49].

5.4.2 Commercial Cellulase Preparation for Various Industrial Uses

With the rise of the global industrial sector, the cellulase market records a high growth over the last few decades. On the commercial enzyme market, the cellulases contribute with nearly 20% [9]. Except in biofuel production, cellulases are widely used in industries such as the laundry and detergent industry, textile, pulp and paper, food and beverage, etc. The main commercial cellulase preparations for industrial use are presented in Table 5.2.

Table 5.2 The main commercial cellulase preparation for various industrial uses.

Commercial name	Industry (description)	Description of cellulases	References
Celluzyme (Novozymes, Denmark)	Laundry and detergent; color brightening, softening, and soil removal from fabric	Fungal cellulase complex; neutral to alkaline cellulase pH	[80]
Cellulclean® 5000 L (liquid) (Novozymes, Denmark)	Laundry and detergent	Works in cold temperatures; keeps white colors white and colors bright; stable in presence of strong chelators; stable in water-rich formulations	[81]
Cellulclean® Evity 4500 (powder) (Novozymes, Denmark)	Laundry and detergent	Works in cold temperatures; keeps white colors white and colors bright	[82]
Biotouch® liquid (FLX1/DCL/FCL75/Duo505) and granules (DCC/FCC45) (AB Enzymes, Germany)	Laundry and detergent	Premium stability at high temperatures	[83]
DeniMax® (Novozymes, Denmark)	Textile industry; denim washing	Acidic and neutral cellulase; low operating temperature (30 °C)	[25, 84, 85]
Cellusoft LT Cellusoft® LT 19500 (Novozymes, Denmark)	Textile industry; biopolishing	pH 5.5–8.0; low operating temperature (30–40 °C)	[86]
CombiPolish (Novozymes, Denmark)	Textile industry; biopolishing, bleaching, and dyeing	Neutral cellulases	[86]
Fiberlife® (Novozymes, Denmark)	Textile industry; biopolishing	Cellulase No available data	[86]
ECOSTONE® (AB Enzymes, Germany)	Textile industry; finishing process of cellulose-based textile fabrics and their blends	Low operating temperature; neutral and acid pH	[87]

(Continued)

Table 5.2 (Continued)

Commercial name	Industry (description)	Description of cellulases	References
FiberCare B FiberCare D FiberCare L FiberCare U FiberCare W (Novozymes, Denmark)	Pulp and paper industry; strength increase	Multicomponent cellulase; pH 5–9 and operating temperature 40–70 °C	[88–90] [91]
Novozyme® 51 008 (Novozymes, Denmark)	Pulp and paper industry; strength increase; polymer replacement and improved runnability	Endoglucanase; pH 5–9 and operating temperature 40–70 °C	[92, 93]
MetZyme® BRILA (MetZyme, Finland)	Pulp and paper industry; improving recycled fibers	No available data	[94, 95]
Celluclast® (Novozymes, Denmark)	Bakery industry; better dough handling during production, increased volume, more uniform crumb structure, softness and bloom	Endoglucanase	[96]
BakeZyme WholeGain (DSM, Netherlands)	Bakery industry; better gluten development, increased dough tolerance, and improved proofing stability	Endoglucanase	[97]
BakeZyme X-pan (DSM, Netherlands)	Bakery industry; improve quality of the bread	Fungal cellulase	[97]
BakeZyme HS10.000 BG BakeZyme HS20.000 BG BakeZyme Multicell BakeZyme® Real-X BakeZyme® X-cell BakeZyme® ARA 10.000 (DSM, Netherlands)	Bakery industry; improve quality of the bread	Hemicellulases	[98]

Pectinex® (Novozymes, Denmark)	Fruit and vegetable juice industry; fruit and vegetable meshes and the maceration of the plant tissues; improving the juice yield in the range of 85%–90% and smoother processing	Pectin transeliminase, polygalacturonase and pectinesterase with small amounts of hemicellulases and cellulases; optimal pH for enzyme mixture 4.5 and working temperature 50 °C	[99]
Rapidase® Fiber (DSM, Netherlands)	Fruit and vegetable juice industry; improving the yield, quality, clarification, and enzymatic filter cleaning	Pectinase enriched in arabinolytic activities from *Aspergillus niger* and cellulolytic activities from *Trichoderma longibrachiatum*	[100]
Rochapect® 10L (AB Enzymes, Germany)	Fruit and vegetable juice industry; depectinization of apple juice and citrus juice; for mash and fruit treatment of acidic fruits	Pectinolytic and cellulolytic enzymes	[101]
Olivex® (Novozymes, Denmark)	Olive oil extraction	Pectinolytic, hemicellulolytic and cellulolytic activities produced by the fungus *Aspergillus aculeatus*; optimal pH 3.5	[102]
Pectinex® Ultra Olio (Novozymes, Denmark)	Olive oil extraction	Pectin lyase, polygalacturonase, side cellulase and hemicellulase activities	[102]
Pectinex® Ultra SPL (Novozymes, Denmark)	Olive oil extraction	Pectolytic activities with side cellulase and hemicellulase activities	[103]
Ultraflo® Max (Novozymes, Denmark)	Brewing industry	β-glucanase and arabinoxylanase	[104]
Rohament CL (AB Enzymes, Germany)	Brewing industry; reducing the viscosity (enzymes for grains)	Cellulase	[105]
Rohament® CEP (AB Enzymes, Germany)	Brewing industry reducing the viscosity (enzymes for grains)	Cellulase	[105]
Roxazyme® G2 (DSM, Netherlands)	Animal feed industry (additive to feed)	β-glucanases, cellulases, and xylanases	[106, 107]

5.4.2.1 Laundry and Detergent Industry

In the laundry and detergent industry, cellulases clean indirectly by gently hydrolyzing certain glycosidic bonds in cotton fibers. In this way, particulate soils attached to microfibrils are removed. A further desirable effect of cellulases is to achieve greater softness and improved color brightness of worn cotton surfaces. Many detergent brands are based on a blend of two or more, even up to eight, different enzyme products.

On the market, cellulases are available under various commercial names. The first commercial enzyme preparation, called Celluzyme®, was developed in 1987 by Novozymes (Denmark) for the detergent industry [108]. This multicomponent enzyme mixture consists of alkaline cellullase, fungal origin, that is active in neutral to alkaline regions. The main benefits of this enzyme mixture are color brightening, softening, and soil removal from fabric. It is ideal for cellulosic cotton fibers and blends [109]. The Novozymes company developed novel cellulase solutions for keeping white colors white and colors bright by preventing dirt particles in the wash water from depositing on otherwise clean clothes. A powder detergent Cellulclean® Evity® 4500 could work at low temperatures in cold water, also as a liquid solution Celluclean® 5000. Celluclean 5000 is developed for stability and consistent performance in detergents with high water content; it is compatible with strong chelators and works well at low temperatures [81].

The company AB Enzymes (Germany) launched in 2018 a liquid cellulase preparation, Biotouch® FLX1, for household laundry detergents. This product has a premium stability at high temperature, and it is stable with the most common proteases and other chemicals in current liquid formulations. The same company produced several liquid and dry cellulase preparations named Biotouch, Biotouch FCL75 and Biotouch Duo505, and granules Biotouch DCC and Biotouch FCC45. All these preparations provide ultimate fabric care (smoother texture and preventing pilling on new cotton fabrics), keep color fresh without anti-graying, show anti-redeposition, color revival, and pilling removal [83].

5.4.2.2 Textile Industry

The first great success of using cellulase in the textile industry was for denim washing. A sustainable way to produce stone-washed jeans is to use enzyme cellulases. Novozymes DeniMax® required 90% less water and up to 50% less energy for abrasion of jeans to get a stone-washed look, compared to classical pumice stones. This enzyme solution could work at low temperatures (30 °C) and thus reduce energy consumption. This innovative enzyme saved water, energy, and heat costs compared to washing with pumice stones. Also, one of the benefits of DeniMax is that it provides a high color contrast finish, high color pull, and a low degree of back staining of the denim [25]. For the abrasion of the denim, neutral cellulases with pH between 6 and 8 and also acidic cellulase with pH optimum between 4 and 6 could be used [25, 84, 85].

Biopolishing is the process that removes dead cotton and protruding fibers from the fabric surface to give a great appearance and to avoid fuzz and pills from forming after wash and wear. By using of cellulase for this process, the fabric is softer, shows a brighter luminosity of colors, and is more resistant to pilling propensity [110]. Biopolishing with cellulase is an environmentally friendly process for extending the lifetime of cotton knits and fabric. The improved formulation comes from the Novozymes in the form of Cellusoft® LT for biopolishing. After the use of this enzyme solution, the fabric is smoother, softer, and stronger and retain their vibrant colors. The Cellusoft LT works at low temperatures, 30–40 °C, and in the pH range between 5.5 and 8. Compared to other conventional neutral and acid cellulases that are used for biopolishing, the Cellusoft LT 19500 has numerous advantages, such as working at low temperatures, pH flexibility, process flexibility, low weight and less strength, high color retention, anti-back staining, and high biopolishing effect with abrasion. This solution could be used for biopolishing before, during, and after dyeing at low temperatures. It could be combined with catalase for bleach removal, biopolishing, and reactive dyeing [86].

The Novozymes company received the award for enzyme innovation called CombiPolish®. This solution is the fastest, gentlest, and most sustainable way to do biopolishing and bleach clean up in the dye bath without any extra water or time consumption. The CombiPolish uses neutral cellulases rather than acidic cellulases, making it possible for textile manufacturers to combine the biopolishing and bleach clean-up process in the dyeing step, saving time and energy. The CombiPolish process saves about 15% of the total water used and also 15% of the time for making cotton knits [86]. The remarkable Fiberlife® solution, with the Livelong® effect, is a brand new enzyme, from the Novozymes, for gentle biopolishing processes. It extends the lifetime of the fibers, allowing fabrics and clothes to stay new-looking for longer, wash after wash. This enzyme mixture can be applied on cotton, modal, viscose and lyocell fabrics. Lyocell is made from wood pulp and has a high tendency to fibrillate when wet. In simple terms, fibrils on the surface of the fiber peel up. If they are not removed, finished garments made of lyocell will end up with an unacceptable pilled look. This is the reason why lyocell fabric is treated with cellulases during finishing. Cellulases also enhance the attractive, silky appearance of lyocell [86].

The company AB Enzymes produced a cellulase ECOSTONE® designed for the finishing process of cellulose-based textile fabrics and their blends. This solution consists of enzymes that work at low temperature and in neutral and acid pH range. This is a very cost-effective solution, and the product is totally biodegradable and harmless to the environment. The ECOSTONE® enzymes save water, time, and energy [87].

5.4.2.3 Pulp and Paper Industry

The application of enzymes in the pulp and paper industry has considerably increased during the last two decades [111]. Cellulases are mainly used for biomechanical pulping processes which include refining and grinding the woody materials, recycling the printed papers, de-inking of paper and as co-additive in bleaching the pulp, fiber enhancement, etc. [12, 111].

To improve the pulp and paper industry, Novozymes developed solutions that are based on a cost reduction, capacity improvements, and sustainability benefits [88]. For fiber modification, Novozymes offer a pallet of solutions named FiberCare® (FiberCare B, FiberCare D, FiberCare L, FiberCare R, FiberCare U, and FiberCare W). All these products contain cellulase with main benefits that are related to fiber savings, refining energy reduction, chemical savings, improved paper and board strength, etc. According to one case study, performed by Novozymes, the multicomponent cellulase in FiberCare readily improves the "beatability" of bleached kraft pulp, reduces up to 49% of refining energy, and increases tear strength by 4.2% and breaking length by 9% [89, 90]. Tang et al. (2012) [112] investigated the effect of cellulase enzyme treatment of the pulp on the reduction of energy consumption and physical strength of old corrugated container fibers. The cellulases from Novozymes, FiberCare® R, FiberCare U and FiberCare D, were used to improve pulp beatability. The results of this study showed that Novozymes FiberCare D was the best choice to save the energy consumption of pulp beating and to improve the paper strength while FiberCare U was better than FiberCare R.

MetGen, a Finland company for the design, manufacturing, and tailoring of industrial enzymes, developed a solution for the pulp and paper and biorefinery industry called MetZyme BRILA™. This is an enzymatic solution, based on cellulase, for improving recycled fibers. These recycled fibers need to return their strength and to be mixed with higher quality fibers [95]. MetZyme BRILA is a liquid enzymatic solution with extraordinary pH and temperature capabilities allowing for adoption to most recycled/de-inked pulp processes [94]. The initial demonstration of MetZyme BRILA was conducted with MetGen's expert team. They used MetZyme BRILA on a customer paper machine and produced 65000 million tons per year of tissue paper (with as raw material a mix of office waste, household waste, and virgin fiber). The tissue paper produced during enzyme application showed (depending on the grade) up to 13% tensile index increase, 3%–5% increase in stretch, 35%–60% refining energy saving and up to 6% increase in paper machine speed without increasing the drying temperature. In addition, paper machine runnability with household grade pulp improved, decreasing break time from 45 min to 15 min per day [94].

5.4.2.4 Bakery Industry

In the bakery industry, cellulases are used for improving the quality of the dough. Dough made from whole grain can be lank and heavy, leading to dry, crumbly bread with poor volume. By using the commercial cellulase preparation Celluclast (Novozymes) these problems can be overcome. This cellulase hydrolyzes the cellulose naturally present in the bran. During hydrolysis, water is released and this water could help form a better gluten networking. The result is better dough handling during production, increased volume, more uniform crumb structure, softness and bloom [96].

BakeZyme® WholeGain is an enzyme solution based on cellulase launched by DSM (Netherlands). This preparation is specially designed for high fiber breads. By using BakeZyme WholeGain the bread has a better volume, texture, and taste. The cellulase in the preparation breaks down the cellulose fibrils and enables better gluten development, increased dough tolerance, and improved proofing stability. Another innovation from DSM is called BakeZyme X-pan. This enzyme is a fungal cellulase solution which partially degrades non-starch polysaccharides—such as cellulose—in dough. This action helps solve common manufacturing problems in white bread-making, such as irregular crumbs and reduced volume—factors that can often dissuade consumers from purchasing. Extensive tests prove that BakeZyme X-pan enhances product quality and texture in a wide variety of white bread applications, including French baguettes and batards as well as American sandwich bread [97].

DSM offers a huge variety of hemicellulases suitable for application in the bakery industry. The commercial preparation can be found under the following commercial names: BakeZyme HS10.000 BG, BakeZyme HS20.000 BG, BakeZyme Multicell, BakeZyme Real-X, BakeZyme X-cell, and BakeZyme ARA 10.000 [98]. The main purpose of these hemicellulases is to improve the quality of the bread.

An adequate choice of enzymes for improving the bread characteristics is very important in the bread-making industry. For example, Tamba-Berehoiu et al. (2004) [113] tested the improved baking qualities of some wheat flour samples by using commercial enzymatic preparations and their mixtures (amylase, lipase, glucose oxydase, different types of hemicellulase). The results showed that the best increase in bread volume was a combination of well-balanced quantities of Bakezyme BXP 5000 (xylanase) and Bakezyme HSP 6000 (hemicellulase) while the quantities of amylase, lipase, glucose oxydase, and ascorbic acid were maintained constant [113].

5.4.2.5 Beverages

For increasing the yield of juice from apples, pears, etc. and the yield for peeling citrus and other fruit, a variety of commercial enzymatic preparations with cellulases are available on the market. Cellulases are highly specific biocatalysts that work in synergy to release sugars, and due to this characteristic they are often found together with pectinases in commercial enzyme preparations [101]. During juice processing, macerating enzymes are used after crushing the fruits and vegetables and after juice extraction. Macerating enzymes can hydrolyze the cell walls and soluble pectins of fruits and vegetables, helping to reduce juice viscosity and release more flavor, proteins, polysaccharides, and enzymes during processing, while maintaining the textures of fruit and vegetable juices [114, 115]. Novozymes Pectinex® is a mixture of several types of pectinases, designed for treating fruit and vegetable meshes and the maceration of plant tissues. The main enzymes are pectintranseliminase, polygalacturonase and pectinesterase. As a side activity, Pectinex also contains small amounts of hemicellulases and cellulases. These enzymes exhibit optimum activity around pH 4.5, at 50 °C. An improved version of Pectinex is Pectinex Ultra Mash, specially designed for improving the juice yield in the range of 85%–90% and smoother processing [99]. This enzymatic solution contains various pectinoltic enyzmes and hemicellulases for obtaining clearer juices. Pectinase treatment of carrot juice processing with Pectinex Smash XXL and Pectinex Ultra SP-L can improve the juice yield by 15% and also enhance total carotene content in the juice by 1.8 times [116]. A combination of Pectinex Ultra Clear and Lallyzme Beta® (a blend of pectinase and β-glucosidase) was tested for Concord grape juice extraction [117]. The synergistic effect between these two commercial preparations improved the juice yield and bioactive compound (phenolic compound) extraction compared with each preparation individually. The complexity of pectinolytic enzymes from Pectinex Ultra Clear allowed better hydrolysis of the pectic complex in the middle lamella, facilitating the access of cellulases from Lallzyme® Beta in the cell wall of the grape cells [117].

DSM developed an enzymatic solution called Rapidase® Fiber for improving yield, quality, clarification, and enzymatic filter cleaning. This is a liquid pectinase enriched in arabinolytic activities from *A. niger* and cellulolytic activities from *Trichoderma longibrachiatum*. This enzymatic preparation is developed to process juice with high colloid content. It improves the filtration rate and maintains and recovers filter membrane performance. Rapidase Fiber could be used for ultrafiltration for filter cleaning in apple juice processing, in colored fruit processing, in stone fruit puree production, vegetable juice extraction, olive oil extraction, stone fruit clear juice production, pineapple mill juice, and pear juice extraction [100].

Rochapect® 10L is another great enzymatic preparation from AB Enzymes. It is commercialized for the extraction and clarification of fruit juices and its main components are pectinolytic (pectinase, pectin lyase, polygalacturonase, and pectin methyl esterase) and cellulolytic enzymes [101]. All these enzymes have maximum activity and stability at pH 4 while at pH 10 only cellulase is reasonably stable [101]. This enzyme is commonly used for the depectinization of apple juice and citrus juice. It is also suitable for mash and fruit treatment of acidic fruits. It improves clarification and filtration. Rohapect 10L can also be used to increase the extraction of juice from raw apples, to increase processing efficiency during pressing, solid settling or removal, and to generate a final product that is clear and visually attractive [118].

In many large breweries, industrial enzymes are added to control the brewing process and produce consistent, high-quality beer [119]. Ultraflo® Max (Novozymes) is a high-performing enzyme mixture that contains β-glucanase (hemicellulases) and arabinoxylanase. This enyzme degrades β-glucans and arabinoxylans into non-viscous polysaccharides and thus increase extract run-off of a more concentrated wort. By using filtration enzymes to break down these components in the malt, the viscosity is reduced and the beer filtration is improved. Ultraflo Max helps to maximize the flexibility of mashing temperature profile, to increase yields during high/very high gravity brewing, and to minimize brew house and beer filtration capacity cost [104]. AB Enzymes launched two commercial preparations Rohament® CL liquid and Rohament CEP dry preparation with cellulases for hydrolyzing non-starch polysaccharides during brewing [105].

5.4.2.6 Edible Oils

For vegetable oil production (olive, kernel, palm, sunflower seed, coconut, etc.), the organic solvent hexane, a potential carcinogenic, that was traditionally used for extraction has been replaced with cell wall degrading enzymes [115]. The addition of enzymes improves the yield and quality of the oil and plant capacity, without negative effects on the oil. The improved quality of the oil was detected by increasing the extraction of phenolic compounds that have antioxidant activity. In olive oil processing, which includes milling, mashing (malaxation), and separation of the obtained phases, the enzymes are applied in the malaxation step. Olivex (Novozymes) is the first commercial enzyme preparation (produced by the fungus *Aspergillus aculeatus*) rich in pectinolytic and hemicellulolytic activities with a low level of cellulase and was used for improving the extraction of olive oil [102]. Pectinex Ultra Olio and Pectinex Ultra SPL, both enzyme solutions from Novozymes, an improved version of Olivex, are also produced by the fungus *Aspergillus*. These enzyme activities are naturally present in small amounts in the olive cells and are responsible for softening of the fruit during maturation. By using these enzymes for oil processing, several benefits are achieved: mixture improved production process, increased oil yield (5%–10%), quicker and better oil clarification, and no change in oil quality [103].

5.4.2.7 Animal Feed Industry

In the animal feed industry, cellulases are used as feed additive proteases for improving the meat quality. Roxazyme® G2 (DSM, Netherlands) is an enzyme complex that improves energy utilization in poultry. This enzyme preparation is used as an additive to feed for broiler chickens, fattening turkeys, ducks, etc., for their fast growth and better meat [106]. Supplementation of animal feed with Roxazyme G2 leads to higher weights of broiler chickens and reduces the consumption of the basic broiler's fed [107].

5.5 Microbial Fermentations for Cellulase Production

In the last few decades there has been a great expansion among researchers for isolating and selecting novel microorganisms, potential producers of industrially important enzymes, especially ones that play a significant role in biofuel production [1, 120–123].

Although fungi are the most frequently reported as hydrolytic enzyme producers in the literature [4, 121, 124], there is an increasing number of newly isolated bacterial strains that produce various hydrolytic enzymes, including cellulases [23, 120, 122, 125–129]. Current research in this area includes constant screening of potential sources of cellulolytic enzymes in the interest of successful bioconversion of lignocellulosic biomass. Some of the bacterial strains isolated in the last 5 years are presented in Table 5.3.

The production of bacterial cellulases can be carried out by solid state fermentation (SsF) or submerged fermentation (SmF). SsF is a fermentation on a solid substrate (usually lignocellulose) without free water, very similar to the natural growth of microorganisms, while SmF fermentation is fermentation in a liquid medium [4, 10, 11, 135].

The soil bacteria that belong to the genera *Bacillus*, *Paenibacillus*, *Streptomyces* are most frequently reported to produce different types of extracellular enzymes, specifically cellulases [23, 120, 122, 123, 125, 130, 131, 133, 136–138]. For example, the soil isolate *Paenibacillus chitinolyticus* CKS1 was able to grow on medicinal herb waste biomass and sawdust, respectively, and to produce exoglucanase (Avicelase), CMCase (endoglucanase), and FPase with various activities (Table 5.3). The strain CKS1 is the first reported *Paenibacillus* sp. that predominantly produces Avicelase. The predominant exoglucanase or Avicelase has a high temperature optimum at 80 °C and good Avicelase thermostability. This characteristic is desirable for harsh industrial processes [23]. Further, the cellulases produced by the same strain, CKS1, were used for immobilization on spent coffee grounds by Buntić et al. [139]. An immobilized cellulase preparation expressed very good Avicelase activity, reaching up to 2.67 U g^{-1}, and revealed an improved storage stability property compared to a free enzyme sample [139]. Also, another bacterial isolate *Sinorhizobium meliloti* 224 was reported as the first strain belonging to the genus *Sinorhizobium* capable of producing endoglucanase and exoglucanase while growing on agro-industrial waste material (waste tobacco) [127, 140]. *Streptomyces* species were also proved to be very potent producers of a wide palette of enzymes simultaneously; they found a purpose in many relevant biotechnological processes, where the marketable cost-effectiveness was particularly emphasized [127, 141, 142]. This group of microorganisms possess minimal nutritional requirements for metabolic activity expressions, while their enzymes exert a strong biodegradation affinity toward certain compounds.

In microorganisms, cellulase production is inducible. It depends on several factors such as the source of carbon, mainly the type of lignocellulose biomass, and physical process parameters (incubation time, inoculum concentration, temperature, pH, agitation speed, etc.). Optimization of these parameters could significantly improve cellulase activity. A current literature survey showed that, in most studies, for optimization of microbial enzyme production a statistical tool–response surface methodology (RSM) was used [1, 125–127, 130, 137, 138, 143, 144]. By this method, a cellulase yield could be increased several times compared to non-optimized media. For instance, a 3-fold increase in cellulase production with optimized media (carboxymethyl cellulose [CMC] 19.21 g L^{-1},

Table 5.3 Some of the recent isolated bacterial strains that produce cellulases.

Bacterial strain and source of isolation	Type of fermentation/substrate	Fermentation conditions	Enzymes	Reference
Paenibacillus chitinolyticus CKS1 from soil sample	SmF/[a]Medicinal herbs waste biomass; [b]sawdust	0.1% (w/v) of medicinal herbs waste in 3 g L^{-1} of yeast extract and 5.0 g L^{-1} of casein hydrolysate in 0.1 M phosphate buffer pH 7, 48 h, 30 °C	Avicelase, U mL^{-1} (80 °C, pH 4.8), [a]1.94 and [b]1.75; CMCase, U mL^{-1} (50 °C, pH 4.8), [a]0.203 and [b]0.145; FPase, U mL^{-1} (50 °C, pH 4.8), [a]0.092 and [b]0.063	[23]
Sinorhizobium meliloti 224, undefined source	SmF/[c]Unmodified tobacco waste; SsF/[d]Tobacco waste modified with NaOH	[a]5 g L^{-1} of the substrate in YMB (without mannitol) medium, 2 days and 9% inoculum [b]1.0 g of sterile tobacco waste (moistured with water in ratio 1:3 w/v) and 10% inoculum, 2 days, 28 °C	Endoglucanase (CMCase) [c]0.405 U mL^{-1} and [d]1.503 U g^{-1} Exoglucanase (Avicelase) [a]0.131 U mL^{-1} and [b]1.615 U g^{-1}	[127]
Bacillus velezensis ASN1 from soil sample from the animal farm	SmF/Waste office paper (WOP)	(g L^{-1}) WOP- 9; $NaNO_3$ 0.35; KH_2PO_4 0.5; $MgSO_4$ 0.2; NaCl 1.5; yeast extract 0.5; KCl 1.5; Tween 80 1; pH 4.72; inoculum load 6.56%, 30 °C, 70 h	FPase 2.46 U mL^{-1}	[130]
Bacillus licheniformis AU01 from marine sediments	SmF/Rice bran	2% rice bran, 0.25% peptone, 0.1% K_2HPO_4, 0.05% $MgSO_4 \cdot 7H_2O$, pH 9	CMCase 188 U mL^{-1}	[131]
Aneurinibacillus aneurinilyticus BKT-9 from urban Himalayan fresh water	SmF/Commercial CMC	Tryptone 1.6%, yeast extract 1%, NaCl 0.6%, 11.1 g L^{-1}, NH_4Cl 0.77 g L^{-1}, $MnSO_4$ 0.49%, CMC 0.5%	CMCase 83.1 U L^{-1}	[129]

Bacillus sp. BMP01 and *Ochrobactrum oryzae* BMP03 from termite gut microbiomes of *Cryptotermes brevis* species	SmF	No available data	CMCase 0.25 U mL^{-1} Xylanase 0.21 U mL^{-1}	[132]
Bacillus cereus JD0404 from muddy sediments of mangrove swamps	SmF/Commercial CMC	CMC broth at 28 °C, 48 h	CMCase, Avicelase β-glucosidase, U mL^{-1} 1.778 ± 0.003, 0.079 ± 0.001, 0.048 ± 0.002	[133]
Geobacillus sp. HTA426 from soil samples from hot spring	SmF/[e]sugarcane bagasse, [f]rice straw, [g]water hyacinth	Mandel and Reese medium +1% substrate	CMCase, U mL^{-1} [e]103.67, [f]74.70, [g]55.10	[134]

Abbreviations: CMC, carboxymethyl cellulose; FP, filter paper; SmF, submerged fermentation; SsF, solid state fermentation; YMB, yeast mannitol broth.

[a] Medicinal herbs waste biomass;

[b] sawdust;

[c] unmodified tobacco waste;

[d] SsF/tobacco waste modified with NaOH;

[e] sugarcane bagasse,

[f] rice straw,

[g] water hyacinth.

$CaCl_2 \cdot 6H_2O$ 25.06 mg L^{-1}, Tween-20 2.96 mL L^{-1}, and temperature 43.35 °C), with a final value of cellulase activity of 40.88 IU mL^{-1} using *Bacillus licheniformis* NCIM 5556, was reported by Shajahan et al. [144]. An approximately 60-fold increase in cellulase activity compared to unoptimized culture medium was reported using bacterial isolate *Aneurinibacillus aneurinilyticus* BKT-9. In this study, the authors used a central composite design (CCD), under the RSM, for medium optimization (glucose 11.1 g L^{-1}, ammonium chloride 0.77 g L^{-1}, manganese sulfate 0.49% w/v, and CMC 0.5% w/v) by the strain BKT-9, reaching a 76.370 IU L^{-1} of cellulase activity [129]. By applying the CCD for optimizing the medium for SsF (ratio wheat bran to water 1:2, inoculum size 14% and temperature 45 °C, 72 h of incubation time), by *Bacillus subtilis* PJK6, cellulase activity resulted in a 58% increase from unoptimized activity (0.98 U mL^{-1} vs. 1.14 IU mL^{-1}) [145]. Moreover, growing *B. subtilis* G_2 in an optimized medium that consisted of banana peel 3.00% (w/v), bamboo leaves 1.50% (w/v), cotton seed cake 1.50% (w/v), and spirulina 1.50% (w/v), Vaid and Bajaj (2017) obtained a cellulase yield of 3.753 IU mL^{-1} which was 119.47% higher than the cellulase activity of 1.710 IU mL^{-1} in the non-optimized medium [146].

The literature data also showed that in-house produced bacterial cellulases are suitable for lignocellulose hydrolysis. Some isolated strains produce only endoglucanase or endoglucanase with exoglucanase, while others could produce cellulosic cocktails (a mixture of crude enzymes). Some of the recently reported bacterial strains, their cellulases activities and conditions for lignocellulose hydrolysis are presented in Table 5.4.

Nevertheless, for involving the cellulases in industrial processes, additional efforts should be conducted by focusing on development of possible techniques for improving of these enzymes functional characteristics. One of the examples is the immobilization of commercial cellulase preparation from *T. reesei* (Sigma Aldrich) on spent coffee. The authors of this study showed that the coffee surface is an excellent base for increasing the cellulase availability to the substrate and enhancing enzyme productivity, offering new perspectives to the industrial sector [148].

5.6 Conclusion

A sustainable biorefinery concept represents a biomass conversion process for producing marketable products and energy within the recycling and remediation approach. For ensuring biorefinery development, the specific goal is to develop a promising biomass conversion technology. Thereby, in recent years, enzymatic conversion-based lignocellulose biorefining is becoming an increasingly attractive model. Many efforts have been made to stimulate and establish a breakthrough for economically viable enzymatic hydrolysis. However, some difficulties still remain, mainly related to the financial feasibility of this process. The costs of enzyme production are among the most meaningful, limiting the sustainability of the overall process, and their reduction is particularly urgent. Because of that, recent advances by researchers within public institutions, but also from leading companies in enzyme manufacturing, are focused on improving the routes for obtaining enzymes. These include constant screening for novel and potent microorganisms, producers of industrially important enzymes and enzyme cocktails, with improved properties, aiming at

Table 5.4 Some of the recent reported bacterial isolates with their cellulase activities involved in lignocellulose hydrolysis.

Bacterial isolate and source	Substrate/fermentation conditions	Cellulases (description)	Hydrolysis conditions	Reducing sugars	Reference
[a]*Bacillus mojavensis* AG21, [b]*Bacillus vallismortis* J77, [c]*Bacillus sonorensis* 7.26, [d]*Bacillus safiensis* AJ49 from desert environments in Tunisia	*Stevia rebaudiana* leaves (SRL) /SmF 50 mL of basal medium with 1%–3% of SRL at various pH (7–10), salinities (0%–10%), and temperatures (24–50 °C), 4 days, 120 rpm	Endoglucanase (CMCase) [a]1890 IU L^{-1} [b]1642 IU L^{-1} [c]1516 IU L^{-1} [d]1455 IU L^{-1}	L/S ratio 50 mL g^{-1}, enzyme concentration 4.4 U g^{-1}, and reaction time 2 h	mg L^{-1} [a]3620 ± 181 [b]3040 ± 164 [c]3656 ± 164 [d]2654 ± 106	[137]
Paenibacillus chitinolyticus CKS1 from soil, foot of the Austrian Alps	Barley bran/SmF casein hydrolysate 5 g L^{-1}; yeast extract 2 g L^{-1}; barley bran 4%	Mix of crude enzymes (U mL^{-1}): CMCase 0.405; Avicelase 0.433	72 h of fermentation at 50 °C with 1.33% of cotton fiber, 37.5 mL of crude bacterial supernatant and 37.5 mL of 0.1 M acetate buffer pH 4.8, 150 rpm	0.117 g glucose g^{-1} of cotton material	[138]
Streptomyces fulvissimus CKS7 from soil, foot of the Austrian Alps	Rye bran/SsF 10 g of rye bran moistened with water (1:2), 10% inoculum, 6 days solid state fermentation at 30 °C	Mix of crude enzymes (U g^{-1}): endoglucanase (CMCase) 8.62 ± 0.08; exoglucanase (Avicelase) 5.98 ± 0.22, pectinase 10.37 ± 0.15; xylanase 11.88 ± 0.23	5% of corn stover, 1% of horsetail, 1% yellow gentian, 1% of cotton fiber, 1% of corona pretreated cotton fiber + mix of crude enzymes obtained during SsF with 0.1 M acetate buffer (1:1 v/v)	mg mL^{-1} 1.89 ± 0.11 corn stover; 2.55 ± 0.09 horsetail (*Equisetum arvense*); 1.81 ± 0.10 yellow gentian (*Gentiana lutea*); 1.87 ± 0.09 cotton fabric; 2.26 ± 0.08 corona treated cotton fabric	[122]
Streptomyces sp. CC48 from submerged vegetation	Commercial CMC/80 mL of minimum medium [1% CMC, 0.05% yeast extract, 0.25% $(NH_4)_2SO_4$, 0.27% KH_2PO_4, 0.53% Na_2HPO_4, 0.02% NaCl, 0.02% $MgSO_4{\cdot}H_2O$]	CMCase 0.76 U mL^{-1} at pH 7.0, 60 °C	0.5 mL of the crude enzyme with 4.5 mL of 50 mM phosphate buffer (pH 7.0) and 4% of alkaline-pretreated substrates	Corn cob 320 mg g^{-1} wheat straw 330 mg g^{-1}	[123]

(*Continued*)

Table 5.4 (Continued)

Bacterial isolate and source	Substrate/fermentation conditions	Cellulases (description)	Hydrolysis conditions	Reducing sugars	Reference
Streptomyces sp. MDS, undefined source	Rice waste biomass (RWB)/5.0 g of RWB moistened with Dubos salt medium, 30 °C, pH 5.5, medium moisture content 1:1.5 and RWB particle size 0.5 mm	$U\ g^{-1}$ endoglucanase (132.6 ± 1.15), exoglucanase (14.6 ± 0.62), cellobiase (125.6 ± 1.75), filter paperase (19.7 ± 0.42), amylase (278.5 ± 3.53), and xylanase (342.5 ± 3.36)	RWB loading 2.0% (w/v) in 50 mM citrate phosphate buffer (pH 5.0) with 0.005% (w/v) sodium azide	465 ± 3.45 $mg\ g^{-1}$	[147]
Galactomyces sp. CCZU11-1, undefined source	Chestnut shell (CNS)/CNS 5 $g\ L^{-1}$, $(NH_4)_2SO_4$ 5 $g\ L^{-1}$, $KHPO_4$ 2 $g\ L^{-1}$, $MgSO_4{\cdot}7H_2O$ 0.5 $g\ L^{-1}$, NaCl 1 $g\ L^{-1}$, Tween-80 8 $g\ L^{-1}$, pH 5.4, 30 °C, 180 rpm, 72 h. Later, sterile-filtered IL 0–30% (w/v) was added and cultured for 72 h	FPase (28.6 $U\ mL^{-1}$), xylanase (186.2 $U\ mL^{-1}$), and CMCase (107.3 $U\ mL^{-1}$)	IL of 20% (w/v), a cellulase loading of 20 FPU g^{-1} CNS	325.28 μg reducing sugars per minute	[128]

Abbreviations: CMC, carboxymethyl cellulose; FP, filter paper; IL, ionic liquid; L/S, liquid/solid; SmF, submerged fermentation; SRL, *Stevia rebaudiana* leaves; SsF, solid state fermentation.

[a] *Bacillus mojavensis* AG21;

[b] *Bacillus vallismortis* J77;

[c] *Bacillus sonorensis* 7.26;

[d] *Bacillus safiensis* AJ49

achievements of high hydrolysis performance and cost lowering. Thanks to those activities, an important biobased product, like cellulosic ethanol, is on the way to large-scale commercialization, primarily in the United States, Europe, and China, but also followed by many other parts of the world.

Acknowledgement

This work was supported by the Ministry of Education, Science and Technological Development of the Republic of Serbia (Contract No. 451-03-68/2022-14/200135).

References

1 Mihajlovski, K., Radovanović, Ž., Carević, M., and Dimitrijević-Branković, S. (2018). Valorization of damaged rice grains: optimization of bioethanol production by waste brewer's yeast using an amylolytic potential from the *Paenibacillus chitinolyticus* CKS1. *Fuel.* 224: 591–599.
2 Ahorsu, R., Medina, F., and Constantí, M. (2018). Significance and challenges of biomass as a suitable feedstock for bioenergy and biochemical production: a review. *Energies.* 11 (12): 3366.
3 Bhatia, L., Garlapati, V.K., and Chandel, A.K. (2019). Scalable technologies for lignocellulosic biomass processing into cellulosic ethanol. *Horizons. Bioprocess. Eng.: Springer* 73–90.
4 Siqueira, J.G.W., Rodrigues, C., de Souza Vandenberghe, L.P. et al. (2020). Current advances in on-site cellulase production and application on lignocellulosic biomass conversion to biofuels: a review. *Biomass Bioenergy* 132: 105419.
5 Chandel, A.K., Albarelli, J.Q., Santos, D.T. et al. (2019). Comparative analysis of key technologies for cellulosic ethanol production from Brazilian sugarcane bagasse at a commercial scale. *Biofuels, Bioprod. Biorefin.* 13 (4): 994–1014.
6 Padella, M., O'Connell, A., and Prussi, M. (2019). What is still limiting the deployment of cellulosic ethanol? Analysis of the current status of the sector. *App. Sci.* 9 (21): 4523.
7 Srivastava, N., Srivastava, M., Mishra, P. et al. (2015). Application of cellulases in biofuels industries: an overview. *J. Biofuel. Bioener.* 1 (1): 55–63.
8 Jayasekara, S. and Ratnayake, R. (2019). *Microbial cellulases: an overview and applications.* Cellulose: Intechopen.
9 Chandel, A.K., Chandrasekhar, G., Silva, M.B., and Silvério da Silva, S. (2012). The realm of cellulases in biorefinery development. *Crit. Rev. Biotechnol.* 32 (3): 187–202.
10 Roth, J.C.G., Hoeltz, M., and Benitez, L.B. (2020). Current approaches and trends in the production of microbial cellulases using residual lignocellulosic biomass: a bibliometric analysis of the last 10 years. *Arch. Microbiol.* 1-17.
11 Mohanram, S., Amat, D., Choudhary, J. et al. (2013). Novel perspectives for evolving enzyme cocktails for lignocellulose hydrolysis in biorefineries. *Sustainable Chem. Processes* 1 (1): 15.

12 Sharma, A., Tewari, R., Rana, S.S. et al. (2016). Cellulases: classification, methods of determination and industrial applications. *Appl. Biochem. Biotechnol.* 179 (8): 1346–1380.

13 Kumar, R., Singh, S., and Singh, O.V. (2008). Bioconversion of lignocellulosic biomass: biochemical and molecular perspectives. *J. Ind. Microbiol. Biotechnol.* 35 (5): 377–391.

14 Abdeshahian, P., Kadier, A., Rai, P.K., and da Silva, S.S. (2020). Lignocellulose as a renewable carbon source for microbial synthesis of different enzymes. *Lignocell. Bioref. Technol.* 185–202.

15 Zeng, Y., Himmel, M.E., and Ding, S.-Y. (2017). Visualizing chemical functionality in plant cell walls. *Biotechnol. Biofuels* 10 (1): 1–16.

16 de Vries, R.P. and Visser, J. (2001). Aspergillus enzymes involved in degradation of plant cell wall polysaccharides. *Microbiol. Mol. Biol. Rev.* 65 (4): 497–522.

17 Shah, A., Patel, H., and Narra, M. (2017). Bioproduction of fungal cellulases and hemicellulases through solid state fermentation. *Fungal. Metabol.* 349–393.

18 Kallioinen, A. (2014). *Development of Pretreatment Technology and Enzymatic Hydrolysis for Biorefineries.* Espoo: VTT Technical Research Centre of Finland.

19 Silveira, M.H.L., Siika-Aho, M., Kruus, K. et al. (2014). The essential role of plant cell wall degrading enzymes in the success of biorefineries: current status and future challenges. In: *Biofuel* (ed. S. da Silva and A. Chandel), 151–172. Brazil: Berlin, Springer.

20 Øbro, J., Hayashi, T., and Mikkelsen, J.D. (2018). Enzymatic modification of plant cell wall polysaccharides. *Ann. Plant. Rev. Online.* 367–387.

21 Sibilla, F. and de María, P.D. (2013). Integrating white biotechnology in lignocellulosic biomass transformations: from enzyme-catalysis to metabolic engineering. In: *The Role of Catalysis for The Sustainable Production of Bio-fuels and Bio-chemicals* (ed. K.S. Triantafyllidis, A.A. Lappas and M. Stöcker), 445–466. Amsterdam: Elsevier.

22 Gupta, M.N. and Bisaria, V.S. (2018). Stable cellulolytic enzymes and their application in hydrolysis of lignocellulosic biomass. *Biotechnol. J.* 13 (6): 1700633.

23 Mihajlovski, K.R., Carević, M.B., Dević, M.L. et al. (2015). Lignocellulosic waste material as substrate for Avicelase production by a new strain of *Paenibacillus chitinolyticus* CKS1. *Int. Biodeterior. Biodegrad.* 104: 426–434.

24 Janusz, G., Pawlik, A., Sulej, J. et al. (2017). Lignin degradation: microorganisms, enzymes involved, genomes analysis and evolution. *FEMS Microbiol. Rev.* 41 (6): 941–962.

25 Novozymes (2013). https://www.novozymes.com/-/media/Project/Novozymes/Website/website/document-library/Brochures_Downloads/General-materials/Enzymes_at_work.pdf (accessed 01 November 2020).

26 Silva, C.O., Vaz, R.P., and Filho, E.X. (2018). Bringing plant cell-wall-degrading enzymes into the lignocellulosic biorefinery concept. *Biofuels, Bioprod. Biorefin.* 12 (2): 277–289.

27 Nahar, N. (2017). Processing trade-offs in a cellulosic biorefinery. Doctoral Dissertation, North Dakota State University, Fargo, North Dakota.

28 Chandel, A.K., Chandrasekhar, G., Silva, M.B., and Silvério da Silva, S. (2012). The realm of cellulases in biorefinery development. Critical reviews. *Biotechnol.* 32 (3): 187–202.

29 Novozymes. http://www.shinshu-u.ac.jp/faculty/engineering/chair/chem010/manual/Ctec3.pdf (accessed 01 Novmber 2020).

30 Chandel, A.K., Garlapati, V.K., Singh, A.K. et al. (2018). The path forward for lignocellulose biorefineries: bottlenecks, solutions, and perspective on commercialization. *Bioresour. Technol.* 264: 370–381.

31 National Renewable Energy Laboratory (Golden, Colorado, United States) (2010). https://www.nrel.gov/docs/fy10osti/47572.pdf. (accessed 13 November 2020).

32 Johnson, E. (2016). Integrated enzyme production lowers the cost of cellulosic ethanol. *Biofuels, Bioprod. Biorefin.* 10 (2): 164–174.

33 IEA Bioenergy. (2018). http://www.ieabioenergy.com/wp-content/uploads/2018/02/3-Crescentino-AdvancedEthanolBiorefinery_IT_Final.pdf. (accessed 15 October 2020).

34 Chandel, A.K., Da Silva, S.S., and Singh, O.V. (2013). Detoxification of lignocellulose hydrolysates: biochemical and metabolic engineering toward white biotechnology. *Bioenergy Res.* 6 (1): 388–401.

35 Zhang, Q. and Bao, J. (2017). Industrial cellulase performance in the simultaneous saccharification and co-fermentation (SSCF) of corn stover for high-titer ethanol production. 4 (1): 1–7.

36 Patel-Predd, P. (2006). *Overcoming the hurdles to producing ethanol from cellulose*, 4052–4053. ACS Publications.

37 NCBE (2016). http://www.ncbe.reading.ac.uk/MATERIALS/Enzymes/celluclast.html (accessed 20 October 2020).

38 Novozymes. https://www.novozymes.com/en/advance-your-business/bioenergy/cellic (accessed 20 October 2020).

39 Novozymes. (2010). https://www.novozymes.com/en/news/news-archive/2010/02/45713. (accessed 20 October 2020).

40 Novozymes (2010). CELLIC Ctec and Htec2—Enzymes for Hydrolysis of Lignocellulosic Materials. Novozymes A/S, Luna. (2010–01668):01.

41 Chiarello, L.M., Ramos, C.E.A., Neves, P.V., and Ramos, L.P. (2016). Production of cellulosic ethanol from steam-exploded *Eucalyptus urograndis* and sugarcane bagasse at high total solids and low enzyme loadings. *Sustainable Chem. Processes* 4 (1): 15–24.

42 Fockink, D.H., Urio, M.B., Sánchez, J.H., and Ramos, L.P. (2017). Enzymatic hydrolysis of steam-treated sugarcane bagasse: effect of enzyme loading and substrate total solids on its fractal kinetic modeling and rheological properties. *Energy Fuels* 31 (6): 6211–6220.

43 Rosales-Calderon, O., Trajano, H.L., and Duff, S.J. (2014). Stability of commercial glucanase and β-glucosidase preparations under hydrolysis conditions. *PeerJ* 2: e402.

44 METGEN. https://www.metgen.com/products/metzyme-suno. (accessed 27 October 2020).

45 Xu, C.C., Mahmood, N., Yuan, Z. et al. (2015). De-polymerization—an engineering approach to valorization of lignin for bio-based chemicals and materials. *NWBC* 2015: 120.

46 Fujii, T., Fang, X., Inoue, H. et al. (2009). Enzymatic hydrolyzing performance of *Acremonium cellulolyticus* and *Trichoderma reesei* against three lignocellulosic materials. *Biotechnol. Biofuels* 2 (1): 24.

47 Dupont (2013). http://www.shinshu-u.ac.jp/faculty/engineering/chair/chem010/manual/accellerase1500_Dupont.pdf (accessed 15 September 2020).

48 GENENCOR (2010). http://www.genencor.com/fileadmin/user_upload/genencor/documents/accellerase_duet_product_info_sheet.pdf. (accessed 28 October 2020).

49 Smit, A. and Huijgen, W. (2017). Effective fractionation of lignocellulose in herbaceous biomass and hardwood using a mild acetone organosolv process. *Green Chem.* 19 (22): 5505–5514.

50 GENENCOR (2011). http://www.genencor.com/fileadmin/user_upload/genencor/documents/TRIO_ProductSheet_LowRes.pdf (accessed 28 October 2020).

51 GENENCOR (2009). http://www.genencor.com/fileadmin/user_upload/genencor/documents/ACCELLERASE_R__BG_0020209.pdf (accessed 28 October 2020).

52 NCBE (2016) http://www.ncbe.reading.ac.uk/MATERIALS/Enzymes/viscozyme.html (accessed 20 October 2020).

53 Novozymes (2005). https://www.novozymes.com/-/media/Project/Novozymes/Website/website/advance-your-business/05_L2_Bioenergy/Benefit-sheets/BS-Viscozyme-2005-34054-03_.pdf (accessed 20 October 2020).

54 Wang, D., Sun, J., Yu, H.-L. et al. (2012). Maximum saccharification of cellulose complex by an enzyme cocktail supplemented with cellulase from newly isolated *Aspergillus fumigatus* ECU0811. *Appl. Biochem. Biotechnol.* 166 (1): 176–186.

55 Prior, B.A. and Day, D.F. (2008). Hydrolysis of Ammonia-Pretreated Sugar Cane Bagasse with Cellulase, β-Glucosidase, and Hemicellulase Preparations. *Appl Biochem Biotechnol.* 146: 151–164.

56 Novozymes (2010). https://repositorio.unal.edu.co/bitstream/handle/unal/51813/2822323.2014%20Anexo.pdf?sequence=1&isAllowed=y (accessed 22 February 2022)

57 Buzała, K., Przybysz, P., Rosicka-Kaczmarek, J., and Kalinowska, H. (2015). Production of glucose-rich enzymatic hydrolysates from cellulosic pulps. *Cellulose.* 22 (1): 663–674.

58 Fehér, C., Gál, B., Fehér, A. et al. (2015). Investigation of commercial enzyme preparations for selective release of arabinose from corn fibre. *J. Chem. Technol. Biotechnol.* 90 (7): 1329–1337.

59 Rajan, K. and Carrier, D.J. (2014). Characterization of rice straw prehydrolyzates and their effect on the hydrolysis of model substrates using a commercial endo-cellulase, β-glucosidase and cellulase cocktail. *ACS Sustainable Chem. Eng.* 2 (9): 2124–2130.

60 Fehér, C., Gál, B., Fehér, A. et al. (2015). Investigation of commercial enzyme preparations for selective release of arabinose from corn fibre. *J. Chem. Technol. Biotechnol.* 90 (7): 1329–1337.

61 Novozymes https://cms.it.gu.se/infoglueDeliverWorking/digitalAssets/1608/1608318_enzyme-information.pdf (accessed 22 February 2022).

62 Kao, M.-R., Kuo, H.-W., Lee, C.-C. et al. (2019). *Chaetomella raphigera* β-glucosidase D2-BGL has intriguing structural features and a high substrate affinity that renders it an efficient cellulase supplement for lignocellulosic biomass hydrolysis. *Biotechnol. Biofuels* 12 (1): 1–18.

63 Rosgaard, L., Andric, P., Dam-Johansen, K. et al. (2007). Effects of substrate loading on enzymatic hydrolysis and viscosity of pretreated barley straw. *Appl. Biochem. Biotechnol.* 143 (1): 27–40.

64 Novozymes https://www.novozymes.com/en/advance-your-business/bioenergy/viscozyme. (accessed 20 October 2020).

65 ISAAA (2009). http://www.isaaa.org/kc/cropbiotechupdate/article/default.asp?ID=3851 (accessed 26 October 2020).

66 Sun, F.F., Zhao, X., Hong, J. et al. (2016). Industrially relevant hydrolyzability and fermentability of sugarcane bagasse improved effectively by glycerol organosolv pretreatment. *Biotechnol. Biofuels* 9 (1): 59.

67 Rodrigues, A.C., Haven, M.Ø., Lindedam, J. et al. (2015). Celluclast and Cellic® CTec2: saccharification/fermentation of wheat straw, solid–liquid partition and potential of enzyme recycling by alkaline washing. *Enzyme Microb. Technol.* 79: 70–77.

68 Álvarez, C., Reyes-Sosa, F.M., and Díez, B. (2016). Enzymatic hydrolysis of biomass from wood. *Microb. Biotechnol.* 9 (2): 149–156.

69 Ascencio, J.J., Chandel, A.K., Philippini, R.R., and da Silva, S.S. (2020;10). Comparative study of cellulosic sugars production from sugarcane bagasse after dilute nitric acid, dilute sodium hydroxide and sequential nitric acid-sodium hydroxide pretreatment. *Biomass. Conv. Bioref.* 813–822.

70 Brar, K., Kaur, S., and Chadha, B. (2016). A novel staggered hybrid SSF approach for efficient conversion of cellulose/hemicellulosic fractions of corncob into ethanol. *Renewable Energy* 98: 16–22.

71 Gao, X., Kumar, R., Singh, S. et al. (2014). Comparison of enzymatic reactivity of corn stover solids prepared by dilute acid, AFEX™, and ionic liquid pretreatments. *Biotechnol. Biofuels* 7 (1): 71–84.

72 Tsai, M.H., Lee, W.C., Kuan, W.C. et al. (2018). Evaluation of different pretreatments of Napier grass for enzymatic saccharification and ethanol production. *Ene. Sci. Eng.* 6 (6): 683–692.

73 Teter, S.A. (2012). *DECREASE Final Technical Report: Development of a Commercial Ready Enzyme Application System for Ethanol*. Novozymes, Inc. Davis, CA, USA.

74 López-Linares, J.C., Romero, I., Cara, C. et al. (2014). Bioethanol production from rapeseed straw at high solids loading with different process configurations. *Fuel.* 122: 112–118.

75 Sun, F.F., Hong, J., Hu, J. et al. (2015). Accessory enzymes influence cellulase hydrolysis of the model substrate and the realistic lignocellulosic biomass. *Enzyme Microb. Technol.* 79: 42–48.

76 Novozymes (2012). https://www.yumpu.com/en/document/view/30214569/novozymes-cellicar-htec3 (accessed 30 October 2020).

77 Flores-Gómez, C.A., Silva, E.M.E., Zhong, C. et al. (2018). Conversion of lignocellulosic agave residues into liquid biofuels using an AFEX™-based biorefinery. *Biotechnol. Biofuels* 11 (1): 1–18.

78 Berlin, A., Gilkes, N., Kilburn, D. et al. (2005). Evaluation of novel fungal cellulase preparations for ability to hydrolyze softwood substrates—evidence for the role of accessory enzymes. *Enzyme Microb. Technol.* 37 (2): 175–184.

79 Rosgaard, L., Pedersen, S., Cherry, J.R. et al. (2006). Efficiency of new fungal cellulase systems in boosting enzymatic degradation of barley straw lignocellulose. *Biotechnol. Progr.* 22 (2): 493–498.

80 Biotech https://www.im-biotech.com/enzymes/detergents. (accessed 03 November 2020).

81 Novozymes https://biosolutions.novozymes.com/en/laundry/products/cellucleanr-5000-l (accessed 22 February 2022).

82 Novozymes https://biosolutions.novozymes.com/en/laundry/products/cellucleanr-evityr-4500-t (accessed 22 February 2022).

83 AB Enzymes. https://www.abenzymes.com/en/your-industry/detergents (accessed 28 October 2020).

84 Vermelho, A.B., Supuran, C.T., and Guisan, J.M. (2012). Microbial enzyme: applications in industry and in bioremediation. *Enzyme Res.* 2012.

85 Heikinheimo, L., Buchert, J., Miettinen-Oinonen, A., and Suominen, P. (2000). Treating denim fabrics with *Trichoderma reesei* cellulases. *Text. Res. J.* 70 (11): 969–973.

86 https://www.novozymes.com/en/advance-your-business/textiles (accessed 28 October 2020).

87 Novozymes https://www.abenzymes.com/en/your-industry/textiles (accessed 28 October 2020).
88 Novozymes (2019). https://www.novozymes.com/-/media/Project/Novozymes/Website/website/document-library/Advance-your-business/Forest-Products/Product_Range_Pulp_and_Paper-Product-Range_2019.pdf (accessed 22 February 2022).
89 Novozymes https://www.novozymes.com/en/advance-your-business/forest (accessed 28 October 2020).
90 Novozymes https://www.novozymes.com/-/media/Project/Novozymes/Website/website/document-library/Advance-your-business/Forest-Products/FiberCare_in_refining_fiber_modification1.pdf (accessed 22 February 2022).
91 Novozymes (2019). Forest products.
92 Sarkar, J., Braden, M., and Shah, J. (2005). Enzyme-assisted clarification and dewatering of wastewater; US 2005/0258095 A1 Google Patents.
93 Poshina, D. and Novozhilov, E. (2015). Modification of spruce sulphite pulp by cellulase treatment. *Cellul. Chem. Technol.* 49: 187–194.
94 METGEN https://www.metgen.com/products/metzyme-brila. (accessed 27 October 2020).
95 De Gonzalo, G. and de María, P.D. (2017). *Biocatalysis: An Industrial Perspective* (ed. G. De Gonzalo and P.D. de María). Royal Society of Chemistry, London, United Kingdom.
96 Novozymes https://biosolutions.novozymes.com/en/baking/products/dough-conditioning/celluclast. (accessed 22 February 2022).
97 DSM http://www.dsm-bakingenzymes.com (accessed 29 October 2020).
98 DSM www.dsm.com›foodandbeverages›enUS›documents›DFS (accessed 29 October 2020).
99 Novozymes (2008). https://www.novozymes.com/en/news/news-archive/2008/12/45179 (accessed 29 October 2020).
100 DSM https://www.dsm.com/food-beverage/en_US/ingredients/beverages-and-brewing/beverage/rapidase.html (accessed 22 February 2022).
101 Dal Magro, L., Kornecki, J.F., Klein, M.P. et al. (2019). Stability/activity features of the main enzyme components of Rohapect 10L. *Biotechnol. Progr.* 35 (6): e2877.
102 Iconomou, D., Arapoglou, D., and Israilides, C. (2010). Improvement of phenolic antioxidants and quality characteristics of virgin olive oil with the addition of enzymes and nitrogen during olive paste processing. *Grasas Aceites* 61 (3): 303–311.
103 Novozymes. https://theolivecentre.com/product/Enzyme-Pectinex-Ultra-SPL-1L-3 (accessed 22 February 2022).
104 Novozymes (2007). https://www.novozymes.com/en/news/news-archive/2007/05/43116 (accessd 30 October 2020).
105 AB Enzymes. https://www.abenzymes.com/media/2375/abenzymes_brochure_grainprocessingproducts_jul2018.pdf (accessed 28 October 2020).
106 DSM https://www.dsm.com/markets/anh/en_US/products/products-feedenzymes/products-feed-carbohydrases/products-feed-carbohydrases-roxazyme-g2.html (accessed 29 October 2020).
107 Ogunsipe, M., Adejumo, J., Agbede, J., and Asaniyan, E. (2015). Effect of Roxazyme G2G supplementation on cassava plant meal fed to broiler chickens. *Bone.* 3 (3.50): 3–50.
108 Novozymes (2001). NOVOZYMES—driven by research and scientists. *Nature.* 409 (6817): 268.
109 Biotech https://www.im-biotech.com (accessed 03 November 2020).

110 Uddin, M.G. (2015). Effects of biopolishing on the quality of cotton fabrics using acid and neutral cellulases. *Tex. Clothing. Sustain.* 1 (1): 9–19.

111 Arora, N.K., Mishra, J., and Mishra, V. (2020). *Microbial Enzymes: Roles and Applications in Industries*. Springer, Berlin.

112 Tang YJ, Huang GW, Xia J, Xue GX, Zhang Y, editors. Effect of cellulase enzyme treatment on the pulp beatability and fiber properties. *Adv. Mater. Res.*; 2012;441:746-49.

113 Tamba-Berehoiu, R., Popa, N., Balan, D. et al. (2004). Testing of some enzymatic mixtures used for the improvement of wheat flours. *Roum. Biotechnol. Lett.* 9 (5): 1871–1878.

114 Toushik, S.H., Lee, K.T., Lee, J.S., and Kim, K.S. (2017). Functional applications of lignocellulolytic enzymes in the fruit and vegetable processing industries. *J. Food Sci.* 82 (3): 585–593.

115 Sharada, R., Venkateswarlu, G., Venkateswar, S., and Rao, M.A. (2014). Applications of cellulases—review. *Int. J. Chem. Biol. Sci.* 4 (2): 424–437.

116 Zhou, P. (2003). Application of Pectinex Smash XXL and Pectinex Ultra SP-L in carrot juice processing. *Sci. Technol. Food Indus.* 9: 62–63.

117 Dal Magro, L., Dalagnol, L.M., Manfroi, V. et al. (2016). Synergistic effects of Pectinex Ultra Clear and Lallzyme Beta on yield and bioactive compounds extraction of Concord grape juice. *LWT Food Sci. Technol.* 72: 157–165.

118 World of Chemicals https://www.worldofchemicals.com/chemicals/chemical-properties/rohapect-10l.html (accessed 05 November 2020).

119 Duza, M.B. and Mastan, S. (2013). Microbial enzymes and their applications—a review. *Am. J. Pharm. Res.* 3 (8): 6208–6219.

120 Mihajlovski, K., Rajilić-Stojanović, M., and Dimitrijević-Branković, S. (2020). Enzymatic hydrolysis of waste bread by newly isolated *Hymenobacter* sp. CKS3: statistical optimization and bioethanol production. *Renewable Energy* 152: 627–633.

121 Verma, N., Kumar, V., and Bansal, M. (2020). Valorization of waste biomass in fermentative production of cellulases: a review. *Waste Biomass Valorization* 12: 613–640.

122 Mihajlovski, K., Buntić, A., Milić, M. et al. (2020). From agricultural waste to biofuel: enzymatic potential of a bacterial isolate *Streptomyces fulvissimus* CKS7 for bioethanol production. *Waste Biomass Valorization* 12: 1–10.

123 Celaya-Herrera, S., Casados-Vázquez, L.E., Valdez-Vazquez, I. et al. (2020). A cellulolytic *Streptomyces* sp. isolated from a highly oligotrophic niche shows potential for hydrolyzing agricultural wastes. *Bioenergy Res.* 14: 1–11.

124 Bhatia, L., Chandel, A.K., Singh, A.K., and Singh, O.V. (2018). Biotechnological advances in lignocellulosic ethanol production. In: *Sustainable Biotechnology—Enzymatic Resources of Renewable Energy* (ed. O. Singh and A. Chandel), 57–82. Springer, Berlin.

125 Mihajlovski, K.R., Radovanović, N.R., Veljović, Đ.N. et al. (2016). Improved β-amylase production on molasses and sugar beet pulp by a novel strain *Paenibacillus chitinolyticus* CKS1. *Ind. Crops Prod.* 80: 115–122.

126 Mihajlovski, K.R., Radovanović, N.R., Miljković, M.G. et al. (2015). β-Amylase production from packaging-industry wastewater using a novel strain *Paenibacillus chitinolyticus* CKS 1. *RSC Adv.* 5 (110): 90895–90903.

127 Buntić, A., Milić, M., Stajković-Srbinović, O. et al. (2019). Cellulase production by *Sinorhizobium meliloti* strain 224 using waste tobacco as substrate. *Int. J. Environ. Sci. Technol.* 16 (10): 5881–5890.

128 He, Y.-C., Liu, F., Gong, L. et al. (2016). Enzymatic in situ saccharification of chestnut shell with high ionic liquid-tolerant cellulases from *Galactomyces* sp. CCZU11-1 in a biocompatible ionic liquid-cellulase media. *Bioresour. Technol.* 201: 133–139.

129 Ahmad, T., Sharma, A., Gupta, G. et al. (2020). Response surface optimization of cellulase production from *Aneurinibacillus aneurinilyticus* BKT-9: an isolate of urban Himalayan freshwater. *Saudi J. Biol. Sci.* 27 (9): 2333–2343.

130 Nair, A.S., Al-Battashi, H., Al-Akzawi, A. et al. (2018). Waste office paper: a potential feedstock for cellulase production by a novel strain *Bacillus velezensis* ASN1. *Waste Manage. (Oxford)* 79: 491–500.

131 Annamalai, N., Rajeswari, M.V., Elayaraja, S. et al. (2012). Purification and characterization of thermostable alkaline cellulase from marine bacterium *Bacillus licheniformis* AU01 by utilizing cellulosic wastes. *Waste Biomass Valorization* 3 (3): 305–310.

132 Tsegaye, B., Balomajumder, C., and Roy, P. (2019). Isolation and characterization of novel lignolytic, cellulolytic, and hemicellulolytic bacteria from wood-feeding termite *Cryptotermes brevis*. *Int. Microbiol.* 22 (1): 29–39.

133 Chantarasiri, A. (2015). Aquatic *Bacillus cereus* JD0404 isolated from the muddy sediments of mangrove swamps in Thailand and characterization of its cellulolytic activity. *Egypt. J. Aqu. Res.* 41 (3): 257–264.

134 Potprommanee, L., Wang, X.-Q., Han, Y.-J. et al. (2017). Characterization of a thermophilic cellulase from *Geobacillus* sp. HTA426, an efficient cellulase-producer on alkali pretreated of lignocellulosic biomass. *PLoS One* 12 (4): e0175004.

135 Carboué, Q., Perraud-Gaime, I., Tranier, M., and Roussos, S. (2017). Production of microbial enzymes by solid state fermentation for food applications. In: *Microbial Enzyme Technology in Food Applications* (ed. R.C. Ray and C.M. Rosell), 437–451. Boca Raton, Florida: CRC Press.

136 Radovanović, N., Milutinović, M., Mihajlovski, K. et al. (2018). Biocontrol and plant stimulating potential of novel strain *Bacillus* sp. PPM3 isolated from marine sediment. *Microbial. Patho.* 120: 71–78.

137 Souii, A., Guesmi, A., Ouertani, R. et al. (2020). Carboxymethyl cellulase production by extremotolerant bacteria in low-cost media and application in enzymatic saccharification of stevia biomass. *Waste Biomass Valorization* 11 (5): 2111–2122.

138 Mihajlovski, K.R., Davidović, S.Z., Veljović, Đ.N. et al. (2017). Effective valorisation of barley bran for simultaneous cellulase and β-amylase production by *Paenibacillus chitinolyticus* CKS1: statistical optimization and enzymes application. *J. Serb. Chem. Soc.* 82 (11): 1223–1236.

139 Buntić, A.V., Pavlović, M.D., Antonović, D.G. et al. (2016). Utilization of spent coffee grounds for isolation and stabilization of *Paenibacillus chitinolyticus* CKS1 cellulase by immobilization. *Heliyon.* 2 (8): e00146.

140 Buntić, A.V., Stajković-Srbinović, O.S., Delić, D.I. et al. (2019). The production of cellulase from the waste tobacco residues remaining after polyphenols and nicotine extraction and bacterial pre-treatment. *J. Serb. Chem. Soc.* 84 (2): 129–140.

141 Buntić, A., Pavlović, M., Šiler-Marinković, S., and Dimitrijević-Branković, S. (2016). Biological treatment of colored wastewater by *Streptomyces fulvissimus* CKS 7. *Water Sci. Technol.* 73 (9): 2231–2236.

142 Buntić, A.V., Pavlović, M.D., Antonović, D.G. et al. (2017). A treatment of wastewater containing basic dyes by the use of new strain *Streptomyces microflavus* CKS6. *J. Cleaner Prod.* 148: 347–354.

143 Tabssum, F., Irfan, M., Shakir, H.A., and Qazi, J.I. (2018). RSM based optimization of nutritional conditions for cellulase mediated saccharification by *Bacillus cereus*. *J. Biol. Eng.* 12 (1): 7.

144 Shajahan, S., Moorthy, I.G., Sivakumar, N., and Selvakumar, G. (2017). Statistical modeling and optimization of cellulase production by *Bacillus licheniformis* NCIM 5556 isolated from the hot spring, Maharashtra, India. *J. King Saud. Univ. Sci.* 29 (3): 302–310.

145 Kaur, P., Bhardwaj, S., Bhardwaj, N.K., and Sharma, J. (2018). Lignocellulosic waste as a sole substrate for production of crude cellulase from *Bacillus subtilis* PJK6 under solid state fermentation using statistical approach. *J. Carbohyd.* 1 (1): 1.

146 Vaid, S. and Bajaj, B.K. (2017). Production of ionic liquid tolerant cellulase from *Bacillus subtilis* G 2 using agroindustrial residues with application potential for saccharification of biomass under one pot consolidated bioprocess. *Waste Biomass Valorization* 8 (3): 949–964.

147 Saratale, G.D., Saratale, R.G., Ghodake, G.S. et al. (2017). Solid state fermentative lignocellulolytic enzymes production, characterization and its application in the saccharification of rice waste biomass for ethanol production: an integrated biotechnological approach. *J. Taiwan Inst. Chem. Eng.* 76: 51–58.

148 Buntić, A., Pavlović, M., Antonović, D. et al. (2018). Customizing the spent coffee for *Trichoderma reesei* cellulase immobilization by modification with activating agents. *Int. J. Biol. Macromol.* 107: 1856–1863.

6

Microbial Production of Biobased Chemicals

Improvements and Challenges

Luana Assis Serra[1,2], *Débora Trichez*[1], *Clara Vida G. C. Carneiro*[1,2], *Letícia M. Mallmann Ferreira*[1,2], *Paula F. Franco*[1], *and João Ricardo M. Almeida*[1,2]

[1] *Laboratory of Genetics and Biotechnology, Embrapa Agroenergia, Brasília, Brazil*
[2] *Graduate Program of Microbial Biology, Department of Cell Biology, Institute of Biology, University of Brasília, Brasília, Brazil*

6.1 Introduction

The dependence on petrochemical resources and environmental and economic factors has intensified the search for cleaner and sustainable energy worldwide. Lignocellulosic biomass is a less polluting and renewable source of energy that can be explored in various industrial sectors (Fatma et al. 2018). In this context, biotechnological solutions for the production of renewable chemicals are under development as an alternative to processes based on fossil fuels. The lignocellulosic-biomass-derived sugars are the most abundant renewable feedstock available in the world and can come from different sources, such as agricultural and agribusiness residues, organic waste (food scraps), and forest residues (wood). These feedstocks lessen the need to expand farming areas, lowering the emission of gases that favor climate change. Lignocellulose is a component of the plant cell wall, a highly organized matrix that consists mainly of cellulose (40%–50%), hemicellulose (25%–30%), and lignin (15%–20%), in which proportions may vary depending on the plant species. Also, it may contain small amounts of other components, for example pectin, and inorganic compounds (Mathews et al. 2015; Bergmann et al. 2019).

Cellulose is the most abundant component of the cell wall, being a linear and homogeneous polymer composed of glucose units. Hemicellulose is a heterogeneous polysaccharide that contains different sugar monomers and is considered the second most abundant constituent of the biomass. This heteropolymer is classified according to the sugar residues present in its chain, which can be either pentoses (C5—xylose and arabinose) or hexoses (C6—glucose, mannose, and galactose). The lignin is a macromolecule formed by phenylpropanoid units. Since the chemical composition of lignocellulose is mainly carbohydrate polymers, it can be used as a final product or raw material for fermentation and production of value-added products (Lee et al. 2014).

Lignocellulose Bioconversion Through White Biotechnology, First Edition. Edited by Anuj Kumar Chandel.

The lignocellulosic biomass must be pretreated to loosen the portions of cellulose and hemicellulose, and after that it is fractionated as monosaccharides and other compounds. The sugars can be set out as a substrate for specific processes that generate several high-value biocompounds. This sugar platform based on biomass is reasoned to be the largest by volume nowadays. Furthermore, the biological fermentation of biomass sugars has been extensively employed in biorefineries for the generation of biochemicals and biofuels, such as organic acids, polymers, alcohols, amino acids, vitamins, and others (de Jong et al. 2020).

The biorefineries are based on the integration of biomass conversion processes for the generation of marketable outcome chemicals, materials, and energy. The biorefinery embraces solutions that involve environmental, economic, and social aspects of sustainability, pointing out a way of making the biochemicals and biofuels production chain more economically viable, valuing waste and co-products. The production of biochemicals in biorefineries can be more market competitive, opening transition pathways to a circular bioeconomy, supported by a biobased and renewable economy. This is further motivated by the global demand for chemicals, which is estimated to be around 90 million tonnes in 2020 (Paes and Almeida 2014; de Jong et al. 2020).

Several biobased fuels and chemicals are being industrially produced by microbial conversion of sugars (de Jong et al. 2020). For instance, *Saccharomyces cerevisiae* has been used in the production of bioethanol, whereas *Aspergillus* sp. in the production of citric acid and lactic acid bacteria (LAB) are the main producers of most of the industrial lactic acid (LA). However, some technological and economic obstacles need to be overcome to make the microbial production of biobased chemicals more economically viable. The ideal microorganism would convert biomass with minimal pre-processing, be able to utilize all the sugars (pentoses and hexoses) contained in the sugar mix, generate maximal product yields, be resilient in the presence of inhibitors, and have the potential for scale-up processes (Elkins et al. 2010).

Most bacteria and fungi do not meet all the required criteria. For this reason, recent discoveries in the field of systems metabolic engineering have been a reliable ally to re-design and create non-native pathways. These new technologies are enabling the integration of evolutionary engineering and synthetic biology with metabolic engineering, which means that the entire microbial process can be optimized, developing more efficient tools to explore the production of chemicals from renewable resources such as lignocellulosic hydrolysates (Lee et al. 2019).

Moreover, as the integration of systems biology has become more attractive, new opportunities of cell engineering aiming to reconstruct artificial biological systems, control microbial consortia and cellular behavior, or increase the natural production of existent metabolites have been extensively explored (McCarty and Ledesma-Amaro 2019). New tools allow the identification of targets for directed genetic modifications and the design of robust microbial cell factories to create entirely new routes, improve the production of desired biobased materials, and even synthesize complete chromosomes (Hutchison et al. 2016). Several studies that implemented synthetic metabolic pathways have demonstrated enhancements in overall efficiency and yield of the final high-value bioproduct (Ro et al. 2006; Galanie et al. 2015; Yadav and Shukla 2019).

In summary, this chapter describes the key challenges and general improvements required for the microbial production of biobased chemicals from lignocellulosic biomass sugars. Initially, we will summarize the main challenges in developing microorganisms for the production of such chemicals from lignocellulosic sugars. Then, the most relevant biobased chemicals in terms of technical development will be identified, presented, and discussed, including prices and market tendencies. Finally, the main challenges and issues for the microbial production of the most relevant biobased chemicals will be addressed.

6.2 Challenges in Developing Microorganisms for Lignocellulosic Sugar Utilization

Lignocellulose-derived sugars are considered the largest platform for the production of fuels and chemicals based on renewable biomass. Using sugars as an intermediary compound, which connects biomass feedstock to final products, has the advantage that it is already the basis for several traditional biorefinery processes and well-established industries (de Jong et al. 2020). Examples are the production of bioethanol from sugarcane and corn sugars or high-value fine chemicals as organic acids (Shahab et al. 2019; Takkellapati et al. 2019; de Jong et al. 2020). However, to reach the industrial biobased production of chemicals three main challenges associated with the conversion of sugars to the products should be considered: (i) the deconstruction of biomass to monosaccharides, (ii) the fermentation of hydrolysates containing sugar mixtures, and (iii) the presence of microbial inhibitory compounds in the biomass.

In a lignocellulosic biorefinery, pentose (C5) and hexose (C6) sugars are the main monosaccharides released from the biomass hydrolysis. Glucose, a C6 sugar, is produced from cellulose, whereas a mixture of C5 (xylose, arabinose) and C6 sugars (glucose, mannose, and galactose) is obtained from the hemicellulose fraction (Agbor et al. 2011; Liu et al. 2020). Thus, a mixed sugar platform, including mainly glucose and xylose, will result from hydrolysis of the lignocellulosic biomass (de Jong et al. 2020) (Table 6.1). Despite the basic composition of the lignocellulose (cellulose, hemicellulose, lignin), the great diversity of plant species and waste residues presents varied proportions of these polymers (Bergmann et al. 2019; Galbe and Wallberg 2019) and consequently a varied content of the individual carbohydrates. Table 6.1 shows the composition of different biomass feedstocks.

Lignocellulose conversion is a critical step for the feasibility and economic efficiency of biotechnological processes. There are several chemical and biological methods that can be used to release the monosaccharides present in the biomass for downstream fermentation by microorganisms (Liu et al. 2020). First, the lignocellulosic material requires a pretreatment step to expose the cellulose and hemicellulose content and facilitate enzymatic or acid hydrolysis of its fractions. Different types of pretreatments have been developed in the past years, including physical, chemical, physicochemical (acid or alkali pretreatment, oxidative, and organosolv methods), biological, or a combination of more methods (Agbor et al. 2011; Galbe and Wallberg 2019; Takkellapati et al. 2019). In the second step, saccharification, or hydrolysis, an enzyme cocktail containing cellulases (endoglucanase, exoglucanase, and β-glucosidase), hemicellulases such as xylanases, ligninases (peroxidases and laccases), and accessory enzymes and proteins may be necessary to reach a more complete deconstruction of the lignocellulosic material. In this sense, either commercial enzymes or enzyme-producing microorganisms can be used (Agbor et al. 2011; Liu et al. 2020).

Table 6.1 Biochemical composition of different biomass feedstock (% dry weight) (Based on Bergmann et al., 2019; E4tech, RE-CORD and WUR, 2015).

Feedstock group	Biomass feedstock	Cellulose	Hemicellulose	Lignin
Forestry	Spruce wood	45.0	21.0	28.0
	Pine	44.0	25.0	26.0
	Oak	40.0	20.0	25.0
	Eucalyptus	43.0	23.0	25.0
Grassy crops	Switchgrass	37.0	31.0	23.0
	Miscanthus	45.0	24.0	21.0
	Arundo donax	33.0	27.0	18.0
Food crops	Sorghum	39.0	24.0	9.0
	Corn	40.0	30.0	12.0
	Wheat	31.0	20.0	7.0
	Sugar beet	26.0	28.0	5.0
	Sugarcane	31.0	12.0	8.0
Agricultural residue	Wheat straw	30.2	21.0	17.0
	Rice straw	31.1	22.3	13.3
	Olive tree prunings	30.0	18.0	21.0
	Sugarcane bagasse	43.1	31.1	11.4
	Corn stover	37.0	22.7	18.6
Agro-industrial residues	Almond shell	39.0	29.0	29.0
	Orange peel	16.0	7.0	5.0
	Olive husk	23.0	25.0	47.0
Others	Macroalgae	9.0	7.0	0.0
	Paper pulp/sludge	59.0	17.0	14.0
	Food industry waste	18.0	21.0	17.0

Despite the advances in biobased chemical production at a commercial scale, one of the major technical challenges is still the bioconversion of biomass into fermentable sugars. The porosity of the material, the crystallinity of the cellulose, the hydrophobicity of lignin, and encapsulation of cellulose microfibrils in the lignin–hemicellulose matrix are some factors that can affect and hinder the hydrolysis process (Agbor et al. 2011; Singhvi and Gokhale 2019). Moreover, high cost and low titers of enzyme production and not optimal biocatalytic efficiency requires high doses and time-consuming treatments, significantly increasing operating costs (Singhvi and Gokhale 2019).

Besides fermentable sugars, pretreatments lead to the release and formation of compounds such as acetic acid, furaldehydes (2-furaldehyde [furfural], 5-hydroxymethylfurfural [HMF]), and phenolic compounds, which inhibit growth and metabolite production by microorganisms, reducing yield and productivity of the overall process (Almeida et al. 2011; Bhatia et al. 2020; Paes et al. 2021). The presence of these inhibitory compounds in lignocellulosic hydrolysates will depend on the nature of the biomass composition and the type of pretreatment process applied. Multiple strategies have been proposed to avoid this negative effect on fermentation, including new pretreatment methods more compatible with downstream fermentation, detoxification of biomass hydrolysates, improvement of plants for reduced lignin content, selection of natural microbial strains, and development of adapted and/or genetically engineered microorganisms with higher inhibitor tolerance (Buschke et al. 2013; Mukherjee et al. 2017; Bhatia et al. 2020).

Taking all these points into consideration, the production of fuels and biochemicals from lignocellulosic sugars will require robust microorganism strains that tolerate harsh environmental conditions and the presence of multiple inhibitory compounds commonly found in lignocellulosic hydrolysates, as well product toxicity, and are capable of utilizing mixed sugars (pentose and hexoses), that is, are able to utilize all carbons present in the biomass (Paes and Almeida 2014; Takkellapati et al. 2019).

Glucose is the preferred carbon source for most microorganisms. As a consequence, this sugar is metabolized and converted into several bioproducts by microorganisms, whereas fermentative conversion of xylose or other C5 sugars is less efficient and reduced to a smaller number of species (Moysés et al. 2016; Bergmann et al. 2019). Furthermore, the presence of glucose induces a carbon catabolite repression response, in which several genes and proteins involved in the metabolism of other carbon sources are suppressed or inactivated (Gao et al. 2019). Several strategies have been reported addressing this issue, including evolutionary engineering or genome shuffling to obtain superior strains capable of using mixed sugars (Lane et al. 2018; Lin et al. 2019), identification and/or engineering of sugar transporters more resistant to glucose inactivation or with improved sugar affinities (Farwick et al. 2014), and disruption of genes involved with glucose signalization in yeast and bacteria (Buschke et al. 2013; Kim et al. 2019).

Microorganisms have different pathways to assimilate pentose sugars. In general, yeast and fungi metabolize pentoses via an oxidoreductive pathway into xylulose 5-phosphate (xylulose-5P), whereas in Eubacteria and some anaerobic fungi the pentoses are isomerized to D-xylulose or L-ribulose by an isomerase and then converted to xylulose-5P (Moysés et al. 2016; Francois et al. 2020). The xylulose-5P formed is further directed to the pentose phosphate pathway or it can also be cleaved into glyceraldehyde-3P and acetyl-phosphate through the phosphoketolase pathway, an alternative route found in lactic bacteria, bifidobacteria, filamentous fungi, and unicellular algae (Francois et al. 2020). In the non-phosphorylating pentose pathway found in Archaea, the oxidation of the pentose sugars and the subsequent dehydration step result in the intermediary compound 2-keto-3-deoxy pentanoate that can follow alternative downstream pathways (Weimberg or Dahms pathways) generating a diversified panel of biobased chemicals (Francois et al. 2020).

Despite the existence of these natural sugar assimilation pathways, usually they are not optimized to reach the yields and productivities that are necessary for industrial application or even to obtain natural or non-natural products of interest with high titers. In this way, metabolic engineering, evolutionary approaches, and synthetic biology are promising tools available to develop microbial cell factories with redesigned carbon metabolism, ensuring cellular homeostasis and optimized performances (Paes and Almeida 2014; Francois et al. 2020).

The reconstruction of artificial biological systems and cellular behavior, or increasing natural production of existent metabolites, even synthesis of complete chromosomes, have been explored (Hutchison et al. 2016; McCarty and Ledesma-Amaro 2019). As an example, the implementation of synthetic metabolic pathways for the production of LA, artemisinin, probiotics, and opioids in yeasts demonstrates enhancements in overall efficiency and yield of the final high-value bioproduct (Ro et al. 2006; Galanie et al. 2015; Yadav and Shukla 2019). In addition to cell engineering strategies, the huge diversity of microorganisms represents a potential source of innovation not yet well exploited. In this respect, bioprospection and physiological characterization of new microorganisms, as well as the identification of new genes, enzymes, and metabolic pathways, may contribute to the development of biotechnological processes, expanding microbial applications in the industry (Paes and Almeida 2014; Mukherjee et al. 2017; Veras et al. 2017, 2019; Trichez et al. 2019).

Although some biobased products are already commercialized, there are several technical, biological, and economic issues to overcome before biomass sugars can be completely exploited. Diversifying the product portfolio, generating co-production of commodities and higher added value bioproducts, is widely accepted to ensure higher competitiveness for the biorefineries (Rosales-Calderon and Arantes 2019; de Jong et al. 2020). Within the broad range of compounds that can be obtained from the renewable lignocellulose biomass, top promising products have been ranked in the last few years. In the following sections, an overview of relevant biobased products for biorefineries and the status of research, technology, and development in relation to commercial applications will be presented as well as the major microorganisms used in these biotechnological processes.

6.3 Relevant Biobased Chemicals from Biomass

Several reports describing the potential use of biomass to produce biobased chemicals have been published by international agencies in the last few years. Based on the analysis of 12 reports elaborated by the European Commission, the United States Department of Agriculture (USDA), the International Energy Agency, and other sources from 2006 to 2020, a list of 50 potential chemicals was identified to be produced from biomass, the so-called biobased chemicals, by chemical and/or biological processes (Table 6.2). These studies established different criteria for considering these chemical compounds as potential targets for production from biomass. Most commonly, they considered technological development (technical readiness level [TRL]), broad application of the product, its market size and value, the level of reported research in the open literature and other things (Braga et al. 2020). Of 50 different chemicals identified in these 12 studies, 15 have been cited in at least 50% of the reports evaluated, whereas this number increases to 18 chemicals when considering 50% citation only in the six most recent reports (2014–2020) (Table 6.2).

These data analyses highlight the potential of using biomass for the production of a diversity of industrially relevant chemicals. The products can be grouped as drop-in or "novel" chemicals. In the first case, the chemicals are identical to existing fossil-based materials and have established markets; in the latter case, chemicals with new or similar properties to petrochemicals are produced. In common, these chemicals usually present at least a functional group that makes them versatile and applicable in different sectors, with the great majority being organic acids (Table 6.2) (Braga et al. 2020).

Table 6.2 Relevant biobased chemicals according to studies from different institutions; chemicals highlighted in bold are the most common among the cited studies.

	Werpy and Petersen (2004)	Patel et al. (2006)	USDA (2008)	Bozell and Petersen (2010)	Bridgwater et al. (2010)	de Jong et al. (2012)	Bain & Company (2014)	E4tech, RE-CORD and WUR (2015)	Nattrass et al. (2016)	Biddy et al. (2016)	E4tech (2017)	de Jong et al. (2020)
Succinic acid	×	×	×	×		×		×	×	×	×	×
Lactic acid	×	×	×	×	×	×				×	×	×
1,3-Propanediol	×	×	×		×	×			×	×		×
Ethanol	×	×	×	×	×	×		×				×
Levulinic acid	×	×		×		×		×	×		×	×
1,4-Butanediol	×	×	×					×	×	×		×
Furfural	×	×				×		×	×	×		×
***n*-Butanol**	×	×					×	×	×		×	×
Propylene glycol	×	×			×	×			×	×		×
2,5-Furandicarboxylic acid (FDCA)	×					×		×	×		×	×
Glycerin	×	×	×			×				×		×
Methanol	×				×	×	×				×	×
Polyhydroxyalkanoates (PHAs)		×	×			×		×	×		×	
Sorbitol	×	×	×			×		×				×
Xylitol		×		×		×		×			×	×
Adipic acid	×	×				×				×	×	
Fatty alcohol	×	×							×	×	×	
Isoprene			×			×		×		×		×

***p*-Xylene**					×		×		×	×	×
Itaconic acid	×				×		×			×	
Polylactic acid	×		×				×	×			
Ethylene	×				×			×			×
Acetic acid	×	×				×	×				
Acrylic acid	×	×			×			×			
Ethyl lactate	×		×		×				×		
Hydroxymethylfurfural (HMF)		×		×				×		×	
Fumaric acid	×	×								×	
Butadiene (1,3-)	×					×			×		
Farnesene					×		×				×
Glucaric acid	×				×					×	
Lysine	×				×						×
Polyethylene (PE)	×		×				×				
Acetone	×						×				
Biodiesel			×		×						
Citric acid	×										×
Epichlorohydrin				×	×						
Ethylene glycol	×							×			
Fatty acid esters			×					×			

(Continued)

Table 6.2 (Continued)

	Werpy and Petersen (2004)	Patel et al. (2006)	USDA (2008)	Bozell and Petersen (2010)	Bridgwater et al. (2010)	de Jong et al. (2012)	Bain & Company (2014)	E4tech, RE-CORD and WUR (2015)	Nattrass et al. (2016)	Biddy et al. (2016)	E4tech (2017)	de Jong et al. (2020)
Polybutylene succinate (PBS)			×						×			
1,3-Butanediol											×	
Biogasoline						×						
D-Mannitol											×	
Dimethyl ether (DME)						×						
Fatty amines									×			
Isobutanol									×			
Isobutene								×				
Levoglucosenone											×	
Muconic acid											×	
Pentamethylene diisocyanate (PDI)												×
Sebacic acid												×

In the next sections, we summarize the main characteristics, applications, and technical developments for the production of 17 relevant chemicals present in Table 6.2. These chemicals were grouped as organic acids (adipic, furan-2,5-dicarboxylic [FDCA], itaconic [IA], LA, polylactic [PLA], levulinic, and succinic acid [SA]), diols (1,3-propanediol [PDO], propylene glycol), polyols (sorbitol, xylitol), alcohols (butanol and ethanol), aldehydes (furfural) and others (polyhydroxyalkanoates [PHAs] and *p*-xylene) (Figure 6.1). Despite being listed as promising chemicals in different studies, glycerin, methanol, fatty alcohol and isoprene will not be discussed in this chapter. These chemicals are obtained exclusively from chemical routes and/or are co-products of industrial processes, like glycerin obtained from biodiesel production.

6.4 Microbial Products from Sugar Fermentation

In this section, an overview is presented of relevant biobased chemicals and the status of research and technology development in relation to their commercial application as well the major microorganisms used in the biotechnological processes.

6.4.1 Organic Acids

6.4.1.1 Adipic Acid

Adipic acid ($C_6H_{10}O_4$) is an aliphatic, straight-chain dicarboxylic acid, with six carbons (Figure 6.1). It is a major component of nylon 66, which is present in carpet fibers, molded plastics, and textile products. Besides nylon, adipic acid is one of the most important commercial acids and can be further converted into polyurethanes, lubricants, and even flavorings (Deng et al. 2016; E4tech 2017).

The global demand for adipic acid is nearly 3 million tonnesa year, with a market value of $8 billion in 2020 (Table 6.3). It is commonly produced from petroleum, by the oxidation of cyclohexane benzene and the nitric acid oxidation of the intermediate, to produce adipic acid. The last step generates almost 10% of the total emission of nitrogen oxide in the atmosphere, which is a toxic greenhouse gas (GHG) (Kruyer et al. 2020; Niu et al. 2020).

The biotechnological production of adipic acid would eliminate several environmental and economic issues regarding its manufacturing. It would eliminate the chemical catalysts and cost-effective purification steps and allow the use of several different renewable lignocellulosic feedstocks, from the simplest sugar to lignin-derived aromatic compounds, helping to increase the value chain of biomass residues (E4tech 2017; de Jong et al. 2020; Kruyer et al. 2020).

The full production of biological adipic acid is being extensively evaluated. Since there is no natural pathway to adipic acid production, several researchers are using metabolic engineering tools to enable microorganisms to produce it. The most investigated microorganism is the bacterium *Pseudomonas putida*; a new pathway was constructed using a series of enzymes to convert lignin into a direct biosynthesis of adipic acid. This pathway enabled *P. putida* to produce adipic acid with a maximum molar yield of 17.4% (Niu et al. 2020), one of the highest reported so far. Recently, another pathway was established in an engineered *Escherichia coli*. The strain was able to produce adipic acid from catechol, a monomer obtained from lignin pretreatment, with a molar yield of 0.241% (Kruyer et al. 2020).

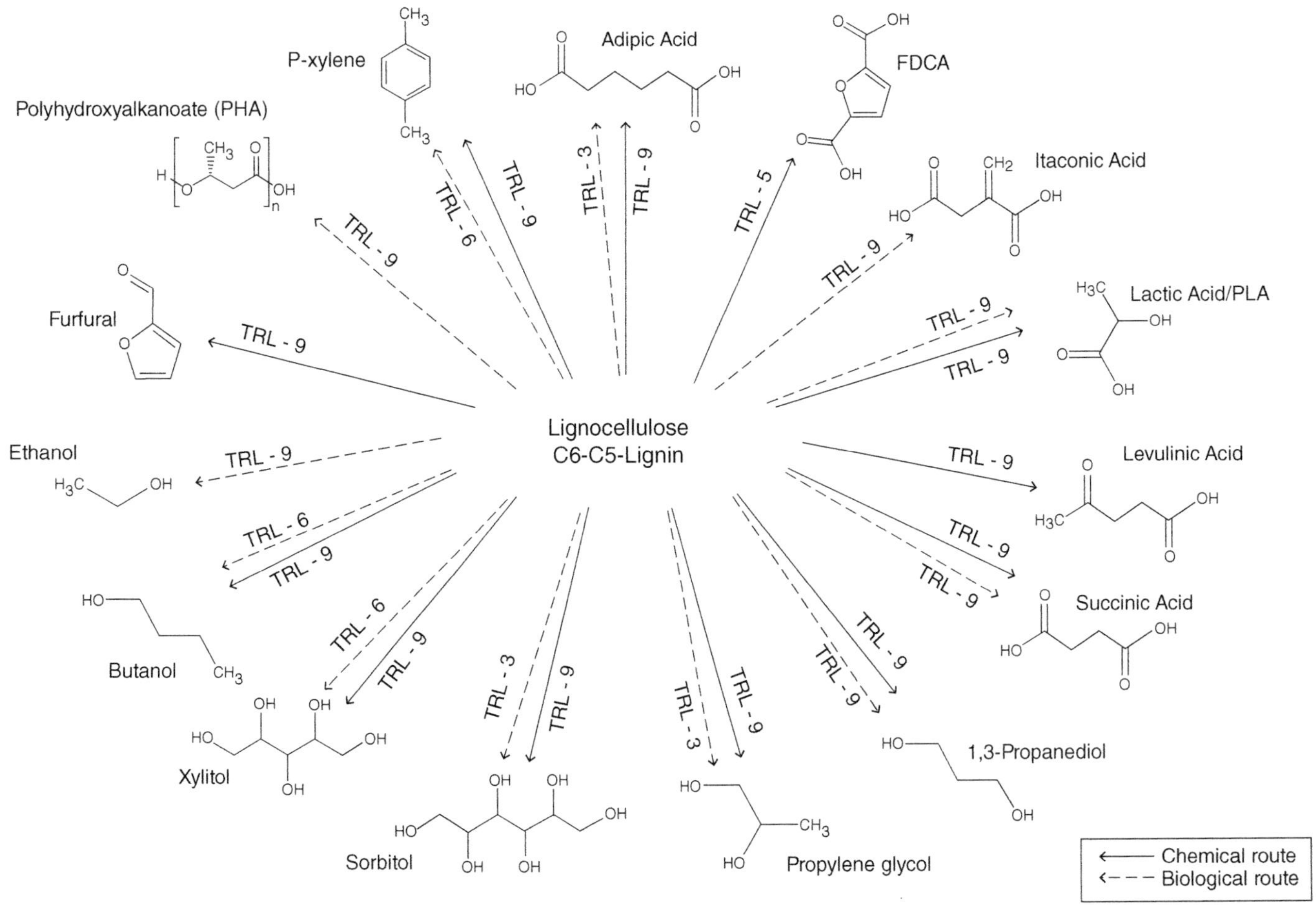

Figure 6.1 Chemical structure of relevant biobased chemicals to be produced by microorganisms. Chemical and biological production processes with their respective technical readiness levels (TRL) are indicated by arrows.

Table 6.3 Application, market, and production process of promising biobased chemicals.

			Production process		
Chemical	Application	Market	Substrate	Industrial process	References
Adipic acid	Nylon 6,6, fibers, molded plastics	3 million tonnes $year^{-1}$ $8 billion (2020)	Lignin	Chemical/semi-biological	Kruyer et al. (2020), Niu et al. (2020)
FDCA	Polymers, PET analogs, polyesters, nylons, and other synthetic fibers	50 kilotonnes per annum (2017) $88.2 million (2014)	C6 sugars	Chemical	Market Analysis Report (2015c), de Jong et al. (2020)
Itaconic acid	Resins, adhesives, coatings, lubricant additives, and synthetic rubber and fibers	41 kilotonnes per annum, $79 million	Commercial scale: glucose and sucrose	Biological	Okabe et al. (2009), E4tech, RE-CORD and WUR (2015)
Lactic acid	Chemicals, cosmetics, food, pharmaceutical	$2.64 billion (2018) 600 000 tonnes (2019)	Corn starch, sugar beets, sugarcane, wheat	Biological/chemical	Market Analysis Report (2019), de Jong et al. (2020)
Polylactic acid	Biomedicine (sutures, prostheses), Biodegradable plastic (fibers, packing)	217 000 tonnes (2018)			
Levulinic acid	Solvents, fuel, pesticides, pharmaceutical applications	20 kilotonnes per annum 26.3 kilotonnes per annum	Sugars derived from corn cob and bagasse	Chemical	de Jong et al. (2020)
Succinic acid	Food, chemicals, cosmetics, pharmaceutical, agricultural	$125 million (2018) 34 kilotonnes per annum	Corn stover Sugarcane Wood	Biological/chemical	Akhtar et al. (2014), Industry ARC (2019), de Jong et al. (2020)
1,3-Propanediol	Polymers, personal care sector, solvents, and lubricants	$484.8 million (2019) 125 000 tonnes (2014)	Glycerol, co-fermentation: glycerol and lignocellulosic substrates	Biological/chemical	Market Analysis Report (2015a)

(*Continued*)

Table 6.3 (Continued)

Chemical	Application	Market	Production process		References
			Substrate	Industrial process	
Propylene glycol or 1,2-propanediol	Unsaturated polyester resins, coolants and antifreeze,	2000 kilotonnes per annum, $3.91 billion	Commercial scale: sorbitol, glycerol hydrogenolysis	Chemical	de Jong et al. (2020), Gérardy et al. (2020)
	paints, and coatings, additive in food and feed products		Research scale: glucose; xylose as co-substrate.	Biological	
Sorbitol	Sweetener, surfactants, polyurethanes, sorbitan, 1,2-propanediol and ascorbic acid	2.4 million tonnes year^{-1}/$4 billion (estimated)	D-glucose	Chemical	de Jong et al. (2020)
Xylitol	Sweetener, polymers, ethylene glycol, propylene glycol, and esters	200 000 tonnes year^{-1}/$880 million	Lignocellulosic biomass or pure D-xylose	Chemical	Market Analysis Report (2017), de Jong et al. (2020)
Butanol	Solvent in the auto industry, "drop-in" fuel, polymers and plastics industries	$7.86 billion (2014)	1st generation: feedstocks that contain sucrose and starch	Biological	Market Analysis Report (2015b), de Jong et al. (2020)
			2nd generation: lignocellulosic biomass		
Ethanol	Solvent, additive to gasoline, fuel	29 000 million gallons (2019)	1st generation: feedstocks that contain sucrose and starch	Biological	Azhar et al. (2017), Bergmann et al. (2019), Market Analysis Report (2020a), Renewable Fuels Association (2020)
		$89.1 billion (2019)	2nd generation: lignocellulosic biomass		

Furfural	Cosmetics, food flavoring, pesticides	360 kilotonnes per annum, $800 million	Xylose and arabinose	Chemical	de Jong et al. (2020)
Polyhydroxyalkanoates (PHAs)	Pellets, powder, and packaging materials, medical implants	17 kilotonnes per annum, $111 million	Commercial plants: sugars derived from corn starch or sugarcane Research and pilot scales: biomass hydrolysates	Biological	E4tech, RE-CORD and WUR (2015), Jiang et al. (2016), Dietrich et al. (2017, 2020)
p-Xylene	Dimethyl terephthalate (DMT), terephthalic acid (PTA), polyethylene terephthalate (PET)	2.7 million tonnes $year^{-1}$/$48.72 billion (estimated)	Petroleum/lignocellulosic biomass	Chemical	Market Analysis Report (2018), de Jong et al. (2020)

Today, the renewable production of adipic acid occurs only by a semi-biological route; the first step is the biological production of muconic acid, from glucose fermentation, followed by chemical hydrogenation to produce adipic acid (Kruyer et al. 2020). Genomatica has patented a synthetic metabolic pathway that enables the production of a fully biological adipate, 6-aminocaproic acid or caprolactam (key intermediates used in nylon production) by microorganisms such as *E. coli*, from renewable carbohydrate feedstocks like glucose. Although there is a high interest in technologies to produce fully biological adipic acid, the conversion yields are still very low compared with the industrial chemical process (Deng et al. 2016; Burgard et al. 2010; de Jong et al. 2020).

6.4.1.2 Furan-2,5-dicarboxylic Acid ($C_6H_4O_5$)

FDCA is characterized by two groups of carboxylic acids attached to a central furan ring (Figure 6.1). FDCA is industrially produced from catalyzed hydrolysis of hexoses in lignocellulosic biomass hydrolysates, via primary formation of HMF, a less stable furanic compound (Sheldon 2014). In this process, HMF can undergo oxidation of the hydroxyl group or oxidation of the aldehyde group followed by oxidation of the hydroxyl group. In both cases, the intermediate products are then oxidized to 5-formyl-2-furancarboxylic acid, and then finally transformed to FDCA (Hameed et al. 2020).

The most important application of FDCA is in the formation of polymers (Table 6.3). FDCA is considered a promising sugar chemical derivative for its potential use in bioplastics, together with PLA and PHAs (de Jong et al. 2020). Global production of bioplastics in 2018 reached 2.11 million tonnes, with polyethylene terephthalate (PET) accounting for 25.6% of all polymers representing more than 50 million tonnes. FDCA is mainly suggested as a replacement for terephthalic acid in the production of polyester polymers. The two key materials for the production of PET are ethylene glycol and *p*-terephthalic acid, derived from ethylene and *p*-xylene, respectively. In 2014, the total FDCA market size was valued at $88.2 million (Table 6.3) and was predicted to grow 33.5% from 2015 to 2020 (Sheldon 2014; Market Analysis Report 2015c).

The Dutch company Avantium, one of the greatest biobased chemical producers at present, is following a "plant bottle" strategy to produce polyethylene furandicarboxylate (PEF) by replacing terephthalic acid with FDCA. The company just bought a new flagship in the Netherlands that is capable of producing 5 kilotonnes of plant-based FDCA. This pilot plant is in development and is to be completed in 2023; also Avantium still needs a license for industrial-scale production (Avantium 2020; de Jong et al. 2020). It is estimated that complete substitution of PET production with biobased PEF could reduce GHG emissions by 20–35 tonnes. They are also on a mission to establish a leading position on FDCA production by a process based on acid catalyzed dehydration of fructose via catalytic oxidation of HMF (Sheldon 2014). Besides the direct application of FDCA as a building block for PEF and PET analogs, it also presents a potential application in furanoic polyesters and polyamides for use in nylons (Werpy and Petersen 2004).

Although there are no microbial routes for the production of furans and their derivatives, the scientific community has shown great interest on behalf of their microbial utilization. Since 1964, a relatively low diversity of microorganisms has been described as capable of metabolizing furanic compounds, such as Gram-negative bacteria (Pseudomonadales, Desulfovibrionales, Rhizobiales, Enterobacteriales, and Burkholderiales), Gram-positive bacteria (Actinomycetales), and fungus (Coniochaetales) (Wierckx et al. 2011). Some studies are suggesting the microbial conversion of HMF and furfural to FDCA. In the work of Wierckx and collaborators (2011), an oxidoreductase from *Cupriavidus basilensis* was expressed in *P. putida* S12, reaching a productivity of 30 g L^{-1} of FDCA from HMF at a yield of 97%. The microbial-derived FDCA can be utilized further in the synthesis of chemical building blocks and a range of other applications (de Jong et al. 2020).

Primary barriers to the production of FDCA are the effective and selective dehydration processes for raw sugars, a technical barrier that is common to several biobased processes. Unstable intermediate molecules can delay the cost-effective production of FCDA as for a variety of other sugar derivatives. Therefore, direct oxidation of HMF remains the most economically efficient route for FDCA production. Different catalyst agents can change conditions and yield of production. Noble metals like gold, platinum, and rhodium, for instance, also appeared to increase the yield of HMF to FDCA conversion to >90% (Hameed et al. 2020). Several furanic products can be explored and invested in new polymers, and biobased FDCA has a great potential for market growth if the productions of big companies, like Avantium, achieve commercialization.

6.4.1.3 Itaconic Acid

IA or 2-methylidenebutanedioic acid ($C_5H_6O_4$) is an unsaturated dicarboxylic acid (Figure 6.1). It can be used in the industrial production of resins, adhesives, coatings, lubricant additives, and synthetic rubber and fibers. Another potential application is as a substitute for petrochemical-based acrylic and methacrylic acid monomers used for plastic production (Table 6.3) (Willke and Vorlop 2001; Steiger et al. 2013). Since the 1960s, *Aspergillus terreus* has been used in IA manufacturing through submerged fermentation of carbohydrates (Figure 6.1) (Okabe et al. 2009). A content of up to 80 g L^{-1} of IA was achieved using an over-producing strain (Yahiro et al. 1995).

The biosynthesis of IA comes from the tricarboxylic acid cycle, in which citrate and *cis*-aconitate are intermediates. In *A. terreus*, IA is synthesized by the decarboxylation of *cis*-aconitate, which is formed in the mitochondrion and exported into the cytosol (Karaffa and Kubicek 2019; Willke and Vorlop 2001). Because *Aspergillus* can produce up to 200 g L^{-1} of citric acid under industrial conditions (Steiger et al. 2013), several efforts have been focused on increasing the IA production by these fungi, including isolation and selection of new strains, optimization of the production process, medium composition, and cell metabolic engineering (Okabe et al. 2009; da Cruz et al. 2018; Karaffa and Kubicek 2019; Xie et al. 2020). These optimizations lead to a wild-type *A. terreus* strain capable of producing 160 g L^{-1} of IA, with yields of 0.56–0.63 g g^{-1}, and overall productivity of 1–1.9 g L^{-1} h^{-1} (Krull et al. 2017).

Another approach emphasizes the utilization of other microbial hosts, with more safety and easy to manage at large-scale fermentation. *Ustilago zeae*, *Ustilago maydis*, and *Candida* sp. were already identified as natural producers of IA. Engineered microorganisms have been developed, including *E. coli*, *Yarrowia lipolytica*, *S. cerevisiae*, and *Pichia kudriavzevii*, although none of them showed better titers than *A. terreus* (Okabe et al. 2009; Sun et al. 2020). Recently, the integration of metabolic engineering of an *U. maydis* strain and optimization of the fermentation process resulted in the highest IA production (220 g L^{-1}) described until now, albeit with a reduction in yield (0.33 g g^{-1}) and productivity (0.45 g L^{-1} h^{-1}) (Tehrani et al. 2019).

Currently, the main companies responsible for global IA production (Qingdao Kehai, Zhejiang Guoguang, and Jinan Huaming Co. Ltd) are located in China and the production capacity seems to be adequate to fulfill the market requirements that are limited by few end-use applications and the relatively high prices compared to cheaper petroleum-based substitutes (de Jong et al. 2020). Since glucose and sucrose are the main substrates used at industrial scale, low-cost feedstocks, such as starch, molasses, and lignocellulosic hydrolysates, have been considered as an alternative to reduce the production cost (da Cruz et al. 2018; Rosales-Calderon and Arantes 2019). However, production of 5–50 $g\,L^{-1}$ of IA with *A. terreus* from lignocellulosic residues (da Cruz et al. 2018) is still small and further technological advances are necessary to ensure industrial interest and the competitiveness of this product.

6.4.1.4 Lactic Acid and Polylactic Acid

LA, also known by the name 2-hydroxypropanoic acid, is an important carboxylic acid with the molecular formula $C_3H_6O_3$ (Figure 6.1). It was purified for the first time in 1780 by Carl Wilhelm. In 1881, the researcher Frémy developed a pioneering industrial production process of LA based on microbial fermentation. Ever since, this acid has been employed in different sectors of industry such as food (as a preservative and pH adjusting), cosmetics (personal products), pharmaceuticals, and detergents. Currently, the interest in this chemical has been growing due to its use in the production of biobased polymers, like PLA, a biodegradable plastic and renewable alternative for the petroleum-based material (Table 6.3) (Shahab et al. 2019; Sin and Tueen 2019).

LA is mainly produced by two routes: chemical synthesis or microbial fermentation. The first is based on petroleum feedstocks and generates a racemic mixture of L- and D-lactic acid (isomers). With the microbial route, different substrates can be explored, like lignocellulosic biomass, and parameters like energy and temperature can be lowered making it an advantageous scenario. Also, choosing the appropriate microbial strain can result in an optically highly pure LA. The pure isomers have elevated value because the optical purity of LA is essential for the physical properties of PLA (Abdel-Rahman et al. 2011). The formation of this polymer can be generated from LA by two routes, direct polycondensation or the ring-opening polymerization of the LA (Sin and Tueen 2019).

L-lactic acid forms poly-L-lactic acid (PLLA), a polymer with great tensile strength suitable for medical products and packing, for example. The D-lactic acid that is the monomer of poly-D-lactic acid (PDLA) is more biodegradable. When both PLLA and PDLA are mixed, a crystalline complex results with higher thermal stability that is desirable for industrial processes (Cubas-Cano et al. 2018). Besides, biobased plastics can offer a better GHG balance than the ones based on fossil fuels. The generation of biobased PLA could generate a GHG saving of 3.3 tonnes of CO_2 per tonne of product (Hermann et al. 2007).

The majority of industrial LA is generated by anaerobic bacterial fermentation of carbohydrates (Figure 6.1). Different genera can produce this compound, especially the LAB like *Lactococcus*, *Lactobacillus*, *Streptococcus*, *Aerococcus*, *Carnobacterium*, *Enterococcus*, and *Weisella*. An important characteristic of LAB is that they have the GRAS (generally recognized as safe) status and can be applied in the food industry (Juturu and Wu 2016). Several studies are reporting the production of LA using lignocellulosic biomasses as substrates, including corn starch, sugar beets, sugarcane, and wheat. Wischral et al. (2019) investigated the fermentation of sugarcane bagasse hydrolysates for LA production by five strains of *Lactobacillus* spp. The *Lactobacillus pentosus* was capable of consuming xylose and glucose reaching a yield of 0.93 g g^{-1} and productivity of 1.01 g L^{-1} h^{-1} (Wischral et al. 2019).

Moreover, genetically modified LAB have been investigated to enhance the yields and productivity of LA. Xylose is a pentose present in the biomass sugar mix that is not assimilated by some microorganisms. To utilize all the sugars from the substrate, Zhang et al. (2016) constructed a recombinant *Lactobacillus plantarum* strain with xylose assimilating genes. The corn stover and soybean meal extract were fermented having an overall yield and productivity of 0.77 g g^{-1} and 0.32 g L^{-1} h^{-1}, respectively (Zhang et al. 2016). Also, some studies have been demonstrating a co-culture of bacteria and fungi to produce LA. Shahab and collaborators (2019) co-fermented *Lactobacilli* sp. and *Trichoderma reesei* from pretreated beech wood reaching 19.8 g L^{-1} of LA (corresponding to 85.2% of the theoretical maximum) with a productivity of 0.1 g L^{-1} h^{-1} (Shahab et al. 2019).

A range of companies are investing in the LA biobased route, following the globally growing demand for this compound; among them are Corbion, NatureWorks, Galactic, and Henan Jindan. The global leader in PLA production is the USA company NatureWorks, which has a commercial scale (TRL 9 level, based on commodity feedstocks) that uses fermentation technology (Biddy et al. 2016; de Jong et al. 2020). Nowadays, the European demand for PLA is 25 000 tonnes per year and market predictions have shown that it could reach 650 000 tonnes per year in 2025 (Juturu and Wu 2016; de Jong et al. 2020). The production of LA and its derivative PLA has been showing constant growth in recent years, showing to be an effective and renewable alternative for the same products based on fossil fuels.

6.4.1.5 Levulinic Acid

Levulinic acid ($C_5H_8O_3$) is an organic acid (Figure 6.1) that can be produced either from furfural or from HMF. Levulinic acid functional groups carbonyl and carboxyl enable it to react as both a carboxylic acid and a ketone, showing a great number of natural and synthetic reactions. Also, this acid presents great solubility in water, alcohols, esters, ethers, and ketones. All these characteristics lead to a great range of derivative compounds, raising the marketability of levulinic acid as a valuable platform chemical (Hayes et al. 1999; Kumaravel et al. 2019).

Levulinic acid derivatives present a series of applications in the industry (Table 6.3). Some of the most valuable derivatives and their applications are levulinate esters (fuel additives, solvents, pesticides), lactones (fuel additive, solvents), 1,4-pentanediol (solvents, polymers, fine chemicals), D-amino levulinic acid (pesticides, herbicides, pharmaceutical applications), methyltetrahydrofuran (fuels, fuel additives, solvents), and SA (polymers, solvents, pesticides).

The yield of levulinic acid production will depend on the production process involved. There are different primary sources of hydrocarbons and various possible intermediate compounds, which tend to cross-react and form unwanted residues (Hayes et al. 1999). Also, different feedstocks have higher and lower proportions of extractive materials—components that can be separated from non-soluble material (Zhang et al. 2012)—and this property affects not only levulinic acid production but all chemical processes from biomass.

One of the most commercially viable lignocellulosic-fractionating technologies currently available is the Biofine Process, which involves the hydrolysis of polysaccharides to their monomeric parts followed by continuous conversion to valuable platform chemicals (Fitzpatrick and Framingham 1990). It consists of two acid-catalyzed stages to give optimal yield with a minimum generation of toxic co-products. This method provides levulinic acid of 75%, which can be purified up to 98% purity.

Some studies have been investigating the bioconversion of levulinic acid into chemicals of interest, with 4-hydroxyvalerate (4-HV) a promising chemical that can be used as a monomer for PHAs. The microbial conversion of levulinic acid is attracting attention since the discovery of catabolic genes and enzymes (especially from *Pseudomonas* sp.) involved in its metabolic pathway (Habe et al. 2020). In this respect, metabolic engineered microorganisms with new routes could be developed to upgrade the production of chemicals of interest derived from levulinic acid. In the study of Rand et al. (2017), a seven-gene operon that enables LA catabolism *in P. putida* KT2440 was identified and characterized. The overexpression of five genes from the *P. putida* operon and deletion of specific genes in *E. coli* LS5218 allowed a strain capable of growing on levulinic acid to be obtained (Rand et al. 2017).

Currently, the global production of biobased levulinic acid is around 20 kilotonnes per annum. The overall levulinic acid market demand was 2606.2 tonnes in 2013 (Table 6.3) and is predicted to be 3820 tonnes by 2020 with an expected growth rate of 4.8%. The prices for the levulinic acid market in 2015 were in the range between \$5 and \$8 per kilo (Market Analysis Report 2015d). The leading market companies are Avantium (Netherlands), GFBiochemicals (Italy), and Circa Group (Australia) with pipeline potential (de Jong et al. 2020).

6.4.1.6 Succinic Acid

SA ($C_4H_6O_4$) is a metabolite produced by different organisms, such as plants, microorganisms, and animals. It is an intermediate of the tricarboxylic acid cycle. Also known by the name of amber or butanedioic acid, it is an important dicarboxylic acid utilized in various marketable sectors like food, chemicals, cosmetics, pharmaceutical, agricultural, and others (Table 6.3) (Ahn et al. 2017). SA was first isolated in 1546 produced by microbial fermentation and the different potential industrial applications of this acid have been proposed since 1980 (Zeikus 1980; Akhtar et al. 2014).

Traditionally, SA is produced by the chemical route based on petrochemical maleic acid. However, the biomass potential for the biological production of SA has gained much attention in the last few years due to the renewability of the process. In addition to being produced by microbial fermentation of biomass-derived sugar, carbon dioxide (CO_2) is assimilated during microbial fermentation to produce SA (Ahn et al. 2017). Studies suggest that biobased SA has the potential to reduce the GHG in 5 tonnes of CO_2 per tonne of product (Hermann et al. 2007). Furthermore, SA is used as a building block chemical for the synthesis of chemicals of interest, such as adipic acid, 1,4-butanediol, tetrahydrofuran, and other compounds (Table 6.3) (Biddy et al. 2016).

Nowadays, biobased SA represents half of the global SA market (de Jong et al. 2020). With the traditional route, SA had a price of \$2.26 per kilogram. Adopting the biobased product using biomass as a substrate the value was reduced to \$1 per kilogram (Efe et al. 2013). A range of companies has been investing in the commercial scale production process (i.e. on the TRL 9 level) of biobased SA by bacteria, for example Myriant and

BioAmber which employ a recombinant *E. coli*. Moreover, other bacteria are capable of using biomass as a substrate such as *Anaerobiospirillum succinoproducens* and *Actinobacillus succinogenes*, and yeasts like *Candida krusei* (Ahn et al. 2017). Also, yeasts like *S. cerevisiae* are utilized by the company Reverdia (Roquette) (de Jong et al. 2020). It is important to note that the players involved in the industrial biobased production of SA have been constantly making new business arrangements.

Genetic engineered microorganisms have been explored to achieve the production of SA in high yields. As an example, Lin et al. (2005) did modifications in the glyoxylate cycle system to enhance SA production and reach the maximum theoretical yield (1 mol mol^{-1} glucose) in *E. coli*. This genetically modified strain under aerobic fermentation of synthetic media containing glucose achieved a yield of 0.94 mol mol^{-1} and productivity of 1.08 g L^{-1} h^{-1} (Lin et al. 2005). Wang et al. (2011) used corn stalk hydrolysates with a mix of sugars (glucose, xylose, and arabinose) to produce SA from a genetically modified *E. coli*. In bioreactor cultivation, the SA production reached a yield of 0.87 g g^{-1} total sugar and a productivity of 0.8 g L^{-1} h^{1} (Wang et al. 2011). Several studies with bacteria presented promising results; Zheng et al. (2010) employed *A. succinogenes* and corn fiber as substrate reaching an SA productivity of 0.99 g L^{-1} h^{-1} and yield of 0.72 g g^{-1} in a bioreactor (Zheng et al. 2010).

As discussed above, SA presents a considerable part in the international chemical market as a final product and chemical building block. In 2017, the companies started to operate using biomass with a capacity (if all projects were completed) of 55 000 tonnes $year^{-1}$. The prediction for production in 2020 was 140 000 tonnes $year^{-1}$ of biobased SA to be available. There is an expectation of SA market growth from $138.61 million in 2017 to $218.14 million by 2025 (Fior Markets 2019). In that way, the global demand for SA is constantly growing, showing it to be a promising chemical for the generation of more sustainable products.

6.4.2 Diols

6.4.2.1 1,3-Propanediol (PDO)

The linear aliphatic diol PDO ($C_3H_8O_2$) is an important chemical utilized as a chemical building block (Figure 6.1). The interest in this chemical has been growing in the last few years because of its range of applications, such as polymers, coatings, the personal care sector, solvents, and lubricants. PDO was identified for the first time by August Freund in a culture of *Clostridium pasteurianum* using glycerol as a substrate (Vilela et al. 2014). One of the main interests in producing PDO is the generation of polytrimethylene terephthalate polymer. It is a useful product for the textile industry, mainly because of the distinguished durability and stain resistance in contrast with nylon (Table 6.3) (Biddy et al. 2016).

PDO can be produced by two routes: chemical synthesis and microbial-based renewable feedstocks (Figure 6.1). The first is based on the hydration of acrelin or the hydroformylation of ethylene oxide and syngas to 1,3-hydroxypropanol with further processing to PDO. On the other hand, the microbial production of PDO has been explored during the past few years, becoming the main route utilized for its generation. The biobased PDO offers a favorable scenario using 40% less energy than the route based on petroleum feedstocks (Biddy et al. 2016). Furthermore, PDO offers some overall environmental advantage presenting a predicted GHG saving of 2.9 tonnes of CO_2 per tonne of product (Hermann et al. 2007).

The extensively investigated natural PDO producers are the facultative anaerobes like *Klebsiella* sp., *Enterobacter* sp., and *Lactobacillus* sp. and the strict anaerobes such as *Clostridium beijerinckii*, *Clostridium butyricum*, and *Clostridium diolis*. These microorganisms use glycerol, which is a by-product of biodiesel, as the main carbon source for PDO production (Wischral et al. 2016). Several studies are exploring the possibilities of utilizing glycerol and lignocellulosic biomass as co-substrate to enhance PDO production. Apiwatanapiwat and collaborators (2018) demonstrated that a strain of *C. butyricum* was capable of using starch-free cassava pulp fiber (CPF) as the co-substrate, boosting the production of PDO. It reached a yield and productivity of 0.662 mole of PDO per mole of glycerol and 0.31 g L^{-1} h^{-1} respectively; these values are higher than fermentation with glycerol alone (Apiwatanapiwat et al. 2018).

Several companies are investing in PDO production and commercialization (TRL 9); as examples are DuPont Tate & Lyle, Glory Biomaterial and Shenghong Group. Currently, among all the companies DuPont Tate & Lyle Bio Products is the biggest producer of biobased PDO with a capacity of 63 500 tonnes $year^{-1}$ (de Jong et al. 2020; DuPont Tate & Lyle Bio Products 2018). The microorganism used by this company in PDO production is a genetically modified *E. coli* that is able to achieve high titers (135 g L^{-1}) by aerobic fermentation (Nakamura and Whited 2003). In 2018, the company announced a plan to increase the annual production of biobased PDO by almost 16 000 tonnes (DuPont Tate & Lyle Bio Products 2018). The estimated global demand for PDO in 2014 was 125 000 tonnes $year^{-1}$ (Harmsen et al. 2013). The market size value in 2019 was estimated at $484.8 million with a growth rate of 13.4% from 2015 to 2022 (Market Analysis Report 2015a). In conclusion, the microbial biobased PDO presents a sustainable and renewable alternative to produce polymers and products of high value-added compared to fossil sources.

6.4.2.2 Propylene Glycol

Propylene glycol, also known as 1,2-propanediol (1,2-PDO) ($C_3H_8O_2$), is a non-toxic and biodegradable chemical with a broad range of industrial applications and a consolidated market (Figure 6.1, Table 6.3). It has been used in the synthesis of plastic and resins, cosmetics and fragrances, as an additive in nutrition products and as a coolant and antifreeze agent (Zhang et al. 2017; Rosales-Calderon and Arantes 2019; de Jong et al. 2020). It has a market volume of 2000 kilotonnes per annum and a value of $3.91 billion (Gérardy et al. 2020).

Currently, commercial 1,2-PDO is mainly produced via chemical processes from petrochemical resources and, to a lesser extent, through catalytic hydrogenolysis of glycerol, a by-product of the biodiesel industry (Rosales-Calderon and Arantes 2019; Gérardy et al. 2020). In addition to glycerol, several other renewable feedstocks can be used in 1,2-PDO production, including cellulosic materials, sugars, sugar alcohols such as xylitol and sorbitol, and LA (Marchesan et al. 2019). These renewable routes, based on chemical processes, are already used for 1,2-PDO production by several companies, such as ADM, Global Biochem, Oleon, HaldorTopsoe/Braskem (Gérardy et al. 2020; de Jong et al. 2020).

In the last decades, microbial production of 1,2-PDO has also gained attention. Some microorganisms, such as *Clostridium thermosaccharolyticum* and *Lactobacillus buchneri*, are capable of producing 1,2-PDO naturally (Saxena et al. 2010; Zhang et al. 2017). Three biochemical pathways have been described so far: (i) the first metabolic pathway uses deoxy sugars like fucose and rhamnose as initial substrates, (ii) the second involves the conversion of the glycolytic intermediate dihydroxyacetone phosphate to methylglyoxal, which is subsequently reduced to yield 1,2-PDO, and (iii) the third route is derived from the anaerobic degradation of LA (Saxena et al. 2010).

Moreover, several attempts to construct metabolic engineered strains have been reported using mainly *E. coli* and *S. cerevisiae* as hosts. Nonetheless, microbial production of 1,2-PDO is still incipient and usual concentrations do not surpass $10\,g\,L^{-1}$ from pure glucose or mixed sugars (Jain et al. 2015; Zhang et al. 2017; Marchesan et al. 2019). Only recently, better results were obtained with the implementation of a synthetic pathway based on the intermediate LA in *E. coli* ($17.3\,g\,L^{-1}$ of the *R*-1,2-PDO stereoisomer, with a molar yield of 42.2% from glucose and a productivity of $0.72\,g\,L^{-1}\,h^{-1}$) (Niu et al. 2019). Also, higher titers of 1,2-PDO were achieved by the wild-type *L. buchneri* strain using whey or lactose as the main substrates supplemented with glucose or xylose ($26–31\,g\,L^{-1}$ of 1,2-PDO) (Veeravalli and Mathews 2018).

The main problems observed with the 1,2-PDO biological production are related to cofactor regeneration, accumulation of toxic intermediates such as methylglyoxal or lactaldehyde, and the formation of by-products that reduce 1,2-PDO yields and can also make the downstream product separation difficult (Zhang et al. 2017). Therefore, biological and technological advances are required to make 1,2-PDO production based on a competitive microbial process possible.

6.4.3 Polyols

6.4.3.1 Sorbitol

Sorbitol (D-sorbitol) is a sugar alcohol with six carbons ($C_6H_{14}O_6$) (Figure 6.1) that can be used in several industry sectors. This compound was discovered in 1872 in a species of wild berry (*Sorbus aucuparia* L.); since then, other natural occurrences in berries and fruits have been found (Park et al. 2016; Rice et al. 2020). Sorbitol is a compound with high solubility and is a potential sweetener; it was found only in nature until 1935, when a pilot plant started to operate and produce $1400\,tonnes\,year^{-1}$ of sorbitol through the electroreduction of D-glucose from corn starch (Park et al. 2016).

Sorbitol demand has increased through the years and currently it is an important commodity with a market expected to achieve 2.4 million tonnes and worth $4 billion in 2023 (Park et al. 2016; Kumar et al. 2020). This important sugar-alcohol has a variety of applications, the most significant being in the food and beverages industry. Besides, sorbitol can be a building block compound for other products such as surfactants, polyurethanes, 1,2-PDO, and ascorbic acid (Table 6.3) (Antunes et al. 2017).

The industrial production of sorbitol occurs by the hydrogenation of glucose using the Raney Nickels catalysis; in this case, the D-glucose is 100% converted to sorbitol. The most common feedstocks for sorbitol production are cassava, wheat, or corn biomass (Antunes et al. 2017; Kumar et al. 2020). The US Department of Energy has included sorbitol as one of the 12 most important building block compounds derived from biomass (Rey-Raap et al. 2019). The most important industries in the field are Roquette Freres, Cargill, and SPI polyols, which together hold 70% of the global market of sorbitol (Antunes et al. 2017; Kumar et al. 2020).

Even though the current production of sorbitol is mainly by chemical processes, several researchers are engaged in exploring biotechnological solutions to produce sorbitol (Figure 6.1). The studies have been focused on the bacterium *Zymomonas mobilis*, which is able to naturally convert fructose and glucose to sorbitol because of a recent glucose–fructose oxidoreductase (GFOR) discovery, which can reduce fructose to sorbitol (Park et al. 2016; Rice et al. 2020). Liu et al. (2010) converted fructose (158.9 g L^{-1}) into sorbitol and showed a yield of 100% and a productivity of 3.06 g g^{-1} h^{-1}. Also, LAB have been genetically modified to produce sorbitol, specifically the species *Lactobacillus casei* and *L. plantarum*, by the oxidation of NADH into NAD^+ in micro or anaerobic conditions (Park et al. 2016; Antunes et al. 2017).

Despite the recent developments in microbial production of sorbitol, an industrial production process will not be reached in the next few years. The reported results are related to the academic fields and to the best of our knowledge no pilot-scale evaluations are being performed.

6.4.3.2 Xylitol

Xylitol ($C_5H_{12}O_5$) is a five carbon polyol commercialized worldwide; it has several applications in different industries, especially in the food, odontological, and pharmaceutical sectors. Xylitol is an insulin-independent low-calorie sweetener, which is safe for patients with diabetes. Also, it can prevent tooth decay and airway inflammation. Besides, as a platform chemical, xylitol can be further converted into several industrial compounds (e.g. polymers, ethylene glycol, propylene glycol, and esters). Xylitol is a chemical with a high global market value, estimated to generate $1.15 billion by 2023, and the global production was 200 000 tonnes in 2019 (Table 6.3) (Antunes et al. 2017; Felipe Hernández-Pérez et al. 2019; de Jong et al. 2020).

Nowadays, the industrial production of xylitol is only by chemical route, by the catalytic hydrogenation of xylose, from lignocellulosic biomass, in high temperature and pressure conditions (180 °C; 50 atm) (Figure 6.1). Even though this reaction can achieve high conversion rates, depending on the feedstock (xylan 50%–60% or pure xylose 98%), it is an expensive process, especially regarding the purification and crystallization of the xylitol, which represents 80% of the total cost of xylitol production (Carneiro et al. 2019; Felipe Hernández-Pérez et al. 2019). The current major producers of xylitol are DuPont Danisco, Roquette Freres, Cargill, Fortress Global, among others (Felipe Hernández-Pérez et al. 2019; de Jong et al. 2020).

The biotechnological production of xylitol can be an eco-friendly alternative to the current chemical routes. This process consists of xylose reduction into xylitol by the enzyme xylose reductase. The xylitol produced can be secreted or further converted into xylulose by a xylitol dehydrogenase enzyme, which may lead to a reduced production. A range of microorganisms can produce xylitol, either naturally or through metabolic engineering processes; among them, yeasts are the most studied microorganisms, for example species of *Candida*, *Meyerozyma*, *Spathaspora*, *Scheffersomyces*, *Debaryomyces*, and *Kluyveromyces* (Carneiro et al. 2019).

Efficient xylitol production in lignocellulosic hydrolysates has been achieved by employing different yeast species and fermentation strategies. *Candida tropicalis* XK12K can produce xylitol with yields as high as 0.98 g g^{-1} when cultivated in corn stover hydrolysate (Hong et al. 2016). In other studies, *C. tropicalis* JA2 was able to produce 109.5 g L^{-1} of xylitol with a yield of 0.86 g g^{-1} of consumed xylose and with a productivity of 2.81 g L^{-1} h^{-1} on sugarcane bagasse hydrolysate containing 8.0 g L^{-1} acetic acid (Morais et al. 2019). Recently, a screening of 960 wild yeasts was reported in which a new strain of *Wickerhamomyces anomalus* was found and was able to convert xylose from sugarcane bagasse hydrolysates into xylitol with a maximum conversion yield of 0.83 g g^{-1}. In the same study, a strain of *Spathaspora* sp. JA1 also presented a high yield on xylitol production (0.74 g g^{-1}) in sugarcane bagasse hydrolysate containing 6 g L^{-1} of acetic acid (Carneiro et al. 2019). The best reports indicate that bioprocesses for xylitol production are under evaluation in relevant operational conditions (bioreactors using biomass hydrolysates) and in a maximum on a pilot scale, characterizing a TRL 6 (Figure 6.1) (E4tech, RE-CORD and WUR 2015; E4tech 2017; de Jong et al. 2020).

Biorefinery integration processes can improve biotechnological xylitol production. Biorefineries aim to optimize the use of substrates and energy and add co-products in the production chain, to increase the value of the production chain and avoid waste (Mountraki et al. 2017). With that in mind, some companies, like Dupont Danisco and Thomson Biotech, that produce xylitol chemically are looking into adopting biotechnological routes in their commercial production (Felipe Hernández-Pérez et al. 2019; de Jong et al. 2020).

6.4.4 Alcohols

6.4.4.1 Butanol

Butanol is a primary alcohol, also known as butyl alcohol, *n*-butanol and butan-1-ol ($C_4H_{10}O$), produced mainly by chemical synthesis (Figure 6.1). The most common process is the Oxo process, in which synthesis gas (a mixture of CO and H_2) is first reacted with propylene to yield butyraldehyde, which is subsequently hydrogenated to butanol (Ranjan and Moholkar 2012). Butanol can be used as a solvent in the auto industry, as a "drop-in" fuel and in the polymers and plastics industries (Table 6.3) (Green 2011; Mariano et al. 2013). Compared to ethanol, butanol has a higher energy density, lower vapor pressure, and lower flashpoint, making its handling safer. It is less volatile and it can be used directly in engines designed for gasoline, due to the similar properties of both (Green 2011).

Biologically, it can be produced from renewable sources (biobutanol), with the acetone-butanol-ethanol (ABE) fermentation process being the most well known and popular, with more than 100 years of history. The first part consists of the production of acetic and butyric acid by bacteria of the genus *Clostridium* and is known as acidogenesis. In the second stage, solventogenesis, acetone, butanol, and ethanol are formed. After fermentation, butanol is recovered and purified (Ranjan and Moholkar 2012). Louis Pasteur was the first to report the discovery about the production of butanol as a product of fermentation by microorganisms, in 1862. He discovered that this compound was normally produced during butyric fermentation of LA (Gabriel and Crawford 1930; Sauer 2016).

During 1912 and 1914, Chaim Weizmann concluded that the production of butanol and isoamyl alcohol by fermentation was essential in the synthetic rubber manufacturing process. He was responsible for isolating and studying some cultures, one of which was called BY, which was later called *Clostridium acetobutylicum*. This microorganism showed the ability to use a variety of substrates, composed mainly of starch, to produce significantly greater amounts of butanol and acetone compared to the original culture isolated by

Fernbach in 1911, capable of fermenting potato starch but not corn starch (Jones and Woods 1986; Sauer 2016). The fermentation process for *C. acetobutylicum* developed by Weizmann yielded a patent in 1919 (US Patent 1315585). Since then, other strains have been discovered, such as *C. beijerinckii*, *Clostridium saccharobutylicum*, and *Clostridium saccharoperbutylacetonicum* (Keis et al. 2001; Jiang et al. 2018).

During the 1920s to 1980s, several countries such as England, the United States, Egypt, India, the former Soviet Union, Japan and China produced butanol on an industrial scale. However, fermentative ABE production of butanol became economically infeasible in the 1960s and could not compete with petrochemically derived butanol due to the development of petrochemical industries (García et al. 2011; Sauer 2016). The production cost caused by problems involved in the process (raw material price, low butanol titers and yields, low volumetric solvent productivity and recovery, high water usage) led to a decline in biobutanol production (Green 2011). However, concerns for environmental problems, such as global warming, pollution, and climate change, and high prices of petrol fuel in the last decades are generating new attempts to raise biobutanol yield (Ibrahim et al. 2017). In this regard, studies have predicted that biobased butanol can reduce the GHG in 3.9 tonnes of CO_2 per tonne of product (Hermann et al. 2007).

Today, many biotechnological companies such as DuPont, BP and GEVO are investing in new alternatives and technologies to enhance the process involved in biobutanol production on an industrial scale, for example the development of superior microbial strains through molecular and metabolic engineering, the use of concentrated sugar solutions, high productivity reactor systems, simultaneous fermentation and product recovery from bioreactors, and production of butanol from less costly sources (Algayyim et al. 2018; Abo et al. 2019; Qureshi et al. 2020). The market for biobutanol was valued at \$7.86 billion in 2014 (Table 6.3) and it is estimated to grow to \$17.78 billion by 2022 (Market Analysis Report 2015b). In this way, biobutanol is an important and competitive biochemical that can be produced as an alternative to petrochemical sources.

6.4.4.2 Ethanol

Ethanol, also known as ethyl alcohol or 1-hydroxyethane (C_2H_5OH) (Figure 6.1), is the main biotechnological commodity globally produced, both volumetrically and economically, as the primary liquid fuel made from a renewable carbon source (Favaro et al. 2019; Rosales-Calderon and Arantes 2019). Ethanol utilization as a biofuel goes back to 1826, when the American inventor Samuel Morey designed an internal combustion engine for a boat, fueled by a mixture of ethanol and turpentine. Later, in 1860, Nicolaus August Otto, a German engineer, developed another internal combustion engine fueled by a blend containing ethanol, followed by American industrialist Henry Ford who constructed tractors that could be powered by ethanol. Since then, ethanol has been identified as the most widely used biofuel in the world (Datta et al. 2011; Parapouli et al. 2020). The interest in using ethanol has been increasing since the 1980s, because of the global concern in reducing consumption of fossil fuels and emissions of GHG (Azhar et al. 2017).

Industrially, ethanol can be produced from a variety of raw materials from the first and second generation. The first-generation ethanol consists of feedstocks rich in sucrose (sugarcane, sugar beet, sweet sorghum, and fruits) and starch (corn, wheat, rice, potato, cassava, sweet potato, and barley). Second-generation ethanol comes from lignocellulosic biomass such as wood, straw, and grasses, and the third generation is derived from algal biomass, including microalgae and macroalgae (Robak and Balcerek 2018; Bergmann et al. 2019;). The United States and Brazil are the largest producers of ethanol worldwide; using corn and sugarcane as raw materials they produced 24 368 million gallons in 2019 from the 29 000 produced worldwide (Table 6.3) (Renewable Fuels Association 2020). In 2019, the international ethanol market was valued at $89.1 million (Table 6.3) with a predicted annual growth rate of 4.8% from 2020 to 2027 (Market Analysis Report 2020a).

The common processes involved in ethanol production are pretreatment and hydrolysis of the substrate for release of the sugars, and conversion of those into ethanol by microbial fermentation (Azhar et al. 2017). Industrial ethanol fermentation is mainly carried out by yeasts, especially *S. cerevisiae*, due to its high ethanol productivity, high ethanol tolerance, and ability to support industrial conditions. Ethanol production from sucrose-containing feedstocks is easier and more efficient than that from starch-based materials, since these feedstocks require pretreatment (liquefaction and saccharification with amylolytic enzymes) before fermentation (Clark et al. 2012). However, there are still some industrial challenges for second-generation ethanol production, especially the presence of inhibitors derived from biomass pretreatment, as discussed previously (Section 6.2), and the inability to ferment pentose sugars. *S. cerevisiae* cannot naturally metabolize xylose and arabinose, sugars present in abundance in several lignocellulosic biomasses. Therefore, industrial pentose-fermenting yeasts have been identified, characterized, and improved through genetic engineering and prolonged evolutionary engineering strategies (Almeida et al. 2011; Huang et al. 2019).

Manufacturing of first-generation ethanol is a relatively simple process, and its main disadvantage is the high price of the sugar/starch feedstocks, which accounts for 40%–70% of the total ethanol cost (Favaro et al. 2019; Rosales-Calderon and Arantes 2019). Besides, sugar- and starch-containing feedstocks compete with their use as food or feed, influencing their supply, which leads to concerns regarding the socio-economic and environmental consequences of large-scale production (Robak and Balcerek 2018). Therefore, lignocellulosic biomass (second generation) represents a promising alternative feedstock for ethanol production due to its low cost, availability and wide distribution, plus it is not competitive with food and feed crops. Other than that, ethanol obtained from this feedstock can considerably reduce the emission of GHG (Bušić et al. 2018). Some studies suggest that bioethanol can reduce GHG annually by 2.7 tonnes of CO_2 per tonne of product (Hermann et al. 2007).

Nevertheless, lignocellulose must be converted to fermentable sugars before it can be fermented to ethanol, and processing this material is still the object of intense research. Feedstock sourcing, pretreatment, production of inhibitory compounds, very high enzyme costs, and harsh fermentation conditions are some of the problems that have to be addressed before second-generation ethanol production can become economically feasible worldwide (Favaro et al. 2019). In 2015, the world had an installed capacity to produce 1390.48 million liters of second-generation ethanol, from the United States 35%, China 24%, Canada 22%, European Union 9%, and Brazil 9%. However, project delays and technical difficulties have led to the operation of few industries so far.

Data show that global ethanol production is constantly increasing. From 2007 to 2019, production went from 49.66 billion to 1107 billion liters (Azhar et al. 2017; Renewable Fuels

Association 2020) The prediction of worldwide ethanol production and its consumption is that it will reach nearly 134.5 billion liters by 2024 (Bušić et al. 2018). As mentioned previously, ethanol is mainly used as fuel, but it is also being used as a substrate for green polyethylene production by Braskem (2020). In this respect, ethanol represents a potential in market growth as the biggest biochemical produced nowadays.

6.4.5 Aldehydes

6.4.5.1 Furfural

Furfural ($C_5H_4O_2$) is a naturally occurring aryl-aldehyde (Figure 6.1) presenting as a colorless to pale oil at room temperature which when exposed to oxygen and light undergoes oxidation, becoming reddish-brown (Machado et al. 2016). Due to the presence of an aldehyde group and a system of double bonds, furfural can go through several reactions (Lange et al. 2012) and thus has broad commercial uses, for example industrial manufacturing, personal care products, food flavoring, and pesticides (Table 6.3) (Reed and Kwok et al. 2014).

About 60%–70% of produced furfural is used for the production of furfuryl alcohol, through C=O hydrogenation (Bhogeswararao and Srinivas 2015). Furfuryl alcohol has emerged in 2019 as the predominant application in the synthesis of furan resins, which are used in the further production of thermoset polymer matrix composites, cement, coatings, and adhesives (Market Analysis Report 2020b). It is also widely used in the production of tetrahydrofurfuryl alcohol, an important intermediate in the food and cosmetic industries (Machado et al. 2016). Through hydrogenation and decarboxylation processes, furfural can also be transformed into secondary building blocks like tetrahydrofurfuryl alcohol, 2-methylfuran, furan, tetrahydrofuran, *n*-butanol, and diols (Bhogeswararao and Srinivas 2015).

Microbial synthesis of furanic compounds occurs rarely in nature. As these molecules are highly reactive, they present potentially toxic effects on both prokaryote and eukaryote cells, including damage to organelles, proteins, and nucleic acids (Allen et al. 2010). However, some studies are investigating genetically engineered microorganisms that are capable of utilizing furfural as a substrate in aerobic processes for the production of biobased fuels and chemicals from biomass-derived sugars. Guarnieri et al. (2017) engineered a *P. putida* KT2440 to utilize furfural as a carbon source, enabling its conversion and enhancing tolerance to hydrolysate inhibitors (Guarnieri et al. 2017).

Furfural has two major functional groups, an aldehyde (CHO) and two olefin groups (CH=CH). It is obtained through acid hydrolysis (a proton-donating acid catalyzes the reaction transforming pentosan into pentose and pentose into furfural) of five-carbon sugars such as xylose and arabinose from the hemicellulose fraction of lignocellulosic waste biomass (Mathew et al. 2018; de Jong et al. 2020). Furfural market growth is driven by the increased use of its derivatives in diverse green chemical industries. The majority of its global manufacturing comes from corn cobs, due to the high pentosan content and fast growth of the crops (Sheldon 2014).

The leading player in the furfural market is China, with the production of approximately 30 million tonnes year^{-1} based on corn cobs, followed by the Dominican Republic, based on sugarcane bagasse (de Jong et al. 2020). The first method for furfural production was developed in 1921 by Quaker Oats Co., which employed sulfuric acid. Since the 1990s, numerous innovative approaches such as those by Biofine®, CIMV®, SupraYield®, MTC®, and Westpro® have been developed to raise furfural yield (Kumar and Wyman 2013). The current furfural market price is around $1 per kilogram and $1.76 per kilogram for furfuryl alcohol; however, it still depends on Chinese supply (Hayes et al. 1999). Nowadays, the market size for furfural is estimated at $814.6 million in 2019 (Table 6.3) and a prospective growth rate of 5.0% from 2020 to 2027 is predicted (Market Analysis Report 2020b).

6.4.6 Polyesters

6.4.6.1 Polyhydroxyalkanoates (PHAs)

PHAs are a class of biobased polymers produced naturally by many Eubacteria and Archaea strains (Figure 6.1). The physical and chemical properties of these polymers are similar to synthetic plastics derived from petroleum. However, unlike those, PHAs are completely biodegradable under several environmental conditions and by different microorganisms (Pagliano et al. 2017). Therefore, PHAs have the potential to replace the main plastic polymers of the global market (polypropylene, polyethylene, and polystyrene polymers) and be applied in the biomedical and pharmaceutical industry, since they are non-toxic compounds, compatible, and naturally absorbed by the body (Table 6.3) (Pagliano et al. 2017; Utsunomia et al. 2020).

Several bacteria have been isolated and identified as PHA producers; *Cupriavidus necator* (formerly *Ralstonia eutropha*), *Burkholderia*, *Pseudomonas*, *Azotobacter*, *Azohydromonas lata*, and recombinant *E. coli* are the major microorganisms used on industrial scales (Jiang et al. 2016; Dietrich et al. 2017; Al-Battashi et al. 2019). In the cells, PHAs are accumulated in intracellular granules in the cytoplasm, which serve as material storage of carbon and energy. Usually, they are synthesized when the bacteria are cultivated under physiological stress conditions such as in excess of a carbon source and limitation of other critical nutrients such as nitrogen and phosphorus (Koller 2018). Different monomers with short, medium, or long chain length can be produced and combined to generate PHA polymers with diversified specificities (such as crystallinity and elasticity). It depends on the bacterial producer strain, the substrate source, and the type of PHA synthase catalyzing the polymerization (Koller 2018; Utsunomia et al. 2020). The most common types are polyhydroxybutyrate (PHB) and the co-polymers polyhydroxybutyrate-*co*-hydroxyvalerate (Dietrich et al. 2017).

The production process in the pilot and industrial plants is conducted in fermentation reactors using glucose, fructose, sucrose, or fatty acids as the main substrates (Cesário et al. 2014; Jiang et al. 2016). *R. eutropha* has been commonly employed in PHA production because it achieves a high cell density (100–200 $g L^{-1}$) containing 75%–80% of PHA (Xu et al. 2010; Dietrich et al. 2017). *E. coli* is also used, since it is easily genetically modified to produce recombinant strains with fast growth and the ability to utilize inexpensive carbon sources. As an example, a recombinant *E. coli* strain showed high final cell density and PHB content (206 $g L^{-1}$ and 73%, respectively) with a productivity of 3.4 $g L^{-1} h^{-1}$ (Dietrich et al. 2017).

To reduce the PHA production cost, several attempts have been focused on the utilization of lignocellulosic feedstock. Fed-batch cultivation of *Burkholderia sacchari* on wheat straw hydrolysate, which was previously pretreated with the AFEX technology, yielded a PHB concentration of 105 $g L^{-1}$, corresponding to 72% cell content, and productivity of 1.6 $g L^{-1} h^{-1}$ (Cesário et al. 2014). Recently, a pilot-scale study demonstrated the feasibility of PHA production from a non-detoxified wood hydrolysate (Dietrich et al. 2020). In this report, a PHB concentration of 34.5 g L^{-1}, with final content of 58% and productivity of 0.72 g L^{-1} h^{-1}, was reached with hydrolysate fermentation by the wild-type *Paraburkholderia sacchari* strain. In another study, improvements in PHB production from xylose were obtained with the over-expression of xylose activated regulator in the bacterium (Guamán et al. 2018). For a more detailed review on PHA production from several substrates, such as lignocellulosic hydrolysates, whey or starch-based materials, molasses and sucrose, see Al-Battashi et al. (2019), Pagliano et al. (2017) and Jiang et al. (2016).

Since 1980, PHAs have been commercially produced as an alternative to replacing various conventional polymers. Nonetheless, over the years, the industrial interest was affected by fluctuations in oil prices (Dietrich et al. 2017). Currently, the PHA manufacturing cost (€2.2–€5.0 kg^{-1}) is still not competitive compared to petrochemical-based polymers (≤€1.0 kg^{-1}) or to other biobased polymers such as PLA (Dietrich et al. 2017; Pagliano et al. 2017). High manufacturing cost is attributed to diverse factors, including common utilization of pure substrates, processing methods that must be adapted to each specific polymer type, and the expensive cost of downstream recovering processes (Utsunomia et al. 2020). Despite this, several companies are producing PHA products worldwide from different substrates (TRL 9). Examples of this are the companies Metabolix/USA (corn sugars), BioOn/Italy (beet sugar), Biomer/Germany (sucrose), TianAn Biopolymer/China (glucose from corn or cassava), Tianjin GreenBio/China (sucrose), and PHB Industrial SA/Brazil (sugarcane) (Jiang et al. 2016; Dietrich et al. 2017).

The diversified PHA properties along with the full biodegradability show the potential application of this polymer. However, the high manufacturing cost is still a barrier to overcome. Promising strategies to reduce the cost of production include the integration of PHA production in biorefineries, usage of cheaper agro-industrial waste residues as substrate, selection of robust native producers and/or development of engineered strains able to use several types of sugars and substrates with high yields and productivities, and technological advances in downstream extraction and purification methods. These future improvements and the development of new biomaterial products can contribute to a PHA market expansion.

6.4.7 Xylenes

6.4.7.1 Para-xylene

p-Xylene is an aromatic hydrocarbon (Figure 6.1). It is one of the three isomers of dimethylbenzene known collectively as xylenes. It is an important chemical commonly used as a building block to produce dimethyl-terephthalate (DMT) and purified terephthalic acid (PTA). *p*-Xylene monomers can be further converted into PET, the major component of plastic bottles, polyester, resins, and films. The global market of *p*-xylene was evaluated at $48.72 billion in 2019, and it is expected to grow 10.5% per year, with predictions of achieving $66.9 billion in 2022; thus, it is one of the largest chemical markets in the world (Table 6.3) (Peters et al. 2011; Williams et al. 2012; E4tech 2017; de Jong et al. 2020).

Today, *p*-xylene is produced by chemical routes, by separating aromatic mixtures from petroleum naphtha or crude oil refining, under high pressure and temperature (Peters et al. 2011; Williams et al. 2012; E4tech 2017). However, its production can be semi-biological (hybrid), as described in the US Patent 2011/0087.000 A1. This work shows that it is possible to produce *p*-xylene with high purity from renewable carbon sources, as lignocellulosic biomass. In the process, the lignocellulosic feedstock (starch, monosaccharides and polysaccharides, pretreated cellulose and hemicellulose, lignin, and pectin) is first converted into isobutanol by one or several microorganisms (yeasts, fungi, bacteria, and archaea), used as biocatalysts to ferment the primary broth to produce *p*-xylene. Then, the isobutanol is purified and goes through a series of chemical reactions (dehydration, dimerization, and aromatization) to finally produce pure *p*-xylene (Peters et al. 2011). Currently, it is known that the hybrid *p*-xylene production is on a pilot or demonstration scale (Figure 6.1) (Peters et al. 2011; E4tech 2017; de Jong et al. 2020).

Several major companies are searching for new metabolic pathways that enable the entire *p*-xylene production to be biological and renewable, removing all petroleum-based chemicals from the process. Since there is a large economic interest in reducing the damage to the environment and the cost of production, industries such as Coca-Cola and PepsiCo are aiming to produce plastic bottles that are entirely biodegradable. Recently a patent was published regarding a fully biological production process of biobased PET polymers (Briois 2019). Also, some other companies are in this search, such as Gevo and Micromidas (de Jong et al. 2020).

6.5 Conclusion

The production of biobased chemicals has gained much attention in the last few years due to economic and environmental aspects. The possibility of using biomass as feedstock instead of fossil resources for the production of industrially relevant chemicals leads to the development of several chemical, biological, and hybrid processes. In this context, the use of microorganisms for the conversion of sugars and lignin of hydrolyzed biomass stands out as an interesting route for the production of chemicals with diverse applications. Different strategies of bioprospecting and genetic improvement of several microbial species have been employed to overcome challenges with biomass hydrolysis and conversion of different types of sugars present in inhibitory hydrolysates. In general, production processes for drop-in chemicals that can be used as building blocks or fuels are more advanced on the development scale than processes for "new" chemicals. Several renewable chemicals are being produced nowadays by microbial conversion of biomass sugars. SA, LA, and PDO are being produced mainly through the bacterial conversion of sugars, whereas ethanol has been widely produced by yeast fermentation. Several other biological processes for the production of xylitol, adipic acid, butanol, and sorbitol are under development, in some cases already at pilot scale. In addition, biobased propylene glycol and IA production have been achieved at the laboratory and commercial scale, respectively, through a hybrid involving a microbial and a chemical step.

References

Abdel-Rahman, M.A., Tashiro, Y., and Sonomoto, K. (2011). Lactic acid production from lignocellulose-derived sugars using lactic acid bacteria: overview and limits. *J. Biotechnol.* 156 (4): 286–301. https://doi.org/10.1016/j.jbiotec.2011.06.017.

Abo, B.O., Gao, M., Wang, Y. et al. (2019). Production of butanol from biomass: recent advances and future prospects. *Environ. Sci. Pollut. Res.* 26 (20): 20164–20182. https://doi.org/10.1007/s11356-019-05437-y.

Agbor, V.B., Cicek, N., Sparling, R. et al. (2011). Biomass pretreatment: fundamentals toward application. *Biotechnol. Adv.* 29 (6): 675–685. https://doi.org/10.1016/j.biotechadv.2011.05.005.

Ahn, J.H., Jang, Y.S., and Lee, Y.L. (2017). Succinic aci. In: *Industrial Biotechnology: Products and Processes* (ed. C. Wittmann and J.C. Liao), 505–544. London: Wiley.

Akhtar, J., Idris, A., and Aziz, R.A. (2014). Recent advances in production of succinic acid from lignocellulosic biomass. *Appl. Microbiol. Biotechnol.* 98 (3): 987–1000. https://doi.org/10.1007/s00253-013-5319-6.

Al-Battashi, H.S., Annamalai, N., Sivakumar, N. et al. (2019). Lignocellulosic biomass (LCB): a potential alternative biorefinery feedstock for polyhydroxyalkanoates production. *Rev. Environ. Sci. Biotechnol.* 18 (1): 183–205. https://doi.org/10.1007/s11157-018-09488-4.

Algayyim, S.J.M., Wandel, A.P., Yusaf, T., and Hamawand, I. (2018). Production and application of ABE as a biofuel. *Renew. Sust. Energ. Rev.* 82: 1195–1214. https://doi.org/10.1016/j.rser.2017.09.082.

Allen, S.A., Clark, W., McCaffery, M. et al. (2010). Furfural induces reactive oxygen species accumulation and cellular damage in *Saccharomyces cerevisiae*. *Biotechnol. Biofuels* 3: 1–10. https://doi.org/10.1186/1754-6834-3-2.

Almeida, J.R.M., Runquist, D., Nogue, V.S. et al. (2011). Stress-related challenges in pentose fermentation to ethanol by the yeast *Saccharomyces cerevisiae*. *Biotechnol. J.* 6 (3): 286–299. https://doi.org/10.1002/biot.201000301.

Antunes, F.A.F., Dos Santos, J.C., Da Cunha, M.A.A. et al. (2017). Biotechnological production of xylitol from biomass. In: *Production of Platform Chemicals from Sustainable Resources.*, 311–342. Springer https://doi.org/10.1007/978-981-10-4172-3.

Apiwatanapiwat, W., Vaithanomsat, P., Thanapase, W. et al. (2018). Xylan supplement improves 1,3-propanediol fermentation by *Clostridium butyricum*. *J. Biosci. Bioeng.* 125 (6): 662–668. https://doi.org/10.1016/j.jbiosc.2017.12.006.

Avantium (2020). *Avantium to build FDCA flagship plant at Chemie Park Delfzijl, Netherlands.* Available at: https://www.avantium.com/2020/avantium-to-build-fdca-flagship-plant-at-chemie-park-delfzijl-netherlands (Accessed: 15 January 2021).

Azhar, S.H.A., Abdulla, R., Jambo, S.A. et al. (2017). Yeasts in sustainable bioethanol production: a review. *Biochem. Biophy. Rep.* 10 (February): 52–61. https://doi.org/10.1016/j.bbrep.2017.03.003.

Bain & Company, Gas Energy (2014). *Diversificação da Indústria Química Brasileira:sumario do estudo.* São Paulo, 30 pp.

Bergmann, J.C., Trichez, D., Junior, W.G.M. et al. (2019). Biotechnological application of non-conventional yeasts for xylose valorization. In: *Non-conventional Yeasts: From Basic Research to Application*, 23–74. Berlin/Heidelberg: Springer https://doi.org/10.1007/978-3-030-21110-3_2.

Bhatia, S.K., Jagtapo, S.S., Bedekar, A.A. et al. (2020). Recent developments in pretreatment technologies on lignocellulosic biomass: effect of key parameters, technological improvements, and challenges. *Bioresour. Technol.* 300: 122724. https://doi.org/10.1016/j.biortech.2019.122724.

Bhogeswararao, S. and Srinivas, D. (2015). Catalytic conversion of furfural to industrial chemicals over supported Pt and Pd catalysts. *J. Catal.* 327 (July): 65–77. https://doi.org/10.1016/j.jcat.2015.04.018.

Biddy, M.J., Scarlata, C.J., Kinchin, C.M. (2016). Chemicals from biomass: a market assessment of bioproducts with near-term potential, *NREL Report*, No. NREL/TP-5100-65509, Golden, CO, USA (March). https://doi.org/10.2172/1244312.

Bozell, J.J. and Petersen, G.R. (2010). Technology development for the production of biobased products from biorefinery carbohydrates—the US Department of Energy's "top 10" revisited. *Green Chem.* 12 (4): 539–555. https://doi.org/10.1039/b922014c.

Braga, M., Ferreira, P.M., and Almeida, J.R.M. (2020). Screening method to prioritize relevant bio-based acids and their biochemical processes using recent patent information. *Biofuels, Bioprod. Biorefining* 15 (1): 231–249. http://doi.org/10.1002/bbb.2156.

Braskem (2020). *I'M GREENTM POLYETHYLENE*. Available at: http://plasticoverde.braskem.com.br/site.aspx/Im-greenTM-Polyethylene (Accessed: 15 January 2021).

Bridgwater, A.V., Chinthapalli, R., Smith, P.W. (2010). *Identification and Market Analysis of Most Promising Added-Value Products to be Co-produced with the Fuels*. [Online] (2010). Available: https://www.bioref-integ.eu/fileadmin/bioref-integ/user/documents/D2total__including_D2.1__D2.2__D2.3_.pdf/ (23 Octuber 2020).

Briois, J.-F. (2019). Process for Producing a Bio-based Polyethylene Terephthalate (PET) Polymer, Entirely from Bio-based Materials, IN201917004769, Nestec S.A.

Burgard, A.P., Pharkya, P. Osterhout, R.E. (2010). Microorganisms for the Production of Adipic Acid and Other Compounds, US7799545B2, Genomatica, Inc.

Buschke, N., Schafer, R., Becker, J., and Wittmann, C. (2013). Metabolic engineering of industrial platform microorganisms for biorefinery applications—optimization of substrate spectrum and process robustness by rational and evolutive strategies. *Bioresour. Technol.* 135: 544–554. https://doi.org/10.1016/j.biortech.2012.11.047.

Bušić, A., Mardetko, N., Kundas, S. et al. (2018). Bioethanol production from renewable raw materials and its separation and purification: a review. *Food Technol. Biotechnol.* 56 (3): 289–311. https://doi.org/10.17113/ftb.56.03.18.5546.

Carneiro, C.V.G.C., de Paula e Silva, F.C., and Almeida, J.R.M. (2019). Xylitol production: identification and comparison of new producing yeasts. *Microorganisms* 7 (11): 1–15. https://doi.org/10.3390/microorganisms7110484.

Cesário, M.T., Raposo, R.S., de Almeida, M.C.D. et al. (2014). Enhanced bioproduction of poly-3-hydroxybutyrate from wheat straw lignocellulosic hydrolysates. *New Biotechnol.* 31 (1): 104–113. https://doi.org/10.1016/j.nbt.2013.10.004.

Clark, J.H., Luque, R., and Matharu, A.S. (2012). Green chemistry, biofuels, and biorefinery. *Annu. Rev. Chem. Biomol. Eng.* 3: 183–207. https://doi.org/10.1146/annurev-chembioeng-062011-081014.

da Cruz, J.C., de Castro, A.M., and Sérvulo, E.E.F. (2018). World market and biotechnological production of itaconic acid. *3 Biotech* 8 (3): 1–15. https://doi.org/10.1007/s13205-018-1151-0.

Cubas-Cano, E., Gonzales-Fernandez, C., Ballesteros, M., and Tomas-Pejo, E. (2018). Biotechnological advances in lactic acid production by lactic acid bacteria: lignocellulose as novel substrate. *Biofuels Bioprod. Biorefin.* 12 (2): 290–303. https://doi.org/10.1002/bbb.1852.

Datta, R., Maher, M.A., Jones, C., and Brinker, R.W. (2011). Ethanol—the primary renewable liquid fuel. *J. Chem. Technol. Biotechnol.* 86 (4): 473–480. https://doi.org/10.1002/jctb.2580.

Deng, Y., Ma, L., and Mao, Y. (2016). Biological production of adipic acid from renewable substrates: current and future methods. *Biochem. Eng. J.* 105: 16–26. https://doi.org/10.1016/j.bej.2015.08.015.

Dietrich, K., Dumont, M., Del Rio, L.F., and Orsat, V. (2017). Producing PHAs in the bioeconomy—towards a sustainable bioplastic. *Sustain. Prod. Cons.* 9: 58–70. https://doi.org/10.1016/j.spc.2016.09.001.

Dietrich, K., Oliveira-Filho, E.R., Dumont, M. et al. (2020). Increasing PHB production with an industrially scalable hardwood hydrolysate as a carbon source. *Ind. Crop. Prod.* 154 (June): 112703. https://doi.org/10.1016/j.indcrop.2020.112703.

DuPont Tate & Lyle (2018). Expanding Bio-PDO Production in Tennessee by 35 million pounds. [Online]. Available at: https://duponttateandlyle.com/dupont-tate-lyle-bio-products-expanding-bio-based-propanediol-production-in-tennessee/ (Accessed: 5 October 2020).

E4tech (2017). UK Top Bio-based Chemicals Opportunities, Report for Biomass Refinery Network (December), pp. 44. Available at: https://www.bbnet-nibb.co.uk/wp-content/uploads/2019/08/LBNet-UKBioChem10_UK-Top-Bio-based-Chemicals-Opportunities_Dec2017.pdf.

E4tech, RE-CORD and WUR (2015). From the sugar platform to biofuels and biochemicals, *Final report for the European Commission Directorate-General Energy*, pp. 183. Contract No. ENER/C2/423–2012/SI2.673791.

Efe, Ç., van der Wielen, L.A.M., and Straathof, A.J.J. (2013). Techno-economic analysis of succinic acid production using adsorption from fermentation medium. *Biomass Bioenergy* 6 (13): 479–492. https://doi.org/10.1016/j.biombioe.2013.06.002.

Elkins, J.G., Raman, B., and Keller, M. (2010). Engineered microbial systems for enhanced conversion of lignocellulosic biomass. *Curr. Opin. Biotechnol.* 21 (5): 657–662. https://doi.org/10.1016/j.copbio.2010.05.008.

Farwick, A., Bruder, S., Schadeweg, V. et al. (2014). Engineering of yeast hexose transporters to transport d-xylose without inhibition by d-glucose. *Proc. Natl. Acad. Sci. USA* 111 (14): 5159–5164. https://doi.org/10.1073/pnas.1323464111.

Fatma, S., Hameed, A., Noman, M. et al. (2018). Lignocellulosic biomass: a sustainable bioenergy source for the future. *Protein Pept. Lett.* 25: 148–163. https://doi.org/10.2174/0929866525666180122144504.

Favaro, L., Jansen, T., and van Zyl, W.H. (2019). Exploring industrial and natural *Saccharomyces cerevisiae* strains for the bio-based economy from biomass: the case of bioethanol. *Crit. Rev. Biotechnol* 39 (6): 800–816. https://doi.org/10.1080/07388551.2019.1619157.

Felipe Hernández-Pérez, A., Jofre, F.M., Queiroz, S.S. et al. (2019). Biotechnological production of sweeteners. In: *Biotechnological Production of Bioactive Compounds*, 261–292. Elsevier https://doi.org/10.1016/B978-0-444-64323-0.00009-6.

Fitzpatrick, S.W., Framingham, M. (1990). Lignocellulose Degradation to Furfural and Levulinic Acid, US4897497A, Biofine Tech LLC.

Francois, J.M., Alkim, C., and Morin, N. (2020). Engineering microbial pathways for production of bio-based chemicals from lignocellulosic sugars: current status and perspectives. *Biotechnol. Biofuels* 13 (1): 1–23. https://doi.org/10.1186/s13068-020-01744-6.

Gabriel, C.L. and Crawford, F.M. (1930). Development of the butyl-acetonic fermentation industry. *Ind. Eng. Chem.* 22 (11): 1163–1165.

Galanie, S., Thodey, K., Trenchard, I.J. et al. (2015). Complete biosynthesis of opioids in yeast. *Science* 349 (6252): 1095–1100.

Galbe, M. and Wallberg, O. (2019). Pretreatment for biorefineries: a review of common methods for efficient utilisation of lignocellulosic materials. *Biotechnol. Biofuels* 12 (1): 1–26. https://doi.org/10.1186/s13068-019-1634-1.

Gao, M., Ploessl, D., and Shao, Z. (2019). Enhancing the co-utilization of biomass-derived mixed sugars by yeasts. *Front. Microbiol.* 10 (Jan): 1–21. https://doi.org/10.3389/fmicb.2018.03264.

García, V., Pakkila, J., Ojamo, H. et al. (2011). Challenges in biobutanol production: how to improve the efficiency? *Renew. Sust. Energ. Rev.* 15 (2): 964–980. https://doi.org/10.1016/j.rser.2010.11.008.

Gérardy, R., Debecker, D.P., Estager, J. et al. (2020). Continuous flow upgrading of selected C2–C6 platform chemicals derived from biomass. *Chem. Rev.* 120 (15): 7219–7347. https://doi.org/10.1021/acs.chemrev.9b00846.

Green, E.M. (2011). Fermentative production of butanol—the industrial perspective. *Curr. Opin. Biotechnol.* 22 (3): 337–343. https://doi.org/10.1016/j.copbio.2011.02.004.

Guamán, L.P., Barba-Ostria, C., Zhang, F. et al. (2018). Engineering xylose metabolism for production of polyhydroxybutyrate in the non-model bacterium *Burkholderia sacchari*. *Microb. Cell Factories* 17 (1): 1–11. https://doi.org/10.1186/s12934-018-0924-9.

Guarnieri, M.T., Franden, M.A., Johnson, C.W., and Beckham, G.T. (2017). Conversion and assimilation of furfural and 5-(hydroxymethyl)furfural by *Pseudomonas putida* KT2440. *Metab. Eng. Commun.* 4 (February): 22–28. https://doi.org/10.1016/j.meteno.2017.02.001.

Habe, H., Sato, Y., and Kirimura, K. (2020). Microbial and enzymatic conversion of levulinic acid, an alternative building block to fermentable sugars from cellulosic biomass. *Appl. Microbiol. Biotechnol.* 104 (18): 7767–7775. https://doi.org/10.1007/s00253-020-10813-7.

Hameed, S., Lin, L., Wang, A. et al. (2020). Recent developments in metal-based catalysts 5-hydroxymethyl-furfural to 2,5-furandicarboxylic acid. *Catalysts* 10 (1): 120. doi: 10.3390/catal10010120.

Harmsen, P.F., Hackmann, M.M., and Bos, H.L. (2013). Green building blocks for bio-based plastics. *Biofuels Bioprod. Biorefin.* 8: 306–324. https://doi.org/10.1002/bbb.

Hayes, D.J., Fitzpatrick, S., Hayes, H.B. et al. (1999). The biofine process: production of levulinic acid, furfural and formic acid from lignocellulosic feedstocks. *Biorefineries Industrial Processecess and Products* 1: 139–164. doi: 10.1002/9783527619849.ch7.

Hermann, B.G., Blok, K., and Patel, M.K. (2007). Producing bio-based bulk chemicals using industrial biotechnology saves energy and combats climate change. *Environ. Sci. Technol.* 41 (22): 7915–7921. https://doi.org/10.1021/es062559q.

Hong, E., Kim, J., Rhie, S. et al. (2016). Optimization of dilute sulfuric acid pretreatment of corn stover for enhanced xylose recovery and xylitol production. *Biotechnol. Bioprocess. Eng.* 21 (5): 612–619. https://doi.org/10.1007/s12257-016-0483-z.

Huang, S., Liu, T., Peng, B., and Geng, A. (2019). Enhanced ethanol production from industrial lignocellulose hydrolysates by a hydrolysate-cofermenting *Saccharomyces cerevisiae* strain. *Bioprocess Biosyst. Eng.* 42 (5): 883–896. https://doi.org/10.1007/s00449-019-02090-0.

Hutchison, C.A., Chuang, R.-Y., Noskov, V.N. et al. (2016). Design and synthesis of a minimal bacterial genome. *Science* 351 (6280): 1–12. https://doi.org/10.1126/science.aad6253.

Ibrahim, M.F., Ramli, N., Bahrin, E.K., and Abd-Aziz, S. (2017). Cellulosic biobutanol by Clostridia: challenges and improvements. *Renew. Sust. Energ. Rev.* 79 (May): 1241–1254. https://doi.org/10.1016/j.rser.2017.05.184.

Industry ARC (2019). *Succinic Acid Market—Forecast (2021–2026)*. Available at: https://www.industryarc.com/Report/17679/succinic-acid-market.html.

Jain, R., Sun, X., Yuan, Q., and Yan, Y. (2015). Systematically engineering *Escherichia coli* for enhanced production of 1,2-propanediol and 1-propanol. *ACS Synth. Biol.* 4 (6): 746–756. https://doi.org/10.1021/sb500345t.

Jiang, G., Hill, D.J., Kowalczuk, M. et al. (2016). Carbon sources for polyhydroxyalkanoates and an integrated biorefinery. *Int. J. Mol. Sci.* 17 (7): 1–21. https://doi.org/10.3390/ijms17071157.

Jiang, Y., Zhang, T., Lu, J. et al. (2018). Microbial co-culturing systems: butanol production from organic wastes through consolidated bioprocessing. *Appl. Microbiol. Biotechnol.* 102 (13): 5419–5425. https://doi.org/10.1007/s00253-018-8970-0.

Jones, D.T. and Woods, D.R. (1986). Acetone-butanol fermentation revisited. *Microbiol. Rev.* 50 (4): 484–524. https://doi.org/10.1128/mmbr.50.4.484-524.1986.

de Jong, E., Higson, A., Walsh, P., Wellisch, M. (2012). *Bio-based Chemicals: Value Added Products from Biorefineries*. [Online]. Available at: http://www.ieabioenergy.com/wp-content/uploads/2013/10/Task-42-Biobased-Chemicals-value-added-products-from-biorefineries.pdf (Accessed: 14 November 2020).

de Jong, E. et al. (2020). *IEA Bioenergy Task 42—Bio-based Chemicals: a 2020 Update*. Available at: http://task42.ieabioenergy.com/wp-content/uploads/2020/02/Bio-based-chemicals-a-2020-update-final-200213.pdf.

Juturu, V. and Wu, J.C. (2016). Microbial production of lactic acid: the latest development. *Crit. Rev. Biotechnol.* 36 (6): 967–977. https://doi.org/10.3109/07388551.2015.1066305.

Karaffa, L. and Kubicek, C.P. (2019). Citric acid and itaconic acid accumulation: variations of the same story? *Appl. Microbiol. Biotechnol.* 103 (7): 2889–2902. https://doi.org/10.1007/s00253-018-09607-9.

Keis, S., Shaheen, R., and Jones, D.T. (2001). Emended descriptions of *Clostridium acetobutylicum* and *Clostridium beijerinckii*, and descriptions of *Clostridium saccharoperbutylacetonicum* sp. nov. and *Clostridium saccharobutylicum* sp. nov. *Int. J. Syst. Evol. Microbiol.* 51 (6): 2095–2103. https://doi.org/10.1099/00207713-51-6-2095.

Kim, S.B., Kwon, D.H., Park, J.B., and Ha, S.J. (2019). Alleviation of catabolite repression in *Kluyveromyces marxianus*: the thermotolerant SBK1 mutant simultaneously coferments glucose and xylose. *Biotechnol. Biofuels* 12 (1): 1–9. https://doi.org/10.1186/s13068-019-1431-x.

Koller, M. (2018). A review on established and emerging fermentation schemes for microbial production of polyhydroxyalkanoate (PHA) biopolyesters. *Fermentation* 4 (2): 1–30. https://doi.org/10.3390/fermentation4020030.

Krull, S., Hevekerl, A., Kuenz, A., and Prüße, U. (2017). Process development of itaconic acid production by a natural wild type strain of *Aspergillus terreus* to reach industrially relevant final titers. *Appl. Microbiol. Biotechnol* 101 (10): 4063–4072. https://doi.org/10.1007/s00253-017-8192-x.

Kruyer, N.S., Wauldron, N., Bommarius, A.S., and Peralta-Yahya, P. (2020). Fully biological production of adipic acid analogs from branched catechols. *Sci. Rep.* 10 (1): 1–8. https://doi.org/10.1038/s41598-020-70158-z.

Kumar, R. and Wyman, C.E. (2013). Enhanced yields of furfural and other products by simultaneous solvent extraction during thermochemical treatment of cellulosic biomass. *RSC Adv.* 3 (25): 9809–9819. https://doi.org/10.1039/C3RA41857J.

Kumar, S., Ali, H., Kansal, S.K. et al. (2020). Sustainable production of sorbitol—a potential hexitol. *Biomass, Biofuels, Biochemicals* 259–281. https://doi.org/10.1016/b978-0-444-64307-0.00009-3.

Kumaravel, S., Thiripuranthagan, S., Radhakrishnan, R. et al. (2019). Liquid phase esterification of levulinic acid into ethyl levulinate over sulphobenzylated nanoporous. *J. Nanosci. Nanotechnol.* 19 (11): 6965–6977. https://doi.org/10.1166/jnn.2019.16637.

Lane, S., Xu, H., Oh, E.J. et al. (2018). Glucose repression can be alleviated by reducing glucose phosphorylation rate in *Saccharomyces cerevisiae*. *Sci. Rep.* 8 (1): 1–12. https://doi.org/10.1038/s41598-018-20804-4.

Lange, J., van Heide, E.V., van Buijtenen, J. et al. (2012). Furfural—a promising platform for lignocellulosic biofuels. *ChemSusChem* 5 (1): 150–166. https://doi.org/10.1002/cssc.201100648.

Lee, H.V., Hamid, S.B.A., and Zain, S.K. (2014). Conversion of lignocellulosic biomass to nanocellulose: structure and chemical process. *Sci. World J.* 2014: 1–21. 631013. https://doi.org/10.1155/2014/631013.

Lee, S.Y., Kim, H.U., Chae, T.U. et al. (2019). A comprehensive metabolic map for production of bio-based chemicals. *Nat. Catal.* 2 (1): 18–33. https://doi.org/10.1038/s41929-018-0212-4.

Lin, H., Bennett, G.N., and San, K. (2005). Metabolic engineering of aerobic succinate production systems in *Escherichia coli* to improve process productivity and achieve the maximum theoretical succinate yield. *Metab. Eng.* 7: 116–127. https://doi.org/10.1016/j.ymben.2004.10.003.

Lin, Y., Cai, Y., Guo, Y. et al. (2019). Development and genomic elucidation of hybrid yeast with improved glucose–xylose co-fermentation at high temperature. *FEMS Yeast Res.* 19 (3): 1–13. https://doi.org/10.1093/femsyr/foz015.

Liu, C., Dong, H., Zhong, J. et al. (2010). Sorbitol production using recombinant *Zymomonas mobilis* strain. *J. Biotechnol.* 148 (2–3): 105–112. https://doi.org/10.1016/j.jbiotec.2010.04.008.

Liu, Y.J., Li, B., Feng, Y., and Cui, Y. (2020). Consolidated bio-saccharification: leading lignocellulose bioconversion into the real world. *Biotechnol. Adv.* 40 (February): 107535. https://doi.org/10.1016/j.biotechadv.2020.107535.

Machado, G., Leon, S., Santos, F. et al. (2016). Literature review on furfural production from lignocellulosic biomass. *Nat. Res. Forum* 7: 115–129.

Marchesan, A.N., Oncken, M.P., Filho, R.M., and Maciel, M.R.W. (2019). A roadmap for renewable C2–C3 glycols production: a process engineering approach. *Green Chem.* 21 (19): 5168–5194. https://doi.org/10.1039/c9gc02949d.

Mariano, A.P., Dias, M.O.S., Junqueira, T.L. et al. (2013). Butanol production in a first-generation Brazilian sugarcane biorefinery: technical aspects and economics of greenfield projects. *Bioresour. Technol.* 135: 316–323. https://doi.org/10.1016/j.biortech.2012.09.109.

Market Analysis Report (2015a). *1,3 Propanediol Market Size, Share and Trends Analysis Report by Application (Polytrimethylene Terephthalate (PTT), Polyurethane, Personal Care and Detergents), by Region, and Segment Forecasts, 2014–2022*. Available at: https://www.grandviewresearch.com/industry-analysis/1-3-propanediol-pdo-market (Accessed: 15 January 2021).

Market Analysis Report (2015b). *Bio-butanol Market Analysis by Application (Acrylates, Acetates, Glycol Ethers, Solvents, Plasticizers), and Segment Forecasts to 2022*. Available at: https://www.grandviewresearch.com/industry-analysis/bio-butanol-industry (Accessed: 15 January 2021).

Market Analysis Report (2015c). *Furandicarboxylic Acid (FDCA) Market Size, Share and Trend Analysis Report by Application (PET, Polyamides, Polycarbonates, Plasticizers, Polyester Polyols), by Region, and Segment Forecast, 2012–2020*. Available at: https://www.grandviewresearch.com/industry-analysis/fdca-industry (Accessed: 15 January 2021).

Market Analysis Report (2015d). *Levulinic Acid Market Analysis and Segment Forecasts to 2020*. Available at: https://www.grandviewresearch.com/industry-analysis/levulinic-acid-market (Accessed: 15 January 2021).

Market Analysis Report (2017). *Xylitol Market Analysis by Application (Chewing Gum, Confectionery, Bakery, Oral Care), by Region (North America, Europe, Asia Pacific, Latin America, MEA), and Segment Forecast, 2018–2025*. Available at: https://www.grandviewresearch.com/industry-analysis/xylitol-market (Accessed: 7 August 2020).

Market Analysis Report (2018). *Paraxylene Market Size, Share and Trends Analysis Report by Application (Dimethyl Terephthalate, Purified Terephthalic Acid), by Region (North America, APAC, Europe, MEA, CSA), and Segment Forecasts, 2012–2022*. Available at: https://www.grandviewresearch.com/industry-analysis/paraxylene-market (Accessed: 7 August 2020).

Market Analysis Report (2019). *Lactic Acid Market Size, Share and Trend Analysis Report by Raw Material (Corn, Sugarcane), by Application (Industrial, Food and Beverages, Polylactic Acid), by Region, and Segment Forecasts, 2019–2025*. Available at: https://www.grandviewresearch.com/industry-analysis/lactic-acid-and-poly-lactic-acid-market (Accessed: 13 January 2021).

Market Analysis Report (2020a). *Ethanol Market Size, Share and Trends Analysis Report by Source (Second Generation, Grain-based), by Purity (Denatured, Undenatured), by Application (Beverages, Fuel and Fuel Additives), and Segment Forecasts, 2020–2027*. Available at: https://www.grandviewresearch.com/industry-analysis/ethanol-market (Accessed: 15 January 2021).

Market Analysis Report (2020b). *Furfural Market Size, Share and Trends Analysis Report by Process, by Raw Material (Corncob, Sugarcane Bagasse), by Application (Furfuryl Alcohol, Solvent), by End-use (Pharmaceutical, Refineries), and Segment Forecasts, 2020–2027*. Available at: https://www.grandviewresearch.com/industry-analysis/furfural-market (Accessed: 15 January 2021).

Mathew, A.K., Abraham, A., Mallapureddy, K.K. et al. (2018). Lignocellulosic biorefinery wastes, or resources? In: *Waste Biorefinery*, 267–297. Elsevier https://doi.org/10.1016/B978-0-444-63992-9.00009-4.

Mathews, S.L., Pawlak, J., and Grunden, A.M. (2015). Bacterial biodegradation and bioconversion of industrial lignocellulosic streams. *Appl. Microbiol. Biotechnol.* 99 (7): 2939–2954. https://doi.org/10.1007/s00253-015-6471-y.

McCarty, N.S. and Ledesma-Amaro, R. (2019). Synthetic biology tools to engineer microbial communities for biotechnology. *Trends Biotechnol.* 37 (2): 181–197. https://doi.org/10.1016/j.tibtech.2018.11.002.

Morais, J.W.G., Pacheco, T.F., Trichez, D. et al. (2019). Xylitol production on sugarcane biomass hydrolysate by newly identified *Candida tropicalis* JA2 strain. *Yeast* 36 (5): 349–361. https://doi.org/10.1002/yea.3394.

Mountraki, A.D., Koutsospyros, K.R., Benjelloun Mlayah, B., and Kokossis, A.C. (2017). Selection of biorefinery routes: the case of xylitol and its integration with an organosolv process. *Waste Biomass Valoriz.* 8 (7): 2283–2300. https://doi.org/10.1007/s12649-016-9814-8.

Moysés, D.N., Reis, V.C.B., Almeida, J.R.M. et al. (2016). Xylose fermentation by *Saccharomyces cerevisiae*: challenges and prospects. *Int. J. Mol. Sci.* 17 (3): 1–18. https://doi.org/10.3390/ijms17030207.

Mukherjee, V., Radecka, D., Aerts, G. et al. (2017). Phenotypic landscape of non-conventional yeast species for different stress tolerance traits desirable in bioethanol fermentation. *Biotechnol. Biofuels* 10 (1): 1–19. https://doi.org/10.1186/s13068-017-0899-5.

Nakamura, C.E. and Whited, G.M. (2003). Metabolic engineering for the microbial production of 1,3-propanediol. *Curr. Opin. Biotechnol.* 14 (5): 454–459. https://doi.org/10.1016/j.copbio.2003.08.005.

Nattrass, L., Clifford, B., Ausilio, B. et al. (2016). The EU bio-based industry: results from a survey. EUR 27736 EN. https://doi.org/10.2791/806858.

Niu, W., Kramer, L., Mueller, J. et al. (2019). Metabolic engineering of *Escherichia coli* for the de novo stereospecific biosynthesis of 1,2-propanediol through lactic acid. *Metab. Eng. Commun.* 8 (June): 1–9. https://doi.org/10.1016/j.mec.2018.e00082.

Niu, W., Willet, H., Mueller, J. et al. (2020). Direct biosynthesis of adipic acid from lignin-derived aromatics using engineered pseudomonas putida KT2440. *Metab. Eng.* 59: 151–161. https://doi.org/10.1016/j.ymben.2020.02.006.

Okabe, M., Lies, D., Kanamasa, S. et al. (2009). Biotechnological production of itaconic acid and its biosynthesis in *Aspergillus terreus*. *Appl. Microbiol. Biotechnol.* 84 (4): 597–606. https://doi.org/10.1007/s00253-009-2132-3.

Paes, B.G. and Almeida, J.R. (2014). Genetic improvement of microorganisms for applications in biorefineries. *Chem. Biol. Technol. Agric.* 1 (1): 1–10. https://doi.org/10.1186/s40538-014-0021-1.

Paes, B.G., Steindorff, A.S., Formighieri, E.F. et al. (2021). Physiological characterization and transcriptome analysis of *Pichia pastoris* reveals its response to lignocellulose-derived inhibitors. *AMB Express* 11 (2): 1–15. https://doi.org/10.1186/s13568-020-01170-9.

Pagliano, G., Ventorino, V., Panico, A., and Pepe, O. (2017). Integrated systems for biopolymers and bioenergy production from organic waste and by-products: a review of microbial processes. *Biotechnol. Biofuels* 10 (1): 1–24. https://doi.org/10.1186/s13068-017-0802-4.

Parapouli, M., Vasileiadis, A., Afendra, A.S., and Hatziloukas, E. (2020). *Saccharomyces cerevisiae* and its industrial applications. *AIMS Microbiology* 6: 1–31. https://doi.org/10.3934/microbiol.2020001.

Park, Y.C., Oh, E.J., Jo, J.H. et al. (2016). Recent advances in biological production of sugar alcohols. *Curr. Opin. Biotechnol* 37: 105–113. https://doi.org/10.1016/j.copbio.2015.11.006.

Patel, M.K., Crank, M., Dornburg, V. et al. (2006). Medium and long-term opportunities and risks of the biotechnological production of bulk chemicals from renewable resources. *European Commission's Growth Programme (DG Reearch)* 5 (1): 30–33.

Peters, M.W., Taylor, J.D., Jenni, M. et al. (2011). Integrated Process to Selectively Convert Renewable Isobutanol to p-Xylene. GEVO, Inc., Englewood, CO, USA. 1 (19), 1–22. US 2011/0087.000 A1

Qureshi, N., Lin, X., Liu, S. et al. (2020). Global view of biofuel butanol and economics of its production by fermentation from sweet sorghum bagasse, food waste, and yellow top presscake: application of novel technologies. *Fermentation* 6 (2): 1–11. https://doi.org/10.3390/FERMENTATION6020058.

Rand, J., Pisithkul, T., Clark, R.L. et al. (2017). A metabolic pathway for catabolizing levulinic acid in bacteria. *Physiol. Behav.* 2 (12): 1624–1634. https://doi.org/10.1038/s41564-017-0028-z.

Ranjan, A. and Moholkar, V.S. (2012). Application of graphene and graphene-based materials in clean energy-related devices Minghui. *Int. J. Energy Res.* 36: 277–323. https://doi.org/10.1002/er.

Reed, N.R. and Kwok, E.S.C. (2014). Furfural. Encyclopedia of Toxicology. *Environ. Fate Behav.* 2: 685–688. https://doi.org/10.1016/B978-0-12-386454-3.00147-0.

Renewable Fuels Association (2020). *Annual Fuel Ethanol Production*. Available at: https://ethanolrfa.org/statistics/annual-ethanol-production (Accessed: 15 January 2021).

Rey-Raap, N., Ribeiro, L.S., Orfao, J.J.M. et al. (2019). Catalytic conversion of cellulose to sorbitol over Ru supported on biomass-derived carbon-based materials. *Appl. Catal. B Environ.* 256 (May): 117826. https://doi.org/10.1016/j.apcatb.2019.117826.

Rice, T., Zannini, E., Arendt, E.K., and Coffey, A. (2020). A review of polyols—biotechnological production, food applications, regulation, labeling and health effects. *Crit. Rev. Food Sci. Nutr* 60 (12): 2034–2051. https://doi.org/10.1080/10408398.2019.1625859.

Ro, D., Paradise, E.M., Ouellet, M. et al. (2006). Production of the antimalarial drug precursor artemisinic acid in engineered yeast. *Nature* 440 (April): 3–6. https://doi.org/10.1038/nature04640.

Robak, K. and Balcerek, M. (2018). Review of second generation bioethanol production from residual biomass. *Food Technol. Biotechnol.* 56 (2): 174–187. https://doi.org/10.17113/ftb.56.02.18.5428.

Rosales-Calderon, O. and Arantes, V. (2019). A review on commercial-scale high-value products that can be produced alongside cellulosic ethanol. *Biotechnol. Biofuels* 12: 240. https://doi.org/10.1186/s13068-019-1529-1.

Sauer, M. (2016). Industrial production of acetone and butanol by fermentation—100 years later. *FEMS Microbiol. Lett.* 363 (13): 1–4. https://doi.org/10.1093/femsle/fnw134.

Saxena, R.K., Anand, P., Saran, S. et al. (2010). Microbial production and applications of 1,2-propanediol. *Indian J. Microbiol.* 50 (1): 2–11. https://doi.org/10.1007/s12088-010-0017-x.

Shahab, R.L., Arias, J.M., Modesto, L.F. et al. (2019). D-lactic acid production from renewable lignocellulosic biomass via genetically modified *Lactobacillus plantarum*. *Biotechnol. Prog* 32 (2): 271–278. https://doi.org/10.1002/btpr.2718.

Sheldon, R.A. (2014). Green and sustainable manufacture of chemicals from biomass: state of the art. *Green Chem.* 16: 950–963. https://doi.org/10.1039/c3gc41935e.

Sin, L.T. and Tueen, B.S. (2019). Synthesis and production of poly(lactic acid). In: *Polylactic Acid* (ed. N. achei), 53–95. Elsevier. Amsterdam, 10.1016/B978-1-4377-4459-0.00002-0

Singhvi, M.S. and Gokhale, D.V. (2019). Lignocellulosic biomass: hurdles and challenges in its valorization. *Appl. Microbiol. Biotechnol* 103 (23–24): 9305–9320. https://doi.org/10.1007/s00253-019-10212-7.

Steiger, M.G., Blumhoff, M.L., Mattanovich, D., and Sauer, M. (2013). Biochemistry of microbial itaconic acid production. *Front. Microbiol.* 4 (Feb): 1–5. https://doi.org/10.3389/fmicb.2013.00023.

Sun, W., Vila-Santa, A., Liu, N. et al. (2020). Metabolic engineering of an acid-tolerant yeast strain *Pichia kudriavzevii* for itaconic acid production. *Metab. Eng. Commun.* 10: e00124. https://doi.org/10.1016/j.mec.2020.e00124.

Takkellapati, S., Li, T., and Gonzalez, M.A. (2019). An overview of biorefinery derived platform chemicals from a cellulose and hemicellulose biorefinery. *Clean Techn. Environ. Policy* 20 (7): 1–30. https://doi.org/10.1007/s10098-018-1568-5.An.

Tehrani, H.H., Becker, J., Bator, I. et al. (2019). Integrated strain- and process design enable production of 220 g L^{-1} itaconic acid with *Ustilago maydis*. *Biotechnol. Biofuels* 12 (1): 1–11. https://doi.org/10.1186/s13068-019-1605-6.

Trichez, D., Steindorff, A., Soares, C.E.V.F. et al. (2019). Physiological and comparative genomic analysis of new isolated 2 yeasts *Spathaspora* sp. JA1 and *Meyerozyma caribbica* JA9 reveal 3 insights into xylitol production. *FEMS Yeast Res.* 19 (4): 1–15. https://doi.org/10.1093/femsyr/foz034.

USDA (2008). Biobased Products Market Potential and Projections Through 2025. Available: http://www.usda.gov/oce/reports/energy/index.htm.

Utsunomia, C., Ren, Q., and Zinn, M. (2020). Poly(4-hydroxybutyrate): current state and perspectives. *Front. Bioeng. Biotechnol.* 8 (April): 1–18. https://doi.org/10.3389/fbioe.2020.00257.

Veeravalli, S.S. and Mathews, A.P. (2018). Exploitation of acid-tolerant microbial species for the utilization of low-cost whey in the production of acetic acid and propylene glycol. *Appl. Microbiol. Biotechnol* 102 (18): 8023–8033. https://doi.org/10.1007/s00253-018-9174-3.

Veras, H.C.T., Parachin, N.S., and Almeida, J.R.M. (2017). Comparative assessment of fermentative capacity of different xylose-consuming yeasts. *Microb. Cell Factories* 16 (1): 1–8. https://doi.org/10.1186/s12934-017-0766-x.

Veras, H.C.T., Campos, C.G., Nascimento, I.F. et al. (2019). Metabolic flux analysis for metabolome data validation of naturally xylose-fermenting yeasts. *BMC Biotechnol.* 19 (1): 1–14. https://doi.org/10.1186/s12896-019-0548-0.

Vilela, C. et al. (2014) The quest for sustainable polyesters—insights into the future, Polym. Chem., 5(9), 3119–3141. 10.1039/c3py01213a.

Wang, D., Li, Q., Yang, M. et al. (2011). Efficient production of succinic acid from corn stalk hydrolysates by a recombinant *Escherichia coli* with ptsG mutation. *Process Biochem.* 46 (1): 365–371. https://doi.org/10.1016/j.procbio.2010.09.012.

Werpy, T. Petersen, G. (2004). *Top Value Added Chemicals From Biomass Volume I : Results of Screening for Potential Candidates*. US Department of Energy (US), United States. https://doi.org/10.2172/15008; https://doi.org/10.2172/15008859859.

Wierckx, N., Koopman, F., Ruijssenaars, H.J., and Winde, J.H. (2011). Microbial degradation of furanic compounds: biochemistry, genetics, and impact. *Appl. Microbiol. Biotechnol.* 92 (6): 1095–1105. https://doi.org/10.1007/s00253-011-3632-5.

Williams, C.L., Chang, C., Do, P. et al. (2012). Cycloaddition of biomass-derived furans for catalytic production of renewable *p*-xylene. *ACS Catal.* 2 (6): 935–939. https://doi.org/10.1021/cs300011a.

Willke, T. and Vorlop, K.D. (2001). Biotechnological production of itaconic acid. *Appl. Microbiol. Biotechnol.* 56 (3–4): 289–295. https://doi.org/10.1007/s002530100685.

Wischral, D., Zhang, J., Cheng, C. et al. (2016). Production of 1,3-propanediol by *Clostridium beijerinckii* DSM 791 from crude glycerol and corn steep liquor: process optimization and metabolic engineering. *Bioresour. Technol* 212, 100–110. https://doi.org/10.1016/j.biortech.2016.04.020.

Wischral, D., Arias, J.M., Modesto, L.F. et al. (2019). Lactic acid production from sugarcane bagasse hydrolysates by *Lactobacillus pentosus*: integrating xylose and glucose fermentation. *Biotechnol. Prog.* 35 (1): 1–9. https://doi.org/10.1002/btpr.2718.

Xie, H., Ma, Q., Wei, D., and Wang, F. (2020). Metabolic engineering of an industrial *Aspergillus niger* strain for itaconic acid production. *3 Biotech* 10 (3): 1–10. https://doi.org/10.1007/s13205-020-2080-2.

Xu, Y., Wang, R.H., Koutinas, A.A., and Webb, C. (2010). Microbial biodegradable plastic production from a wheat-based biorefining strategy. *Process Biochem.* 45 (2): 153–163. https://doi.org/10.1016/j.procbio.2009.09.001.

Yadav, M. and Shukla, P. (2019). Efficient engineered probiotics using synthetic biology approaches : a review. *Biotechnol. Appl. Biochem.* 67: 1–8. https://doi.org/10.1002/bab.1822.

Yahiro, K., Takahama, T., Park, Y.S., and Okabe, M. (1995). Breeding of *Aspergillus terreus* mutant TN-484 for itaconic acid production with high yield. *J. Ferment. Bioeng.* 79 (5): 506–508. https://doi.org/10.1016/0922-338X(95)91272-7.

Zeikus, J.G. (1980). Chemical and fuel production by anaerobic bacteria. *Annu. Rev. Microbiol.* 34: 423–464.

Zhang, C., Roland, J.Y.Z., and Sessions, J. (2012). Fractionation of forest residues of Douglas-fir for fermentable sugar production by SPORL pretreatment. *Bioenergy Research* 5: 978–988. https://doi.org/10.1007/s12155-012-9213-3.

Zhang, Y., Kumar, A., Hardwidge, P.R. et al. (2016). D-lactic acid production from renewable lignocellulosic biomass via genetically modified *Lactobacillus plantarum*. *Biotechnol. Prog.* 32 (2): 271–278. https://doi.org/10.1002/btpr.2212.

Zhang, Y., Liu, D., and Chen, Z. (2017). Production of C2–C4 diols from renewable bioresources: new metabolic pathways and metabolic engineering strategies. *Biotechnol. Biofuels* 10 (1): 1–20. https://doi.org/10.1186/s13068-017-0992-9.

Zheng, P., Fang, L., Xu, Y. et al. (2010). Succinic acid production from corn stover by simultaneous saccharification and fermentation using *Actinobacillus succinogenes*. *Bioresour. Technol.* 101 (20): 7889–7894. https://doi.org/10.1016/j.biortech.2010.05.016.

7

Molecular Biology Based Innovations in Lignocellulose Biorefinery

Nilesh Kumar Sharma[1] *and Mohit Bibra*[2]

[1] *Department of Chemical Engineering, Worcester Polytechnic Institute, Worcester, MA, USA*
[2] *Prairie AquaTech LLC, South Dakota, Brookings, SD, USA*

7.1 Introduction

Energy sustainability is the idea of managing and maintaining the availability of resources by finding an alternative bioresource for the depleting conventional sources such as the fossil fuel reserve. Increasing the human population and altering daily practice will increase the energy demand, which will not be accomplished with only conventional sources of fuel. Moreover, fossil fuels lead the increased emission of greenhouse gases and global warming in the environment. Almost one-third of the world crude oil supply is fulfilled by Middle East countries only and, in a situation of political crisis, it can be hampered and cause an upsurge in price (Zabed et al. 2016; Chettri et al. 2020). In this view, several potential alternative fuels have been projected such as bioethanol, biobutanol, biodiesel, methanol, hydrogen, compressed natural gas, biogas, Fischer–Tropsch fuel, electricity, and solar fuels. Yet only 3% of the transportation fuels requirement can be contributed by liquid biofuels (Arora et al. 2015). Bioethanol has been considered as an alternative transportation fuel due to its huge potential to replace conventional fuels (MacLean 2004; Hoover and Abraham 2009; Ben-Iwo et al. 2016; Bibra et al. 2020). Moreover, fuel combustion in vehicles and the reduction of carcinogen emission could also be improvised through the use of bioethanol (Khuong et al. 2016; Chandel and Sukumaran 2017). At present the United States and Brazil contribute approximately 80% of total world bioethanol demand by using corn or sugarcane as feedstock (Saini et al. 2015; Popp et al. 2016). However, in developing countries, edible feedstocks cannot be used for bioethanol production due to food versus fuel issues (Ben-Iwo et al. 2016; Sharma et al. 2016; Bibra et al. 2020). Therefore, second-generation bioethanol, that is, lignocellulosic bioethanol, has been considered a more economical and sustainable biofuel as a long-term goal. Lignocellulosic biomass (LCB) is composed of approximately 50% cellulose, 30% hemicellulose, and 15% lignin, generally with pectin, lipid, protein, and minerals (Bibra et al. 2014). Cellulose and hemicellulose can be easily degraded into simple sugars such as glucose and xylose, respectively, by using

Lignocellulose Bioconversion Through White Biotechnology, First Edition. Edited by Anuj Kumar Chandel.

carbohydrate-active enzymes. Furthermore, these simple forms of sugars can be further converted into ethanol and other valuable molecules using various wild-type microorganisms including *Saccharomyces cerevisiae*, *Zymomonas mobilis*, *Pachysolen tennophilus*, *Candida shehatae*, *Scheffersomyces stipitis*, *Kluyveromyces marxianus*, *Debaryomyces hansenii* and so forth (Chandel et al. 2011; Sharma et al. 2018). Conventional yeast *S. cerevisiae* is well studied for the conversion of hexose sugars but is not able to utilize pentose sugars. Alternatively, some native pentose utilizing strains were identified such as *S. stipitis*, *K. marxianus*, and *D. hansenii*. However, low ethanol productivity has been reported for the co-utilization of hexose and pentose. Also, xylose utilizing strains are lacking a robust nature for industrial usage due to sensitivity and low inhibitor and solvent tolerance (Caspeta et al. 2015). Therefore, it is still a major challenge to efficiently breakdown complex sugars into simple sugars and further conversion of all reducing sugars into ethanol economically and sustainably. This chapter is dedicated to exploring the progress in the biomass to biofuel conversion process including biomass pretreatment together with the different approach of strain engineering for sugar conversion.

7.2 Lignocellulosic Biomass Potential

According to the sources of availability, LCBs can be categorized as primary, secondary, and tertiary biomass. Crops and products are considered as the primary source of LCB such as sugarcane, corn, sugar beet, etc. (Maity 2015), followed by the agricultural, forestry residues or residues generated after processing which can be considered as secondary sources of biomass including bagasse, paddy straw, paper pulp, and so on. Further, the residues produced after or during application like an organic fraction of municipal solid waste, sludge, sawdust, and so forth are tertiary sources of biomass (Fischer and Schrattenholzer 2001; Kang et al. 2014; Roy et al. 2020). Agricultural residues have the most potential for biofuel application which have been estimated as up to 70 eta-joules per year and cover by-products of agriculture or a related industry such as cotton stalk, paddy, rice, and wheat straw, banana stem, rice husk, and maize cobs. However, the low density of biomass, centralized production, and transportation cost make it less economically sustainable for fuel production. Total LCB has significant potential as an estimated 150 eta-joules per year. However, only a few parts of these resources have been explored as feedstock for bioethanol production.

7.3 Biomass Pretreatment

LCBs possess a complex cell wall structure composed of lignin, cellulose, and hemicellulose. The lignin layer needs to be broken before the cellulose and hemicellulose components can be accessed for hydrolysis and sugar production. This process of making biomass accessible for enzymatic hydrolysis to obtain monomeric sugars is called pretreatment. The removal of lignin and hydrolysis of cellulose and hemicellulose into monomeric sugar is the most costly step in any bioprocesses (Bibra et al. 2014; Kumar et al. 2015). Several decades of research have been put into developing new mechanical, physico-chemical, chemical, and biological methods of pretreatment and optimizing and reducing their associated cost to make the step more economical (Behera et al. 2014). The pretreatment step usually needs a combination of one or more types of pretreatment methods due to economic constraints and environmental concerns. A combination pretreatment step is most commonly employed. Table 7.1 also presents examples of various methods used for the pretreatment. This section will discuss the different pretreatment types most commonly used.

7.3.1 Mechanical Pretreatment

Mechanical pretreatment aims to increase the surface area and porosity of LCBs to reduce the crystallinity of cellulose present in the cell wall (Olofsson et al. 2008). Almost all the lignocellulosic substrates undergo some sort of mechanical pretreatment. Various mechanical methods like milling (ball milling, two-roll milling, hammer milling, and colloid milling), chipping, and grinding has been used to improve the pretreatment and further the enzymatic accessibility of LCBs (Behera et al. 2014). Although size reduction plays a major role to augment enzymatic accessibility of biomass, high energy consumption makes overall pretreatment cost increase (Arora et al. 2021). Extrusion is another commonly used method where the biomass is pretreated through simultaneous heat, mix, and shear followed by passing through an extruder which changes the materialistic properties of biomass (Behera et al. 2014). More development is needed to improve the energy use for mechanical pretreatment.

7.3.2 Chemical Pretreatment

Chemical pretreatment is a widely used method for degradation of the crystalline structure of cellulose and separate lignin/hemicellulose from it (Brodeur et al. 2011). This method has also been applied to process the cellulosic waste of the paper and pulp industries. Various chemicals including acids, alkalis, salts, oxidizing agents, ionic liquids, eutectic solvents, supercritical fluids, etc. have been extensively used for the separation of lignin and hemicellulose from LCBs (Brodeur et al. 2011; Jaswal et al. 2019; Rai et al. 2019; Roy et al. 2020; Arora et al. 2021; Hans et al. 2021). The most commonly used acid and alkali for chemical-based pretreatment are H_2SO_4 and NaOH, respectively (Amin et al. 2017; Wang et al. 2019). One of the greatest advantages chemical pretreatment provides is the short time and energy required to complete the process of pretreatment compared to mechanical and biological pretreatment methods. However, their environmental, sugar loss and worker safety concerns sometimes outweigh the time and energy benefits. Concentrated acids give the best results during pretreatment but the corrosive nature, environmental and worker safety concerns, sugar loss and economic feasibility of the process outweigh their benefits in several instances. Hence, researchers have been trying to develop green chemicals that can alleviate these concerns. Organic acids such as salicylic acid, oxalic acid, and acetylsalicylic acid have been used as a catalyst in the process or a cocktail of organic solvents and acids, for example hydrochloric acid and sulfuric acid, may breakdown lignin and hemicellulose bonds (Behera et al. 2014; Gunny et al. 2015). Usually, the chemical pretreatment step is preceded by a mechanical pretreatment step and followed by a biological pretreatment step. Several studies have successfully used this combinational pretreatment strategy for the pretreatment of the LCB (Chen et al. 2012; Hoşgün et al. 2017; Rahayu et al. 2017; Bay et al. 2020). The research is still far from where it can be claimed that the process is economical and optimum.

Table 7.1 The various pretreatment methods and combination thereof used for the pretreatment of lignocellulosic biomass for bioethanol production.

Lignocellulosic biomass	Solid loading	Pretreatment conditions	Results	Ethanol	References
		Acidic pretreatment			
Forest residue	10% (w/v)	H_2SO_4 (1.5% w/v) at 121 °C for 90 min		0.64 g L^{-1}	Rahayu et al. (2017)
Wheat straw		0.05 M HCl, 4 h, and 85 °C	Total 85.71% lignin and 4.30% hemicellulose were removed by solubilization	-	Sun et al. (2005)
Fresh softwood sawdust (*Picea abies*)	-	Two-step H_2SO_4 pretreatment Step 1: H_2SO_4 (0.5% w/w) at 180 °C for 10 min Step 2: H_2SO_4 (2% w/w) at 210 °C for 2 min	21% (w/w) glucose was obtained after step 2 of the acidic pretreatment	40.8 g L^{-1}	Söderström et al. (2003)
		Alkali pretreatment			
Pinewood	5% (w/v)	Milled to obtain size of 177–833 μm NaOH 8% (w/v) at 93 °C for 2 h followed by enzymatic hydrolysis, cellulase 20 FPU g^{-1} cellulose at 45 °C for 72 h	Glucose yield increased to 49.3% compared to 8.4% when no NaOH treatment was given	2.02 g L^{-1}	Bay et al. (2020)
		Organosolv/ionic liquid/eutectic solvent pretreatment			
Corn stover	10% (w/v)	40% (v/v) acetic acid at 80 °C for 2 h	Decreased cellulose, hemicellulose, and lignin concentration by 8.8%, 4.7%, and 10.3%, respectively	-	Gill et al. (2021)
Switchgrass	10% (w/w)	Ionic liquid $[C_2C_1Im][OAc]$	Yeast strain capable of producing >10 g L^{-1} of ethanol with 0.2% (w/w) of residual $[C_2C_1Im][OAc]$ found	7.5 g L^{-1}	Sitepu et al. (2019)

		Enzyme pretreatment			
Corn stover	5% (w/v)	In situ enzyme treatment with *Geobacillus* sp. strain DUSEL R13 (10% v/v)	48.6% of corn stover was hydrolyzed	3.72 g L^{-1}	Bibra et al. (2018)
Napier grass	10% (w/w)	Physical size reduction to obtain powder with particle size 150 μm followed by enzymatic hydrolysis by cellulase (Acremozyme) 10% at 45 °C for 48 h	61.4% of the substrate was hydrolyzed	8.8 g L^{-1}	Yasuda et al. (2011)
		Combinational pretreatment			
Cotton stalk	4% (w/v) for alkali treatment 2% (w/v) of dried alkali treated substrate for enzymatic pretreatment	NaOH (7% w/v) at 30 °C for 72 h Cellulase (30 FPU g^{-1} solid)	241 mg g^{-1} reducing sugar	3.8 g L^{-1}	Malik et al. (2020)
Common reed	20 g substrate for γ-irradiation and 6% (w/v) glucan for enzyme	20 g of substrate irradiated with γ-irradiation dose of 800 kGy followed by detoxification with $NaBH_4$ and enzymatic (cellulose, β-glucanase, and xylanase) hydrolysis with loading of 30% (w/w) at 50 °C for 102 h	1710 g L^{-1} of sugar from the lignocelluosic biomass after combined pretreatment	60% increase in ethanol yield for detoxified substrate compared to undetoxified	Liu et al. (2016)

(Continued)

Table 7.1 (Continued)

Lignocellulosic biomass	Solid loading	Pretreatment conditions	Results	Ethanol	References
Corn stover	16.67% (w/v)	Corn stover treated with magnesium bisulfite solution (3.5%, w/v) at 170 °C for 60 min followed by enzymatic (cellulase 15 FPU g^{-1} cellulose and cellobiohydrolase 30 U g^{-1} cellulose) hydrolysis at 50 °C for 48 h	Magensium bisulfite removed 90% of lignin and 80% of hemicellulose	10 g L^{-1}	Yu et al. (2016)
Wheat straw	12% (w/v)	Physical size reduction followed by steam explosion at 200 °C for 2.5 min followed by enzymatic hydrolysis laccase 1% (w/w) at 50 °C for 8 h, cellulase (15 FPU g^{-1} substrate) and β-glucosidase (15 IU g^{-1} substrate), respectively, at 50 °C for 8 h	Laccase treatment increased the ethanol yield by 55 times compared to non-laccase treated substrate	16.7 g L^{-1}	Moreno et al. (2013)
Sugarcane bagasse	10% (w/v)	NaOH (10% w/w) at 90 °C for 1.5 h followed by 10% (v/w) peracetic acid 75 °C for 2.5 h followed by cellulase (15 FPU g^{-1} solid) treatment for 120 h	95.62% material delignified Reducing sugars yield 92.04%		Zhao et al. (2009)

Abbreviation: FPU, filter paper unit.

7.3.3 Biological Pretreatment

Biological pretreatment involves the use of enzymes or microbial strains, namely *Trichoderma*, *Aspergillus*, *Geobacillus*, etc., with the ability to produce ligninolytic enzymes to carry out the pretreatment, hydrolysis, and saccharification (Bibra et al. 2018; Kumar et al. 2018; Jiang et al. 2019; Rai et al. 2019; Bibra et al. 2020). Some of the fungal species like brown fungi, white fungi, soft rot fungi, and actinomycetes have been reported to have ligninolytic activities with very little consumption of cellulose. Several white rot fungi such as *Phanerochaete chrysosporium*, *Ceriporia lacerate*, *Cyathus stercolerus*, *Ceriporiopsis subvermispora*, *Pycnoporus cinnarrbarinus*, and *Pleurotus ostreaus* have shown high delignification efficiency on different LCBs (Cong et al. 2017). The biological pretreatment method is advantageous compared to other pretreatment methods including mechanical and chemical pretreatment as the method is more specific, has low energy requirements, no generation of toxic compounds, and is environmentally friendly with less sugar loss and higher yield (Chandel and Sukumaran 2017). Despite all these advantages, the biological pretreatment method is not the method of choice due to the economical constraints of enzyme production and the longer time required to complete the pretreatment. Also, there are concerns of microbial contamination and sugar consumption during in situ enzyme production that reduce the overall sugar yield. Several efforts are being made to produce recombinant enzymes, organisms with higher substrate specificity, use extremophilic enzymes, and hosts for pretreatment to make this process more economical and of short duration to alleviate the concerns.

7.3.4 Other Methods

Several other methods are being developed and used for LCB pretreatment, namely microwave (Tsubaki et al. 2012; Biller et al. 2013; Klein et al. 2016), plasma treatment (Rathinam et al. 2019), electrochemical (Rathinam et al. 2020), and use of specific metals, for example Cu (Cheng et al. 2020), to further optimize the pretreatment process.In the current scenario, chemical pretreatment is being used widely and is more promising than the rest of the methods. Several chemicals such as ammonia, hydrogen peroxide, sodium hydroxide are efficient during the structural breakdown of the cell wall and separation of cellulose and lignin/hemicellulose (Behera et al. 2014; Arora et al. 2021; Hans et al. 2021). Besides the importance of chemical pretreatment, it requires a massive amount of chemicals and water during the process, which must be recycled and treated. Hence, high facility investment and high treatment cost for potential environmental pollution is needed. Further, some proportion of glucose and xylose get converted into aromatic inhibitors such as hydroxymethylfurfural (HMF) and furfural, respectively, which hampers the survival of microbes during the fermentation process.

7.4 Different Approaches to Enhance Xylose Utilization

Several key issues have been faced during xylose fermentation as a sole source of carbon or co-fermentation with glucose for ethanol conversion including slow metabolism and low productivity. There are two major steps in primary xylose metabolism which are the reduction of xylose to xylitol followed by dehydrogenation of xylitol to xylulose mediated by xylose reductase (XR) and xylitol dehydrogenase (XDH), respectively. In this phenomenon, cofactors play major roles to metabolize xylose as NADPH and NAD^+ are used by XR and XDH, respectively. Because it lacks these enzymes, *S. cerevisiae* cannot metabolize xylose conventionally but grows in xylulose and its isomers (Hahn-Hägerdal et al. 2007a). Therefore *S. cerevisiae* has been heterologously engineered with *xyl1* and *xyl2* genes encoding for XR and XDH, respectively, for xylose fermentation into ethanol (Matsushika et al. 2009; Sharma et al. 2018). However, cofactor differences between XR and XDH raise redox imbalance and encourage xylitol accumulation instead of ethanol production. Moreover, it has been attributed that the difference in cofactor preference of XR and XDH restricts xylose conversion and produces xylitol instead of ethanol (Krahulec et al. 2010). Furthermore, slow xylose uptake also plays a role in xylose to ethanol conversion and yield. Various transporters have been identified and engineered to enable xylose uptake and increase ethanol production. Numerous genome engineering approaches including random and targeted mutagenesis have been approached to resolve the aforementioned issues for two decades and enable conventional and non-conventional yeast strains for enhanced ethanol production.

7.4.1 Random Mutagenesis

7.4.1.1 Evolutionary Adaptation

Evolutionary adaptation is the classic approach for evolving strains for specific conditions and work on the principle of natural selection. It is a systematic algorithm based on a selection procedure to develop a biological system with an expected phenotype that does not requires prior genetic or biochemical information. Therefore, this approach is quite popular for non-conventional yeast strains which lack the targeted genome engineering tools. Besides, knockout or overexpression of genes is not enough to get desired results; hence, extensive transcription reprogramming is needed to balance metabolic flux. Adaptation is also useful for initial strain improvement and to explore the molecular basis of desired phenotypes. Evolutionary adaptation was also used for enhanced ethanol production via increased tolerance to inhibitors, stabilizing the growth during multiple sugar utilization, and increased substrate uptake (Sharma et al. 2016, 2017). It has also been applied to stabilize engineered microbes through sequential mutations to tolerate environmental changes.

Young et al. (2012) engineered *S. cerevisiae* with sugar transporter genes *gxs1* and *xut3* followed by applied directed evolution. The evolved strain showed a 70% enhanced growth rate on xylose as the parent strain. Similarly, adaptive evolution was applied on *S. cerevisiae* harboring the XR/XDH pathway after deletion of cytochrome C oxidase subunit IV gene, which resulted in increased ethanol yield of $0.39\,g\,g^{-1}$ respectively on xylose as carbon source (Peng et al. 2012). Similarly, upregulation of xylose metabolic genes *xyl1*, *xyl2*, and *xks1* was reported in recombinant *S. cerevisiae* after evolutionary adaptation and resulted in 78.3% reduced xylitol production together with increased ethanol production of 2.6-fold more than the parental strain, during xylose fermentation (Hasunuma et al. 2014). Ethyl methyl sulfonate (EMS) was used to induce random mutagenesis in engineered *S. cerevisiae* harboring *xylA* from *Clostridium phytofermentans*, followed by evolutionary adaptation in a xylose-containing medium. Adaptation of the recombinant strain resulted in higher xylose fermentation capability in the evolved strain (Demeke et al. 2015). Reznicek et al. (2015) reported a higher affinity for xylose in GAL2 transporter after directed evolution of *S. cerevisiae* followed by increased xylose uptake rate. Non-conventional yeasts *Kluyveromyces marxianus* NIRE-K1 and NIRE-K3 were also approached for evolutionary adaptation on xylose as the carbon source and resulted in a 4.84- and 5.07-fold higher xylose consumption rate, respectively, in the evolved strains *K. marxianus* NIRE-K1.1 and NIRE-K3.1 (Sharma et al. 2016, 2017). Moreover, there was also reported to be increased xylosidase activity in both evolved strains (Behera et al. 2016).

Processing of biomass after the pretreatment is recommended to remove the inhibitors generated including furans, furfurals, and weak acids for optimal growth of microbes and ethanol fermentation, which increases the cost of the process (Zhu et al. 2009). Therefore, evolutionary adaptation has been approached to increase the tolerance in sugar fermenting yeast strains (Ibraheem and Ndimba 2013). The adaptation was applied for *S. stipitis* CBS 6054 on solid agar and broth medium, separately, and resulted in higher ethanol titer of 19.4 g L^{-1} and 18.4 g L^{-1} compared to a non-adapted strain (16.3 g L^{-1}) (Agbogbo et al. 2008). Increased ethanol yield was reported in the fermentation of undetoxified hydrolysates when cells of adapted *S. stipitis* and *C. shehatae* were used (Zhu et al. 2009). Similarly, *S. stipitis* CBS 5776 was adapted on steam-exploded prehydrolysate of corn stover and reported 80.34% of the maximum theoretical ethanol yield (Zhu et al. 2009). In another study, 11% ethanol (v/v) tolerance was reported in *S. cerevisiae* after 2 years of adaptation on selective media supplemented with a gradually increased ethanol concentration of 6%–12% (v/v) (Voordeckers et al. 2015). The *S. cerevisiae* strain was also developed through adaptive laboratory evolution for enhancing thermotolerance and the growth of the adapted strain at 40 °C was reported (Caspeta and Nielsen 2015). Further, some more examples of evolutionary adaptation have been included in Table 7.2.

7.4.1.2 Strain Hybridization

Intergeneric strain hybridization is an effective, simple, and cheap technique to improve the performance of yeast strains via an exchange of gene segments from genomic DNA. It is also useful for genetically uncharacterized organisms and involves an exchange of characteristic advantages between the organisms. Fusogenic agents including polyethylene glycol (PEG) have been used for the induction of protoplast fusion for the development of hybrids. Several studies have been reported to develop strain for efficient xylose utilization through protoplast fusion (Kuhad et al. 2011). Gera et al. (1997) developed a fusant of *K. marxianus* and *S. cerevisiae* by protoplasmic hybridization, which can grow at up to 45 °C and produces ethanol from 6% molasses (v/v) at 30 °C. Similarly, *D hansenii* and *S. cerevisiae* have been fused and resulted in 2-fold higher xylose uptake in fusant compared to parental strains (Loray et al. 1997). In the pipeline, protoplast fusion was applied on *S. cerevisiae* and *C. shehatae* and the developed fusant resulted in 0.43 g g^{-1} ethanol yield on hydrolysates of *Prosopis juliaflora* biomass (Pasha et al. 2007). Further, UV-mutagenized ethanol-tolerant *C. shehatae* and *S. cerevisiae* have been fused and resulted in a 28% increase in ethanol yield on xylose as a carbon source (Li et al. 2008). In the advancement of strain hybridization, strategic genome shuffling has been approached on *S. stipitis* strain by cross-mating between UV-mutagenized *S. stipitis* strains to develop a tolerance for liquid hardwood spent sulfite liquor (HW SSL) (Bajwa et al. 2010). The evolved strain was able to grow in more than 80% of HW SSL and utilized 4% (w/v) xylose present in HW SSL. Further, GS301 and GS302 fusant yeast strains were developed by genome shuffling in UV-mutagenized *S. stipitis*

Table 7.2 Evolutionary adaptation of yeast strains under different conditions.

			Phenotypical changes		
Strain	**Stress type**	**Method**	**Effect of stress tolerance**	**Effect on ethanol production**	**Reference**
Scheffersomyces stipitis C4	Hardwood spent sulfite liquor (HSSL)	Evolutionary adaptation on sequentially 20%, 40%, and 60% (v/v) undetoxified eucalyptus HSSL	*S. stipitis* was successfully adapted to 60% (v/v) undetoxified eucalyptus HSSL	7% higher ethanol efficiency and improved ethanol yield	Pereira et al. (2015)
Saccharomyces cerevisiae	Nickel	Evolutionary engineering by batch selection under gradually increasing nickel stress levels	Nickel hyper-resistant mutants that could resist up to 5.3 mM $NiCl_2$	–	Küçükgöze et al. (2013)
Corynebacterium glutamicum ATCC 13032	Cellobiose and xylose	Strain was sequentially cultivated in minimal medium containing 2% (w/v) cellobiose as sole carbon source for 48 h each time	Strain utilized cellobiose as sole carbon and co-fermented cellobiose and xylose	–	Lee et al. (2016)
S. cerevisiae ISO12	Thermotolerant and hydrolysate-derived compounds	Long-term adaptation on combined stress, i.e., non-detoxified spruce hydrolysate at 39 °C	Evolved yeast was capable of growing and fermenting the liquid fraction of non-detoxified spruce hydrolysate at 39 °C	Ethanol yield was 0.39 $g\,g^{-1}$ sugars present in non-detoxified spruce hydrolysate	Wallace-Salinas and Gorwa-Grauslund (2013)
S. cerevisiae TMB3500	Lignocellulosic inhibitors and low pH	Short-term adaptation on the defined medium pH 5.0 under aerobic conditions, followed by gradual decrease in pH to 3.7 for 3600 h	Robust population was obtained after 3600 h with an ability to grow and ferment at pH 3.7 with inhibitors	3-fold increase in ethanol yield at pH 3.7	Narayanan et al. (2016)
S. cerevisiae	Hyperosmotic stress	Adaptive experimental evolution on KCl (hyperosmotic stress) for 200 generations followed by intrastrain hybridization	Higher glycerol production	1.3% (v/v) less ethanol during wine production	Tilloy et al. (2014)
Pichia guilliermondii	Corncob residues (CCR) hydrolysate	Adaptation in CCR hydrolysate without external nutrients with gradual increase in hydrolysate fraction from 25% to 100%	Tolerant to detoxified CCR hydrolysate and could ferment the sugars anaerobically	Ethanol productivity and titer obtained by evolved strain were reported as 56.3 $g\,L^{-1}$ and 0.47 $g\,L^{-1}\,h^{-1}$, respectively	Fan et al. (2013)
Kluyveromyces marxianus MTCC 1389	Osmotolerance	Adaptation in 200 $g\,L^{-1}$ lactose for 65 days	Improved osmotolerance for lactose in adapted *K. marxianus* cells	17.5% higher ethanol titer than the parental strain	Saini et al. (2017)

strains and resulted in improved tolerance for HW SSL; they produced 0.39%–1.4% (w/v) ethanol on a different type of wood hydrolysate (Bajwa et al. 2011). A hybrid strain was developed using multiple rounds of genome shuffling of *S. cerevisiae* W5 strain and resulted in strain GS3-10 which showed 50% greater efficacy to ferment glucose and xylose (Jing-Ping et al. 2012). Another hybrid strain developed through genome shuffling between *S. cerevisiae* and *S. stipitis* showed a higher xylose uptake rate (1.28 g L^{-1} h^{-1}) than that of native *S. stipitis* (0.93 g L^{-1} h^{-1}) (Zhang and Geng 2012). Kahar and Tanaka (2014) developed xylose-fermenting yeast FSC1 with successful fusion between *S. cerevisiae* and *Candida intermedia*. The fusant FSC1 strain produced 0.38 g g^{-1} ethanol yield on a mixture of glucose and xylose. A fusant yeast strain has been developed through protoplast fusion between *S. cerevisiae* and *C. shehatae*, which showed high fermentation performance with 0.424 g g^{-1} ethanol yield on a glucose/xylose mixture (Ge et al. 2014). Shi et al. (2014) developed a xylose-fermenting strain TJ2-3 using genome shuffling in *S. stipitis* which showed a 1.5-fold higher ethanol titer on xylose than wild *S. stipitis*. Cheng et al. (2015) developed diploid recombinants PMM1, PMM3, and PMM4 using genome shuffling of wild-type strain with haploids PG-132, PG-259, PG-353, PR-164, and PR-438 strain. As a result, PMM1 showed higher tolerance for furfural, whereas PMM3 and PMM4 showed increased tolerance for HMF. Similarly, a hybrid strain of *S. cerevisiae* was developed through genome shuffling between eight differentially isolated *S. cerevisiae* and reported 0.45 g g^{-1} ethanol yield on an 8 L pilot-scale test (Snoek et al. 2015). Further, some more phenotypical changes in the hybrid strain are included in Table 7.3,

7.4.2 Site-specific Engineering

7.4.2.1 Targeting Sugar Transporters

Several sugar transporters have been reported to facilitate the assimilation of hexose and pentose sugars in conventional and non-conventional yeast strains (Sharma et al. 2018). However, co-assimilation of glucose and xylose is inhibited due to a higher affinity for glucose in transporters than xylose (Lee et al. 2002). Some hexose transporters (HXTs) were engineered in the conventional yeast *S. cerevisiae* to increase xylose assimilation and promote co-utilization of glucose and xylose (Cai et al. 2012; Moysés et al. 2016). Moreover, overexpression of sugar transporter genes together with evolutionary engineering may improve the expression of endogenous transporters (Vilela et al. 2015; Moysés et al. 2016). Some HXTs such as HXT1p, HXT4p, HXT5p, HXT7p, and GAL2p have also been reported to be facilitated the influx of both sugars including glucose and xylose (Hamacher et al. 2002; Sedlak and Ho 2004). During co-fermentation of glucose and xylose, some of the HXTs including HXT5p and HXT7p show dual behavior so that they prefer glucose when the concentration is more than 200 mM but change preferences or majorly transport xylose when the glucose level gets lower than about 0.2% (Hamacher et al. 2002; Sedlak and Ho 2004). Similarly, transcriptome analysis also revealed the higher level of gene expression in *hxt1* in the presence of ≥2% glucose in a medium but *hxt7* shows a higher expression in the presence of xylose in a medium (Gonçalves et al. 2014).

S. cerevisiae has always been preferred as a synthetic biology model because of established molecular tool techniques such as transformation tools and well-annotated whole genome information. In this pipeline, heterologous genes *hxt* and *xylt* encoding for HXT and

Table 7.3 Phenotypic changes in hybrid strains developed through intergeneric hybridization.

Hybridization				
Strain 1	**Strain 2**	**Method**	**Phenotype**	**References**
Saccharomyces cerevisiae	*Spathaspora passalidarum* (UV mutagenized)	Genome shuffling	Tolerance to inhibitors produced by lignocellulosic substrate and enhanced ethanol yield	Hou and Yao (2012)
S. cerevisiae	*Scheffersomyces stipitis*	Genome shuffling	Enhanced xylose uptake rate	Zhang and Geng (2012)
S. cerevisiae	*Pacchysolen tennophilus*	Protoplast fusion	Enhanced ethanol titer as $76.8 \pm 0.31\,g\,L^{-1}$ by using mixture of glucose and xylose sugar ($167.5 \pm 0.82\,g\,L^{-1}$)	Kumari and Pramanik (2012)
S. cerevisiae	*S. stipitis*	Protoplast fusion	Enhanced ethanol titer as $71.0 \pm 0.28\,g\,L^{-1}$ by using mixture of glucose and xylose sugar ($162.4 \pm 1.23\,g\,L^{-1}$)	Kumari and Pramanik (2012)
S. cerevisiae	*Candida shehatae*	Protoplast fusion	Enhanced ethanol titer as $63.4 \pm 0.14\,g\,L^{-1}$ by using mixture of glucose and xylose sugar ($146.6 \pm 1.14\,g\,L^{-1}$).	Kumari and Pramanik (2012)
Kluyveromyces marxianus TY-22	*S. cerevisiae* AY-5	Protoplast fusion	About 3.8% (v/v) ethanol in 72 h by fermenting cheese whey powder	Guo et al. (2012)
S. cerevisiae	*P. tennophilus*	Protoplast fusion	Enhanced ethanol yield as $0.46\,g\,g^{-1}$ on fermentation of detoxified *Ipomoea carnea* biomass hydrolysate	Kumari and Pramanik (2013)
S. cerevisiae	*Candida intermedia*	Protoplast fusion	Enhanced ethanol yield on xylose as sole carbon source	Kahar and Tanaka (2014)
S. cerevisiae W5	*C. shehatae* 20 335	Protoplast fusion	About 43.4% and 16.7% higher ethanol yield than that of *S. cerevisiae* and *C. shehatae*, respectively, on glucose/xylose mixture	Ge et al. (2017)
S. cerevisiae	*Saccharomyces mikatae*	Protoplast fusion	Improved xylose consumption AFEX-pretreated corn stover hydrolysate	Peris et al. (2017)
S. cerevisiae	*Saccharomyces kudriavzevii*	Protoplast fusion	Improved xylose consumption AFEX-pretreated corn stover hydrolysate	Peris et al. (2017)
S. cerevisiae YIIc17_E5	*S. cerevisiae* UWOPS87-2421	Protoplast fusion	Grew faster in the presence of acetic acid, levulinic acids, and 2-furaldehyde (10 mM)	Štafa et al. (2019)

glucose/xylose transporter from *Talaromyces emersonii* were transformed into *hxt* null *S. cerevisiae* and no growth was observed on either hexose and/or pentose sugars (Saloheimo et al. 2007). The possible reason for no growth was depicted as the transporter proteins were not localized to the plasma membrane or the structural conformation was changed due to the composition of the phospholipid sterols in *S. cerevisiae* (Fernandes and Murray 2010). Furthermore, the kinetic characterization of heterologous xylose transporters *GXF1*, *SUT1*, and *AT5G59250* in *S. cerevisiae* reported a higher affinity of xylose for *GXF1* (Hector et al. 2008; Runquist et al. 2009). Fonseca et al. (2011) performed heterologous expression of *gxf1* in different *S. cerevisiae* strains, in which TMB3400 showed 2-fold higher growth in xylose compared to the parental strain. In another report, *S. cerevisiae* co-utilized glucose and xylose 80 and 40 g L^{-1}, respectively, within 24 h, when an expression of heterologous *xylA* of *Piromyces* sp., *gxf1* of *C. intermedia* and overexpression of endogenous *xks1* (xylulokinase), *rki1* (ribose-5-phosphate ketol-isomerase), *rpe1* (ribulose 5-phosphate epimerase), *tkl1* (transketolase), and *tal1* (transaldolase) were tried (Kuhad et al. 2011). Xylose utilizing *S. cerevisiae* strain was approached for individual overexpression of endogenous transporters *hxt1* and *hxt7* from *S. cerevisiae* and heterologous *gxf1* and *gxs1* of *C. intermedia* (Tanino et al. 2012). The resultant strains, overexpressing *hxt1*, *gxs1*, and *gxf1*, showed a significant uptake rate of 0.399, 0.367 and 0.378 g xylose (g^{-1} dry cell weight) h^{-1} respectively on xylose as the carbon source,. Similarly, heterologous expression of transporter gene *AT5G17010* of *Arabidopsis thaliana* in *Candida tropicalis* resulted in a 37%–73% increase in xylose uptake than isogenic *C. tropicalis* (Jeon et al. 2013). Transporter gene *MGT05196* of *Meyerozyma guilliermondii* which is similar to *gal2* was expressed in *S. cerevisiae* and resulted in a 26.3-fold higher growth on xylose compared to the isogenic strain (Wang et al. 2015a). Moreover, the substitution of asparagine residue at 360 loci to phenylalanine (N360F) of MgT05196p increase xylose transport. Wang et al. (2015b) also reported that Asp72 and Arg164 of MgT05196p are specific xylose binding residues. Heterologous expression of transporter gene *hxt2*, *xut1*, and *qup2* of *S. stipitis* individually into *S. cerevisiae* harboring overexpressed *xyl1*, *xyl2*, and *xks1* resulted in a significant ethanol yield of 0.28, 0.30, and 0.26 g g^{-1}, respectively, on glucose and xylose co-fermentation (de Sales et al. 2015).

7.4.2.2 Targeting Xylose Metabolic Pathway

As mentioned earlier, *S. cerevisiae* was always studied as a model organism to explore the loopholes in pentose sugar utilization for sustainable biofuel production (Liu and Hu 2010; Rao et al. 2011). In this process, limitations in the xylose metabolic pathway were also investigated via a comparison between a native xylose utilizing strain and *S. cerevisiae* either by regulating the expression of endogenous metabolic genes or heterologous expression (Weber et al. 2010). Conversion of xylose occurs in two routes: the first is two steps using XR and XDH and the second is one step of isomerization using xylose isomerase (XI) (Jeffries 2006; Chu and Lee 2007). In the XR/XDH pathway, XR prefers NADPH and XDH prefers NAD^+ as cofactors to convert xylose into xylitol and further into xylulose, respectively (Hahn-Hägerdal et al. 2007b). In this pathway, researchers have observed redox imbalance. Unlike the XR/XDH pathway, the XI pathway converts xylose to xylulose in a single step and bypasses the cofactor imbalance (Matsushika et al. 2009). The first time, heterologous expression of *xylA* gene encoding for XI of *Thermus thermophilus* in *S. cerevisiae* showed significant production of ethanol at 40 °C and maximum activity at 85 °C (Walfridsson et al. 1996; Lönn et al. 2003). Similarly, Madhavan et al. (2009a), reported the higher activity of XI as 1.91 U mg^{-1} in *S. cerevisiae* when heterologous gene *xylA* of *Orpinomyces* was expressed. Further, an improved specific growth rate of 0.13 h^{-1} and ethanol yield of 0.48 g g^{-1} was achieved when *sut1* (encoding the xylose transporter) of *S. stipitis* was heterologously expressed together with overexpression of endogenous *xks1* (encoding xylulose kinase) and evolutionary adaptation (Madhavan et al. 2009a, 2009b). Recombinant *S. cerevisiae* harboring heterologous *xylA* of *Clostridium phytofermentas* resulted in significant ethanol yield of 0.43 g g^{-1} on xylose as the carbon source (Brat et al. 2009). Heterologous expression of *xyl*A from *Piromyces* sp. and *xks* of *S. stipitis* in *S. cerevisiae*, in a combination of adaptation in xylose-containing medium, resulted a significant growth rate of 0.203 h^{-1} with xylose uptake rate and ethanol rate of 1.866 g g^{-1} h^{-1} and 0.41 g g^{-1}, respectively, on xylose as the carbon source (Zhou et al. 2012). Similarly, the heterologous expression of *xylA* and *xks1* of *Prevotella ruminicola* in *S. cerevisiae* produced a 3.8-fold higher specific growth rate on xylose (Hector et al. 2013). The heterologous expression of *xylA* and the non-oxidative pentose phosphate pathway (PPP) from *C. phytofermentans* in *S. cerevisiae* resulted in complete utilization of xylose, that is, 35 g L^{-1} with 1.1 g $L^{-1}h^{-1}$ xylose consumption rate (Demeke et al. 2013). The heterologous XI pathway was integrated into different *S. cerevisiae* mutants having knocked out the *gre3* and *pho13* (encoding phosphatase) genes individually and in the combination of overexpressed *xks1* and *tal1*, which resulted in a significantly higher ethanol yield on xylose (Lee et al. 2014). In another study, the expression of *xylA* in *S. cerevisiae* showed a higher specific growth rate, that is, $0.12 \pm 0.01\,h^{-1}$, and shorter lag phase (4 h versus 30 h) than the control strain during anoxic chemostat growth (Smith et al. 2014). Further, an engineered *S. cerevisiae* harboring heterologous *xylA* gene of *Burkholderia cenocepacia* showed the ability to co-utilize glucose and xylose, and significant ethanol productivity of 0.42 g g^{-1} h^{-1} on xylose without accumulation of xylitol under anaerobic conditions (Vilela et al. 2013, 2015). Similarly, *S. cerevisiae* having the XI pathway of *Bacteroides vulgatus* showed enhanced XI activity during xylose fermentation after adaptation on xylose as the carbon source (Hasunuma et al. 2014). A comparative transcriptomic study during xylose and glucose fermentation of engineered *S. cerevisiae* reported a higher expression of ergosterol genes during glucose fermentation. However, a higher transcriptome level of genes related to the central metabolic pathway was observed during xylose fermentation, which indicates the higher preference of fermentation for glucose over xylose in *S. cerevisiae* (Matsushika et al. 2014). The expression of the heterologous gene encoding NADH oxidase from *Lactococcus lactis* was optimized using different promotors including GPD2 and TEF1 in xylose utilizing *S. cerevisiae* harboring the XI and XR/XDH pathways individually and resulted in 84% reduced glycerol accumulation with 8% and 12% increased ethanol yield in the XI and XR/XDH harboring strains, respectively (Hou et al. 2014). Two-stage transcriptional reprogramming, including screening of stage-specific promotors and flux balance, was approached to engineer *S. cerevisiae* and resulted in 0.48 g g^{-1} ethanol yield on xylose (Cao et al. 2014). Comparative XI activity was also studied in a *S. cerevisiae* strain having overexpressed *XKS1* and non-oxidative PPP genes with knocked out *GRE3*, and individually expressing heterologous *xylA* gene of different mammal gut Bacteroidetes including *B. vulgatus*, *Tannerella* sp., *Paraprevotella xylaniphila*, and *Alistipes* sp. (Peng et al. 2015). Moreover, *S. cerevisiae* expressing *xylA* of *B. vulgatus* showed significantly higher XI activity up to 1.2 U mg^{-1}.

7.4.2.3 Targeting Non-oxidative Pathway

Engineering only the xylose metabolic pathway, that is, XR/XDH or XI, is not enough to efficiently convert xylose into ethanol until a follow-up PPP is considered. Tuning the expression of *xks1* together with the non-oxidative pathway may increase the conversion rate of xylose or the flux balance by increasing the cofactor regeneration for the oxide reduction pathway during co-utilization of xylose and glucose (Senac and Hahn-Hägerdal 1991; Jin et al. 2003). The non-oxidative pathway in *S. cerevisiae* is not efficient for xylose conversion but helps to convert glycolytic intermediates from xylulose (Matsushika et al. 2009). In this way, reduction in xylitol was observed when *gre3*, encoding for predictive aldose reductase, was knocked out in xylose utilizing *S. cerevisiae* (Senac and Hahn-Hägerdal 1991; Traff et al. 2001). Similarly, overexpression of *gdp1* and deletion of *zwf1* gene encoding for $NADP^+$-dependent D-glyceraldehyde-3-phosphate dehydrogenase and glucose-6-phosphate dehydrogenase also facilitates the NADPH availability of a yeast strain harboring the XR/XDH pathway (Verho et al. 2003). Overexpression of *xks1*, *rpi*, *rpe*, *tkl*, and *tal* genes of PPP encoding for xylulose kinase, ribulose-5-phosphate isomerase, ribulose-5-phosphate epimerase, transketolase, and transaldolase, respectively, in *S. cerevisiae* RWB202-AFX with knocked out *gre3* resulted in significant growth at $0.12\,h^{-1}$ specific growth rate on xylose with $0.41\,g\,g^{-1}$ ethanol yield (Kuyper et al. 2005a, 2005b). In another study, heterologous expression *gapN* gene, encoding for non-phosphorylating $NADP^+$-dependent glyceraldehyde-3-phosphate dehydrogenase in xylose utilizing *S. cerevisiae* increased ethanol yield by 25% above the control strain during glucose xylose co-utilization (Bro et al. 2006). Heterologous expression of *xks1* from *S. stipitis* and knockout of endogenous *pho13* in xylose utilizing *S. cerevisiae* reported enhanced ethanol titer on xylose as the carbon source (Van Vleet et al. 2008). Almeida et al. (2009) predicted an *in silico* model for xylose metabolism and proposed an increased redox balance with reduced xylitol accumulation via overexpression of *gdh2* and knockout of *gdh1* genes encoding for NAD^+-dependent glutamate dehydrogenase and $NADP^+$-dependent glutamate dehydrogenase, respectively. It has been depicted that tuning genes of non-oxidative PPP variably affect the ethanol yield in yeast strains (Chu and Lee 2007; Matsushika et al. 2009; Bera et al. 2011). Heterologous expression of *pho13* and deletion of *ald6* increases the ethanol yield in xylose metabolizing *S. cerevisiae* (Kim et al. 2013). Moreover, reduced acetate accumulation was observed when *ald3* was deleted and increases ethanol production.

7.4.2.4 Engineering Non-conventional Yeast

Instead of engineering a conventional yeast *S. cerevisiae*, several non-conventional, xylose-utilizing yeast strains such as *S. stipitis*, *C. intermedia*, *C. shehatae*, *D. hensenii*, *C. tropicalis*, and *K. marxianus* have been developed for a sustainable LCB conversion process. Developing engineering tools for non-conventional yeast strains such as *S. stipitis* and *D. hansenii* is not as easy as for *S. cerevisiae*. *S. stipitis* has been found to be resistant to common antibiotics and auxotrophic markers. Moreover, it also decodes CUG for serine residue rather than leucine; therefore a heterologous expression of genes requires codon optimization and the process is complex (Ruchala et al. 2020). However, various tools have been developed to engineer the strain including sexual conjugation and loxP/Cre recombination together with auxotrophic and antibiotic selection. In the series, *S. stipitis* has been reported as an efficient converter of xylose to ethanol together with utilizing all the sugars present in woody biomass (Ruchala et al. 2020). Moreover, the major issue of catabolite repression during co-utilization of glucose and xylose was also fixed by deletion of *hxk1* gene (encoding for hexokinase). Engineered strain *S. stipitis* SS6 showed a rapid and simultaneous utilization of glucose and xylose whereas a wild-type strain started consumption of xylose after completed utilization of glucose (Dashtban et al. 2015). Some non-conventional naturally xylose-utilizing yeasts have also approached alteration in a xylose metabolic pathway including *xyl1* and *xyl2* with altered co-enzyme preferences for enhanced xylose utilization. Heterologous expression of *Candida perapsilosis xyl1* gene encoding for NADH preferring XR was done in *C. tropicalis* (Lee et al. 2003), and the improved strain resulted in an increased level of XR activity than that of the type strain (Rao et al. 2006). The *D. hansenii* strain having knocked out the *xyl2* gene was observed with 2.5-fold increased xylitol accumulation when co-substrates were utilized, that is, xylose and glycerol were used as the source of carbon (Pal et al. 2013). Biswas et al. (2013) studied *xyl2* of *D. hansenii* and reported that DhXDHp is thermosensitive and highly active against xylitol compared to XDH of the other yeasts. Similarly, Zhang et al. (2014) expressed heterologous transporters *gxf1* and *gxs1* of *C. intermedia* together with overexpression of endogenous *fps1* in thermotolerant *K. marxianus* resulting in xylitol productivity of 2.81 $g\,L^{-1}\,h^{-1}$ on xylose as a carbon source.

7.5 Conclusion and Future Prospects

The worldwide energy demand requires an alternative and sustainable source of energy, which can be fulfilled using natural sources such as LCB. In the series various approaches have been made to convert LCB into potential biofuel. This process includes novel techniques for biomass pretreatment, breakdown of complex molecules into simple sugars and potential microbes for sugar metabolism. Bioethanol is one of the possible solutions to lend a hand for the worldwide fuel crisis. This chapter summarized promising methods of biomass pretreatment together with various efforts to engineer conventional and non-conventional yeast strains. Moreover, improvisation in the potential strains was also explored using random mutagenesis and site-specific engineering. However, there are still bottlenecks to be resolved. It is recommended to explore more naturally pentose metabolizing species together with the development of genetic tools for non-conventional microbes to make a sustainable process.

References

Agbogbo, F.K., Haagensen, F.D., Milam, D., and Wenger, K.S. (2008). Fermentation of acid pretreated corn stover to ethanol without detoxification using *Pichia stipitis*. *Appl. Biochem. Biotechnol.* 145: 53–58.

Almeida, J.R.M., Bertilsson, M., Hahn-Hägerdal, B. et al. (2009). Carbon fluxes of xylose-consuming *Saccharomyces cerevisiae* strains are affected differently by NADH and NADPH usage in HMF reduction. *Appl. Microbiol. Biotechnol.* 84: 751–761.

Amin, F.R., Khalid, H., Zhang, H. et al. (2017). Pretreatment methods of lignocellulosic biomass for anaerobic digestion. *AMB Express* 7: 72.

Arora, R., Behera, S., Sharma, N.K. et al. (2015). A new search for thermotolerant yeasts, its characterization and optimization using response surface methodology for ethanol production. *Front. Microbiol.* 6: 889.

Arora, R., Behera, S., Sharma, N.K. et al. (2021). Bioprospecting saccharification of alkali pretreated paddy straw through statistically designed parameters for biofuel production. *Ind. Biotechnol.* 16: 375–385.

Bajwa, P.K., Pinel, D., Martin, V.J.J. et al. (2010). Strain improvement of the pentose-fermenting yeast *Pichia stipitis* by genome shuffling. *J. Microbiol. Methods* 81: 179–186.

Bajwa, P.K., Phaenark, C., Grant, N. et al. (2011). Ethanol production from selected lignocellulosic hydrolysates by genome shuffled strains of *Scheffersomyces stipitis*. *Bioresour. Technol.* 102: 9965–9969.

Bay, M.S., Karimi, K., Esfahany, M.N., and Kumar, R. (2020). Structural modification of pine and poplar wood by alkali pretreatment to improve ethanol production. *Ind. Crop. Prod.* 152: 112506.

Behera, S., Arora, R., Nandgopal, N., and Kumar, S. (2014). Importance of chemical pretreatment for bioconversion of lignocellulosic biomass. *Renew. Sust. Energ. Rev.* 36: 91–106.

Behera, S., Sharma, N.K., Arora, R., and Kumar, S. (2016). Effect of evolutionary adaption on xylosidase activity in thermotolerant yeast isolates *Kluyveromyces marxianus* NIRE-K1 and NIRE-K3. *Appl. Biochem. Biotechnol.* 179: 1143–1154.

Ben-Iwo, J., Manovic, V., and Longhurst, L. (2016). Biomass resources and biofuels potential for the production of transportation fuels in Nigeria. *Ren. Sust. Energy Rev.* 63: 172–192.

Bera, A.K., Ho, N.W., Khan, A., and Sedlak, M. (2011). A genetic overhaul of *Saccharomyces cerevisiae* 424A(LNH-ST) to improve xylose fermentation. *J. Ind. Microbiol. Biotechnol.* 38: 617–626.

Bibra, M., Wang, J., Squillace, P. et al. (2014). Biofuels and value-added products from extremophiles. In: *In Advances in Biotechnology* (ed. N.N. Nawani, M. Khetmalas, P.N. Razdan and A. Pandey), 268. New Delhi: I.K. International Publishing House.

Bibra, M., Kumar, S., Wang, J. et al. (2018). Single pot bioconversion of prairie cordgrass into biohydrogen by thermophiles. *Bioresour. Technol.* 266: 232–241.

Bibra, M., Rathinam, N.K., Johnson, G.R., and Sani, R.K. (2020). Single pot biovalorization of food waste to ethanol by *Geobacillus* and *Thermoanaerobacter* spp. *Renew. Energy* 155: 1032–1041.

Biller, P., Friedman, C., and Ross, A.B. (2013). Hydrothermal microwave processing of microalgae as a pre-treatment and extraction technique for bio-fuels and bio-products. *Bioresour. Technol.* 136: 188–195.

Biswas, D., Datt, M., Aggarwal, M., and Mondal, A.K. (2013). Molecular cloning, characterization, and engineering of xylitol dehydrogenase from *Debaryomyces hansenii*. *Appl. Microbiol. Biotechnol.* 97: 1613–1623.

Brat, D., Boles, E., and Wiedemann, B. (2009). Functional expression of a bacterial xylose isomerase in *Saccharomyces cerevisiae*. *Appl. Environ. Microbiol.* 75: 2304–2311.

Bro, C., Regenberg, B., Forster, J., and Nielsen, J. (2006). In silico aided metabolic engineering of *Saccharomyces cerevisiae* for improved bioethanol production. *Metab. Eng.* 8: 102–111.

Brodeur, G., Yau, E., Badal, K. et al. (2011). Chemical and physicochemical pretreatment of lignocellulosic biomass: a review. *Enzyme Res.* 2011: 787532.

Cai, Z., Zhang, B., and Li, Y. (2012). Engineering *Saccharomyces cerevisiae* for efficient anaerobic xylose fermentation: reflections and perspectives. *Biotechnol. J.* 7: 34–46.

Cao, L., Tang, X., Zhang, X. et al. (2014). Two-stage transcriptional reprogramming in *Saccharomyces cerevisiae* for optimizing ethanol production from xylose. *Metab. Eng.* 24: 150–159.

Caspeta, L. and Nielsen, J. (2015). Thermotolerant yeast strains adapted by laboratory evolution show trade-off at ancestral temperatures and preadaptation to other stresses. *MBio* 6: e00431–e00415.

Caspeta, L., Castillo, T., and Nielsen, J. (2015). Modifying yeast tolerance to inhibitory conditions of ethanol production processes. *Front. Bioeng. Biotechnol.* 3: 184.

Chandel, A.K. and Sukumaran, R.K. (2017). *Sustainable Biofuels Development in India*. Cham: Springer.

Chandel, A.K., Chandrasekhar, G., Radhika, K. et al. (2011). Bioconversion of pentose sugars into ethanol: a review and future directions. *Biotechnol. Mol. Biol. Rev.* 6: 8–20.

Chen, C., Boldor, D., Aita, G., and Walker, M. (2012). Ethanol production from sorghum by a microwave-assisted dilute ammonia pretreatment. *Bioresour. Technol.* 110: 190–197.

Cheng, C., Almario, M.P., and Kao, K.C. (2015). Genome shuffling to generate recombinant yeasts for tolerance to inhibitors present in lignocellulosic hydrolysates. *Biotechnol. Lett.* 37: 2193–2200.

Cheng, H.H., Pien, T.T., Lee, Y.C. et al. (2020). Effects of copper on biological treatment of NMF- and MDG-containing wastewater from TFT-LCD industry. *Chemosphere* 258: 127125.

Chettri, D., Verma, A.K., and Verma, A.K. (2020). Innovations in CAZyme gene diversity and its modification for biorefinery applications. *Biotechnol. Rep.* 28: e00525.

Chu, B.C.H. and Lee, H. (2007). Genetic improvement of *Saccharomyces cerevisiae* for xylose fermentation. *Biotechnol. Adv.* 25: 425–441.

Cong, B., Wang, N., Liu, S. et al. (2017). Isolation, characterization and transcriptome analysis of a novel antarctic *Aspergillus sydowii* strain MS-19 as a potential lignocellulosic enzyme source. *BMC Microbiol.* 17: 129.

Dashtban, M., Wen, X., Bajwa, P. et al. (2015). Deletion of *hxk1* gene results in derepression of xylose utilization in *Scheffersomyces stipitis*. *J. Ind. Microbiol. Biotechnol.* 42: 889–896.

Demeke, M.M., Dietz, H., Li, Y. et al. (2013). Development of a d-xylose fermenting and inhibitor tolerant industrial *Saccharomyces cerevisiae* strain with high performance in lignocellulose hydrolysates using metabolic and evolutionary engineering. *Biotechnol. Biofuels* 6: 89. https://doi.org/10.1186/1754-6834-6-89.

Demeke, M.M., Foulquie-Moreno, M.R., Dumortier, F., and Thevelein, J.M. (2015). Rapid evolution of recombinant *Saccharomyces cerevisiae* for xylose fermentation through formation of extra-chromosomal circular DNA. *PLoS Genet.* 11: e1005010.

Fan, C., Qi, K., Xia, X.X., and Zhong, J.J. (2013). Efficient ethanol production from corncob residues by repeated fermentation of an adapted yeast. *Bioresour. Technol.* 136: 309–315.

Fernandes, S. and Murray, P. (2010). Metabolic engineering for improved microbial pentose fermentation. *Bioeng. Bugs* 1: 424–428.

Fischer, G. and Schrattenholzer, L. (2001). Global bioenergy potentials through 2050. *Biomass Bioenergy* 20: 151–159.

Fonseca, C., Olofsson, K., Ferreira, C. et al. (2011). The glucose/xylose facilitator Gxf1 from *Candida intermedia* expressed in a xylose fermenting industrial strain of *Saccharomyces cerevisiae* increases xylose uptake in SSCF of wheat straw. *Enzym. Microb. Technol.* 48: 518–525.

Ge, J., Zhao, J., Zhang, L. et al. (2014). Construction and analysis of high-ethanol-producing fusants with co-fermentation ability through protoplast fusion and double labeling technology. *PLoS One* 9: e108311.

Ge, J., Du, R., Song, G. et al. (2017). Metabolic pathway analysis of the xylose-metabolizing yeast protoplast fusant ZLYRHZ7. *J. Biosci. Bioeng.* 124: 386–391.

Gera, R., Dhamija, S.S., Gera, T., and Singh, D. (1997). Intergeneric ethanol producing hybrids of thermotolerant *Kluyveromyces* and non-thermotolerant *Saccharomyces cerevisiae*. *Biotechnol. Lett.* 19: 189–194.

Gill, M.K., Kocher, G.S., and Panesar, A.S. (2021). Optimization of acid-mediated delignification of corn stover, an agriculture residue carbohydrate polymer for improved ethanol production. *Carbohydr. Polym. Technol. Appl.* 2: 100029.

Gonçalves, D.L., Matsushika, A., Belisa, B. et al. (2014). Xylose and xylose/glucose co-fermentation by recombinant *Saccharomyces cerevisiae* strains expressing individual hexose transporters. *Enzym. Microb. Technol.* 63: 13–20.

Gunny, A.A.N., Arbain, D., Nashef, E.M., and Jamal, P. (2015). Applicability evaluation of deep eutectic solvents–cellulase system for lignocellulose hydrolysis. *Bioresour. Technol.* 181: 297–302.

Guo, X., Wang, R., Chen, Y., and Xiao, D. (2012). Intergeneric yeast fusants with efficient ethanol production from cheese whey powder solution: construction of a *Kluyveromyces marxianus* and *Saccharomyces cerevisiae* AY-5 hybrid. *Eng. Life Sci.* 12: 656–661.

Hahn-Hägerdal, B., Karhuma, K., Jeppssonm, M., and Gorwa-Grauslund, M.F. (2007a). Metabolic engineering for pentose utilization in *Saccharomyces cerevisiae*. *Adv. Biochem. Eng. Biotechnol.* 108: 147–177.

Hahn-Hägerdal, B., Karhumaa, K., Fonseca, C. et al. (2007b). Towards industrial pentose-fermenting yeast strains. *Appl. Microbiol. Biotechnol.* 74: 937–953.

Hamacher, T., Becker, J., Gárdonyi, M. et al. (2002). Characterization of the xylose-transporting properties of yeast hexose transporters and their influence on xylose utilization. *Microbiology* 148: 2783–2788.

Hans, M., Garg, S., Pellegrini, V.O.A. et al. (2021). Liquid ammonia pretreatment optimization for improved release of fermentable sugars from sugarcane bagasse. *J. Clean. Prod.* 281: 123922.

Hasunuma, T., Ismail, K.S., Nambu, Y., and Kondo, A. (2014). Coexpression of TAL1 and ADH1 in recombinant xylose-fermenting *Saccharomyces cerevisiae* improves ethanol production from lignocellulosic hydrolysates in the presence of furfural. *J. Biosci. Bioeng.* 117: 165–169.

Hector, R.E., Qureshi, N., Hughes, S.R., and Cotta, M.A. (2008). Expression of a heterologous xylose transporter in a *Saccharomyces cerevisiae* strain engineered to utilize xylose improves aerobic xylose consumption. *Appl. Microbiol. Biotechnol.* 80: 675–684.

Hector, R.E., Dien, B.S., Cotta, M.A., and Mertens, J.A. (2013). Growth and fermentation of d-xylose by *Saccharomyces cerevisiae* expressing a novel d-xylose isomerase originating from the bacterium *Prevotella ruminicola* TC2-24. *Biotechnol. Biofuels* 6: 84.

Hoşgün, E.Z., Berikten, D., Kıvanç, M., and Bozan, B. (2017). Ethanol production from hazelnut shells through enzymatic saccharification and fermentation by low-temperature alkali pretreatment. *Fuel* 196: 280–287.

Hoover, F. and Abraham, J.A. (2009). Comparison of corn-based ethanol with cellulosic ethanol as replacements for petroleum-based fuels: a review. *Int. J. Sust. Energy* 28: 171–182.

Hou, L. and Yao, S. (2012). Improved inhibitor tolerance in xylose fermenting yeast *Spathaspora passalidarum* by mutagenesis and protoplast fusion. *Appl. Microbiol. Biotechnol.* 93: 2591–2601.

Hou, J., Suo, F., Wang, C. et al. (2014). Fine-tuning of NADH oxidase decreases byproduct accumulation in respiration deficient xylose metabolic *Saccharomyces cerevisiae. BMC Biotechnol.* 14: 13.

Ibraheem, O. and Ndimba, B.K. (2013). Molecular adaptation mechanisms employed by ethanologenic bacteria in response to lignocellulose-derived inhibitory compounds. *Int. J. Biol. Sci.* 9: 598–612.

Jaswal, R., Shende, A., Nan, W. et al. (2019). Hydrothermal liquefaction and photocatalytic reforming of pinewood (*Pinus ponderosa*)-derived acid hydrolysis residue for hydrogen and bio-oil production. *Energy Fuel* 33: 6454–6462.

Jeffries, T.W. (2006). Engineering yeasts for xylose metabolism. *Curr. Opin. Biotechnol.* 17: 320–326.

Jeon, W.Y., Shim, W.Y., Lee, S.H. et al. (2013). Effect of heterologous xylose transporter expression in *Candida tropicalis* on xylitol production rate. *Bioprocess Biosyst. Eng.* 36: 809–817.

Jiang, L.Q., Wu, Y.X., Wang, X.B. et al. (2019). Crude glycerol pretreatment for selective saccharification of lignocellulose via fast pyrolysis and enzyme hydrolysis. *Energy Convers. Manag.* 199: 111894.

Jin, Y.S., Ni, H., Laplaza, J.M., and Jeffries, T.W. (2003). Optimal growth and ethanol production from xylose by recombinant *Saccharomyces cerevisiae* require moderate d-xylulokinase activity. *Appl. Environ. Microbiol.* 69: 495–503.

Jing-Ping, G., Hongbing, S., Gang, S. et al. (2012). A genome shuffling-generated *Saccharomyces cerevisiae* isolate that ferments xylose and glucose to produce high levels of ethanol. *J. Ind. Microbiol. Biotechnol.* 39: 777–787.

Kahar, P. and Tanaka, S. (2014). A xylose-fermenting yeast hybridized by intergeneric fusion between *Saccharomyces cerevisiae* and *Candida intermedia* mutants for ethanol production. *Sustain. Chem. Process.* 2: 17.

Kang, Q., Appels, L., Tan, T., and Dewil, R. (2014). Bioethanol from lignocellulosic biomass: current findings determine research priorities. *Sci. World J.* 2014: 1–13.

Khuong, L.S., Zulkifli, N.W.M., Masjuki, H.H. et al. (2016). A review on the effect of bioethanol dilution on the properties and performance of automotive lubricants in gasoline engines. *RSC Adv.* **6**: 66847–66869.

Kim, S.R., Skerker, J.M., Kang, W. et al. (2013). Rational and evolutionary engineering approaches uncover a small set of genetic changes efficient for rapid xylose fermentation in *Saccharomyces cerevisiae. PLoS One* 8: e57048.

Klein, M., Griess, O., Pulidindi, I.N. et al. (2016). Bioethanol production from *Ficus religiosa* leaves using microwave irradiation. *J. Environ. Manag.* 177: 20–25.

Krahulec, S., Petschacher, B., Wallner, M. et al. (2010). Fermentation of mixed glucose-xylose substrates by engineered strains of *Saccharomyces cerevisiae*: role of the coenzyme specificity of XR, and effect of glucose on xylose utilization. *Microb. Cell Factories* 9: 16.

Küçükgöze, G., Alkım, C., Yılmaz, Ü. et al. (2013). Evolutionary engineering and transcriptomic analysis of nickel-resistant *Saccharomyces cerevisiae. FEMS Yeast Res.* 8: 731–746.

Kuhad, R.C., Gupta, R., Khasa, Y.P. et al. (2011). Bioethanol production from pentose sugars: current status and future prospects. *Renew. Sust. Energ. Rev.* 15: 4950–4962.

Kumar, S., Bhalla, A., Bibra, M. et al. (2015). Thermophilic biohydrogen production: challenges at the industrial scale. In: *Bioenergy: Opportunities and Challenges* (ed. N. Krishnara and J.-S. Yu), 34. New York: Apple Academic Press.

Kumar, S., Dheeran, P., Taherzadeh, M., and Khanal, S. (2018). *Fungal Biorefineries*. Cham: Springer https://doi.org/10.1007/978-3-319-90379-8.

Kumari, R. and Pramanik, K. (2012). Improved bioethanol production using fusants of *Saccharomyces cerevisiae* and xylose-fermenting yeasts. *Appl. Biochem. Biotechnol.* 167: 873–884.

Kumari, R. and Pramanik, K. (2013). Bioethanol production from *Ipomea carnea* biomass using a potential hybrid yeast strain. *Appl. Biochem. Biotechnol.* 171: 1771–1785.

Kuyper, M., Hartog, M.M., Toirkens, M.J. et al. (2005a). Metabolic engineering of a xylose-isomerase expressing *Saccharomyces cerevisiae* strain for rapid anaerobic xylose fermentation. *FEMS Yeast Res.* 5: 399–409.

Kuyper, M., Toirkens, M.J., Diderich, J.A. et al. (2005b). Evolutionary engineering of mixed-sugar utilization by a xylose-fermenting *Saccharomyces cerevisiae* strain. *FEMS Yeast Res.* 5: 925–934.

Lee, W.J., Kim, M.D., Ryu, Y.W. et al. (2002). Kinetic studies on glucose and xylose transport in *Saccharomyces cerevisiae*. *Appl. Microbiol. Biotechnol.* 60: 186–191.

Lee, J.K., Koo, B.S., and Kim, S.Y. (2003). Cloning and characterization of the xyl1 gene, encoding an NADH preferring XR from *Candida parapsilosis*, and its functional expression in *Candida tropicalis*. *Appl. Environ. Microbiol.* 69: 6179–6188.

Lee, S.M., Jellison, T., and Alper, H.S. (2014). Systematic and evolutionary engineering of a xylose isomerase-based pathway in *Saccharomyces cerevisiae* for efficient conversion yields. *Biotechnol. Biofuels* 7: 122.

Lee, J., Saddler, J.N., Um, Y., and Woo, H.M. (2016). Adaptive evolution and metabolic engineering of a cellobiose- and xylose-negative *Corynebacterium glutamicum* that co-utilizes cellobiose and xylose. *Microb. Cell Factories* 15: 20.

Li, J., Zhao, X., Liu, C. et al. (2008). Construction of yeast strains for efficient ethanol fermentation from xylose by protoplast fusion. *J. Biotechnol.* 136: 402–459.

Liu, E. and Hu, Y. (2010). Construction of a xylose-fermenting *Saccharomyces cerevisiae* strain by combined approaches of genetic engineering, chemical mutagenesis and evolutionary adaptation. *Biochem. Eng. J.* 48: 204–210.

Liu, Y., Zhou, H., Wang, L. et al. (2016). Improving *Saccharomyces cerevisiae* growth against lignocellulose-derived inhibitors as well as maximizing ethanol production by a combination proposal of γ-irradiation pretreatment with in situ detoxification. *Chem. Eng. J.* 287: 302–312.

Lönn, A., Träff-Bjerre, K.L., Cordero Otero, R.R. et al. (2003). Xylose isomerase activity influences xylose fermentation with recombinant *Saccharomyces cerevisiae* strains expressing mutated xylA from *Thermus thermophilus*. *Enzym. Microb. Technol.* 32: 567–573.

Loray, M.A., De Figueroa, L.I., and Höfer, M. (1997). Monosaccharide uptake in a yeast hybrid produced by protoplast fusion. *Folia Microbiol. (Praha)* 42: 239–241.

MacLean, H.L. (2004). Alternative transport fuels for the future. *Int. J. Vehicle Design* 35: 27–49.

Madhavan, A., Tamalampudi, S., Srivastava, A. et al. (2009a). Alcoholic fermentation of xylose and mixed sugars using recombinant *Saccharomyces cerevisiae* engineered for xylose utilization. *Appl. Microbiol. Biotechnol.* 82: 1037–1047.

Madhavan, A., Tamalampudi, S., Ushida, K. et al. (2009b). Xylose isomerase from polycentric fungus *Orpinomyces*: gene sequencing, cloning, and expression in *Saccharomyces cerevisiae* for bioconversion of xylose to ethanol. *Appl. Microbiol. Biotechnol.* 82: 1067–1078.

Maity, S.K. (2015). Opportunities, recent trends and challenges of integrated biorefinery: part I. *Renew. Sust. Energ. Rev.* 43: 1427–1445.

Malik, K., Salama, E.S., Kim, T.H., and Li, X. (2020). Enhanced ethanol production by *Saccharomyces cerevisiae* fermentation post acidic and alkali chemical pretreatments of cotton stalk lignocellulose. *Int. Biodeterior. Biodegrad.* 147: 104869.

Matsushika, A., Inoue, H., Kodaki, K., and Sawayama, S. (2009). Ethanol production from xylose in engineered *Saccharomyces cerevisiae* strains: current state and perspectives. *Appl. Microbiol. Biotechnol.* 84: 37–53.

Matsushika, A., Goshima, T., and Hoshino, T. (2014). Transcription analysis of recombinant industrial and laboratory *Saccharomyces cerevisiae* strains reveals the molecular basis for fermentation of glucose and xylose. *Microb. Cell Factories* 28: 13–16.

Moreno, A.D., Ibarra, D., Ballesteros, I. et al. (2013). Ethanol from laccase-detoxified lignocellulose by the thermotolerant yeast *Kluyveromyces marxianus*—effects of steam pretreatment conditions, process configurations and substrate loadings. *Biochem. Eng. J.* 79: 94–103.

Moysés, D.N., Reis, V.C.B., de Almeida, J.R.M. et al. (2016). Xylose fermentation by *Saccharomyces cerevisiae*: challenges and prospects. *Int. J. Mol. Sci.* 17: 207. https://doi.org/10.3390/ijms17030207.

Narayanan, V., Sànchez i Nogué, V., van Niel, E.W.J., and Gorwa-Grauslund, M.F. (2016). Adaptation to low pH and lignocellulosic inhibitors resulting in ethanolic fermentation and growth of *Saccharomyces cerevisiae*. *AMB Express* 6: 59.

Olofsson, K., Bertilsson, M., and Lidén, G. (2008). A short review on SSF—an interesting process option for ethanol production from lignocellulosic feedstocks. *Biotechnol. Biofuels* 1: 7.

Pal, S., Choudhary, V., Kumar, A. et al. (2013). Studies on xylitol production by metabolic pathway engineered *Debaryomyces hansenii*. *Bioresour. Technol.* 147: 449–455.

Pasha, C., Kuhad, R.C., and Rao, L.V. (2007). Strain improvement of thermotolerant *Saccharomyces cerevisiae* VS strain for better utilization of lignocellulosic substrates. *J. Appl. Microbiol.* 103: 1480–1489.

Peng, B., Shen, Y., Li, X. et al. (2012). Improvement of xylose fermentation in respiratory deficient xylose fermenting *Saccharomyces cerevisiae*. *Metab. Eng.* 14: 9–18.

Peng, B., Huang, S., Liu, T., and Geng, A. (2015). Bacterial xylose isomerases from the mammal gut Bacteroidetes cluster function in *Saccharomyces cerevisiae* for effective xylose fermentation. *Microb. Cell Factories* 14: 70.

Pereira, S.R., Sànchez i Nogué, V., Frazão, C.J.R. et al. (2015). Adaptation of *Scheffersomyces stipitis* to hardwood spent sulfite liquor by evolutionary engineering. *Biotechnol. Biofuels* 8: 50.

Peris, D., Moriarty, R.V., Alexander, W.G. et al. (2017). Hybridization and adaptive evolution of diverse *Saccharomyces* species for cellulosic biofuel production. *Biotechnol. Biofuels* 10: 28.

Popp, J., Harangi-Rákos, M., Gabnai, Z. et al. (2016). Biofuels and their co-products as livestock feed: global economic and environmental implications. *Molecules* 21: 285.

Rahayu, F., Kawai, Y., Iwasaki, Y. et al. (2017). Thermophilic ethanol fermentation from lignocellulose hydrolysate by genetically engineered *Moorella thermoacetica*. *Bioresour. Technol.* 245: 1393–1399.

Rai, R., Bibra, M., Chadha, B.S., and Sani, R.K. (2019). Enhanced hydrolysis of lignocellulosic biomass with doping of a highly thermostable recombinant laccase. *Int. J. Biol. Macromol.* 137: 232–237.

Rao, R.S., Cherukuri, P.J., Reddy, P.S. et al. (2006). Strain improvement of *Candida tropicalis* for the production of xylitol: biochemical and physiological characterization of wild-type and mutant strain CT-OMV5. *J. Microbiol.* 44: 113–120.

Rao, R.P., Dufour, N., and Swana, J. (2011). Using microorganisms to brew biofuels. *In vitro Cell. Dev. Biol. Plant* 47: 637–649.

Rathinam, N.K., Bibra, M., Rajan, M. et al. (2019). Short term atmospheric pressure cold plasma treatment: a novel strategy for enhancing the substrate utilization in a thermophile, *Geobacillus* sp. strain WSUCF1. *Bioresour. Technol.* 278: 477–480.

Rathinam, N.K., Bibra, M., Salem, D.R., and Sani, R.K. (2020). Bioelectrochemical approach for enhancing lignocellulose degradation and biofilm formation in *Geobacillus* strain WSUCF1. *Bioresour. Technol.* 295: 122271.

Reznicek, O., Facey, S.J., and de Waal, P.P. (2015). Improved xylose uptake in *Saccharomyces cerevisiae* due to directed evolution of galactose permease Gal2 for sugar co-consumption. *J. Appl. Microbiol.* 5: 46.

Roy, R., Rahman, M.S., and Raynie, D.E. (2020). Recent advances of greener pretreatment technologies of lignocellulose. *Curr. Res. Green Sust. Chem.* 3: 100035.

Ruchala, J., Kurylenk, O.O., Dmytruk, K.V., and Sibirny, A.A. (2020). Construction of advanced producers of first- and second-generation ethanol in *Saccharomyces cerevisiae* and selected species of non-conventional yeasts (*Schefersomyces stipitis, Ogataea polymorpha*). *J. Ind. Microbiol. Biotechnol.* 47: 109–132.

Runquist, D., Fonseca, C., Rådström, P. et al. (2009). Expression of the Gxf1 transporter from *Candida intermedia* improves fermentation performance in recombinant xylose-utilizing *Saccharomyces cerevisiae*. *Appl. Microbiol. Biotechnol.* 82: 123–130.

Saini, J.K., Saini, R., and Tewari, L. (2015). Lignocellulosic agriculture wastes as biomass feedstocks for second-generation bioethanol production: concepts and recent developments. 3. *Biotech.* 5: 337–353.

Saini, P., Beniwal, A., Kokkiligadda, A., and Vij, S. (2017). Evolutionary adaptation of *Kluyveromyces marxianus* strain for efficient conversion of whey lactose to bioethanol. *Proc. Biochem.* 62: 69–79.

de Sales, B.B., Scheid, B., Gonçalves, D.L. et al. (2015). Cloning novel sugar transporters from *Scheffersomyces* (Pichia) *stipitis* allowing d-xylose fermentation by recombinant *Saccharomyces cerevisiae*. *Biotechnol. Lett.* 37: 1973–1982.

Saloheimo, A., Rauta, J., Stasyk, O.V. et al. (2007). Xylose transport studies with xylose-utilizing *Saccharomyces cerevisiae* strains expressing heterologous and homologous permeases. *Appl. Microbiol. Biotechnol.* 74: 1041–1052.

Sedlak, M. and Ho, N.W.Y. (2004). Characterization of the effectiveness of hexose transporters for transporting xylose during glucose and xylose co-fermentation by a recombinant *Saccharomyces* yeast. *Yeast* 21: 671–684.

Senac, T. and Hahn-Hagerdal, B. (1991). Effects of increased transaldolase activity on d-xylulose and d-glucose metabolism in *Saccharomyces cerevisiae* cell extracts. *Appl. Environ. Microbiol.* 57: 1701–1706.

Sharma, N.K., Behera, S., Arora, R., and Kumar, S. (2016). Enhancement in xylose utilization using *Kluyveromyces marxianus* NIRE-K1 through evolutionary adaptation approach. *Bioprocess Biosyst. Eng.* 39: 835–843.

Sharma, N.K., Behera, S., Arora, R., and Kumar, S. (2017). Evolutionary adaptation of *Kluyveromyces marxianus* NIRE-K3 for enhanced xylose utilization. *Front. Energy Res.* 5: 32.

Sharma, N.K., Behera, S., Arora, R. et al. (2018). Xylose transport in yeast for lignocellulosic ethanol production: current status. *J. Biosci. Bioeng.* 125: 259–267.

Shi, J., Zhang, M., Zhang, L. et al. (2014). Xylose-fermenting *Pichia stipitis* by genome shuffling for improved ethanol production. *Microb. Biotechnol.* 7: 90–99.

Sitepu, I.R., Enriquez, L.L., Nguyen, V. et al. (2019). Ethanol production in switchgrass hydrolysate by ionic liquid-tolerant yeasts. *Bioresour. Technol. Rep.* 7: 100275.

Smith, J., van Rensburg, E., and Görgens, J.F. (2014). Simultaneously improving xylose fermentation and tolerance to lignocellulosic inhibitors through evolutionary engineering of recombinant *Saccharomyces cerevisiae* harbouring xylose isomerase. *BMC Biotechnol.* 15: 14–41.

Snoek, T., Picca Nicolino, M., Van den Bremt, S. et al. (2015). Large-scale robot-assisted genome shuffling yields industrial *Saccharomyces cerevisiae* yeasts with increased ethanol tolerance. *Biotechnol. Biofuels* 8: 32.

Söderström, J., Pilcher, L., Galbe, M., and Zacchi, G. (2003). Two-step steam pretreatment of softwood by dilute H_2SO_4 impregnation for ethanol production. *Biomass Bioenergy* 24: 475–486.

Štafa, A., Žunar, B., Pranklin, A. et al. (2019). Novel approach in the construction of bioethanol-producing *Saccharomyces cerevisiae* hybrids. *Food Technol. Biotechnol.* 57: 5–16.

Sun, X.F., Sun, R., Fowler, P., and Baird, M.S. (2005). Extraction and characterization of original lignin and hemicelluloses from wheat straw. *J. Agric. Food Chem.* 53: 860–870.

Tanino, T., Ito, T., Ogino, C. et al. (2012). Sugar consumption and ethanol fermentation by transporter-overexpressed xylose metabolizing *Saccharomyces cerevisiae* harboring a xylose isomerase pathway. *J. Biosci. Bioeng.* 114: 209–211.

Tilloy, V., Ortiz-Julien, A., and Dequin, S. (2014). Reduction of ethanol yield and improvement of glycerol formation by adaptive evolution of the wine yeast *Saccharomyces cerevisiae* under hyperosmotic conditions. *Appl. Environ. Microbiol.* 80: 2623–2632.

Traff, K.L., Otero Cordero, R.R., van Zyl, W.H., and Hahn-Hagerdal, B. (2001). Deletion of the GRE3 aldose reductase gene and its influence on xylose metabolism in recombinant strains of *Saccharomyces cerevisiae* expressing the xylA and XKS1 genes. *Appl. Environ. Microbiol.* 67: 5668–5674.

Tsubaki, S., Oono, K., Onda, A. et al. (2012). Microwave-assisted hydrothermal hydrolysis of cellobiose and effects of additions of halide salts. *Bioresour. Technol.* 123: 703–706.

Van Vleet, J.H., Jeffries, T.W., and Olsson, L. (2008). Deleting the para-nitrophenyl phosphatase (pNPPase), PHO13, in recombinant *Saccharomyces cerevisiae* improves growth and ethanol production on d-xylose. *Metab. Eng.* 10: 360–369.

Verho, R., Londesborough, J., Penttila, M., and Richard, P. (2003). Engineering redox cofactor regeneration for improved pentose fermentation in *Saccharomyces cerevisiae*. *Appl. Environ. Microbiol.* 69: 5892–5897.

Vilela, L.F., de Mello, V.M., Reis, V.C. et al. (2013). Functional expression of *Burkholderia cenocepacia* xylose isomerase in yeast increases ethanol production from a glucose-xylose blend. *Bioresour. Technol.* 128: 792–796.

Vilela, L., de Araujo, V.P.G., Paredes, R. et al. (2015). Enhanced xylose fermentation and ethanol production by engineered *Saccharomyces cerevisiae* strain. *AMB Express* 5: 16.

Voordeckers, K., Kominek, J., Das, A. et al. (2015). Adaptation to high ethanol reveals complex evolutionary pathways. *PLoS Genet.* 11: e1005635.

Walfridsson, M., Bao, X., Anderlund, M. et al. (1996). Ethanolic fermentation of xylose with *Saccharomyces cerevisiae* harboring the *Thermus thermophilus* xylA gene, which expresses an active xylose (glucose) isomerase. *Appl. Environ. Microbiol.* 62: 4648–4651.

Wallace-Salinas, V. and Gorwa-Grauslund, M.F. (2013). Adaptive evolution of an industrial strain of *Saccharomyces cerevisiae* for combined tolerance to inhibitors and temperature. *Biotechnol. Biofuels* 6: 151.

Wang, C., Bao, X., Li, Y. et al. (2015a). Cloning and characterization of heterologous transporters in *Saccharomyces cerevisiae* and identification of important amino acids for xylose utilization. *Metab. Eng.* 30: 79–88.

Wang, C., Bao, X., Li, Y. et al. (2015b). Data set for cloning and characterization of heterologous transporters in *Saccharomyces cerevisiae* and identification of important amino acids for xylose utilization. *Data Brief* 4: 119–126.

Wang, Q., Wang, W., Tan, X. et al. (2019). Low-temperature sodium hydroxide pretreatment for ethanol production from sugarcane bagasse without washing process. *Bioresour. Technol.* 291: 121844.

Weber, C., Farwick, A., Benisch, F. et al. (2010). Trends and challenges in the microbial production of lignocellulosic bioalcohol fuels. *Appl. Microbiol. Biotechnol.* 87: 1303–1315.

Yasuda, M., Miura, A., Yuki, R. et al. (2011). The effect of TiO_2-photocatalytic pretreatment on the biological production of ethanol from lignocelluloses. *J. Photochem. Photobiol. A* 220: 195–199.

Young, E.M., Comer, A.D., Huang, H.S., and Alper, H.S. (2012). A molecular transporter engineering approach to improving xylose catabolism in *Saccharomyces cerevisiae*. *Metab. Eng.* 14: 401–411.

Yu, H., Ren, J., Liu, L. et al. (2016). A new magnesium bisulfite pretreatment (MBSP) development for bio-ethanol production from corn stover. *Bioresour. Technol.* 199: 188–193.

Zabed, H., Sahu, J.N., Boyce, A.N., and Faruq, G. (2016). Fuel ethanol production from lignocellulosic biomass: an overview on feedstocks and technological approaches. *Renew. Sust. Energ. Rev.* 66: 751–774.

Zhang, W. and Geng, A. (2012). Improved ethanol production by a xylose fermenting recombinant yeast strain constructed through a modified genome shuffling method. *Biotechnol. Biofuels* 5: 46.

Zhang, J., Zhang, B., Wang, D. et al. (2014). Improving xylitol production at elevated temperature with engineered *Kluyveromyces marxianus* through overexpressing transporters. *Bioresour. Technol.* 175: 642–645.

Zhao, X., Peng, F., Cheng, K., and Liu, D. (2009). Enhancement of the enzymatic digestibility of sugarcane bagasse by alkali–peracetic acid pretreatment. *Enzym. Microb. Technol.* 44: 17–23.

Zhou, H., Cheng, J.S., Wang, B.L. et al. (2012). Xylose isomerase overexpression along with engineering of the pentose phosphate pathway and evolutionary engineering enable rapid xylose utilization and ethanol production by *Saccharomyces cerevisiae*. *Metab. Eng.* 14: 611–622.

Zhu, J.J., Yong, Q., Xu, Y. et al. (2009). Adaptation fermentation of *Pichia stipitis* and combination detoxification on steam exploded lignocellulosic prehydrolyzate. *Nat. Sci.* 1: 47–54.

8

Recent Developments in Synthetic Biology and their Role in Uplifting Lignocellulose Bioeconomy

Nayanika Sarkar[1,2], *Adhinarayan Vamsidhar*[1], *Pratham Khaitan*[1], *and Samuel Jacob*[1]

[1] *Department of Biotechnology, School of Bioengineering, College of Engineering and Technology, Faculty of Engineering and Technology, SRM Institute of Science and Technology, SRM Nagar, Chengalpattu Dist., Kattankulathur, 603203, Tamil Nadu, India*
[2] *Department of Biotechnology, Bhupat and Jyoti Mehta School of Biosciences, Indian Institute of Technology Madras, Chennai, Tamil Nadu, India*

8.1 Introduction

In order to meet the ever-increasing demand for non-renewable sources of energy like fossil fuels, to reduce the problems related to environmental pollution and the dearth of energy sources, lignocellulosic biorefineries have been developed as an effective "green" alternative for the production of chemicals and fuels from various lignocellulosic biomass (Li et al. 2019; Banner et al. 2021). These lignocellulosic biomass, as the name suggests, are mainly composed of lignin, which are complex phenolic polymers, cellulose, and hemicellulose. Being a major part of plant cell wall, these are abundantly available on earth, with a worldwide production of around 181.5 billion tonnes according to the data provided by Paul and Dutta (2018). Synthetic biology can be defined as the composition and fabrication of pristine biological constituents such as cells, enzymes, and genetic circuits, or refurbishing the pre-existing living systems (Nieves et al. 2015). It is based on the knowledge of systemic biology, and for the assembly of high throughput pathways, multigene constructs, and high precision gene editing it uses top-notch tools with excellent precision (Kalluri et al. 2014). Synthetic biology has surfaced as a prominent discipline which has immense potential to develop novel biotic systems in a cost-efficient manner (Nieves et al. 2015). It has the ability to combine with transgenes in a combinatorial and modular manner and offers improved tolerance of solvent which will eventually result in effective conversion of biomass into fruitful products (French 2009). The adoption of precise engineering of designer enzymes and metabolic pathways in synthetic biology will substantially increase lignocellulosic biomass properties (Kalluri et al. 2014). The principle behind synthetic biology is to maneuver biology by fabricating reusables and standard biology parts (Figure 8.1) (Liu 2012). These parts are known as bio-bricks, which are the modular

Lignocellulose Bioconversion Through White Biotechnology, First Edition. Edited by Anuj Kumar Chandel.

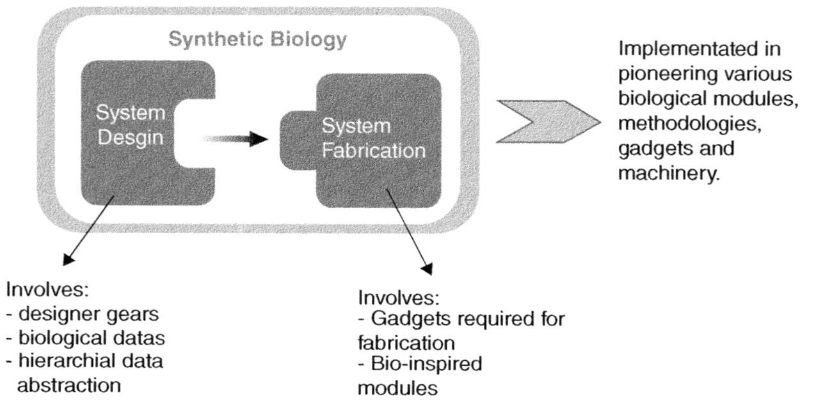

Figure 8.1 Components of synthetic biology (Based on Liu 2012).

and interchangeable bricks of DNA that enable the engineering of organisms rapidly. The bricks consist of transcriptional terminators, coding sequences, ribosome binding sites, promoters; thus they can be identified as naturally occurring nucleic acid sequences which encode a distinct purpose to conform into one or more defined standards (Lakhundi 2012).

Being one of the most abundant in nature, the lignocellulosic biomass, which is often a part of agricultural wastes, serves as an attractive, low-cost feedstock that can effectively replace conventional fossil fuels, as well as the glucose-based sources of carbon. The complex polymeric network of lignin and cellulose can be degraded (delignification) to release monomeric phenols and sugars respectively. This degradation process can be achieved through mechanical or thermo-chemical processing methods which, however, have several drawbacks and safety concerns. Microbes, with their inherent ability to secrete various enzymes capable of degrading these complex polymers, can be exploited for this purpose.

The major limitation of using such biological processes is the low levels of enzyme production, and thus low degradation rates. Many research groups have focused on ways to engineer microorganisms using synthetic biology tools to achieve increased delignification rates, and further produce various value-added products (Rajeswari et al. 2021). While many different genetic tools are available for rewiring the genome of the model organisms, the cost of such tools is often very high, thus limiting their utility. Many of the recent research groups are focusing on elucidating the genomes and roles played by non-model organisms, which often have unique and complex evolved phenotypes. Synthetic biology could be a good way of overcoming such issues for implementation in a circular bioeconomy. In this chapter, the application of synthetic biology routes pertaining to delignification of lignocellulosic biomass, hydrolysis of cellulose, and synthesis of high and low value alcohols from the sugars obtained, thereby establishing an overall biorefinery framework, have been described.

8.1.1 Synthetic Biology Routes for the Delignification of Lignocellulosic Biomass for Biorefinery

Lignocellulosic biomasses are often by-products from various agro-industries. Its composition provides appropriate substrates from biofuel production through fermentation by various microorganisms through conversion of carbohydrates into sugars, and finally to biofuels. However, for the microbes to access the inner cellulose layer, the outer, hard lignin layers need to be depolymerized or separated. This process is called delignification, and the purpose is to improve the sugar digestibility by the microbes (Sánchez et al. 2011). It was also found that most of the biorefineries target the cellulosic part due to its ease of degradation and conversion, leaving behind the recalcitrant part, lignin (Li et al. 2019). Lignin can be an excellent source for various value-added chemicals; however, it is often neglected by industry because of its complex, organic nature which makes its degradation a very difficult task. For an effective valorization and conversion into other chemicals, the lignin needs to be depolymerized into smaller fragments of monomeric or short-chained oligomeric aromatic compounds (Li et al. 2019).

This process of depolymerization of lignin is conventionally carried out by various physico-chemical methods that involve extreme conditions, in turn negatively affecting the enzyme production capacity and biofuel production efficiency of microorganisms. This can be easily overcome by exploiting the ability of certain bacteria and fungi to produce lignin-degrading enzymes like laccase, manganese peroxidase, etc. (Linares et al. 2018). However, one reason why the biorefinery industries are reluctant to use such biological methods is due to the slow rates of lignin degradation owing to limited enzyme production which makes lignin degradation the rate limiting step, and adds to the cost. Apart from optimization studies for enhancing the production of such enzymes, another powerful method would be using the various synthetic biology tools for creating recombinant strains capable of overproducing the desired enzymes. This part of the chapter will focus on the various synthetic biology methods and tools that have been used for the purpose of improving the lignin depolymerization process.

8.1.2 The Key Players of Delignification

The complex aromatic rings syringyl (S), guaiacyl (G) and hydroxyphenyl (H) groups, with a wide distribution of C–O and C–C linkages, bring complexity to the structure of lignin and make the process of degradation particularly difficult. Certain enzymes are capable of breaking down the structure of lignin to release the smaller aromatic and monomeric groups. The following points about the key players, that is, the enzymes of the lignin degradation, are important (Li et al. 2019):

- Fungal peroxidase belongs to the family of peroxidase-catalase; these enzymes utilize hydrogen peroxide as co-substrates. The most significant and well-known fungal peroxidases include lignin, manganese and versatile peroxidases, and these have a high degree of sequence homogeneity.
- Dye-decolorizing peroxidases belong to a heme group containing peroxidases with unique functional and structural features. These peroxidases can be classified into decolorizing peroxide (DyP) A, B and C which, unlike the fungal peroxidases, are found in much higher levels in bacteria.
- Laccases are enzymes that belong to a class of metalloproteins. These utilize dioxygen as e-acceptor and have a major advantage of releasing water as the by-product. White rot fungi are a well-known producer of laccase. Certain bacteria can also produce laccase, with a good amount of structural homogeneity with the fungal laccases.

- Of the β-etherases, one of the most abundant constituents in lignin is the β-aryl ether, comprising up to 62% of all the inter-monomer bonds in lignin. It can be present is four possible stereoisomers and two diastereomers. The cleavage of this bond is catalyzed by the β-etherases.
- Ring-cleaving dioxygenases are crucial enzymes that catalyze the cleaving of the aromatic rings into a linear form, which may in turn be converted into intermediates of the Krebs cycle (or tricarboxylic acid cycle). Such strategies are often used by bacteria to get the carbon sources from lignin.

8.1.3 Case Studies

8.1.3.1 Fungi as Expression Host

Fungal systems can be efficient for this purpose because of their inherent ability of delignification (Li et al., 2018). Using the various available synthetic biology tools for the process of delignification is an active topic of research. One such study was carried out by the research group of Linares et al. (2018), where they aimed to achieve an enhanced delignification process through overexpression of the lignocellulosic enzymes in *Phanerochaete chrysosporium,* a white rot fungi. For this, they used four different strains overexpressing the *lipH8* (lignin peroxidase), *mnp1* (manganese peroxidase), *vpl2* (versatile peroxidase), and *lacIIIb* (laccase) genes. Plasmids (expression vectors) of each of these genes were prepared (Linares et al. 2015), and the fungal strain was co-transformed by using shock waves under water. The simultaneous over-expression of all the four enzymes in one fungal strain was thought to improve the efficiency of lignin degradation.

Results from this study showed that the recombinant strains could improve the rate of the delignification activity and around 20% reduction in degradation time, compared with the non-recombinant control strains. The recombinant strains were capable of drastically enhancing the levels of the soluble sugars. The reason is that the improved lignin breakdown exposed the hemicelluloses and cellulose, which could be readily hydrolyzed, resulting in increased sugar levels (Linares et al. 2015). This is particularly important for analyzing the efficiency of lignocellulosic biomass to biofuel conversion. However, more analysis is needed to get insights on the enzymes that are responsible in this hydrolysis process. Another interesting finding was that the versatile peroxidase activity synergistically enhances lignin peroxidase and manganese peroxidase. Results from this study could achieve a 25% hike in delignification, around 4- and 2.6-fold increase in laccase and peroxidase activity respectively. In another study by Li et al. (2019) aiming to achieve improved laccase production, a fungal host system and a rational design approach were used. The *pslcc* gene from parent strain *Pycnoporus sanguineus* was expressed in another fungi *Aspergillus nidulans* and controlled by RsmA transcription factor. This was coupled to the knocking out of two proteases, *DppV* and *PepA I*. The results show a successful 15-fold enhanced expression and 13 times improvement in the activity of the recombinant strain (Li et al. 2019). In another study by Pardo et al. (2018) and his group, they targeted the OB1 from *Pycnoporus cinnabarinus* and interchanged the D2 cupredoxin domain. The resulting enzyme was an acid-tolerant enzyme retaining 90% of the total activity in pH as low as 2.

8.1.3.2 Yeast as Expression Host

For yeasts, although they have many benefits as an expression host compared to other fungi and bacteria, the expression of the lignin degrading enzymes may be particularly challenging, one of the challenges being a lack of understanding of the post translational modification in these fungal systems and how these effect the enzymatic efficiency. Studies by Song et al. (2020) show that expressing the *Lac2* gene in *Pichia pastoris* (X33) yeast from parent strain *Pleurotus ostreatus* (white rot fungi) resulted in laccase with higher catalytic efficiency and greater ability to degrade 2,6-dimethoxyphenol and 2,2′-azino-bis(3-ethylbenzo thiazoline-6-sulfonic acid). When corn stover was utilized as the lignin source, a 4.33% increase in degradation rate was observed in the recombinant laccase compared to the wild strain. The reason, as inferred by the authors, was the lower production rate of the enzyme in the native strain.

Another study by Xu et al. (2019) expressed *Lcc9* laccase protein from *Pleurotus ostreatus* in *Pichia pastoris*. Results indicated that the catalytic ability of the recombinant enzyme had much more improved specific activity (around 3.03-fold higher) in the heterologously expressed host organism. The catalytic efficiency also increased to around 1.38-fold. The engineered enzyme was also superior in terms of thermal and alkaline pH tolerance. The authors inferred that one possible reason for the enhancement could be the hyperglycosylation (a type of post transcriptional modification where extra oligosaccharide residues are added to the translated protein). The occurrence of hyperglycosylation was supported by the fact that the molecular weight of the recombinant protein was higher than the native one. Similar results were shown by Otterbein et al. (2000) where they expressed the *lac1* gene from parent organism *Pycnoporus cinnabarinus* (white rot fungus) I-937 into *P. pastoris*. To enhance the secretion of laccase, α-factor signal peptide was also expressed, along with a laccase leader peptide. The results showed that the molecular weight of the recombinant protein was about 1.35 times that of the native one (81 kDa), suggesting that hyperglycosylation has taken place. Unlike the results from Xu et al. (2019), in this study the activity of laccase did not improve and remained almost the same as the wild-type strain. A similar study was done by Nitheranont et al. (2018), where the *lac1* gene from *Grifola frondosa* was expressed in *P. pastoris* culture. An enhancing agent, Cu^{2+}, was used. However, even though the recombinant laccase was found to be hyperglycosylated (supported by the above two studies), the catalytic efficiency was much less and the K_m values were higher compared to the native laccase, in contrast to the results by Xu et al. (2019) and Otterbein et al. (2000). The recombinant enzyme had a higher thermal and pH tolerance, as observed in studies by Xu et al. (2019). The authors inferred that the reason could be the improvement in rigidity due to hyperglycosylation. A novel approach was carried out by Mateljak et al. (2019); they used the SCHEMA-RASPP structure- guided recombination approach to generate superior laccase chimeras with higher thermal and pH stability, while maintaining around 70% of sequence homogeneity. The *lac3* laccase protein from parent strain *Trametes* species (Basidiomycete) was isolated and expressed in *Saccharomyces cerevisiae* host. A leader peptide and α-factor signal peptide was also expressed. The results showed that the recombinant proteins were much more robust with high thermal stability as well as enhanced activity. Table 8.1 represents some of the synthetic biology routes adopted for delignification of lignocellulosic biomass.

Table 8.1 Synthetic biology approaches for enhancing delignification and the results achieved in each of the studies.

Host	Parent strain (origin)	Gene	Strategy used	Result	References
Pseudomonas putida	*Amycolatopsis* species	DyP2 (dye-decolorizing peroxidase)	Substituting the wild-type promoter with a strong constitutive promoter to overexpress genes	A 2.1-fold improvement in the growth rates were observed compared to the wild-type strains	Lin et al. (2016)
Pichia pastoris	*Pleurotus ostreatus*	*POXA 1B* (laccase)	Native enzyme subjected to photocatalysis, compared with recombinant strain	Higher chemical oxygen demand (COD) removal and decolorization by treated native strain compared to the recombinant strain	Rivera-Hoyos et al. (2018)
Escherichia coli BL21	*Phanerochaete chrysoporium* (white rot fungus)	*LiPH8* (lignin peroxidase)	Used *in silico* design strategies to achieve enhanced activity	Increase in tolerance towards acidic pH, around 9-fold improved catalytic efficiency	Pham et al. (2018)
Pichia pastoris	*Moniliophthora rareri*	*MrMnP1* (manganese peroxidase)	Used *in silico* design approach	Enhanced ability to degrade a range of substrates by the new strains, in contrast to the wild-type which were incapable of degrading	Bronikowski et al. (2018)
Escherichia coli Rosetta	*Bacillus pumilus*	*cotA* (laccase)	DNA shuffling method and structure guided engineering	Enhanced catalytic efficiency and thermal tolerance was achieved	Ihssen et al. (2017)
Trichoderma reesei	*Pycnoporus sanguineus*	*laC* (laccase)	Random genome integration approach	The thermal tolerance was improved. Fed-batch strategy was used and highest enzyme production of around 17.7 U mL^{-1} could be achieved	Zhao et al. (2018)

8.1.3.3 Bacteria as Expression Host

A study by Xie et al. (2019) used a bacterial host *Rhodococcus opacus* for the expression of these fungal laccase genes. The parent strain for the laccase was *Streptomyces coelicolor*. This could be a challenging task because bacteria lack the inherent ability to secrete enzymes, unlike the fungus. This research group had overcome this problem by using a proteomics guided-design to completely redesign the flow of central dogma, protein folding and secretion. Signal peptide S258 was added, along with a Tat secretary system whose expression was enhanced by overexpressing *TatA* and *TatC*. Along with it, an appropriate promoter, ribosome binding site, could result in a drastic improvement in extracellular laccase levels in comparison to the wild strain which was completely incapable of producing the enzyme. Around 81% of β-5 phenolic groups was achieved, along with 26% guaiacyl OH groups and 41% of OH groups in aliphatic chains. A successful heterologous expression and secretion of laccase enzyme could be achieved.

The group of Brissos et al. (2017) used the approach of random mutagenesis aimed at enhancing the activities of the dye-decolorizing peroxidases using *Pseudomonas putida* MET94. A high throughput screening resulted in selection of a variant with an enhanced oxidizing capacity and about a hundred-fold improvement in the catalytic efficiency for degrading 2,6-dimethoxyphenol (DMP) compared to the wild-type bacteria. The efficiency was almost at par with the fungal peroxidases. The new strain was also more robust with increased alkaline pH tolerance. Another advantage of the new strain was its increased resistance to substrate inhibition, which allows treatment of higher phenolic concentrations. A high hydrogen peroxide concentration tolerance of the new strain is particularly advantageous from the point of view of industrial use, where its accumulation is a major limitation. The increased ability to degrade the β-aryl ether groups was also observed with about 70% breakdown, in contrast to only 20% breakdown in the native wild strains. Optimal pH was also increased.

8.2 Synthetic Biology Routes for Cellulose Degradation in Lignocellulosic Biomass

8.2.1 Cellulose—a Major Plant Component

Biomass mainly comprises long cellulose fibers which are encapsulated in an indeterminate matrix of lignin and hemicellulose. The cellulose fibers have β-(1,4)-linked D-glucose residues in the form of long chains in an opposite orientation (Figure 8.2). These chains have crystalline regions bonded with insoluble hydrogen which gives the cellulose

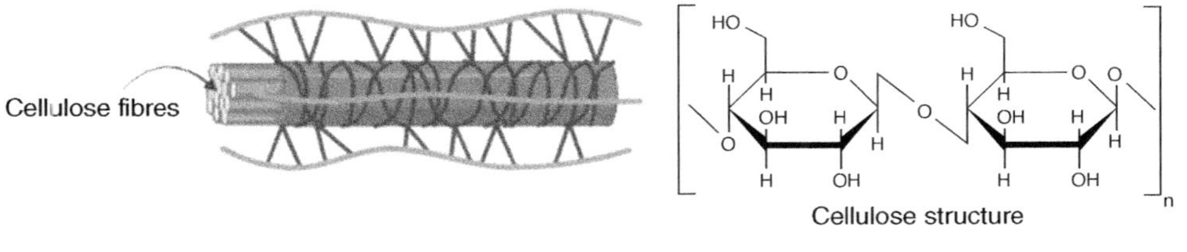

Figure 8.2 Cellulose fibers and molecular structure.

tremendous tensile strength. The hydrolyzation of cellulose would produce D-glucose, which can be used widely in the production of feedstock, chemicals, and biofuels as it is a competent carbon source. However, the proximity of lignin and the insoluble crystalline structure of cellulose renders the enzymatic or non-enzymatic conversion of the cellulose to its sugar component. The potential degradation of crystalline cellulose is comparatively difficult compared to amorphous and soluble cellulose, but synthetic biology with its novel approaches has made it possible with microorganisms that degrade cellulose. Through synthetic biology, the assembly of interchangeable and modular DNA components is possible in any desirable number and order for the generation of multigene complex systems, by generating sticky ends which are both incompatible and compatible due to the conjunction of various restriction sites. High recalcitrance and abundance of cellulose acts as a prolific carbon source for which some of the microorganisms compete. The efficacious degrading of crystal cellulose involves the Gram-negative aerobic bacteria *Cytophaga* and the anaerobic *Fibrobacter*, followed by the Gram-positive aerobic bacteria with high G-C content *Cellulomonas* and the anaerobic low G-C content spore forming *Clostridium* (French 2009). Cellulose is the most vital constituent of the plant cell wall and is superabundant. The cell wall comprises polysaccharides and the cellulose fibers are immersed in a matrix which contain these polysaccharides and glycoproteins. Cellulose is a part of both primary and secondary walls where it acts as a platform for other cell wall constituents.

The cellulose microfibrils have 36 parallel β-1,4-glucan chains and have a diameter of 3 nm. Cellulose biosynthesis is performed by enzyme complexes which are membrane bound by cellulose synthase and are the most profuse biomachines available. The complexes when examined under an electron microscope using a freeze fracture technique appear as hexameric rosettes. The rosettes in plasma membranes have CESA proteins which are coded by cellulose synthase catalytic subunit (CESA) genes. The cellulose present in the primary and secondary cell walls is synthesized by different sets of CESA proteins. There are six rosette subunits present in each hexameric rosette and there are six CESA proteins in each rosette subunit, which accounts for 36 CESA proteins per rosette. CESA proteins have intrinsic membrane proteins with eight transmembrane domains and a sizable hydrophilic domain facing cytosol. These domains have active sites and conserved residues. The proteins have zinc finger motifs on the amino terminal. The zinc finger sequences help in the assembly of CESA monomers into the rosette complexes by acting as multimerization domains that are redox regulated (Lerouxel et al. 2006).

8.2.2 Synthetic Biology for Hydrolysis of Cellulose

Significant advancements have been formulated for the efficient degradation of lignocellulosic biomass using the fungal and bacterial enzymatic complexes. In the past few years, a batch of redox enzymes for cellulose degradation has been considered and suggested as an essential advancement in the degradation of cellulose. Several research workers have shown that the enzymes known as lytic polysaccharide mono-oxygenases (LPMOs) are proficient enough to cleave the polysaccharide chains of the crystalline cellulose using an oxidative mechanism. The efficiency of cellulose conversion increases greatly due to the improved activity of the hydrolytic cellulases after the inclusion of these enzymes (Chakraborty et al. 2020). LPMOs have a reaction mechanism that involves the oxidation of the carbon-C1 of a sugar molecule which results in a crack in the cellulose chain and formation of aldonic acid. LPMOs being exclusively available in aerobic organisms do not profit from the aids provided by the cellulosomal system. The polysaccharide mono-oxygenases are devised in such a way that they acquire the potential to assemble in the cellusome complex. When the enzymes disintegrate in these complexes, they result in an enhanced rate of cellulose degradation in contrast to enzymes. LPMOs have been regarded as "cellulase boosters" and in three carbohydrate active enzyme (CAZy) families it has been categorized as "auxiliary activities (AA)," AA11, bacterial-AA10 and fungal-AA9, previously known as CBM33 and GH61. Several studies present that, by using big multi-enzyme complexes known as cellulosomes, the degradation of biomass has a higher efficiency (Arfi et al. 2014).

Designer cellulosomes were initiated to subsume non-cellulosomal enzymes into the complexes and further utilize the cellusome architecture. The conceptualization of the designer cellulosomes are based on the interlocking modular characteristics of the cellulosomal complex and its constituents (Arfi et al. 2014). These cellulosomes have laid out a bottom-up strategy for (i) assessing the synergistic interaction among the various CAZymes, (ii) affluently depicting the carbohydrate-binding domain mediated "targeting" effects of cellulosomes and the positive additions of enzyme "proximity," (iii) analyzing the function of several domain modules. Various free enzymes have been devised for the construction of designer cellulosomes (Elkins et al. 2010; Rajeswari et al. 2021). A chimeric scaffolding having several chimeric enzymes with dockerin integrants and cohesin integrants can be designed by using the high specificity amidst dockerin and cohesin from the same species. The incorporation of LPMOs into the designer cellulosomes have shown enhanced enzymatic integration of the crystalline cellulose. The systematic degradation of lignocellulosic biomass requires detachment of lignin from the polysaccharide backbone or disintegration of its components. Thus, an endeavor is required for the incorporation of a wide range of enzymatic activities into the cellulosomes (Arfi et al. 2014).

The fabrication of the convoluted metabolic pathways is of major pertinence for synthetic biology. The accurate blending of different enzymes holds the potential for enhanced activity of the enzymes for better degradation of biomass. Predicting the perfect blending of enzymes is quite difficult without the knowledge of detailed characteristics of the enzymes. An alternative to this could be performing parallel screening on a huge scale. A wide-reaching library of genes encoded with cellulase is developed and needs to be screened in parallel for the release of sugars from various cellulosic substrates. Bio-bricks are most suitable for such experiments. A library must be prepared which includes biomass degrading enzymes from distinct organisms and then organisms expressing assortments of synthesis are assembled in parallel and finally screened for their aptness to efficiently degrade biomass. This method is beneficial for empirically determining the most efficacious mixture of enzymes for enhanced biomass degradation (French 2009).

Cellulose synthases are classified as CESAs; these are the catalytic enzymes which are responsible for building β-glucan chains from the substrate, uridine diphosphate glucose, UDP-glucose. The microfibrils extended from the CSC or CESA functional complexes appear as rosettes when viewed under an electron microscope. Some of the factors for the progressive regulation and degradation of cellulose in plants include (i) pH, Ca^{2+} levels and H^{+} ions, (ii) associated wall proteins such as COBRA, KOBITO and KORRIGAN, (iii) the arrangement, composition, and number of rosettes, and (iv) membrane dynamics such as alteration in lipid response to environmental stress and the activity of cell wall enzymes (Kalluri et al. 2014). The cellulose extraction from plants depends on two divisions of nanomaterials. CNCs—cellulose nanocrystals—are derived from plants by acid hydrolysis. The hydrolysis removes the amorphous regions while the higher resistance of the crystalline regions keep them intact and they form nanocrystals shaped like rods. CNFs—cellulose nanofibrils—are obtained when the plants undergo mechanical shear force and the disintegrated fibrils return back to structural fibrils (Yang et al. 2019).

In synthetic biology, cellulose with its structural integrity is both an impediment in the production of biofuels and an exceptional insight for tailoring pristine biomaterials. The structural breakdown of cellulose is arduous, and so spatial organization plays a vital role in its breakdown. Cellulose can be disintegrated by only a few enzymatic systems and cellulose is one such mega Dalton level complex which is extracellular and found on the surface of *Clostridium thermocellum*, an anaerobic bacterium. A non-catalytic scaffold protein with carbohydrate binding and modular scaffolding domains is present central to the complex. The dockerin domains present in the cellulosomes bind to the primary scaffolding domains and these scaffolds have two crucial tasks to perform: (i) attaching the bacterial cell wall directly to the plant cell wall and (ii) co-localizing the enzymes, hemicellulases, and cellulases. In this fashion, the host cell and the plant cell sandwich the cellulosome and provide the host with favorable ingress to the degradation products (Agapakis et al. 2012). The protein complexes used as molecular scaffolds were to construct designer cellulosomes. It was done using SP1, a stress-responsive protein taken from *Populus tremula*. Cel6B exoglucanase and Cel5A endoglucanase from recombinant *Thermobifida fusca* were accumulated on the protein scaffold SP1 with a dockerin module. The endoglucanase complex had 2-fold higher activity than the free enzymes on the carboxymethyl cellulose because of their dense packaging in the scaffold (Elkins et al. 2010). The noteworthy modularity of the cellulosomes marshaled the structuring of synthetic cellulosomes. The construction of mint cellulosomes tagged with xylanase and cellulase enzymes can be targeted to specific sites using divergent dockerin-scaffolding pairs on the synthetic cellulosomes for enhanced cellulose degradation (Agapakis et al. 2012).

The organisms which degrade cellulose possess cellulosomes and cellulases. An association of at least three enzymes, β-glucosidase, exoglucanase, and endoglucanase, are required for the hydrolysis of cellulose. Cellulose disintegrates when endoglucanase hydrolyzes the β-1,4-glycosidic bonds present between the joined glucose residues. The attributes influencing cellulose biodegradation include the essence of cellulose-binding lignin, degree of crystallinity, size of cellulase compared to the cellulose-capillary structure and face of the crystalline cellulose microfibril. Cellulose binding domains (CBDs) facilitate the hitching of the insoluble cellulose with the cellulases. These cellulose binding domains prevent the cellulases from being wiped off the cellulose surface, and also corroborate confined cellulose hydrolysis. Each discrete cellulase has this domain present in it which acts on complex cellulosomes and cellulose substrates. The discovered novel cellulases can disintegrate the cellulose directly without extricating chains from the cellulose crystals. These cellulases are fungal glycoside hydrolase 61 (GH61) and bacterial carbohydrate-binding modules (CBM33). GH61 and CBM33 possess substrate binding domains and the disintegration of cellulose takes place through an oxidative mechanism. The microorganisms accountable for biomass conversion have GH61 and CBM33 genes profusely available in them (Munawar 2014). The complex hydrolytic enzymes in anaerobic microorganisms have a cellulose system which is utilized for enhanced degradation of the biomass components. Several catalytic constituents are accumulated concurrently on the scaffolding subunits by strong non-covalent type I protein interactions. These interactions occur between dockerin modules and cohesin modules which are present on every single cellulosome enzyme. These cellulosome enzymes collectively bind to the lignocellulosic substrates using the cellulose-specific carbohydrate-binding domains (CBMs) of the scaffolds whereas the free enzymes have individual CBMs available. For efficient biomass hydrolysis, cellulosome-based complexes target the enzymes towards the insoluble substrates through CBM domains available on the scaffold (Elkins et al. 2010). Oxidative enzymes have offered a new paradigm for future cellulose degradation. The conjugation of nanoparticles and cellulases has manifested enhanced cellulolytic aptness for crystalline cellulose degradation (Munawar 2014).

8.2.3 Degradation using Nanoparticles

Silver nanoparticles (AgNPs) have recently been used for enhanced cellulose degradation. These nanoparticles are environmentally friendly, cost efficient and can easily be synthesized. They are less toxic and have high electron transfer efficiency. AgNPs can speed up the reaction rate and act as nano-catalysts in the hydrolysis of cellulose. The AgNPs synthesized by plant and cellulose after interacting with cellulase can degrade the cellulose complex. In contrast to free cellulose, these AgNPs can attach the enzymes on their surfaces and render themselves immobilized, which results in faster degradation of cellulose. The rigid reinforcement of the solid particles along with immobilized enzymes limits the steric orientation and collision frequency which results in increased rate of reaction and the cellulose degradation occurs at a much higher speed. The increased rate of cellulose hydrolysis in the presence of AgNPs could be due to various reasons, one of which can be illustrated as follows. The interplay between plant-cellulose coalesced AgNPs and cellulase results in cellulose degradation into reducing sugars by enzymes and immobilization of the cellulase present on the surface of silver particles. Cellulose when degraded in the presence of AgNPs produces higher quantities of reducing sugars. The precise workings backing the enhanced cellulose hydrolyzation using the AgNPs is not yet clear; the particles behaving as a catalyst using some mysterious mechanisms need to be further investigated. Fascinatingly, the smaller-sized nanoparticles manifested superior catalytic activity in contrast to the larger nanoparticles. Thus, it can be said that AgNPs are suitable for nanocatalysis and help in rapid degradation of cellulose into smaller components (Salunke et al. 2015).

8.3 Synthetic Biology Routes for the Production of Low-value and High-value Alcohols

With the rise in the need for an alternative source of fuel in the world due to the rapid decline in crude oil sources, biofuels have been presented to be an excellent substitute and additive along with the fuels that are currently in use (Rao et al. 2016). The production of fuel from lignocellulosic biomass provides a viable replacement. Nevertheless, the recalcitrance of lignocellulosic materials and the inability of native microorganisms to efficiently ferment lignocellulosic hydrolysates pose numerous challenges for the production of bio-alcohols (Colin et al. 2011).

Breeding techniques and genetic engineering are notable methods to enhance the productivity of microorganisms exhibiting high alcohol yield (Atsumi and Connor 2010). In this section, the various synthetic biological routes that are being utilized to increase the rate of production of low-value and high-value alcohols from lignocellulosic biomass are briefly discussed.

8.3.1 Low-value Alcohols

8.3.1.1 Ethanol

Bioethanol production from lignocellulosic biomass or non-edible feedstock (second generation) provides us with an opportunity to tackle the growing environmental concerns related to the consumption of fossil fuels (Weber et al. 2010). Hence, to make the process of production more economically viable, microorganisms have been synthetically designed to be utilized more efficiently. Altering the yeast strains through mutation has been done for many years to increase ethanol production (Chandel et al. 2011). Although this method has proved to be a viable method to increase the production of ethanol, there have been difficulties in extracting the stable ethanol titers after the fermentation from the mutagenic strains. Various studies have been conducted in the past, to depict strain improvement on mutation (Maleszka et al. 1983). *Candida* sp. produced five times more ethanol when subjected to mutation through UV mutagenesis, compared to the native strain. The anaerobic thermophiles are sensible targets to be genetically improved, as they have the capacity to produce ethanol from a wide range of substrates. One of the major drawbacks of using thermophiles for the production of ethanol is their low substrate and ethanol tolerance. Adaptation methods such as non-specific mutagenesis and artificial selection are used to improve specific traits of microorganisms for the production of bioethanol on a large scale. Examples of the adaptation methods are as follows (Scully and Orlygsson 2019):

- Mutant strains of *Thermoanaerobacter ethanolicus* are obtained by selection after they are deprived of iron and pyruvate, which leads to increased ethanol tolerance (10% v/v) at substrate concentrations above 10 g L^{-1} (He et al. 2009).
- *C. thermocellum* showed an increase in ethanol tolerance up to 5% v/v by the stepwise increase and transfer of cultures, which increases the concentration of ethanol produced (Shao et al. 2011).
- *Thermoanaerobacter pentosaceus* has gradually adapted for fuel ethanol production from sugarcane to higher substrate concentrations and demonstrated higher ethanol tolerance and substrate utilization (Sittijunda et al. 2013).

8.3.1.2 Methanol

Methanol is attributed as a next-generation source of carbon as it is available in large quantities and at affordable prices (Olah 2005). Additionally, the risk of contamination is less on the usage of methanol and residual methanol can be removed (Zhang et al. 2018) with ease at higher temperatures. The gasification of lignocellulosic materials provides a method to convert biomass into a mixture of gases, which are H_2, CO, CO_2, and CH_4, known as synthetic gas or syngas (Frazão and Walther 2020). Chemical process research has mainly focused on transportation fuel production from syngas, such as Fischer–Tropsch liquid fuels, hydrogen, methanol, dimethyl ether, etc. The Wood–Ljungdahl metabolic pathway is the most commonly employed process by acetogens for the production of methanol. Methanotrophs have been employed to produce biofuels through oxidation reactions with the help of enzymes such as methane monooxygenase, methane dehydrogenase, formaldehyde dehydrogenase. This was successfully achieved by using the genetically modified methanotrophic bacterium *Methylosinus trichosporium* IMV 3011 to produce methanol through a series of reactions catalyzed by enzymes. The optimum production was about 20% along with other gases without loss of productivity (Bjorck et al. 2018). To increase the production of methane monooxygenase enzymes can be expressed in hosts such as *E. coli*, which have a higher growth rate. Nevertheless, this approach does not have any report to support its possibility of increased methanol production (Gautam et al. 2020).

8.3.2 High-value Alcohols

8.3.2.1 Xylitol

Xylitol is a sugar alcohol with various areas of applications, mainly used in the pharmaceutical and food industries. Generally, xylitol is produced by the process of purification of xylose, obtained from the acid-based hydrolysis of lignocellulosic materials in the presence of a metal catalyst. The use of synthetic biology routes utilizes hemicellulosic hydrolysates which are obtained from biomass, which are converted to pentose sugars by the action of bacteria and yeast. This process is an environmentally friendly process that can be performed under mild conditions such as atmospheric pressure and ambient temperature, thus avoiding the need to purify xylose, which makes the process more economically viable (Prakasham et al. 2009). For the production of xylitol, a microorganism must be one that can be easily cultivated, shows a high capacity of xylitol production, and must have excellent tolerance to stress and toxic compounds (Guo et al. 2006). In similar areas, various studies in the field of molecular engineering are being conducted to discover and formulate ideal conditions for the economic production of xylitol. *S. cerevisiae* was first identified as an ideal candidate to be genetically modified to optimize the production of xylitol. Optimization of metabolic pathways and simultaneous utilization of sugars are some of the genetic modifications that are being studied (Su et al. 2013). Among the naturally occurring D-xylose fermenting yeasts, *Candida* sp. is attributed as one of the best producers of xylitol. Unfortunately, due to its pathogenic nature, it is not commercially viable. The improvement of xylitol production by the overexpression of homologous xylose reductase (XR) in *Kluyveromyces marxianus* IIPE453 (KSIIPE453) was considered bearing in mind that the deletion of the xylitol dehydrogenase (XDH) gene would result in the accumulation of xylitol (Figure 8.3). Nevertheless, the use of genetically modified strains poses some challenges to be overcome such as an increase in their metabolic stability and tolerance to the presence of inhibitors (Pal et al. 2016).

8.3.2.2 Butanol

Butanol is a renewable fuel that is produced by a process known as the alcoholic fermentation of biomass (Birgen et al. 2019; Hansen et al. 2005). Butanol has a better capability to be blended with gasoline, due to the presence of longer hydrocarbon chains, resulting in a

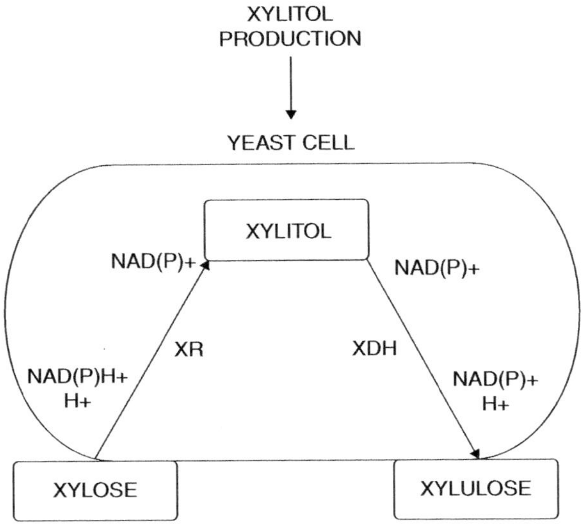

Figure 8.3 Metabolic pathway of xylose production in yeasts. XK, xylulose kinase; XR, xylose reductase; XDH, xylitol dehydrogenase.

lower polarity. Butanol's energy density is almost identical to that of gasoline (Jin et al. 2011). The acetone–butanol–ethanol (ABE) fermentation process is used to produce biobutanol from various substrates, including lignocellulosic materials. The transcriptional repression of so1R and the overexpression of aldehyde dehydrogenase (aad) are recognized to produce the best results in terms of higher production rates (Tomas et al. 2003). There was an increase of 8.2 g L^{-1} acetone, 2.2 g L^{-1} ethanol, and 17.6 g L^{-1} butanol, which in turn were 66%, 194%, and 51% higher than what was obtained using the wild-type strain. Increasing the production of selective butanol production was unsuccessful by the overexpression of enzymes which are directly related to its rate of production. Hence an antisense RNA strategy was employed for the knockdown of CoA transferase (ctfB), which is an acetone producing enzyme, in addition to the overexpression of the butanol producing enzyme alcohol/aldehyde dehydrogenase (aad). This resulted in a 2-fold increase of the butanol/acetone increase with a minimal decrease in butanol titer of approximately 9.8 g L^{-1} compared to the wild-type which resulted in a decrease of approximately 10.2 g L^{-1} (Weber et al. 2010). *Clostridium cellulovorans* can be genetically engineered by introducing the heterologous bifunctional aldehyde/alcohol dehydrogenase gene (adhE2) to produce *n*-butanol and ethanol directly from cellulose (Figure 8.4).

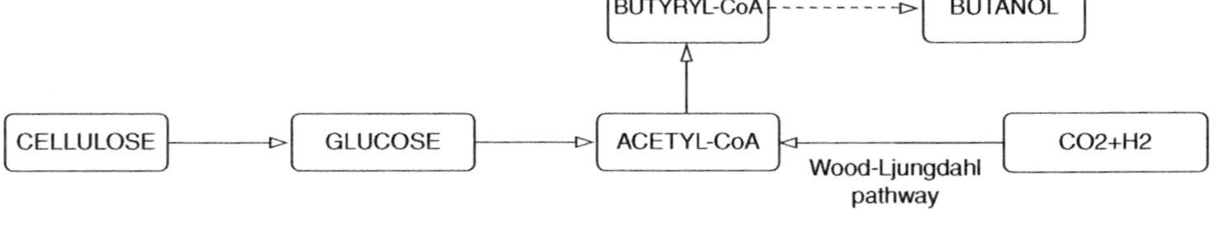

Figure 8.4 Butanol production from lignocellulosic mass.

8.4 Conclusion

We anticipate that in the years to come many more unknown, novel genomes in model as well as non-model organisms will be studied, thus creating a highly efficient and cost effective platform for achieving effective depolymerization and valorization of lignocellulosic biomass. These engineered organisms could provide an efficient way to ensure that the produced alcohols are both economically viable and of high purity. However, most of these methods are still in their development phases and are in use in pilot-scale production units in various countries around the world. There is still scope for improvement in this field and it will require extensive research and analysis to ensure these bio-alcohols and green chemicals will be an effective substitute to fossil fuels and chemicals produced from lignocellulosic biomass.

References

Agapakis, C.M., Boyle, P.M., and Silver, P.A. (2012). Natural strategies for the spatial optimization of metabolism in synthetic biology. *Nat. Chem. Biol.* 8 (6): 527–535. https://doi.org/10.1038/nchembio.975.

Arfi, Y., Shamshoum, M., Rogachev, I. et al. (2014). Integration of bacterial lytic polysaccharide monooxygenases into designer cellulosomes promotes enhanced cellulose degradation. *Proc. Natl Acad. Sci.* 111 (25): 9109–9114.

Atsumi, S. and Connor, M.R. (2010). Synthetic biology guides biofuel production. *J. Biomed. Biotechnol.* 2010: 541698. https://doi.org/10.1155/2010/541698.

Banner, A., Toogood, H.S., and Scrutton, N.S. (2021). Consolidated bioprocessing: synthetic biology routes to fuels and fine chemicals. *Microorgan.* 9 (5): 1079.

Birgen, C., Dürre, P., Preisig, H.A., and Wentzel, A. (2019). Butanol production from lignocellulosic biomass: revisiting fermentation performance indicators with exploratory data analysis. *Biotechnol. Biofuels* 12 (1): https://doi.org/10.1186/s13068-019-1508-6.

Bjorck, C.E., Dobson, P.D., and Pandhal, J. (2018). Biotechnological conversion of methane to methanol: evaluation of progress and potential. *AIMS Bioeng.* 5 (1): 1–38. 2375-1495.

Brissos, V., Tavares, D., Sousa, A.C. et al. (2017). Engineering a bacterial DyP-type peroxidase for enhanced oxidation of lignin-related phenolics at alkaline pH. *ACS Catal.* 7 (5): 3454–3465.

Bronikowski, A., Kodchorreck, K., and Urlacher, V.B. (2018). Redesign of a new manganese peroxidase highly expressed in *Pichia pastoris* towards a lignin-degrading versatile peroxidase. *Chem. Bio. Chem.* 19: 2481–2489.

Chakraborty, D., Sarkar, N., Biswas, I., and Jacob, S. (2020). Molecular aspects of prokaryotic and eukaryotic cellulases and their modulation for potential application in biofuel production. In: *Genetic and Metabolic Engineering for Improved Biofuel Production from Lignocellulosic Biomass* (ed. A. Kuila and V. Sharma), 81–95. Oxford: Elsevier https://doi.org/10.1016/B978-0-12-817953-6.00006-3.

Chandel, A.K., Chandrasekhar, G., Radhika, K. et al. (2011). Bioconversion of pentose sugars into ethanol: a review and future directions. *Biotechnol. Mol. Biol. Rev.* 6 (1): 8–20.

Colin, V.L., Rodríguez, A., and Cristbal, H.A. (2011). The role of synthetic biology in the design of microbial cell factories for biofuel production. *J. Biomed. Biotechnol.* 2011: 601834. https://doi.org/10.1155/2011/601834.

Elkins, J., Raman, B., and Keller, M. (2010). Engineered microbial systems for enhanced conversion of lignocellulosic biomass. *Curr. Opin. Biotechnol.* 21 (5): 657–662.

Frazão, C.J.R. and Walther, T. (2020). Syngas and methanol-based biorefinery concepts. *Chemie-Ingenieur-Technik* 92 (11): 1680–1699.

French, C. (2009). Synthetic biology and biomass conversion: a match made in heaven? *J. Royal Soc. Interf.* 6 (suppl_4): S547–S558. https://doi.org/10.1098/rsif.2008.0527.focus.

Gautam, P., Neha, Upadhyay, S.N. et al. (2020). Bio-methanol as a renewable fuel from waste biomass: current trends and future perspective. *Fuel* 273: 117783. https://doi.org/10.1016/j.fuel.2020.117783.

Guo, C., Zhao, C., He, P. et al. (2006). Screening and characterization of yeasts for xylitol production. *J. Appl. Microbiol.* 101 (5): 1096–1104.

Hansen, A.C., Zhang, Q., and Lyne, P.W.L. (2005). Ethanol-diesel fuel blends—a review. *Bioresour. Technol.* 96 (3): 277–285.

He, Q., Lokken, P.M., Chen, S., and Zhou, J. (2009). Characterization of the impact of acetate and lactate on ethanolic fermentation by *Thermoanaerobacter ethanolicus*. *Bioresour. Technol.* 100 (23): 5955–5965.

Ihssen, J., Jankowska, D., Ramsauer, T. et al. (2017). Engineered *Bacillus pumilus* laccase-like multi-copper oxidase for enhanced oxidation of the lignin model compound guaiacol. *Prot. Eng, Design. Select.* 30 (6): 449–453.

Jin, C., Yao, M., Liu, H. et al. (2011). Progress in the production and application of *n*-butanol as a biofuel. *Renew. Sust. Energy Rev.* 15 (8): 4080–4106.

Kalluri, U., Yin, H., Yang, X., and Davison, B. (2014). Systems and synthetic biology approaches to alter plant cell walls and reduce biomass recalcitrance. *Plant Biotechnol. J.* 12 (9): 1207–1216.

Lakhundi, S.S. (2012). Synthetic biology approach to cellulose degradation. Doctoral dissertation, University of Edinburgh. (Viewed 22 June 2012.) Available from: https://era.ed.ac.uk/handle/1842/9887.

Lerouxel, O., Cavalier, D., Liepman, A., and Keegstra, K. (2006). Biosynthesis of plant cell wall polysaccharides—a complex process. *Curr. Opin. Plant Biol.* 9 (6): 621–630.

Li, W., Yu, J., Li, Z., and Yin, W.B. (2018). Rational design for fungal laccase production in the model host *Aspergillus nidulans*. *Sci. China Life Sci.* 62 (1): 84–94.

Li, C., Chen, C., Wu, X. et al. (2019). Recent advancement in lignin biorefinery: with special focus on enzymatic degradation and valorization. *Bioresour. Technol.* 291: 121898.

Lin, L., Cheng, Y., Pu, Y. et al. (2016). Systems biology-guided biodesign of consolidated lignin conversion. *Green Chem.* 18: 5536–5547.

Linares, N.C., Ortiz-Vázquez, E., Fernándezc, F. et al. (2015). Recombinant expression of four oxidoreductases in *Phanerochaete chrysosporium* improves degradation of phenolic and non-phenolic substrates. *J. Biotechnol.* 209: 76–84.

Linares, N.C., Fernández, F., Lokse, A.M., and Gómez-Lim, M.A. (2018). Enhanced delignification of lignocellulosic biomass by recombinant fungus *Phanerochaete chrysosporium* overexpressing laccases and peroxidases. *J. Mol. Microbiol. Biotechnol.* 28 (1): 1–13.

Liu, C.K. (2012). Cellulose degradation system of *Cytophaga hutchinsonii*. Doctoral dissertation, University of Edinburgh. (Viewed 30 November 2012.) Available from: http://hdl.handle.net/1842/7657.

Maleszka, R., James, A.P., and Schneider, H. (1983). Ethanol production from various sugars by strains of *Pachysolen tannophilus* bearing different numbers of chromosomes. *J. Gen. Microbiol.* 129 (8): 2495–2500.

Mateljak, I., Rice, A., Yang, K. et al. (2019). The generation of thermostable fungal laccase chimeras by SCHEMA-RASPP structure-guided recombination *in vivo*. *ACS Synth. Biol.* 8 (4): 833–843.

Munawar, A.H. (2014). Developing synthetic biology tools for ligno cellulose degradation. Master's thesis (33726), Tampere University of Technology. (Viewed 5 November 2014.) Available from: http://urn.fi/URN:NBN:fi:tty-201410301533.

Nieves, L., Panyon, L., and Wang, X. (2015). Engineering sugar utilization and microbial tolerance toward lignocellulose conversion. *Front. Bioengineer. Biotechnol.* 3: 17. https://doi.org/10.3389/fbioe.2015.00017.

Nitheranont, T., Watanabe, A., and Asada, Y. (2018). Molecular cloning of cDNA encoding a major laccase isozyme from *Grifola frondosa* and its expression in *Pichia pastoris*. *Eur. J. Biochem.* 25 (4): 134–140.

Olah, G.A. (2005). Beyond oil and gas: the methanol economy. *Angew. Chem. Int. Ed.* 44 (18): 2636–2639.

Otterbein, L., Recoed, E., Longhi, S. et al. (2000). Molecular cloning of the cDNA encoding laccase from *Pycnoporus cinnabarinus* I-937 and expression in *Pichia pastoris*. *European J. Biochem.* 267: 1619–1625.

Pal, S., Mondal, A.K., and Sahoo, D.K. (2016). Molecular strategies for enhancing microbial production of xylitol. *Process Biochemistry* 51 (7): 809–819.

Pardo, I., Rodríguez-Escriban, D., Aza, P. et al. (2018). A highly stable laccase obtained by swapping the second cupredoxin domain. *Sci. Rep.* 8 (1): 15669.

Paul, S. and Dutta, A. (2018). Challenges and opportunities of lignocellulosic biomass for anaerobic digestion. *Resour. Conserv. Recycl.* 130: 164–174.

Pham, L.T.M., Seo, H., Kin, K.J., and Kim, Y.H. (2018). In silico-designed lignin peroxidase from *Phanerochaete chrysosporium* shows enhanced acid stability for depolymerization of lignin. *Biotechnol. Biofuels* 11: 325.

Prakasham, R.S., Sreenivas Rao, R., and Hobbs, P.J. (2009). Current trends in biotechnological production of xylitol and future prospects. *Curr. Trend. Biotechnol. Pharm.* 3 (1): 1–38, 0973-8916.

Rajeswari, G., Jacob, S., Chandel, A.K., and Kumar, V. (2021). Unlocking the potential of insect and ruminant host symbionts for recycling of lignocellulosic carbon with a biorefinery approach: a review. *Microb. Cell Fact.* 20: 107. https://doi.org/10.1186/s12934-021-01597-0.

Rao, A., Sathiavelu, A., and Mythili, S. (2016). Genetic engineering in biobutanol production and tolerance. *Braz. Arch. Biol. Technol.* 59: https://doi.org/10.1590/1678-4324-2016150612.

Rivera-Hoyos, C.M., Morales-Alcarez, E.D., Abelló-Esparza, J. et al. (2018). Detoxification of pulping black liquor with *Pleurotus ostreatus* or recombinant *Pichia pastoris* followed by CuO/TiO_2/visible photocatalysis. *Sci. Rep.* 8: 3503.

Salunke, B., Sawant, S., Kang, T. et al. (2015). Potential of biosynthesized silver nanoparticles as nanocatalyst for enhanced degradation of cellulose by cellulase. *J. Nanomater.* 2015: 1–8.

Sánchez, O., Sierra, R., and Alméciga-Díaz, C. J. (2011). Delignification Process of Agro-Industrial Wastes an Alternative to Obtain Fermentable Carbohydrates for Producing Fuel. In (Ed.), Alternative Fuel. IntechOpen. https://doi.org/10.5772/ 22381

Scully, S.M. and Orlygsson, J. (2019). Progress in second generation ethanol production with thermophilic bacteria. In: *Fuel Ethanol Production from Sugarcane* (ed. T.P. Basso and L.C. Basso). IntechOpen.

Shao, X., Raman, B., Zhu, M. et al. (2011). Mutant selection and phenotypic and genetic characterization of ethanol-tolerant strains of *Clostridium thermocellum*. *Appl. Microbiol. Biotechnol.* 92 (3): 641–652.

Sittijunda, S., Tomás, A.F., Reungsang, A. et al. (2013). Ethanol production from glucose and xylose by immobilized *Thermoanaerobacter pentosaceus* at 70°C in an up-flow anaerobic sludge blanket (UASB) reactor. *Bioresour. Technol.* 143: 598–607.

Song, Q., Deng, X., and Song, U.Q. (2020). Expression of *Pleurotus ostreatus* laccase gene in *Pichia pastoris* and its degradation of corn stover lignin. *Microorgan.* 8 (4): 601.

Su, B., Wu, M., Lin, J., and Yang, L. (2013). Metabolic engineering strategies for improving xylitol production from hemicellulosic sugars. *Biotechnol. Lett.* 35 (11): 1781–1789.

Tomas, C.A., Welker, N.E., and Papoutsakis, E.T. (2003). Overexpression of groESL in *Clostridium acetobutylicum* results in increased solvent production and tolerance, prolonged metabolism, and changes in the cell's transcriptional program. *Appl. Environ. Microbiol.* 69 (8): 4951–4965.

Weber, C., Farwick, A., Benisch, F. et al. (2010). Trends and challenges in the microbial production of lignocellulosic bioalcohol fuels. *Appl. Microbiol. Biotechnol.* 87 (4): 1303–1315.

Xie, S., Sun, S., Lin, F. et al. (2019). Mechanism-guided design of highly efficient protein secretion and lipid conversion for biomanufacturing and biorefining. *Adv. Sci.* 6 (3): 1801980.

Xu, G., Wang, J., Yin, Q. et al. (2019). Expression of a thermo- and alkali-philic fungal laccase in *Pichia pastoris* and its application. *Pro. Purif. Exp.* 154: 16–24.

Yang, D., Li, Z., Liu, M. et al. (2019). Biomass-derived carbonaceous materials: recent progress in synthetic approaches, advantages, and applications. *ACS Sustain. Chem. Eng.* 7 (5): 4564–4585.

Zhang, W., Song, M., Yang, Q. et al. (2018). Current advance in bioconversion of methanol to chemicals. *Biotechnol. Biofuels* 24 (11): 260. https://doi.org/10.1186/s13068-018-1265-y.

Zhao, J., Zeng, S., Xia, Y., and Xia, L. (2018). Expression of a thermotolerant laccase from *Pycnoporus sanguineus* in *Trichoderma reesei* and its application in the degradation of bisphenol A. *J. Biosci. Bioeng.* 125: 371–376.

9

Lignocellulose Bioconversion through Chemical Methods, Platform Chemicals, and New Chemicals

Manoela Martins, Patrícia F. Ávila, Marcos Fellipe da Silva, Allan Henrique Felix de Melo, Alberto M. Moura Lopes, and Rosana Goldbeck

Bioprocess and Metabolic Engineering Laboratory, School of Food Engineering, University of Campinas (UNICAMP), Campinas, SP, Brazil

9.1 Introduction

The modern concept of biorefineries of valuable bioproducts taking over the traditional petrochemical route has become popularized and has been seen as the future direction for industrialization in terms of enhanced carbohydrate-based products. The sustainable production of commercially valuable chemicals and biofuels using lignocellulose-based processes has been considered one of the hot areas and needs the innovation of new emerging technologies (Patel et al. 2019). This has the potential to decrease greenhouse gas emissions, bringing benefits to rural economies and promoting energy security (Bhatia et al. 2020). Lignocellulosic biomass is an inexpensive renewable source that can be used to produce biofuels and bioproducts. It is composed of three major interwoven components (cellulose, hemicellulose, and lignin) (Bhatia et al. 2018). However, its recalcitrant nature presents a technical challenge for releasing fermentable sugars, and a major hurdle in its use in biorefineries (Sindhu et al. 2017; Bhatia et al. 2020).

Pretreatment of lignocellulosic biomass is the initial step in exposing the cellulose and hemicellulose content for hydrolysis using enzymatic or chemical methods to increase the surface area and provide easily accessible binding sites for enzymes (Bhatia et al. 2020). Lignin content, hemicellulose chemistry, and acetyl groups affect the enzymatic conversion of cellulose into glucose. These pretreatments include mechanical pretreatment, biological pretreatment, alkaline pretreatment, dilute acid hydrolysis, ammonia fiber explosion (AFEX) pretreatment, hydrothermal treatment (steam explosion and hot water pretreatment), and novel green processes (ionic liquids [ILs] and sub/supercritical fluids) (Agbor et al. 2011; Brodeur et al. 2011; Singh et al. 2014).

Enzymatic hydrolysis is mainly focused on biomass degradation with various potentials (Goldbeck et al. 2014), and it has endless capacity to improve productivity and reduce costs (Ávila et al. 2018). The discovery and production of new enzymes through bio-prospection,

Lignocellulose Bioconversion Through White Biotechnology, First Edition. Edited by Anuj Kumar Chandel.

genetic engineering, and strain improvements have led to the identification of thousands of enzymes. A better understanding of which enzymes and proportions are essential for lignocellulosic degradation could lead to the rational design of more efficient and less expensive mixtures (Ávila et al. 2020). This chapter provides an update on recent reports and an overview of the advancements in biochemical methods and strategies for lignocellulosic biomass conversion in bioethanol and other chemical products.

9.2 Lignocellulosic Biomass

Lignocellulosic biomass is the most abundant plant material, available in much higher quantities and at lower cost than starch and sucrose-based materials. This biomass makes up the cell walls of woody plants such as trees, shrubs, and grasses. Its industrial use can come from various sources, such as agriculture, forest residues, and municipal wastes. The use of agricultural residues, for example cereal straws and corn cobs, as biorefinery feedstock allows food and fuels to be produced using the same land (Bhatia et al. 2018). However, its structural complexity has become a technical challenge for releasing fermentable sugars from biomass and a major hurdle in its use in biorefineries (Bhatia et al. 2020). The aspects about chemical composition and recalcitrance structure properties are covered in Sections 9.2.1 and 9.2.2.

9.2.1 Chemical Composition of Lignocellulosic Biomass

Lignocellulose biomass is composed of carbohydrate polymers of cellulose (40%–50%), hemicellulose (20%–30%), lignin 10%–25%, and small amounts of pectin, proteins, and extractives (chlorophyll, waxes, and nonstructural sugars) (Bhatia et al. 2020). The composition of lignocellulosic biomass varies with plant species, age, stage of growth, and season (Sindhu et al. 2017).

9.2.1.1 Cellulose

Cellulose is a linear polymer composed of D-glucose units linked together by β-(1–4) glycosidic bonds (Wyman et al. 2005). The degree of polymerization is approximately 4000–6000 in woody biomass. Polymers of cellulose are interlinked through hydrogen and van der Waals bonds to form a microfibril, and present crystalline and amorphous forms. Microfibrils are covered by hemicellulose and lignin. Crystalline cellulose fibers attach to each other by non-covalent hydrogen bonding, which provides 3–30 times lower degradability compared to the amorphous part. Cellulase is readily able to hydrolyze amorphous cellulose but is not effective at degrading the less accessible crystalline portion (Taherzadeh and Karimi 2008).

9.2.1.2 Hemicellulose

Hemicellulose is the second largest polysaccharide in lignocellulosic biomass. It supports cellulose microfibrils by hydrogen bonding and binds to lignin by covalent bonds. Unlike cellulose, hemicellulose is a heterogeneous and amorphous polysaccharide with a degree of polymerization of approximately 200 and is composed of pentoses (xylose and arabinose), hexoses (glucose, galactose, and mannose) and sugar acids such as D-glucuronic acid, 4-*O*-methyl-D-glucuronic acid, and D-galacturonic acid. The composition and structure of hemicellulose vary depending on the species. Moreover, considerable differences in hemicellulose composition exist between the stems, branches, roots, and bark. Xylan is the main component of hemicellulose in grasses and hardwood, while glucomannan is the dominant hemicellulose component in softwood (Saha 2003).

9.2.1.3 Lignin

Lignin is the most complex amorphous polyphenolic polymer and is composed of three *o*-methoxylated *p*-hydroxyphenyl propanoid units, varying according to source. These monomer units give rise to *p*-hydroxyphenyl (H), guaiacyl (G), and syringyl (S) subunits when incorporated into a lignin polymer (Bhatia et al. 2020). Gymnosperms (softwood) and fern lignin are generally composed of G, followed by a small H content. In contrast, in angiosperm (hardwood) plants, lignin is mainly composed of S units followed by G units; in herbaceous crops, G is followed by H and S units. Various monomer units are linked through (β-O-4) aryl ether bonds (Brandt et al. 2013). Lignin acts as a glue around cellulose and hemicellulose, providing mechanical strength and resistance against microbial attack. It makes biomass recalcitrant by restricting the accessibility of cellulase and by preventing the deactivation of enzymes by various lignin-derived compounds (Bichot et al. 2018).

9.2.2 Biomass Types and Recalcitrance Properties

Biomass recalcitrance to bioprocessing is directly related to the inherent properties of the biomass source. Lignin content, cellulose accessibility to cellulase, and cellulose crystallinity determine the overall digestibility of the biomass. The complexity of a given biomass type is reflected in the relationship between its structural and carbohydrate components (Bhatia et al. 2020). The factors that contribute to recalcitrance include crystallinity and degree of polymerization of cellulose; accessible surface area (or porosity); protection of cellulose by lignin; cellulose sheathing by hemicellulose; and fiber strength. The removal of lignin enhances biomass digestibility up to the point where the effect of the presence of lignin no longer limits it. Highly crystalline cellulose is less accessible to cellulase attack than amorphous cellulose, and cellulose accessibility to cellulase is one of the most important factors in enzymatic hydrolysis when the effect of lignin is minimized (Ávila et al. 2018, 2020; Bhatia et al. 2020).

Ultrafine grinding leads to smaller particle sizes, changing polymerization and porosity, which enhances enzymatic hydrolysis. Biobased research efforts targeting woody biomass should be focused on upstream activity such as size reduction and physicochemical pretreatment to improve direct microbial utilization of polysaccharides and sugar yields following enzymatic saccharification (Zhu et al. 2010).

9.3 Pretreatment and Fractionation of Lignocellulosic Materials

The pretreatment of lignocellulosic materials is a crucial step for disorganizing the recalcitrant structure of the lignocellulosic cell wall, enabling the release of fermentable sugars and greater accessibility of enzymes to the substrate during enzymatic hydrolysis

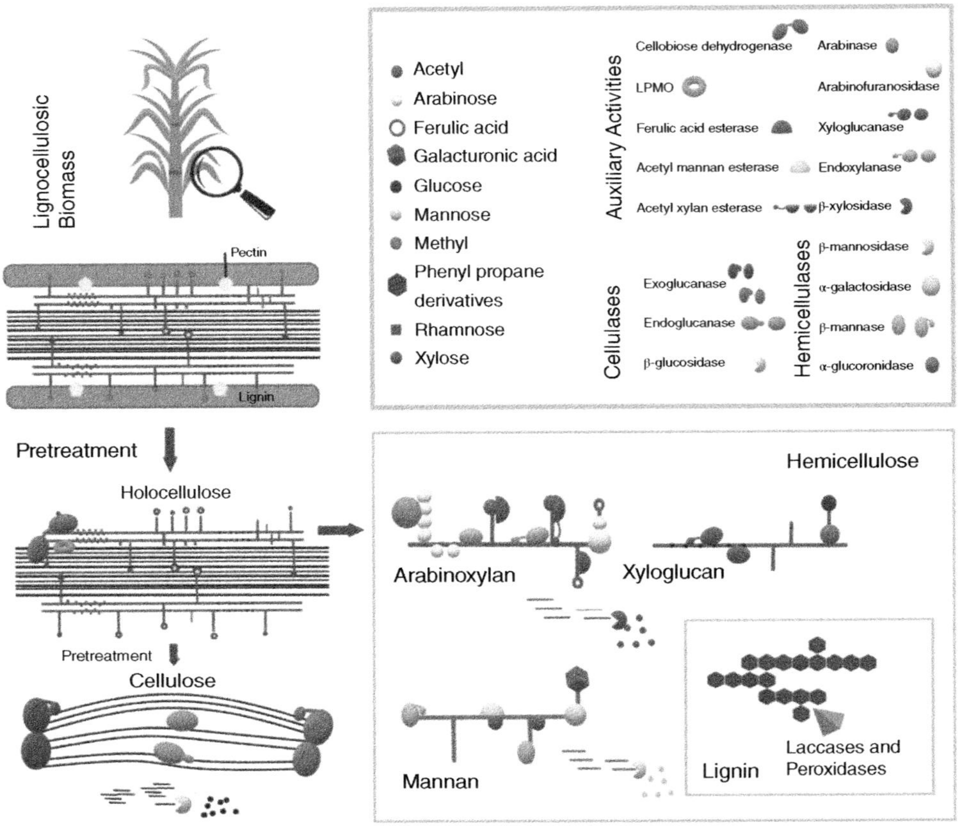

Figure 9.1 Lignocellulosic biomass structure, pretreatment and enzymes enrolled in the deconstruction of the lignocellulosic biomass. Based on Shill et al. 2011.

(Figure 9.1) (Shill et al. 2011). Therefore, pretreatment is essential for the removal of lignin, reduction of cellulose crystallinity and increased porosity of the material. The ideal biomass pretreatment method should have the following characteristics: no need for biomass size reduction, preservation of the cellulosic and hemicellulose part, no toxic inhibitors generated, lower energy demands, low cost, and recycling of chemicals (Singh et al. 2014).

Different lignocellulosic biomasses require different pretreatments based on their composition to convert these into free sugars. A generalized classification of pretreatment methods groups them into physical, chemical, biological, and multiple or combinatorial pretreatments. In combinatorial pretreatment methods, physical parameters such as temperature, pressure, or a biological step are combined with chemical treatments and are termed physicochemical or biochemical pretreatment methods (Agbor et al. 2011). Combinatorial pretreatment strategies are generally more effective in enhancing biomass digestibility and are often employed in designing leading pretreatment technologies (Brandt et al. 2013). This section covers concepts and advances of chemical and physicochemical pretreatment methods (Sections 9.3.1 and 9.3.2) to overcome the lignocellulosic biomass intrinsic recalcitrant nature prior to enzymatic hydrolysis treatment for bioethanol production, as well as treatments for fractionating lignocellulosic biomass (Section 9.3.3) to produce other high value-added biobased products.

9.3.1 Chemical Pretreatments

9.3.1.1 Alkaline Pretreatment

Pretreatment with alkali such as NaOH, KOH, $Ca(OH)_2$, hydrazine, and anhydrous ammonia causes swelling of the biomass, which increases the internal surface area of the biomass and decreases both the degree of polymerization and cellulose crystallinity (Rocha et al. 2009). Alkaline pretreatment disrupts the lignin structure and breaks the linkage between lignin and the other carbohydrate fractions in lignocellulosic biomass, increasing accessibility to the carbohydrates in the lignocellulosic matrix structure (Agbor et al. 2011). Acetyl and other uronic acid substitutions on hemicellulose that lessen the accessibility of enzymes to the cellulose surface are also removed by alkali pretreatments. Alkali pretreatment is most effective with low lignin content biomass such as agricultural residues, but becomes less effective as the lignin content of the biomass increases (Ali et al. 2020). Several studies have combined alkaline pretreatment with other pretreatment methods, such as wet oxidation (WO), steam explosion, AFEX, and ammonia recycled percolation (ARP), which are discussed in later sections.

9.3.1.2 Acidic Pretreatment

Acidic pretreatment involves the use of concentrated and diluted acids to break the rigid structure of the lignocellulosic material. The most commonly used acid is dilute sulfuric acid (H_2SO_4), which has been commercially used to pretreat a wide variety of biomass types: switchgrass (Sun et al. 2015), corn stover (Zhu et al., 2005), and sugarcane biomass (Goldbeck et al. 2014; Ávila et al. 2018). The acid pretreatment enhances the hydrolysis of hemicelluloses and a portion of amorphous cellulose, thus increasing the recovery of hemicelluloses as monomers in liquid fraction and digestible cellulose in the solid fraction. However, concentrated acids are toxic, corrosive, and hazardous in nature (Singh et al. 2014).

Diluted acid pretreatment (0.5%–1.5%, temperature above 160 °C) is more favorable for industrial applications since a high sugar yield is obtained from hemicelluloses (Singh et al. 2014). A key advantage of acid pretreatment is that a subsequent enzymatic hydrolysis step is sometimes not needed, as the acid itself hydrolyzes biomass to yield fermentable sugars. Hemicellulose and lignin are solubilized with minimal degradation, and hemicellulose is converted to sugars. However, extensive washing and/or a detoxification step is required to remove the acid before a fermentation step. Due to the corrosive nature and toxicity of most acids, an adequate material for the reactor is required to withstand the required experimental conditions and corrosiveness of the acids. Another drawback is the production of fermentation inhibitors such as furfural and hydroxymethyl furfural that reduce the effectiveness of the pretreatment method and further processes (Agbor et al. 2011; Brodeur et al. 2011; Bhatia et al. 2020).

9.3.1.3 Ionic Liquids

A new approach to chemical pretreatment of biomass is the use of ILs. ILs are a new class of solvents with low melting points (below 100 °C), high polarities, high thermal stability, and negligible vapor pressure, consisting entirely of ions (cations and anions). Most ILs used in biomass fractionation are imidazonium salts. Studies show that 1-allyl-3-methylimidazonium chloride and 1-butyl-3-methylimidazonium chloride can effectively be used as solvents for the dissolution of cellulose at temperatures below 100 °C (Brandt et al. 2013). A possible dissolution mechanism suggests that these ILs compete with lignocellulosic components for hydrogen bonding, thus disrupting their three-dimensional network. Through this dissolution mechanism, ILs are employed in the dissolution of cellulose, presenting good solubility (5%–20%). ILs can remove lignin from lignocellulosic biomass in the range of 15%–92% and partial hemicellulose, while the cellulose part remains intact (Brandt et al. 2013).

The cost, recovery, and reuse of ILs are important considerations for industrial applications. An almost full recovery of ILs was achieved by adding glycerol to carboxylate ILs (Clough et al. 2016). The high cost of imidazolium cations is a major obstacle limiting the industrial application of ILs for biomass pretreatment. Therefore, novel ILs based on cations have been developed. Ferrari et al. (2019) have studied strategies to obtain low-cost and easily synthesizable ethanolammonium-based ILs to be applied as a pretreatment step of sugarcane straw biomass. The IL mixtures improved enzymatic digestibility at higher solids loading with water addition in the reaction medium under mild temperature conditions (90 °C), which resulted in a cellulose conversion of 98% after enzymatic hydrolysis.

ILs can also be synthesized using vanillin, *p*-anisaldehyde and furfural derived from lignin and hemicellulose as starting materials. The techno-economic analysis suggested that the raw materials cost \$13.71 kg^{-1} for the biomass-derived ILs. However, the modification of the reductive amination step in the synthesis of ILs could significantly lower the raw material cost to \$4 kg^{-1} (Socha et al. 2014).

9.3.1.4 Wet Oxidation

Wet oxidation (WO) utilizes oxygen dissolved in water. There are two reactions that occur during this process. One is low temperature hydrolysis, and the other is high temperature oxidation (Brodeur et al. 2011). Typically, the procedure for WO consists of drying and milling lignocellulosic biomass to obtain particles that are 2 mm in length, to which 1 liter of water is added for every 6 g of biomass. This method of pretreatment is performed at 195 °C for a range of 10–20 min (Pedersen and Meyer 2009). WO has been shown to be effective in pretreating a variety of biomasses such as wheat straw, corn stover, sugarcane bagasse, cassava, peanuts, rye, canola, fava beans, and reed to obtain glucose and xylose after enzymatic hydrolysis (Martín and Thomsen 2007; Pedersen and Meyer 2009). During WO, lignin is decomposed to carbon dioxide, water, and carboxylic acids. The amount of lignin removed after pretreatment ranges from 50% to 70% depending on the type of biomass pretreated and the conditions used. For bagasse with a pretreatment time of 15 min 50% lignin was removed, which resulted in 57.4% conversion of cellulose compared to only 35% lignin removal and 48.9% cellulose conversion for steam explosion, explained later, under the same conditions (Martín and Thomsen 2007).

9.3.2 Physicochemical Pretreatment

9.3.2.1 Steam Explosion

Steam explosion pretreatment is one of the most commonly used pretreatment options, as it uses both chemical and physical techniques to break the structure of the lignocellulosic material. This method subjects the material to high pressures and temperatures for a short duration of time after which it rapidly depressurizes the system, disrupting the structure of the fibrils (Brodeur et al. 2011).

A two-step steam explosion pretreatment option has also been examined to solubilize the structure in stages, creating an optimum cellulose fraction during hydrolysis. The first step involved temperatures of 180 °C to solubilize and remove the hemicellulose fraction. The second stage used a high-temperature pressurized pretreatment with temperatures up to 210 °C, not exceeding 240 °C, in which the cellulose fraction was subjected to breaking down of its carbohydrate linkage (Agbor et al. 2011; Brodeur et al. 2011). This two-step process increased the downstream ethanol yield by increasing accessibility to cellulose structure because of the reduction of the hemicellulose fraction. Operation costs also decreased as less enzyme dosage was required due to increased accessibility of the cellulose fraction (Bhatia et al. 2020). However, increased costs of equipment are required for processing, and the additional energy usage of a second steam explosion process is required (Brodeur et al. 2011).

9.3.2.2 Liquid Hot Water

Liquid hot water pretreatment uses water at elevated temperatures and high pressures to maintain its liquid form to promote disintegration and separation of the lignocellulosic matrix (Singh et al. 2014). Temperatures can range from 160 °C to 240 °C and over length soft immersing from a few minutes up to an hour with temperatures dominating the types of sugar formation and time dominating the amount of sugar formation (Agbor et al. 2011). Xylose recovery is high (88%–98%) and no acid or chemical catalysis is needed in this process, which makes it economically interesting and environmentally attractive (Mosier et al. 2005). However, it requires higher energy due to high pressure and a large amount of water.

9.3.2.3 Ammonia Fiber/Freeze Explosion (AFEX), Ammonia Recycle Percolation (ARP) and Soaking Aqueous Ammonia (SAA)

AFEX, ARP, and SAA are alkaline pretreatment methods that use liquid ammonia to pretreat biomass. As a physicochemical process, AFEX is similar to steam explosion for operating at high pressure, but it is conducted at ambient (<90 °C) temperatures, while ARP is carried out at higher temperatures. SAA is a modified version of AFEX utilizing aqueous ammonia to treat biomass in a batch reactor at moderate temperatures (25–60 °C) to reduce the liquid throughput during pretreatment. At ambient temperatures, the duration could be up to 10–60 days while at higher temperatures (150–190 °C) the effect of ammonia is rapid, and the duration of pretreatment is reduced to minutes (Alizadeh et al. 2005).

In both mentioned processes, the lignocellulosic material is exposed to ammonia at a given temperature and high pressure, which causes swelling and phase change in cellulose crystallinity in addition to the alteration and removal of lignin. This increases the reactivity of the remaining carbohydrates after pretreatment. The pretreated biomass is easily hydrolysable with close to theoretical yields after enzymatic hydrolysis at low enzyme loadings compared to pretreated biomass from other pretreatment methods (Agbor et al. 2011). To optimize the process and establish a continuous process, by dropping the liquid ammonia throughput Kim et al. (2006) reported a low liquid ARP using one reactor void volume with 15 wt% ammonia. The digestibility of the low liquid ARP pretreated biomass under optimum conditions was at least 86% of the theoretical maximum even at the lowest enzyme loading of 7.5 FPU g^{-1} glucan (FPU is filter paper unit).

9.3.2.4 Supercritical Fluid

A supercritical fluid is a material that can be either liquid or gas, used in a state above the critical temperature and critical pressure where gases and liquids can coexist (Agbor et al. 2011). A supercritical fluid can penetrate the crystalline structure of lignocellulosic biomass overcoming the mass transfer limitations encountered in other pretreatments (King and Srinivas 2009).

Supercritical carbon dioxide (CO2) with a critical temperature (Tc) of 31°C and a critical pressure (Pc) of 7.4 MPa is typically also used as an extraction solvent (Schacht et al. 2008). Currently it has gained importance as a solvent for the pretreatment of different varieties of biomass. When used in combination with water, it forms carbonic acid which favors polymer hydrolysis (Brodeur et al. 2011). Once the biomass is pretreated, the explosive release of CO2 disrupts the cellulose and hemicellulose structure, thus increasing the access for enzyme hydrolysis. The lower temperature used in the process aids in the stability of the sugars and prevents the degradation observed in other pretreatments (Agbor et al. 2011). Even though supercritical fluids are being investigated by a significant number of researchers for pretreatment, the whole process has not proven to be economically viable, with the high pressures involved being a deterrent. Improvements need to be made to implement the process on a large scale (King and Srinivas 2009).

9.3.3 Fractionating Treatments of Lignocellulosic Compounds

Another approach to overcoming the cost barrier is the production of high value-added co-products from lignocellulose. Prior to producing these co-products, the separation of each component in the lignocellulosic biomass is needed. Each component can be utilized by itself or converted into intermediates of various products (Zheng et al. 2009). The fractionation process can contribute to reducing the cost in the downstream process, because a significant amount of energy, chemicals, and water is consumed in the downstream process (Zhang et al. 2014).

Alkaline treatment also releases other polysaccharides, such as lignin, thus decreasing arabinoxylan purity. Alkaline peroxide is an effective reagent for delignification and improving the whiteness of hemicelluloses from cereals. In alkaline media, hydrogen peroxide forms the highly active hydroperoxide anion (HOO^{-}–), which oxidizes lignin structures by reacting with the hydrophilic (carboxyl) groups, breaking some interunit bonds, and resulting in the separation of lignin and hemicelluloses. A process that is gaining prominence is organosolv, which consists of the fractionation of lignocellulosic biomass using a mixture of organic solvents and water at high temperatures to separate cellulose, hemicellulose, and lignin. There are a variety of solvents but the most commonly used are acetone, ethanol, methanol, and organic acids (acetic acid and formic acid). The choice of the solvent is defined according to the studied process, the costs, and the ease of acquisition of the solvent (Zhang et al., 2014).

The organosolv process is generally carried out without the addition of catalysts; however, a catalyst can be used to accelerate reactions. The commonly used catalysts are acetic acid, H_2SO_4, HCl, NaOH, $CaCl_2$, NH_3, $AlCl_3$, $Fe_2(SO_4)_3$, $Al_2(SO_4)_3$, anthraquinone, oxalic acid, and sulfonic acid. The main disadvantages of using catalysts are the possibility of equipment corrosion, additional catalyst recovery and recycling steps, contamination of the extracted lignin, and the consequent increase in the cost of the process (Skrebets et al. 2004). Direct extraction of lignin in biomass without prior removal of cellulose and hemicellulose implies that more than 60% of mass added to the extraction reactor is polysaccharides (cellulose and hemicellulose), which can suffer degradation during the extraction process, reducing the degree of purity of the extracted lignin (Yamamoto et al. 2014). Generally, during the lignin extraction process hemicellulose is also extracted in large quantities (>70%) (Esteves et al. 2008).

9.4 Enzymatic Hydrolysis of Lignocellulosic Biomass

Although chemical and physical catalysis can hydrolyze lignocellulosic biomass to products of interest, enzymatic reactions have some advantages since they occur at moderate temperatures, have stereospecificity, and eliminate the use of chemicals and toxic reagents that hinder purification and are difficult to recycle, which is particularly beneficial for the synthesis of biomolecules for food or pharmaceutical use. The use of enzymes is ecologically friendly and industrially well established, and is applied in the manufacture of biofuels, resins, processed meats, cheeses, juices, oils, alcoholic beverages, etc. Currently, purification techniques and genetic engineering have expanded the performance of enzymatic processes (Lopes et al. 2018). Due to the variety and complexity of lignocellulose structures, coordinated action of an enzymatic complex is required to achieve complete degradation. Information about lignocellulosic enzymes can be checked in the carbohydrate-active enzyme (CAZy) database (www.cazy.org). The heterogeneity of biomass results in a multiplicity of enzymes that differ in their structural and physicochemical properties, mode of action, and substrate specificity (Obeng et al. 2017).

9.4.1 Cellulases

Key enzymes required for hydrolysis of microcrystalline cellulose, the main polysaccharide in the cell wall, are the glycoside hydrolases endoglucanases, exoglucanases, and β-glucosidases, which work cooperatively to enforce hydrolysis of the complex microfibrils of cellulose. These enzymes commonly have a catalytic domain that cleaves the glycosidic bond and a carbohydrate binding module that targets the catalytic domain to the substrate (Obeng et al. 2017).

Endo-1,4-β-D-glucanases (EC 3.2.1.4) are short loops that randomly hydrolyze the internal bonds of cellulose fibers present in the amorphous part of cellulose, releasing cello-oligosaccharides and cellobiose, which is hydrolyzed to glucose by the action of β-glucosidase or cellobiases (EC 3.2.1.21) (Wyman et al. 2005; Polizeli et al. 2017).

β-glucosidases possess a rigid structure with the active site inside a large cavity, a site pocket formed by four hydrophilic loops with different conformations to enhance carbohydrate binding, which favors the entry of cellobiose. They are also essential to relieve the feedback inhibition from the products of glucanases. However, β-glucosidases also face inhibition from their products and thus the search for glucose-tolerant β-glucosidases is being developed (Obeng et al. 2017).

Exo-1,4-β-D-glucanases have long loops and affinity for crystalline sites along external cellulose chains. The loops form a tunnel around the catalytic site to where the substrate is directed. They include the 1,4-β-D-glucan glucohydrolases (EC 3.2.1.74), which liberate glucose from 1,4-β-D-glucans and slowly hydrolyze cellobiose, and 1,4-β-D-glucan cellobiohydrolases (CBHs), which liberate cellobiose from 1,4-β-glucans from the reducing (CBHI) or non-reducing ends (CBHII). Moreover, CBH can hydrolyze highly crystalline cellulose (Wyman et al. 2005; Obeng et al. 2017; Polizeli et al. 2017).

Cellobiose dehydrogenase (EC 1.1.99.18), capable of oxidizing cellobiose, belongs to the auxiliary activities family and has a synergistic auxiliary activity, activating hydrolytic cellulases by relieving enzymatic inhibition by cellobiose product, similarly to β-glucosidases (Moreira et al. 2011; Obeng et al. 2017).

Hydrolytic enzymes that also belong to the auxiliary activities family that help in cellulose degradation are lytic polysaccharide monooxygenases (LPMOs) (Moreira et al. 2011; Obeng et al. 2017). These copper-dependent enzymes have received attention because of their synergic effect on cellulase activity and their activity toward a range of substrates in addition to cellulose: chitin, starch, xyloglucan, and glucomannan (Filho et al. 2020). The active sites of LPMOs are located in the center of a flat face. This flat structure promotes efficient interaction with oligomers. Unlike other cellulases that require the presence of a loose substrate, the strong mechanism of LPMO seems to allow oxidation at the surface of a cellulose crystalline microfibril (Levasseur et al. 2013). The catalysis occurs from the binding of an active oxygen molecule from the atmosphere to monomeric Cu^{2+}, which culminates in the interaction of the active site with polysaccharide chains (Obeng et al. 2017). It is believed that LPMOs constitute the first attack on lignocellulose polymers, cleaving bonds by oxidizing C1 and C4, and improving the degradation of this recalcitrant polymer, making the material more accessible to other enzymes, although the exact mechanism of action is still unclear (Johansen 2016).

9.4.2 Ligninolytic Enzymes

Lignin is the most abundant aromatic polymer, consisting of phenolic and non-phenolic compounds. There are two families of ligninolytic enzymes that play an important role in enzymatic degradation: phenol oxidase (laccase) and peroxidases (lignin peroxidase, versatile peroxidase, and manganese peroxidase), both from the auxiliary activities category (Baruah et al. 2018). The delignification of phenylpropane polymers involves laccases (EC 1.10.3.2), also called benzenediol oxygen oxidoreductase, a multicopper enzyme that oxidizes phenolic compounds and aromatic amines, boosting cellulase activity by liberating cellulose from binding sites on lignin (Obeng et al. 2017); lignin peroxidase (EC 1.11.1.14) and manganese peroxidase (EC 1.11.1.13) are enzymes of low specificity capable of oxidizing lignin (Crestini et al. 2011). Versatile peroxidase (EC 1.11.1.16) also oxidizes phenolic and non-phenolic aromatic compounds. These enzymes have the catalytic activities of both manganese and lignin peroxidase and can oxidize Mn^{2+} as well as high-redox potential non-phenolic compounds (Baruah et al. 2018). Vanillyl alcohol oxidase (EC 1.1.3.38) catalyzes the conversion of phenolic compounds released during the lignin breakdown, and 1,4-benzoquinone reductase (EC 1.6.5.6) degrades aromatic compounds (Levasseur et al. 2013).

9.4.3 Pectic Enzymes

While abundant in fruits, very little pectin is present in traditional biomass such as wood, sorghum, or sugarcane bagasse. Pectic substances are deconstructed by endopoligalacturonase and exopoligalacturonase (EC 3.2.1.15 and 3.2.1.67), cleaving the chain formed by polygalacturonic acid and releasing oligogalacturonate and monogalacturonate, respectively. Carbohydrate esterase CE8 pectinesterases (EC 3.1.1.11) deesterify the ester bonds of the methyl group, releasing methanol (Jayani et al. 2005). Additional rhamnogalacturonases (EC 3.2.1.-) and rhamnogalacturonan lyases (EC 4.2.2.-) are responsible for the degradation of the "hairy regions" in pectin, and CE16 pectin acetylesterases (EC 3.1.1.6) remove acetyl groups from acetylated homogalacturonan and rhamnogalacturonan (Van Dyk and Pletschke 2012).

9.4.4 Mannases

Mannans, hemicellulosic polysaccharides, are degraded by the synergistic action of β-mannanase (EC 3.2.1.78), which acts on the main mannan chain, β-mannosidase (EC 3.2.1.25), which releases mannose from the non-reducing ends, and β-glucosidase (EC 3.2.1.21), which cleaves β-D-1,4-glucopyranose in the final non-reducing region of the oligosaccharides released from glucomannan and galactoglucomannan by β-mannanase, releasing monosaccharides. Acetyl mannan esterase (EC 3.1.1.6) and α-galactosidase (EC 3.2.1.22) cleave the bonds of the acetyl and galacturonic acid groups, respectively (Moreira et al. 2011; Lopes et al. 2018).

9.4.5 Xylanases

Xylan, due to its complex and heterogeneous nature, has more complicated enzymatic hydrolysis than other plant polysaccharides. Different enzymes are required depending on the specific type of xylan to be hydrolyzed and the product to be obtained. The enzymes involved in the xylan depolymerization process can be categorized into three groups: (i) enzymes that hydrolyze backbones; (ii) enzymes that remove side chains; and (iii) accessory enzymes that remove substituted residues, such as acetyl groups. Due to the insolubility and recalcitrance of plant cell walls, many hemicellulases are modular proteins, pursuing other functional modules in addition to their catalytic domain, such as carbohydrate binding modules. A detailed explanation of these enzymes is described hereafter (Wyman et al. 2005).

9.4.5.1 Backbone Enzymes

The core for xylan degradation is endo-1,4-β-xylanases (EC 3.2.1.8), which cleave glycosidic bonds between xylose units in the main xylan chain, releasing oligomers of lesser degree of polymerization and xylo-oligosaccharides. Only two families belong to this group of enzymes: GH10 and GH11. While the GH10 family, exhibiting strand-loop-helix-turn $(\beta/\alpha)_8$ folding with a deep active site groove at the loops $\beta \rightarrow \alpha$, can hydrolyze xylan chains closer to the branching points, releasing shorter chains of xylo-oligosaccharides, the GH11 family, with β-jellyroll folding, is able to cleave into more distant regions of the side chains.

Thus, branching residue distribution interferes with the action of these enzymes on the xylan backbone. β-xylosidases (EC 3.2.1.37) hydrolyze small xylo-oligosaccharides and xylobiose from their non-reducing ends to xylose. They are usually not able to hydrolyze xylan but can cleave artificial substrates such as *p*-nitrophenyl-β-D-xylopyranoside (Polizeli et al. 2005).

9.4.5.2 Side Chain Enzymes

α-glucuronidase (EC 3.2.1.131) hydrolyzes α-1,2 bonds between glucuronic acid and β-D-xylopyranosyl residues found in glucoxylans; arabinose (EC 3.2.1.99) removes residues of L-arabinose from positions 2 and/or 3 in residues of β-D-xylopyranosyl; arabinofuranosidases (EC 3.2.1.55) hydrolyze α-1,2-, α-1,3-, and α-1,5-L-arabinofuranosidic bonds in L-arabinose-containing polysaccharides, removing residues substituted at positions 2 and 3. They are active on the synthetic substrate *p*-nitrophenyl α-L-arabinofuranoside; xyloglucanases (EC 3.2.1.151) hydrolyze the xyloglucan backbone (glucose β-1,4 linkages) in glucose residues (Polizeli et al. 2005; Segato et al. 2014).

9.4.5.3 Accessory Enzymes

Other enzymes needed for the complete depolymerization of xylan are acetyl xylan esterase (EC 3.1.1.72), which removes the *O*-acetyl groups from positions 2 and/or 3 in the residues β-D-xylopyranosyl, facilitating the action of endoxylanases, and ferulic acid esterase (EC 3.1.1.73) and *p*-coumaric acid esterase (EC 3.1.1.-), which cleave ester bonds from the xylan side chains and pectin (Segato et al. 2014). The additional auxiliary enzyme glyoxal oxidase (EC 1.1.3.-) converts aldehydes to carboxylic acids, and galactose oxidase (EC 1.1.3.9) oxidizes carbohydrates and primary alcohols (Levasseur et al. 2013).

9.4.6 Enzyme Synergism

Lignocellulosic biomass is an exceedingly complex substrate, requiring several enzymes to degrade. It is therefore expected that there is synergism among these enzymes in nature, producing much higher yields when acting in a cooperative manner. In addition, the more effective cooperation reduces enzymatic loadings, and the more effective the production costs. Great efforts have been spent on improving enzyme cocktails for specific substrates (Van Dyk and Pletschke 2012; Lopes et al. 2018).

Cellulosomes are a natural means of harnessing the synergistic nature of certain hydrolytic enzymes to degrade lignocellulose. A cellulosome is an exocellular enzymatic complex found on the cell surface of the extracellular medium growth of some anaerobic fungi (e.g. *Neocallimastix frontalis*) and bacteria (e.g. *Clostridium thermocellum*) where several saccharolytic enzymes are anchored to scaffolding proteins (Wyman et al. 2005; Schuster and Chinn 2013). Since its discovery, many investigators have used a variety of preparations to demonstrate the synergy among various lignocellulosic enzymes. More recently, Goldbeck et al. (2014) enhanced hemicellulose depolymerization by investigating synergic enzymatic mixtures, Ávila et al. (2020) reported synergy between endoxylanase and arabinofuranosidase, and Martins et al. (2020) applied this synergy to optimize xylo-oligosaccharide production. In another approach, Furtado et al. (2013) designed a bifunctional enzyme showing synergism between β-glucanase (EC 3.2.1.73) and the copper-dependent oxidase laccase (EC 1.10.3.2) that catalyzes the oxidation of aromatic compounds, enhancing the decrease in sugar production. Inspired by nature, artificial chimeric proteins, which are domains of two or more enzymes expressed in a single peptide sequence, have been designed to reduce production costs and improve enzymatic performance (Saadat 2017).

9.5 Biorefinery–Biobased Chemicals Platform

9.5.1 Contextualization–Bioeconomic and Biorefinery

World development over the years has undergone several transformations to meet global dynamics (Francois et al. 2020). Oil exploration, in the mid-twentieth century, generated a low-cost energy source that triggered global industrialization. Since then, oil has been applied not only as a fuel source but also as a base for most chemicals currently available (Hess et al. 2016). Burning fossil fuels releases more than 21.3 billion tons of carbon dioxide (CO_2) per year (Shinde et al. 2020), which is associated with the greenhouse effect, posing serious risks to human health and the environment. Thus, current economic development requires the preservation and management of the various resources available to promote a profitable and biobased productive system. Therefore, innovative materials, methods and technologies have been the main challenge for researchers to provide alternative and viable subsidies for energy and chemicals to the economy. Bio-economics is the term given to the economic sector to produce value-added products through the sustainable conversion of biobased raw materials. More recently, the term circular bioeconomy was defined since it was realized that circularity and efficiency are not automatically present in bio-economic mechanisms (D'Adamo et al. 2020). In practice, the main motivation of the circular bioeconomy refers to the transformation of conventional industrialists through the implementation of efficient biorefineries, to enable economic growth unrelated to environmental degradation (Mohan et al. 2020; Escobar and Laibach 2021). The American National Renewable Energy Laboratory (NREL) defines biorefineries as installations of integrated biomass conversion processes to produce fuels, energy, and chemicals.

The chemical market based on biological platforms has an expected annual growth of 12.6%, with an expectation of reaching $22.75 billion in 2025 (López-Porfiri et al. 2020). Biofuels represent an important part of that market. Bioethanol, for example, has an extensive production volume, contributing to energy sufficiency. However, the economic viability of biorefineries is associated with the ability to manage multiple processes. Analogous to an oil refinery, a biorefinery can take advantage of different fractions of biomass to produce various compounds and maximize their gains (Bozell and Petersen 2010). More than 200 value-added chemicals can be produced by refining lignocellulosic biomass. Chemical products and platform intermediates, also called building blocks, represent the key to the growth of bio-economics (Isikgor and Becer 2015). Advances in bioethanol production and the main biologically based chemicals will be covered in Sections 9.5.2 and 9.5.3, respectively.

In general, biomass has more oxygen and lower fractions of hydrogen and carbon than oil. Due to this differentiated composition, a wider range of products can be obtained. However, a greater number of processes and technologies must be developed. To produce biofuels, biomass needs to be deoxygenated. In general, the presence of oxygen increases the polarity of the biofuel, making it difficult to mix with fossil fuels, in addition to decreasing energy capacity (Cherubini and Strómman 2010). Oxygen is rejected in the typical form of CO_2 or H_2O. Both forms have disadvantages. Whereas in the first case half a mole of carbon is lost per mole of oxygen (loss of energy), in the second, two moles of hydrogen are unwantedly removed per mole of oxygen. The removal of oxygen can be favored by the elimination of undesirable elements, especially sulfur and nitrogen (SO_2 and NO_2, respectively) (Lange 2007). Other chemicals may require much less deoxygenation, because the presence of oxygen generally provides greater diversification of products with interesting physical and chemical properties (Wang and Liu 2020).

9.5.2 Bioethanol

Bioethanol is produced from the fermentation of sugars obtained from biomass. Bioethanol feedstock must contain either sucrose (e.g. sugarcane, sugar beet) or starch (e.g. corn, wheat) or be a lignocellulosic material (e.g. sugarcane bagasse, wood, and straw). Sugarcane is a feedstock used in Brazil and, thus far, the most efficient raw material for bioethanol production. CAZymes can be naturally expressed by a high number of microorganisms in the medium such as aerobic and anaerobic prokaryotes and eukaryotes (Himmel et al. 2010; Champreda et al. 2019) or can be expressed in microbial cell factories (Claes et al. 2020). The cost of these enzymes could constitute up to 30% of the bioethanol production from lignocellulosic biomass (Valdivia et al. 2016); thus, reducing costs is paramount.

Typically, during sugarcane processing, large amounts of bagasse are produced (~240 kg per ton of sugarcane), which are burned in boilers for steam and electricity generation. However, better technologies for co-generation and optimization of bioethanol production allow it to be used as a raw material for producing second-generation bioethanol (E2G) and other biobased products. E2G can be produced from plant biomass by the use of microorganisms in four different bioprocesses: separate hydrolysis and fermentation, where enzymatic hydrolysis and fermentation occur in separate vessels, simultaneous saccharification and fermentation, simultaneous saccharification and co-fermentation, and consolidated bioprocessing, where enzymatic hydrolysis is associated with fermentation in the same compartment (Salehi Jouzani and Taherzadeh 2015). The combination of hydrolysis and fermentation in the same compartment could reduce costs associated with a reduced number of compartments and energy (Olofsson et al. 2008; Salehi Jouzani and Taherzadeh 2015). The separate hydrolysis and fermentation, simultaneous saccharification and fermentation and simultaneous saccharification and co-fermentation processes use exogenous enzymes (Figure 9.2) obtained from specialized companies or produced on-site at the same refinery, with an additional cost. However, consolidated bioprocessing could remove this additional cost in bioethanol production by allowing the microorganism to perform the production of CAZymes and fermentation simultaneously, also allowing lower contamination and inhibition of cellulases (Salehi Jouzani and Taherzadeh 2015).

To develop an ideal microorganism for bioethanol production in a consolidated bioprocessing system, several authors performed modifications in *Saccharomyces cerevisiae* strains, such as the expression of heterologous cellulases from other microorganisms (Claes et al. 2020) or modification of genes associated with improved phenotypes (Caspeta et al. 2015).

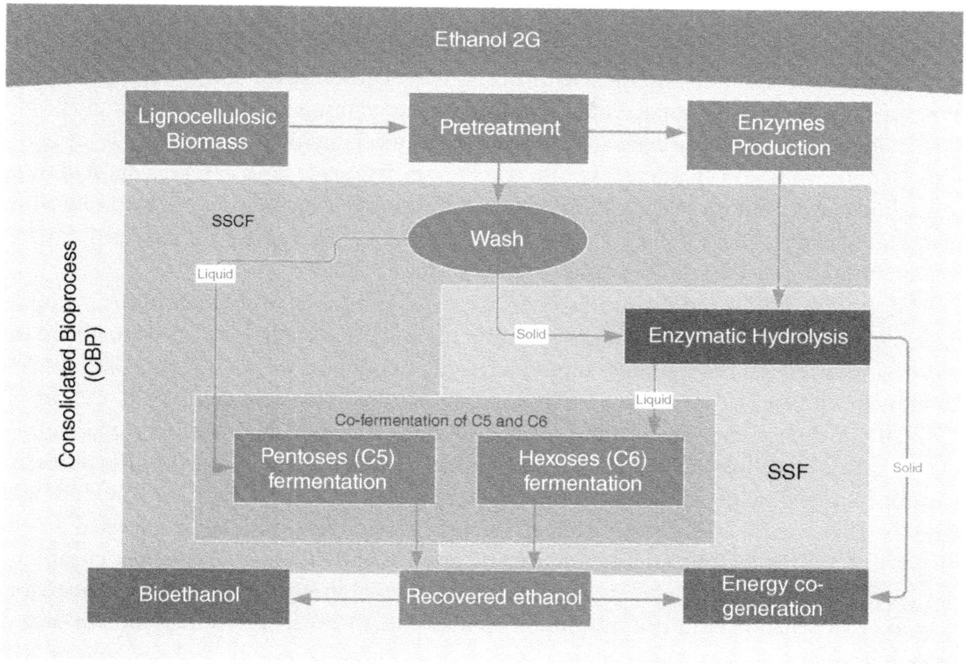

Figure 9.2 Strategies used for hydrolysis of the pretreated biomass and fermentation at the same compartment using simultaneous saccharification and fermentation (SSF), simultaneous saccharification, and co-fermentation (SSCF) in a consolidated bioprocess (CBP).

9.5.3 Other Value-added Bioproducts Obtained from Lignocellulosic Biomass

In the context of biorefineries, one of the main challenges is the application of the most diverse biotechniques to optimize bioprocesses to obtain the most diversified compounds with high added value from agro-industrial residues, aiming at application in other industrial sectors (Silva et al. 2020). From each fraction of lignocellulosic biomass and by means of various strategies, the most diverse bioproducts can be obtained. Figure 9.3 shows a scheme of the main compounds obtained from each fraction.

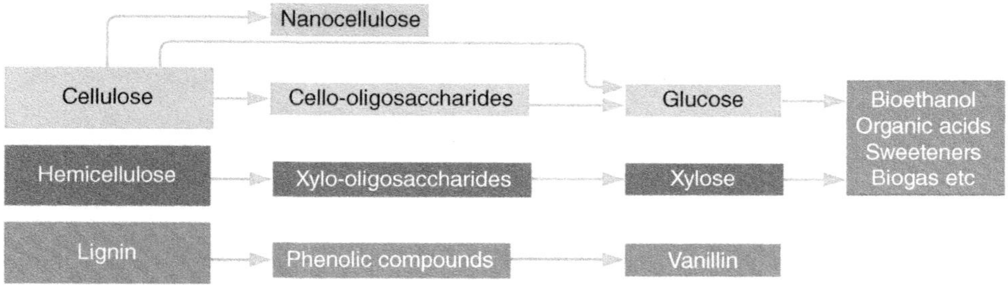

Figure 9.3 Main bioproducts obtained from the different fractions of lignocellulosic biomasses.

9.5.3.1 Nanocellulose

In the face of advances in nanotechnology, researchers worldwide have been looking for the development of nanomaterials with properties of interest for multiple applications (Pires et al. 2019). In this respect, nanocellulose has stood out, in view of its multiple properties and applications. It is defined as a cellulosic material with at least one of its dimensions (length, width, or diameter) within the nanometric scale, and can be classified into crystalline nanocellulose (NCC), fibrillated nanocellulose (NFC) and bacterial nanocellulose (Thomas et al. 2020). In this chapter we focus on the first two types since they can be obtained from lignocellulosic biomass.

Other names for NCC are cellulose nanocrystals or nanocrystalline cellulose, which denote nanostructures of a crystalline nature up to 500 nm in length and up to 70 nm in width (Moon et al. 2011). They stand out for their chemical purity, rigid structure, resistance, ability to disperse easily, and typically being amphiphilic (Thomas et al. 2020). Currently, NCC is commonly obtained from acid hydrolysis of lignocellulosic biomass or cellulosic isolate. Sulfuric acid is the main reagent used in this process, usually in high concentrations, mainly guaranteeing good colloidal stability of the crystals. Other acids such as phosphoric, hydrochloric, and citric acid are used as an alternative approach. To promote a more sustainable and ecologically correct commercial production of NCC, researchers have been developing recovery techniques for re-using the acids used in the treatment, in addition to studying applications of other less abrasive and more ecological techniques such as enzymatic hydrolysis, and application of ultrasound. Crystalline nanocellulose has multiple industrial applications (Cui et al. 2016). In the food industry it has been used mainly in coatings, films, and food packaging, in view of promoting improvements in the mechanical properties of polymers, in addition to being able to be used as a functional ingredient (dietary fiber), emulsifying agent, and stabilizer. In the medical field, NCC has been applied in the development of materials with great potential in therapies that involve the repair, replacement, and regeneration of tissues and organs. In the materials industry, cellulose nanocrystals have an important application in the development of high resistance polymers, serving to minimize the excessive dependence on plastics and other synthetic polymers (Thomas et al. 2020).

Other names for NFC are cellulose nanofibers or nanocellulose fibers, which designate individual cellulosic fibers in nano dimensions, usually with a diameter less than 100 nm in width and a few micrometers in length (Moon et al. 2011). Cellulose nanofibers have interesting properties, such as low thermal expansion, high aspect ratio, strengthening effect, and good mechanical and optical properties. Due to these specific characteristics, nanocellulose fibers have been used in composites, food packaging, coating additives, airgels, and membranes, and as a gas barrier materials, fillers, flocculants, emulsifiers, food thickeners, and reinforcement materials (Czaikoski et al. 2020). Mechanical breakdown or defibrillation is one of the main methods used to produce NFC from the amorphous regions of cellulose present in lignocellulosic biomass. Other methods have also been used together to optimize the production of possible nanofibers, such as electrostatic induction and enzymatic and oxidative treatments (Thomas et al. 2020).

9.5.3.2 Prebiotics

Currently, in the face of changes in the population's eating habits, both the consumption and the development of functional foods have gained considerable prominence. Among the main functional ingredients, prebiotics are selective substrates of microorganisms naturally present in the organism, such as bifidobacterial and lactobacillus, thus promoting several health benefits (Hurtado-Romero et al. 2020). In the context of biorefineries, involving the use of lignocellulosic biomass, there is the possibility of obtaining two important classes of prebiotics: the cello-oligosaccharides (COS) and the xylo-oligosaccharides (XOS).

COS or cello-oligomers are defined as oligomers that consist of two to six glucose units linked by β-1,4 type bonds. COS can be obtained from the cellulosic fraction of biomass through different treatments, such as acid hydrolysis, hydrolysis on carbon catalysts, a hydrothermal process, and controlled enzymatic hydrolysis. The latter is considered the most selective method in view of the greater control regarding the generation of monomers, thus providing higher yields of COS. COS are functional molecules of great interest to the food, feed, and pharmaceutical industries, since they have a prebiotic effect, favoring the growth of probiotic strains in the large intestine, which inhibits the development of pathogenic bacteria and helps to combat intestinal constipation, maintaining intestinal health. In addition, COS can be used to produce bioethanol (Barbosa et al. 2020).

XOS or xylo-oligomers are defined as oligomers consisting of xylose units, interconnected through β-1,4 bonds (Samanta et al. 2015). The XOS can be obtained from the xylan found in the hemicellulosic fraction of biomasses by different treatments and enzymatic procedures, similar to the COS. In addition to their prebiotic properties, XOS present an important antioxidant activity, helping the capture of free radicals and inhibiting the development of cancer cells, especially in the small intestine, colon, and liver tissues; they promote the reduction of blood glucose, triglycerides, and cholesterol levels, and help to fight diabetes mellitus and obesity. They can also act as immunomodulating agents, strengthening the immune system, especially in the fight against infectious diseases . In the pharmaceutical industry, these prebiotics are mainly used in the production of drugs to treat obesity and gastrointestinal diseases. In the food industry, XOS can be used as sugar and/or fat substitutes, acting as preserving, stabilizing, and gelling agents (Zhang et al. 2017).

9.5.3.3 Organic Acids

Organic acids are one of the main products in the chemical platform of biorefineries. The biotechnological production of these compounds is preferable to traditional methods, in view of the high selectivity and purity, cost–benefit, working in the optimized conditions, and being more ecological. Organic acids have a wide industrial application, standing out in the sectors of food and beverages, pharmaceutical products, textiles, detergents, solvents, dyes, adhesives, rubbers, etc. Many organic acids such as levulinic acid, acetic acid, lactic acid, and succinic acid, can be produced from renewable raw materials, such as lignocellulosic biomasses, which provide sugars that act as a source of carbon for the most diverse microorganisms that, through different metabolic routes, produce these organic molecules (Rajendra Singh et al. 2017).

Levulinic acid is a chain fatty acid linked with a ketone carbonyl and a carboxylic acid group. This organic molecule is recognized as one of the chemicals with the highest added value because it is a promising precursor to other valuable chemicals such as γ-valerolactone (GVL), methyltetranhydrofuran (MTHF), hydroxymethylfurfural (HMF), ethyl levulinate, succinic acid, 2-pentanol, and 2-butanol (Kumar et al. 2019). Levulinic acid can be produced from glucose, fructose, sucrose, and xylose, which can be found in the biomass of agricultural residues, such as rice straw, rice husk, wheat straw, corn straw, bagasse, grass, etc. On an industrial scale, levulinic acid is produced through various routes, such as synthesis, biological, and extraction (Eda et al. 2018). The biological route through fermentation is preferred by researchers, mainly because it avoids the use of large amounts of chemical conductors. This organic acid compound is essential for several industries, such as the oil industry; the pharmaceutical industry due to its applications in biofuels, capsules, tablets and injections; the food industry, as a flavor regulator; and the agricultural industry in plant cultivation (Kumar et al. 2019).

Lactic acid is an important bioproduct used in a wide range of applications in the food, pharmaceutical, cosmetic, and leather industries. It has great potential to produce biodegradable and biocompatible polylactic acid polymer, which has vital applications in the pharmaceutical industry for the construction of prosthetic devices, sutures, and internal dosage of medicines. The acquisition of lactic acid can be carried out by chemical methods or by fermentation. Fermentative production is more advantageous because, in addition to producing a racemic mixture of lactic acid, this type of production uses sugars from low-cost renewable substrates and minimizes energy consumption. The lactic acid bacteria most used for the commercial production of lactic acid belong to the genus *Lactobacillus*, such as *Lactobacillus rhamnosus, Lactobacillus casei, Lactiplantibacillus plantarum, Lactobacillus pentosus*, and *Lactobacillus delbrueckii*. These bacteria consist of one of the most industrially important groups, since most of them are Generally Considered Safe (GRAS) and are used to efficiently convert sugars derived from various raw materials into lactic acid (Karnaouri et al. 2020).

Succinic acid is defined as a dicarboxylic acid composed of four carbons. Today, this important molecule is being recognized as the foundation for many other important chemicals applied in the food, chemical, and pharmaceutical industries. It has great application potential in the manufacture of polybutylene succinate, surfactants, detergents, flavors and fragrances, herbicides, fungicides, and food additives. Currently, most of this acid is produced from fossil fuels. However, recent studies have demonstrated the efficient production of this important acid from renewable raw materials, making the process more economical and environmentally friendly (Kumar et al. 2020). Naturally, bio-succinic acid is formed by plants, animals, and microorganisms as a fermentation product. However, the greatest production is achieved by anaerobic microbial fermentation. Several microorganisms can be used to produce this acid, among them the bacterium *Actinomycetemcomitans succinogenes* is known as one of the most promising microorganisms for industrial production, for producing relatively large quantities and for being able to use a wide range of carbon sources including lactose, xylose, arabinose, cellobiose, and other reduced sugars, to produce succinic acid as the main end product. In addition, several renewable and inexpensive resources have been used by *Actinobacillus succinogenes* to produce bio-succinic acid, especially fruit peels and husks, sugarcane bagasse, corn straw and fiber, rapeseed flour, wheat bran, bread residues, corn stalk, cotton stalk, cane molasses, fresh cassava root lentils, and cheese whey, among others (Putri et al. 2020).

9.5.3.4 Sweeteners

Currently, a significant portion of the population is seeking healthier lifestyle habits. As a result, researchers are looking for new ingredients that can be applied to food products to meet these requirements (Hong et al. 2019). In this scenario, the production of sugar substitutes, such as xylitol and arabitol, has become a target of research by scientists worldwide.

Xylitol is a type of alcohol pentahydrate (polyol) that is characterized as a white, crystalline, odorless, non-toxic powder with a sweet taste. It is mainly used as a sweetener, replacing or reducing the amount of sucrose in formulations, but it can also be used as a preservative and stabilizing agent in foods. Another aspect of the great relevance of this substance is its high anticariogenic action, in addition to helping combat various physiological dysfunctions of the human organism. Due to all these characteristics, xylitol has a high added value, presenting great interest in food and pharmaceuticals (Dasgupta et al. 2017). Biotechnological production of xylitol on a large scale usually occurs through fermentative processes, using microorganisms such as bacteria, yeasts, and filamentous fungi capable of assimilating pentoses, as a substrate. Among these microorganisms, yeasts of the genus *Candida* sp. present great potential to be used in these bioprocesses. This large-scale production is influenced by operational parameters such as pH, temperature, initial xylose concentration, and oxygen transfer rate (Tamburini et al. 2015).

Arabitol is another important polyol that is widely used in the food, chemical, and pharmaceutical industries. It has a sweetness similar to sucrose, low in calories ($0.2\,kcal\,g^{-1}$) and can be absorbed slowly by the human digestive tract. Like xylitol, it also has an important anticariogenic action for oral health, in addition to being able to compose products and medicines for diabetics. In addition, D-arabitol has been used as a promising raw material for the biosynthesis of some value-added products, including xylitol, arabinoic acid, xylonic acid, propylene, ethylene glycol, enantiopure compounds, immunosuppressive glycolipids, herbicide formulations, etc. In the last few decades, arabitol has been biosynthesized from different substrates or hydrolysates derived from lignocellulosic biomasses, composed of glucose, xylose, crude glycerol, arabinose, fructose, xylose, and galactose. A wide variety of microorganisms, mainly osmophilic yeasts, have been reported to produce arabitol from different substrates, including *Candida*, *Pichia*, *Debaryomyces*, and *Zygosaccharomyces* species (Guo et al. 2019).

9.5.3.5 Biogas

Biogas is produced by anaerobic digestion of organic raw material through a process mediated by microorganisms, which occurs in four stages: hydrolysis, acidogenesis, acetogenesis, and methanogenesis. It is mainly composed of methane (40%–65% v/v) and CO_2 (35%–55% v/v) with small amounts of hydrogen sulfide (H_2S) (0.1%–3% v/v), moisture, and other traces of contaminants depending on the raw material (Kapoor et al. 2020). Several lignocellulosic biomasses can be used as an organic material for this anaerobic digestion process, such as rice straw, corn straw, sunflower residues, wheat straw, etc., requiring some pretreatment, such as diluted sulfuric acid, sodium hydroxide, and ILs in order to reduce the recalcitrance of biomass, removing lignin and consequently increasing access to cellulosic and hemicellulosic fractions (Mirmohamadsadeghi et al. 2021). Choosing the inoculum as the right microbiological community is also important to reduce the process time (15–40 days) and increase biogas production. Furthermore, it is important to have a balance between the abundance of lignocellulolytic and methanogenic bacteria in the inoculum to achieve improved productivity (Vu et al. 2020).

9.5.3.6 Vanillin

Vanillin (4-hydroxy-3-methoxybenzaldehyde) is the main component of the vanilla aroma, presenting a sweet and creamy odor. Its flavor and fragrance properties have made this compound important worldwide, especially for the food and pharmaceutical industries. Vanillin can be obtained from vanilla orchid grains via extraction. However, vanillin yield is very low because it accumulates at low levels in the plant. The biotechnological production of vanillin using microorganisms has attracted much attention in recent years because the higher yields are obtained using lignocellulosic biomasses, in the optimized conditions (Furuya et al. 2017). The biochemical route for obtaining vanillin has a precursor ferulic acid (4-hydroxy-3-methoxycinnamate), which is a phenolic compound that is linked to the cell wall of plant matrices through ester bonds and can be released through enzymatic reactions. Some strains of microorganisms, such as the bacterium *Bacillus subtilis* and the yeast *Streptomyces sannanensis*, as well as agro-industrial residues, such as wheat bran, have shown success in the production of vanillin from ferulic acid esters in bioreactors (de Albuquerque et al. 2019).

Acknowledgment

The authors thank Espaço da Escrita, Pró-Reitoria de Pesquisa, UNICAMP, for the language services provided.

References

Agbor, V.B., Cicek, N., Sparling, R. et al. (2011). Biomass pretreatment: fundamentals toward application. *Biotechnol. Adv.* 29 (6)): 675–685. https://doi.org/10.1016/j.biotechadv.2011.05.005.

de Albuquerque, T.L., de Silva, J.S., de Macedo, A.C. et al. (2019). Biotechnological strategies for the lignin-based biorefinery valorization. In: *Reference Module in Chemistry, Molecular Sciences and Chemical Engineering*. Elsevier https://doi.org/10.1016/b978-0-12-409547-2.14570-6.

Ali, N., Zhang, Q., Liu, Z.Y. et al. (2020). Emerging technologies for the pretreatment of lignocellulosic materials for bio-based products. *Appl. Microbiol. Biotechnol.* 104 (2): 455–473. https://doi.org/10.1007/s00253-019-10158-w.

Alizadeh, H., Teymouri, F., Gilbert, T.I. et al. (2005). Pretreatment of switchgrass by ammonia fiber explosion (AFEX). *Appl. Biochem. Biotechnol.* **124**: 1133–1141. https://doi.org/10.1385/ABAB:124:1-3:1133.

Anuj, K., Shende, D.Z., and Wasewar, K.L. (2019). Production of levulinic acid: a promising building block material for pharmaceutical and food industry. *Mat. Today: Proc.* 29: 790–793. https://doi.org/10.1016/j.matpr.2020.04.749.

Ávila, P.F., Forte, M.B.S., and Goldbeck, R. (2018). Evaluation of the chemical composition of a mixture of sugarcane bagasse and straw after different pretreatments and their effects on commercial enzyme combinations for the production of fermentable sugars. *Biomass Bioenergy* 116: 180–188. https://doi.org/10.1016/j.biombioe.2018.06.015.

Ávila, P.F., Cairo, J.P.L.F., Damasio, A. et al. (2020). Xylooligosaccharides production from a sugarcane biomass mixture: effects of commercial enzyme combinations on bagasse/straw hydrolysis pretreated using different strategies. *Food Res. Int.* 128 (September 2019): 108702. https://doi.org/10.1016/j.foodres.2019.108702.

Barbosa, F.C., Martins, M., Brenelli, L.B. et al. (2020). Screening of potential endoglucanases, hydrolysis conditions and different sugarcane straws pretreatments for cello-oligosaccharides production. *Bioresour. Technol.* 316: 1–8. https://doi.org/10.1016/j.biortech.2020.123918.

Baruah, J., Nath, B.K., Sharma, R. et al. (2018). Recent trends in the pretreatment of lignocellulosic biomass for value-added products. *Front. Ener. Res.* 6 (141): https://doi.org/10.3389/fenrg.2018.00141.

Bhatia, S.K., Joo, H.S., and Yang, Y.H. (2018). Biowaste-to-bioenergy using biological methods—a mini-review. *Energy Conv. Manage.* 177: 640–660. https://doi.org/10.1016/j.enconman.2018.09.090.

Bhatia, S.K., Jagtap, S.S., Bedekar, A.A. et al. (2020). Recent developments in pretreatment technologies on lignocellulosic biomass: effect of key parameters, technological improvements, and challenges. *Bioresour. Technol.* 300: 1–13. https://doi.org/10.1016/j.biortech.2019.122724.

Bichot, A., Delgenès, J.P., Méchin, V. et al. (2018). Understanding biomass recalcitrance in grasses for their efficient utilization as biorefinery feedstock. *Rev. Environ. Sci. Biotechnol.* 17 (4)): 707–748. https://doi.org/10.1007/s11157-018-9485-y.

Bozell, J.J. and Petersen, G.R. (2010). Technology development for the production of biobased products from biorefinery carbohydrates—the US Department of Energy's "top 10" revisited. *Green Chem.* 12 (4): 539–555. https://doi.org/10.1039/b922014c.

Brandt, A., Gräsvik, J., Hallett, J.P., and Welton, T. (2013). Deconstruction of lignocellulosic biomass with ionic liquids. *Green Chem.* 15 (3)): 550–583. https://doi.org/10.1039/c2gc36364j.

Brodeur, G., Yau, E., Badal, K. et al. (2011). Chemical and physicochemical pretreatment of lignocellulosic biomass: a review. *Enzyme Res.* 2011 (1): 1–17. https://doi.org/10.4061/2011/787532.

Caspeta, L., Castillo, T., and Nielsen, J. (2015). Modifying Yeast Tolerance to Inhibitory Conditions of Ethanol Production Processes. *Front. Bioeng. Biotechnol.* 3: 184. https://doi.org/10.3389/fbioe.2015.00184.

Champreda, V., Mhuantong, W., Lekakarn, H. et al. (2019). Designing cellulolytic enzyme systems for biorefinery: from nature to application. *J. Biosci. Bioeng.* 128 (6): 637–654. https://doi.org/10.1016/j.jbiosc.2019.05.007.

Cherubini, F. and Strómman, A.H. (2010). Production of biofuels and biochemicals from lignocellulosic biomass: estimation of maximum theoretical yields and efficiencies using matrix algebra. *Energy Fuels* 24 (4): 2657–2666. https://doi.org/10.1021/ef901379s.

Claes, A., Deparis, Q., Foulquié-Moreno, M.R., and Thevelein, J.M. (2020). Simultaneous secretion of seven lignocellulolytic enzymes by an industrial second-generation yeast strain enables efficient ethanol production from multiple polymeric substrates. *Metabol. Eng.* *59*: 131–141. https://doi.org/10.1016/j.ymben.2020.02.004.

Clough, M.T., Griffith, J.A., Kuzmina, O., and Welton, T. (2016). Enhancing the stability of ionic liquid media for cellulose processing: acetal protection or carbene suppression? *Green Chem.* *18* (13): 3758–3766. https://doi.org/10.1039/C6GC00027D.

Crestini, C., Melone, F., and Saladino, R. (2011). Novel multienzyme oxidative biocatalyst for lignin bioprocessing. *Bioorg. Med. Chem.* 19 (16): 5071–5078. https://doi.org/10.1016/J.BMC.2011.05.058.

Cui, S., Zhang, S., Ge, S. et al. (2016). Green preparation and characterization of size-controlled nanocrystalline cellulose via ultrasonic-assisted enzymatic hydrolysis. *Ind. Crops Prod.* 83: 346–352. https://doi.org/10.1016/j.indcrop.2016.01.019.

Czaikoski, A., da Cunha, R.L., and Menegalli, F.C. (2020). Rheological behavior of cellulose nanofibers from cassava peel obtained by combination of chemical and physical processes. *Carbohydr. Polym.* 248: 116744. https://doi.org/10.1016/j.carbpol.2020.116744.

D'Adamo, I., Falcone, P.M., Imbert, E., and Morone, P. (2020). A socio-economic indicator for EoL strategies for bio-based products. *Ecol. Econ.* 178: 106794. https://doi.org/10.1016/j.ecolecon.2020.106794.

Dasgupta, D., Bandhu, S., Adhikari, D.K., and Ghosh, D. (2017). Challenges and prospects of xylitol production with whole cell bio-catalysis: a review. *Microbiol. Res.* 197: 9–21. https://doi.org/10.1016/j.micres.2016.12.012.

Eda, S., Borra, A., Parthasarathy, R. et al. (2018). Recovery of levulinic acid by reactive extraction using tri-*n*-octylamine in methyl isobutyl ketone: equilibrium and thermodynamic studies and optimization using Taguchi multivariate approach. *Sep. Purif. Technol.* 197: 314–324. https://doi.org/10.1016/j.seppur.2018.01.014.

Escobar, N. and Laibach, N. (2021). Sustainability check for bio-based technologies: a review of process-based and life cycle approaches. In: *Renew. Sustain. Energ. Rev.*, vol. 135. Amsterdan: https://doi.org/10.1016/j.rser.2020.110213.

Esteves, B., Graca, J., and Pereira, H. (2008). Extractive composition and summative chemical analysis of thermally treated eucalypt wood. *Holzforschung 62* (3): 344–351. https://doi.org/10.1515/HF.2008.057.

Ferrari, F.A., Pereira, J.F.B., Witkamp, G.J., and Forte, M.B.S. (2019). Which variables matter for process design and scale-up? A study of sugar cane straw pretreatment using low-cost and easily synthesizable ionic liquids. *ACS Sust. Chem. Eng.* *7* (15): 12779–12788. https://doi.org/10.1021/acssuschemeng.9b01385.

Filho, E.X.F., de Souza Moreira, L.R., de Aquino Ximenes, E., and Farinas, C.S. (2020). *Recent Advances in Bioconversion of Lignocellulose to Biofuels and Value Added Chemicals Within the Biorefinery Concept*. Amsterdam: Elsevier https://books.google.com.br/books?id=bYfiDwAAQBAJ.

Francois, J.M., Alkim, C., and Morin, N. (2020). Engineering microbial pathways for production of bio-based chemicals from lignocellulosic sugars: current status and perspectives. *Biotechnol. Biofuels* 13 (1): https://doi.org/10.1186/s13068-020-01744-6.

Furtado, G.P., Ribeiro, L.F., Lourenzoni, M.R., and Ward, R.J. (2013). A designed bifunctional laccase/β-1,3-1,4-glucanase enzyme shows synergistic sugar release from milled sugarcane bagasse. *Pro. Eng, Des. Sel.* 26 (1): 15–23. https://doi.org/10.1093/protein/gzs057.

Furuya, T., Kuroiwa, M., and Kino, K. (2017). Biotechnological production of vanillin using immobilized enzymes. *J. Biotechnol.* 243: 25–28. https://doi.org/10.1016/j.jbiotec.2016.12.021.

Goldbeck, R., Damásio, A.R.L., Gonçalves, T.A. et al. (2014). Development of hemicellulolytic enzyme mixtures for plant biomass deconstruction on target biotechnological applications. *Appl. Microbiol. Biotechnol.* 98 (20): 8513–8525. https://doi.org/10.1007/s00253-014-5946-6.

Guo, Q., Zabed, H., Zhang, H. et al. (2019). Optimization of fermentation medium for a newly isolated yeast strain (*Zygosaccharomyces rouxii* JM-C46) and evaluation of factors affecting biosynthesis of d-arabitol. *LWT* 99: 319–327. https://doi.org/10.1016/j.lwt.2018.09.086.

Hess, J.R., Lamers, P., Stichnothe, H., and Beermann, M. (2016). Chapter 1. Bioeconomy strategies. In: *Developing the Global Bioeconomy*, 1–9. Amsterdam: Elsevier https://doi.org/10.1016/B978-0-12-805165-8.00001-X.

Himmel, M.E., Xu, Q., Luo, Y. et al. (2010). Microbial enzyme systems for biomass conversion: emerging paradigms. *Biofuels 1* (2): 323–341. https://doi.org/10.4155/bfs.09.25.

Hong, C., Corbett, D., Venditti, R. et al. (2019). Xylooligosaccharides as prebiotics from biomass autohydrolyzate. *LWT* 111: 703–710. https://doi.org/10.1016/j.lwt.2019.05.098.

Hurtado-Romero, A., Del Toro-Barbosa, M., Garcia-Amezquita, L.E., and García-Cayuela, T. (2020). Innovative technologies for the production of food ingredients with prebiotic potential: modifications, applications, and validation methods. *Trends Food Sci. Technol* 104: 117–131. https://doi.org/10.1016/j.tifs.2020.08.007.

Isikgor, F.H. and Becer, C.R. (2015). Lignocellulosic biomass: a sustainable platform for the production of bio-based chemicals and polymers. *Polym. Chem.* 6 (25): 4497–4559. https://doi.org/10.1039/c5py00263j.

Jayani, R.S., Saxena, S., and Gupta, R. (2005). Microbial pectinolytic enzymes: a review. *Process Biochem.* 40 (9): 2931–2944. https://doi.org/10.1016/J.PROCBIO.2005.03.026.

Johansen, K.S. (2016). Lytic polysaccharide monooxygenases: the microbial power tool for lignocellulose degradation. *Trends Plant Sci.* 21 (11): 926–936. https://doi.org/10.1016/j.tplants.2016.07.012.

Kapoor, R., Ghosh, P., Tyagi, B. et al. (2020). Advances in biogas valorization and utilization systems: a comprehensive review. *J. Cleaner Prod.* 273: 123052. https://doi.org/10.1016/j.jclepro.2020.123052.

Karnaouri, A., Asimakopoulou, G., Kalogiannis, K.G. et al. (2020). Efficient d-lactic acid production by *Lactobacillus delbrueckii* subsp. *bulgaricus* through conversion of organosolv pretreated lignocellulosic biomass. *Biomass Bioenergy* 140: 105672. https://doi.org/10.1016/j.biombioe.2020.105672.

Kim, T.H., Lee, Y.Y., Sunwoo, C., and Kim, J.S. (2006). Pretreatment of corn stover by low-liquid ammonia recycle percolation process. *Appl. Biochem. Biotechnol. 133* (1): 41–57. https://doi.org/10.1385/ABAB:133:1:41.

King, J.W. and Srinivas, K. (2009). Multiple unit processing using sub- and supercritical fluids. *J. Supercrit. Fluids 47* (3): 598–610. https://doi.org/10.1016/j.supflu.2008.08.010.

Kumar, R., Basak, B., and Jeon, B.H. (2020). Sustainable production and purification of succinic acid: a review of membrane-integrated green approach. *J. Cleaner Prod.* 277: 123954. https://doi.org/10.1016/j.jclepro.2020.123954.

Lange, J. (2007). Lignocellulose conversion: an introduction to chemistry. *Biofuels, Bioprod. Biorefin.* 1: 39–48. https://doi.org/10.1002/bbb.7.

Levasseur, A., Drula, E., Lombard, V. et al. (2013). Expansion of the enzymatic repertoire of the CAZy database to integrate auxiliary redox enzymes. *Biotechnol. Biofuels* 6 (1): 41. https://doi.org/10.1186/1754-6834-6-41.

Lopes, A.M., Filho, E.X.F., and Moreira, L.R.S. (2018). An update on enzymatic cocktails for lignocellulose breakdown. *J. Appl. Microbiol.* 125: 632–645. https://doi.org/10.1111/jam.13923.

López-Porfiri, P., Gorgojo, P., and Gonzalez-Miquel, M. (2020). Green solvent selection guide for biobased organic acid recovery. *ACS Sust. Chem. Eng.* 8 (24): 8958–8969. https://doi.org/10.1021/acssuschemeng.0c01456.

Martín, C. and Thomsen, A.B. (2007). Wet oxidation pretreatment of lignocellulosic residues of sugarcane, rice, cassava and peanuts for ethanol production. *J. Chem. Technol. Biotechnol. 82* (2): 174–181. https://doi.org/10.1002/jctb.1648.

Martins, M., Ávila, P.F., de Andrade, C.C.P., and Goldbeck, R. (2020). Synergic recombinant enzyme association to optimize xylo-oligosaccharides production from agricultural waste. *Biocatal. Agricult. Biotechnol. 28*: 101747. https://doi.org/10.1016/j.bcab.2020.101747.

Mirmohamadsadeghi, S., Karimi, K., Azarbaijani, R. et al. (2021). Pretreatment of lignocelluloses for enhanced biogas production: a review on influencing mechanisms and the importance of microbial diversity. *Renew. Sust. Energ. Rev.* (135): –110173. https://doi.org/10.1016/j.rser.2020.110173.

Mohan, S.V., Varjani, S., Pant, D. et al. (2020). Bioresource technology. *Bioresour. Technol.* 318 (September): 124084. https://doi.org/10.1016/j.biortech.2020.124084.

Moon, R.J., Martini, A., Nairn, J. et al. (2011). Cellulose nanomaterials review: structure, properties and nanocomposites. *Chem. Soc. Rev.* 40 (7): 3941–3994. https://doi.org/10.1039/c0cs00108b.

Moreira, L.R.S., Milanezi, N.vG., and Filho, E.X.F. (2011). Enzymology of plant cell wall breakdown: an update. In: *Routes to Cellulosic Ethanol* (ed. M. Buckeridge and G. Goldman), 73–96. New York: Springer https://doi.org/10.1007/978-0-387-92740-4_6.

Mosier, N., Wyman, C., Dale, B. et al. (2005). Features of promising technologies for pretreatment of lignocellulosic biomass. *Bioresour. Technol.* 96 (6): 673–686. https://doi.org/10.1016/j.biortech.2004.06.025.

Obeng, E.M., Adam, S.N.N., Budiman, C. et al. (2017). Lignocellulases: a review of emerging and developing enzymes, systems, and practices. *Bioresour. Bioprocess.* 4 (1): 16. https://doi.org/10.1186/s40643-017-0146-8.

Olofsson, K., Bertilsson, M., and Lidén, G. (2008). A short review on SSF—an interesting process option for ethanol production from lignocellulosic feedstocks. *Biotechnol. Biofuels* **1**: 7. https://doi.org/10.1186/1754-6834-1-7.

Patel, A.K., Singhania, R.R., Sim, S.J., and Pandey, A. (2019). Thermostable cellulases: current status and perspectives. *Bioresour. Technol.* 279: 385–392. https://doi.org/10.1016/j.biortech.2019.01.049.

Pedersen, M. and Meyer, A.S. (2009). Influence of substrate particle size and wet oxidation on physical surface structures and enzymatic hydrolysis of wheat straw. *Biotechnol. Prog. 25* (2): 399–408. https://doi.org/10.1002/btpr.141.

Pires, J.R.A., Souza, V.G.L., and Fernando, A.L. (2019). Valorization of energy crops as a source for nanocellulose production—current knowledge and future prospects. *Ind. Crops Prod.* 140: 111642. https://doi.org/10.1016/j.indcrop.2019.111642.

Polizeli, M.L.T.M., Rizzatti, A.C.S., Monti, R. et al. (2005). Xylanases from fungi: properties and industrial applications. *Appl. Microbiol. Biotechnol.* 67: 577–591. https://doi.org/10.1007/s00253-005-1904-7.

Polizeli, M.L.T.M., Somera, A.F., de Lucas, R.C. et al. (2017). Enzymes involved in the biodegradation of sugarcane biomass: challenges and perspectives. In: *Advances of Basic Science for Second Generation Bioethanol from Sugarcane* (ed. M. Buckeridge and A. De Souza), 55–79. Denmark: Springer International Publishing https://doi.org/10.1007/978-3-319-49826-3_5.

Putri, D.N., Sahlan, M., Montastruc, L. et al. (2020). Progress of fermentation methods for bio-succinic acid production using agro-industrial waste by *Actinobacillus succinogenes*. *Ener. Rep.* 6: 234–239. https://doi.org/10.1016/j.egyr.2019.08.050.

Rocha, M.V.P., Rodrigues, T.H.S., De MacEdo, G.R., and Gonçalves, L.R.B. (2009). Enzymatic hydrolysis and fermentation of pretreated cashew apple bagasse with alkali and diluted sulfuric acid for bioethanol production. *Appl. Biochem. Biotechnol.* 155 (1–3): 407–417. https://doi.org/10.1007/s12010-008-8432-8.

Saadat, F. (2017). A review on chimeric xylanases: methods and conditions. *3 Biotech.* 7 (1): 1–7. https://doi.org/10.1007/s13205-017-0660-6.

Saha, B.C. (2003). Hemicellulose bioconversion. *J. Ind. Microbiol. Biotechnol.* 30 (5): 279–291. https://doi.org/10.1007/s10295-003-0049-x.

Salehi Jouzani, G. and Taherzadeh, M.J. (2015). Advances in consolidated bioprocessing systems for bioethanol and butanol production from biomass: a comprehensive review. *Biofuel Res. J. 2* (1): 152–195. https://doi.org/10.18331/BRJ2015.2.1.4.

Samanta, A.K., Jayapal, N., Jayaram, C. et al. (2015). Xylooligosaccharides as prebiotics from agricultural by-products: production and applications. *Bioact. Carbohydr. Dietary Fibre* 5 (1): 62–71. https://doi.org/10.1016/j.bcdf.2014.12.003.

Schacht, C., Zetzl, C., and Brunner, G. (2008). From plant materials to ethanol by means of supercritical fluid technology. *J. Supercrit. Fluids 46* (3): 299–321. https://doi.org/10.1016/j.supflu.2008.01.018.

Schuster, B.G. and Chinn, M.S. (2013). Consolidated bioprocessing of lignocellulosic feedstocks for ethanol fuel production. *Bioenergy Res.* 6 (2): 416–435. https://doi.org/10.1007/s12155-012-9278-z.

Segato, F., Damásio, A.R.L., De Lucas, R.C. et al. (2014). Genomics review of holocellulose deconstruction by aspergilli. *Microbiol. Molecul. Biol. Rev.* 78 (4): 588–613. https://doi.org/10.1128/MMBR.00019-14.

Shill, K., Padmanabhan, S., Xin, Q. et al. (2011). Ionic liquid pretreatment of cellulosic biomass: enzymatic hydrolysis and ionic liquid recycle. *Biotechnol. Bioeng.* 108 (3): 511–520. https://doi.org/10.1002/bit.23014.

Shinde, S.H., Hengne, A., and Rode, C.V. (2020). Lignocellulose-derived platform molecules. In: *Biomass, Biofuels, Biochemicals* (ed. S. Saravanamurugan, A. Pandey, H. Li and A. Riisager), 1–31. Amsterdam, Elsevier https://doi.org/10.1016/b978-0-444-64307-0.00001-9.

Silva, M.F., Menis-Henrique, M.E., Felisberto, M.H. et al. (2020). Bamboo as an eco-friendly material for food and biotechnology industries. In: *Curr. Opin. Food Sci.*, S. Saravanamurugan, A. Pandey, H. Li, and A. Riisager (eds.), vol. 33, 124–130. Amsterdan https://doi.org/10.1016/j.cofs.2020.02.008.

Sindhu, R., Binod, P., Mathew, A.K. et al. (2017). Development of a novel ultrasound-assisted alkali pretreatment strategy for the production of bioethanol and xylanases from chili post harvest residue. *Bioresour. Technol.* 242: 146–151. https://doi.org/10.1016/j.biortech.2017.03.001.

Singh, R., Shukla, A., Tiwari, S., and Srivastava, M. (2014). A review on delignification of lignocellulosic biomass for enhancement of ethanol production potential. *Renew. Sustain. Energ. Rev.* 32: 713–728. https://doi.org/10.1016/j.rser.2014.01.051.

Singh, R., Mittal, A., Kumar, M., and Mehta, P.K. (2017). Organic acids: an overview on microbial production. *Int. J. Adv. Biotechnol. Res.* 8 (1): 104–111.

Skrebets, T.E., Bogolitsyn, K.G., Kosyakov, D.S., and Verbitskaya, S.A. (2004). An IR study of organosolvent lignin. *Russian J. Appl. Chem. 77* (9): 1536–1539. https://doi.org/10.1007/s11167-005-0067-8.

Socha, A.M., Parthasarathi, R., Shi, J. et al. (2014). Efficient biomass pretreatment using ionic liquids derived from lignin and hemicellulose. *Proc. Natl Acad. Sci. 111* (35): E3587–E3595. https://doi.org/10.1073/pnas.1405685111.

Sun, Q., Pu, Y., Meng, X. et al. (2015). Structural transformation of isolated poplar and switchgrass lignins during dilute acid treatment. *ACS Sust. Chem. Eng.* 3 (9): 2203–2210. https://doi.org/10.1021/acssuschemeng.5b00426.

Taherzadeh, M.J. and Karimi, K. (2008). Pretreatment of lignocellulosic wastes to improve ethanol and biogas production: a review. *Int. J. Mol. Sci.* 9 (9)): 1621–1651. https://doi.org/10.3390/ijms9091621.

Tamburini, E., Costa, S., Marchetti, M.G., and Pedrini, P. (2015). Optimized production of xylitol from xylose using a hyper-acidophilic *Candida tropicalis*. *Biomol.* 5 (3): 1979–1989. https://doi.org/10.3390/biom5031979.

Thomas, P., Duolikun, T., Rumjit, N.P. et al. (2020). Comprehensive review on nanocellulose: recent developments, challenges and future prospects. *J. Mech. Behav. Biomed. Mater.* 110: 103884. https://doi.org/10.1016/j.jmbbm.2020.103884.

Valdivia, M., Galan, J.L., Laffarga, J., and Ramos, J.L. (2016). Biofuels 2020: biorefineries based on lignocellulosic materials. *Microbial Biotechnol. 9* (5): 585–594. https://doi.org/10.1111/1751-7915.12387.

Van Dyk, J.S. and Pletschke, B.I. (2012). A review of lignocellulose bioconversion using enzymatic hydrolysis and synergistic cooperation between enzymes—factors affecting enzymes, conversion and synergy. *Biotechnol. Adv.* 30 (6): 1458–1480. https://doi.org/10.1016/j.biotechadv.2012.03.002.

Vu, H.P., Nguyen, L.N., Vu, M.T. et al. (2020). A comprehensive review on the framework to valorise lignocellulosic biomass as biorefinery feedstocks. *Sci. Total Environ.* 743: 140630. https://doi.org/10.1016/j.scitotenv.2020.140630.

Wang, Y. and Liu, Q. (2020). Manganese-catalyzed dehydrogenative/deoxygenative coupling of alcohols. *Synlett.* 31 (15): 1464–1473. https://doi.org/10.1055/s-0040-1707126.

Wyman, C.E., Decker, S.R., Himmel, M.E. et al. (2005). Hydrolysis of cellulose and hemicellulose. In: *Polysaccharides: Structural Diversity and Functional Versatility* (ed. S. Dumitriu), 1023–1062. Boca Raton, FL: Taylor & Francis https://doi.org/10.1.1.519.9024.

Yamamoto, M., Iakovlev, M., and van Heiningen, A. (2014). The effect of chemical and physical characteristics of spruce SEW pulps on enzymatic hydrolysis. *Cellulose 21* (5): 3395–3407. https://doi.org/10.1007/s10570-014-0396-y.

Zhang, Z., Smith, C., and Li, W. (2014). Extraction and modification technology of arabinoxylans from cereal by-products: a critical review. *Food Res. Int. 65*: 423–436. https://doi.org/10.1016/j.foodres.2014.05.068.

Zhang, H., Xu, Y., and Yu, S. (2017). Bioresource technology co-production of functional xylooligosaccharides and fermentable sugars from corncob with effective acetic acid prehydrolysis. *Bioresour. Technol.* 234: 343–349. https://doi.org/10.1016/j.biortech.2017.02.094.

Zheng, Y., Pan, Z., and Zhang, R. (2009). Overview of biomass pretreatment for cellulosic ethanol production. *Int. J. Agric. Biol. Eng. 2* (3): 51–68. https://doi.org/10.3965/j.issn.1934-6344.2009.03.051-068.

Zhu, Y., Lee, Y.Y., and Elander, R.T. (2005). Optimization of Dilute-Acid Pretreatment of Corn Stover Using a High-Solids Percolation Reactor. *Appl. Biochem. Biotechnol.* 124 (1–3): 1045–1054.

Zhu, W., Zhu, J.Y., Gleisner, R., and Pan, X.J. (2010). On energy consumption for size-reduction and yields from subsequent enzymatic saccharification of pretreated lodgepole pine. *Bioresour. Technol.* 101 (8): 2782–2792. https://doi.org/10.1016/j.biortech.2009.10.076.

10

Lignin Conversion through Biological and Chemical Routes

Marcos H. L. Silveira[1]*, Alain E. M. Mera*[2]*, Anuj Kumar Chandel*[2]*,*
and Eduardo A. Ribeiro[3]

[1] *Chemnnova Mhl Silveira Consultoria–Inovacao Para Novos Negocios LTDA, Centro Santa Rosa de Lima/SC, 88763-000, Brazil*
[2] *Department of Biotechnology, Engineering School of Lorena (EEL), University of São Paulo, Lorena/SP, 12.602-810, Brazil*
[3] *Instituto Federal de Santa Catarina, IFSC, Criciúma/SC, 88813-600, Brazil*

10.1 Introduction

Lignin is an organic macromolecule formed principally by polyphenols, which are crosslinked in different ways; it is synthesized by plants and constitutes a key structural material of vegetable support tissues. The three principal lignin precursors are *p*-hydroxyphenyl (H), guaicyl (G) and syringyl (S), which have different units of the methoxy radical substituted in the aromatic ring in positions 4 or 6.

Considering that lignin is the second most abundant entity on earth (Ikeda et al. 2002; Hoffmann et al. 2019) its conversion into valuable products represents a relevant economic and environmental opportunity to be attained, and to achieve that currently two routes exist: the chemical and the biological approach. However, in both cases, the extraction of lignin is realized by cleavage of bonds in the lignin structure.

Lignin can be extracted from the biomass via the sulfite or kraft process. The first one includes the employment of solutions of sulfite or bisulfite ions, which cleave the bonds that link cellulose to lignin; and the second one implies the treatment of wood with NaOH, water, sodium sulfite, among other mechanical as well as chemical steps. Nowadays, about 90% of pulp production has been generated from the kraft pulping process (Dessbesell et al. 2020).

At an industrial level, the chemical removal of lignin from black liquor implies the acidification of black liquor to precipitate lignin that was alkali solubilized in the wood cooking, the filtration or dewatering of the solution to recover liquid side streams and some chemicals for cooking the liquor, and finally the extraction or purification of lignin and its by-products. At present there are two main technological approaches for lignin extraction from black liquor in industry, LignoBoost™ from Valmet and LignoForce™ from FPInnovations and NORAM.

Lignocellulose Bioconversion Through White Biotechnology, First Edition. Edited by Anuj Kumar Chandel.

On the other hand, the biological approach presents the advantage of integrating the processes of depolymerization and production of end-use products using lignin as feedstock, a process named Lignin Consolidate Bioprocessing (LCB) (Salvachúa et al. 2015). The biological approach implies the use of microbes or their derived chemicals to depolymerize lignin and to produce end-use products; in this sense, as fungi are the natural and principal degraders of lignin, they were mainly used to study the LCB of lignin. However, nowadays, bacteria have been more extensively studied due to various advantages compared to fungus. Despite all this, many aspects to fully utilize lignin still remain to be solved, and in this line the aim of the present chapter is to describe lignin conversion through current biological and chemical routes.

10.1.1 Lignin Availability

Currently, lignin is solubilized in the pulp mills by the use of alkaline cooking liquor in the presence of sulfur-containing molecules. By doing so, lignin is thus mostly solubilized to constitute black liquor.

If the industrial process is focused on dissolving pulp production instead of market pulp, a higher cellulose content is required, and therefore hemicellulose is extracted from the final pulp either by pre-hydrolysis or by alkaline extraction after delignification. In this case, alkaline extraction leads to a side stream of C5 enriched liquids that can be further recovered for use in biorefinery. On the other hand, when using a pre-hydrolysis, black liquor will contain also higher amounts of carbohydrates and furan compounds such as furan and 5-hydroxymethylfurfural from hexose and pentose degradation along with lignin carbohydrate complex solubilization.

Another important factor is the chemical structure of biomass macromolecules (cellulose, hemicellulose, and lignin), which varies according to the species and the growth environment of the plant. A practical example is observed in biorefineries; compared to other species, the levels of extractives produced by conifers and leafy plants are lower and reach values between 5% and 8% and 3%–10% respectively (de D'Almeida 1988).

10.1.2 Lignin Structure

Lignin is a macromolecule formed mostly by polyphenols. It is synthesized by plants and when associated with other polysaccharides forms a complex of high stability. The term lignin was coined by Anselme Payen when describing the soluble residue of the wood after treatment with concentrated nitric acid (Dos Santos 2001; Gutiérrez Villanueva et al. 2020).

Biomass represents the largest source for obtaining lignin, reaching around 20–30 wt%; the remaining fractions are essentially composed of hemicelluloses (20–30 wt%) and cellulose (40–50 wt%) (Shenoy et al. 2011; Ramsurn and Gupta 2012). However, these compositions may change according to the nature of the biomass. In plant tissue, lignin formation originates from precursors such as *p*-hydroxyphenyl (H), guaicyl (G) and syringyl (S), which have different units of the methoxy radical substituted in the aromatic ring in positions 4 and/or 6, as represented in Figure 10.1a.

From the polymerization of alcohols, cumaric ((E)-4-(3-hydroxyprop-1-en-1-yl) phenol), coniferyl ((E)-4-(3-hydroxyprop-1-en-1-yl)-2-methoxyphenol) and sinapyl ((E)-4-(3-hydroxyprop-1-en-1-yl)-2,6-dimethoxyphenol), as shown in Figure 10.1b (Howard

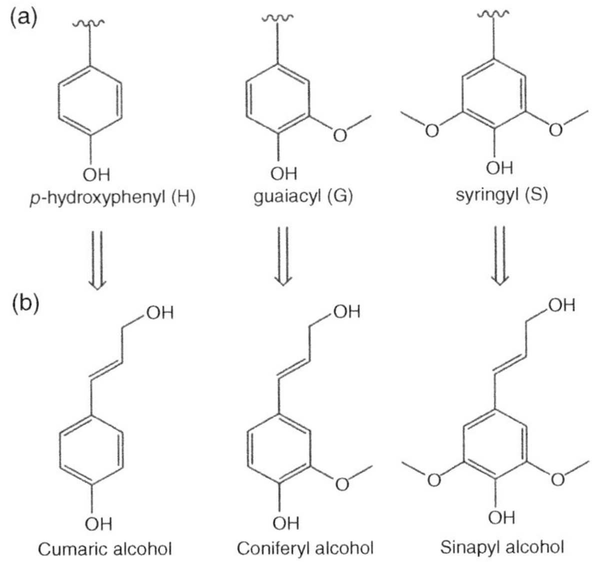

Figure 10.1 Chemical structures of alcohol precursors of lignin. *Source:* adapted from Sobrinho (2017).

et al. 2003; Fang et al. 2015), a lignin is formed. Therefore, due to the enormous possibilities of the combination of chemical linkages, a heterogeneous structure is formed, and it changes all along its distribution in the plant cell wall. That is why lignin-based biorefinery is such a challenge.

As mentioned above, in more detail the polymerization reaction occurs at random and leads to the formation of a complex three-dimensional branched structure and the availability of lignin precursors during lignin biosynthesis is the crucial driver that leads to the different types of lignin, classified as HGS, GS, and G.

The HGS-type lignin is mostly found in plants such as grasses (monocotyledonous angiosperm), in which there is a reduction in the non-methoxylated aromatic units that are derived from coumaryl alcohol. In more complex plants such as dicotyledonous or leafy angiosperm, also known as hardwood, lignin is classified as GS type and comes mostly from synaphyl alcohol. Type G lignin is found mostly in conifers, also called softwoods; they are made up of coniferous alcohol derivatives and phenylpropanoids that give rise to guaiacyl units (Ramos 2003).

The presence of lignin in the cell wall gives rigidity to the tissue, and thus acts as a protective barrier; the hydrophobic nature of lignin displaces water molecules which favors the formation of hydrogen bonds between hemicelluloses and cellulose (Northcote 1972), avoiding degradation by chemical and biological agents (Rubik 2013) besides, hampering its mechanical deconstruction.

In biomass, the increase in lignin concentration as well as the presence of G-type lignin increases the plant's resistance. In the case of G-type lignin, the hydroxyl present in the benzene ring acts as an ortho advisor on the ring since the para position is occupied by the aliphatic portion in the molecule, thus favoring the formation of C–C and C–O in these positions (Brunow and Lundquist 2010).

Other important evidence that corroborates the statement of G-type lignin resistance is due to the fact that its industrial removal from biomass is carried out by the use of strong nucleophiles, such as sulfide (S^{2-}) in the well-known processes kraft and sulfite (Zhu et al. 2010). The process of delignifying softwoods, such as Lodgepole pine wood and conifers, occurs with greater difficulty due to the high degree of G-type lignin present in the structure of these plants, which can present levels close to 95% (Brunow and Lundquist 2010).

The composition and contents of lignin present in these species are strongly related to physical characteristics such as color, physical and microbiological resistance, odor, in addition to abrasive properties (de D'Almeida 1988).

Another very expressive characteristic shown by the lignin is related to the structuring of their monomer units. While in celluloses and hemicelluloses the monomeric chains have a better arranged linear ordering, in lignin this is not observed. Lignin is an amorphous polymer; its three-dimensional structure is irregular (Lu et al. 2017) and thus the wide variety of functional groups and the different possibilities of C–C or C–O bonds present in the polymer reflect their greater stability. Figure 10.2 shows the main types of bonds present in lignin.

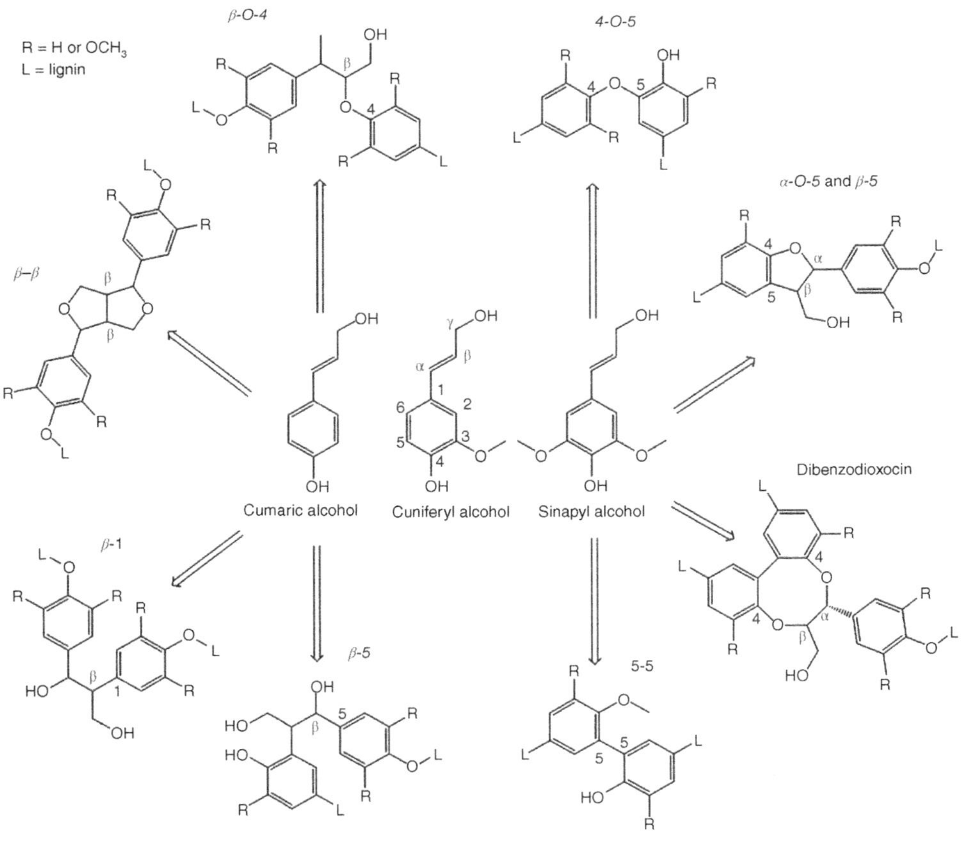

Figure 10.2 Main types of C–C and C–O bonds present in lignin. *Source:* adapted from Lu et al. (2017).

Table 10.1 Fractions of each type of bond.

Linkage	Fraction (%)
β–O–4	45–50
Dibenzodiocin, 5–5'	18–25
β–5	9–12
β–1	7–10
α–O–4	6–8
4–O–5	4–8
β–β	0–3

Source: adapted from Lu et al. (2017).

Among the different possibilities of C–C bonds formed in lignin are C3 of the aromatic ring (β–β', α–α', α–β') that occurs between aliphatic chains, links between aliphatic chains and aromatic rings of the type (β–5', β–1', α–1', β–6'), and finally between the aromatic ring carbons (5–5') (Higuchi 1985; Lu et al. 2017). The C–C bonds have less polarization compared to C–O bonds and thus have greater stability (Leal 2014).

The β–O–4 etheric bonds are the most expressive in lignin and represent about 50% of all the bonds present in lignin (Lee 1997); however, other bonds such as α–O–4' (Higuchi 1985) and β–1 (Lee 1997) have also been reported. In the cleavage of the β–O–4 bond there is formation of the hydroxyl group in the aliphatic structure of lignin (Du et al. 2020). In more recent studies the fractions of each type of bond have been determined in the lignin polymer. These values are shown in Table 10.1.

Due to the great heterogeneity and polydispersity demonstrated in the processing of lignin in the different industrial sectors, the application as a more attractive polymer still represents a great challenge to be overcome.

10.1.3 Chemical Transformation Routes

Lignin is the second most abundant polymer on earth (Ikeda et al. 2002; Hoffmann et al. 2019). The extraction of lignin is realized by cleavage of bonds in the lignin structure. There are many reports in the literature about this process, and one of the greatest reported difficulties is related to the large number of products that are generated in this step.

Black liquor, therefore, has been the target of research all over the world in the field of lignin recovery and its use as a precursors in new chemical routes, in which molecules with greater commercial value are desired.

Basically, from the initial studies proposed by Anselme Payen in the late 1830s (Dos Santos 2001) to the most recent, the industrial process has been based on three crucial stages: initially (i) black liquor is acidified in order to precipitate lignin that was alkali solubilized in the wood cooking; (ii) thus, filtration/dewatering takes place in order to recover a liquid side stream and possibly chemicals for cooking liquor; and finally (iii) a third stage where the extraction and purification of lignin and its derivatives occurs.

Today, about 90% of pulp production has been generated from the kraft pulping process (Dessbesell et al. 2020). Currently, there are two main technological approaches for lignin

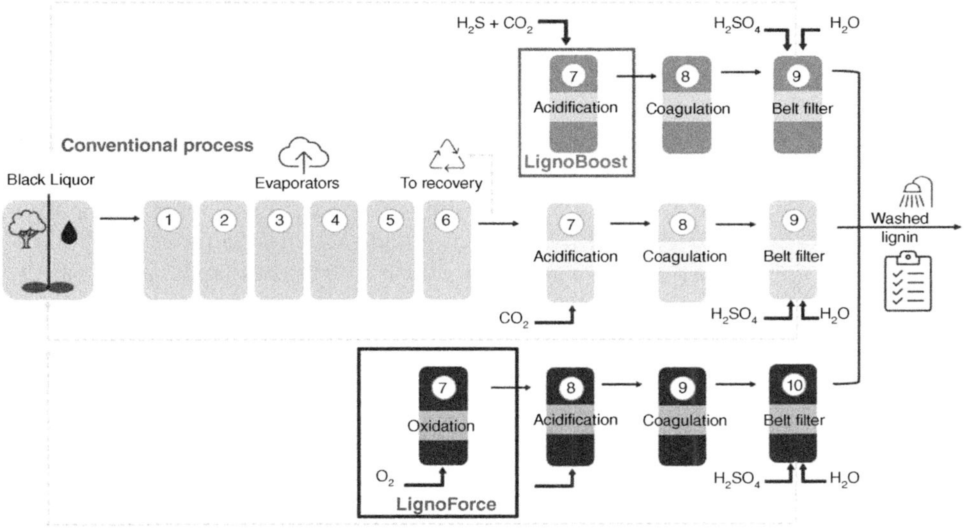

Figure 10.3 Comparison of principal routes for the obtention of lignin: conventional, LignoForce and LignoBoost process. Based on WO2009104995A1 and US 8771464.

extraction from black liquor: LignoBoost (WO2009104995A1) from Valmet and LignoForce from FPInnovations and NORAM (US 8771464), as briefly shown in Figure 10.3.

Both technologies are focused on the acid precipitation and filtration of lignin, but by using different agents. The LignoForce process is based on an initial use of oxygen to oxidize and remove sulfur compounds, followed by sulfuric acid precipitation. By doing so, this process tentatively avoids the formation of mercaptan-derived compounds, therefore reducing the recovered lignin toxicity for final use. On the other hand, LignoBoost is based on two successive acidifications with carbon dioxide and sulfuric acid, respectively.

Although black liquor is the same, these processes will deliver different compounds in terms of chemical functionalities. More than that, filtration and mass recovery efficiency are supposedly different.

10.1.4 Lignin Conversion by Biological Routes

The biological function of lignin is related to the formation of cell walls from vascular plants and some algae, and to fill space between cellulose, hemicellulose, and pectin components (Tumuluru et al. 2011). In the same way, many organisms have evolved mechanisms to degrade lignin and use the resulting aromatic compounds as sources of carbons and energy (Salvachúa et al. 2015). Considering that lignin is the second most abundant fount of aromatic compounds on the planet (Lee et al. 2019) its incorporation in the industrial production of energy and biomaterials is quite advantageous, both environmentally and economically; however, some technical challenges still remain, such as overcoming lignin recalcitrance, which is considered a hindrance for its exploitation. In this regard, biological, thermal, and chemical approaches have been proposed as solutions (Lee et al. 2019). The interest in lignin has increased because of the increased interest in biomass-based products due to the energy crisis and environmental concerns (Lee et al. 2019). Formerly, lignin was considered as an impediment to producing cellulose of higher quality; however, with the advent of the energy and environmental crisis, the demand for biomass-based products has increased dramatically, and thus lignin must be fully utilized.

Lignin can be depolymerized via thermal, chemical, or biological approaches (Akash 2016). The biological approach presents the advantage of simultaneous depolymerization and the production of end-use products from lignin LCB (Salvachúa et al. 2015). Second-generation ethanol is produced with cellulosic microbes, and thus a similar approach could be used to integrate depolymerization and product generation using lignin as source at large scales (Salvachúa et al. 2015). As fungi are the natural and principal degraders of lignin, they were mainly used to study the LCB of lignin; however, nowadays, bacteria have been more extensively studied due to various advantages compared to fungus: they are easier to grow, their metabolic routes are better known, and they are versatile organisms, that is, their metabolic routes can be engineered to produce a huge range of end-use products.

The natural depolymerization of lignin is produced by white or brown rot fungi using oxidative enzymes like laccase and peroxidases, and other enzymes which aid with the coupling or decoupling of monolignols (Lee et al. 2019). In addition, many bacteria secrete lignin oxidizing enzymes to degrade lignin, use lignin-derived and other chemicals presented in lignin-enriched culture and accumulate intracellular sources that can be used to produce industry relevant materials, chemicals, and fuels (Salvachúa et al. 2015). Considering the abundance of lignin on the planet (Lee et al. 2019) new biological methods are necessary to improve the utilization of lignin and the improvement of the present methods. It has been recommended that there must be a particular solution for every type of lignin (Balakshin et al. 2021).

Enzymes that break lignin are currently named lignin-modifying enzymes, formerly lignases or ligninases because they use oxidative rather than hydrolyzing mechanisms. These enzymes are produced by fungi and bacteria and are principally peroxidases, like lignin peroxidase, versatile peroxidase, and manganese peroxidase, and phenoloxidase laccases. Laccases (EC 1.10.3.2.) are multicopper oxidative enzymes which oxidize a variety of phenolic compounds via one-electron oxidation. Fungus laccases are principally monomers and dimers and are involved in lignin degradation through direct action in collaboration with enzymes such as peroxidases, or by detoxification by degradation of toxic phenols produced during lignin degradation.

Three functions have been attributed to laccases: (i) the formation of pigments, (ii) lignin degradation, and (iii) detoxification (Solomon et al. 1996). Fungal laccases are mainly monomers or dimers, glycosylated but to a more minor degree than plant laccases (10%–25% against 22%–45%), and they are extracellular. They have been associated with the functions of pigment formation, lignin degradation, and detoxification (Solomon et al. 1996). Laccases present low specificity and can act over *o*-quinols, *p*-quinols, phenylenediamine, aminophenols, and the semiquinone, which may react further enzymatically or non-enzymatically (Gasteiger et al. 2003).

On the other hand, peroxidases are enzymes that catalyze the breakdown or formation of peroxide bonds (R–O–O–R′) to ROH and R′OH, using an electron donor. Lignin peroxidase

(EC 1.11.1.14) catalyzes two reactions: (i) 1-(3,4-dimethoxyphenyl)-2-(2-methoxyphenoxy) propane-1,3-diol + H_2O_2 = 3,4-dimethoxybenzaldehyde + 2-methoxyphenol + glycolaldehyde + H_2O; and (ii) 2-(3,4-dimethoxyphenyl)methanol + H_2O_2 = 2-(3,4-dimethoxyphenyl) methanol radical + $2H_2O$.

Lignin peroxidase is a hemoprotein, where the heme iron is oxidized by hydrogen peroxide and later reduced by the substrate, thus forming an intermediate which decomposes spontaneously. This enzyme can act on a variety of aromatic substances; however, it cannot act directly over lignin because it does not fit in the active site (Iubmb 2021).

One of the most extensively studied catabolic pathways to utilize the heterogeneous mixture of aromatic compounds liberated by extracellular lignin oxidizing enzymes is the β-ketoadipate pathway, which proceeds via oxidative ring opening of central aromatic intermediates (Salvachúa et al. 2015). Alkaline pretreated liquor can be "biologically funneled" up to a central intermediate like acetyl coenzyme A (acetyl-CoA): the constituents *p*-coumaric acid, vanillic acid, and ferulic acid can be converted into a common intermediate like protocatechuic acid, which can be converted via the β-ketoadipate pathway to β-ketoadipate acid, and later to acetyl-CoA. Acetyl-CoA, as well as residual glucose and acetate, present in the culture, may be metabolized to medium chain length polyhydroxyalkanoate (*mcl*-PHA) via the fatty acid pathway, or to other end-use products (see Table 10.2) (Linger et al. 2014). Biomass-derived lignin can be used to produce chemicals, fuels, and material through natural aromatic catabolic pathways (Linger et al. 2014).

10.1.5 Potential Chemicals from Lignin

In order to reach an industrial application of lignin, it is required to remove it from biomass (fractionation). Therefore, in addition to being used as a raw material for pyrolysis of chemical resin, it is possible to use the chemical framework of lignin for future applications. In this case, it is necessary to fractionate black liquor, as well as upgrade lignin and target molecules.

To make a biorefinery viable, the lignin component of wood must be addressed. Lignin decomposes over a wider temperature range than cellulose and hemicellulose. Currently, residual lignin from paper pulping is burned off to generate heat and electricity (Stöcker 2008). In terms of upgrading, the outcome of the process will strongly depend on the method used, which can be (i) thermal, (ii) base-catalyzed, (iii) acid-catalyzed, (iv) reductive or (v) oxidative. There are numerous studies investigating the potential application of lignin (Chio et al. 2019).

Lignin is used as a potential feedstock to produce biobased polymers such as biodegradable (polyhydroxyalkanoates [medium chain PHAs], polyhydroxybutyrates [short chain PHAs], polylactic acid) and non-biodegradable (polyurethane, polyolefin, polyester, polyamide, and polyol) (Rajesh Banu et al. 2019) polymers, shown in Figure 10.4.

10.2 Conclusions

Around 90% of pulp production has been generated from the kraft pulping process, and lignin is extracted from black liquor by two technologies: LignoBoost and LignoForce. Lignin extraction from liquor implies its acidification, precipitation, filtration, extraction,

Table 10.2 Organisms that metabolize lignin and generate industrially relevant products. Based on LINGER et al., 2014.

Microorganism	Cultivation conditions	Effect on lignin structure	Products synthesized from lignin	References
Pseudomonas putida KT2440	Alkaline pretreated liquor (APL): lignin (32%), extractives (23%), inorganic components (11%), and acetate (8%), wt/wt	Transformation to the intermediate protocatechuic acid, then via the β-ketoadipate (β-KA) pathway to β-KA, next to acetyl-CoA, and finally to *mcl*-PHAs through the fatty acid pathway	Medium chain length polyhydroxyalkanoates (*mcl*-PHAs)	Linger et al. (2014)
Pseudomonas putida A514 carrying pTDV and pGJ4C1	Commercial insoluble kraft lignin	Lignin depolymerization and use of aromatic compounds	PHA	Lin et al. (2016)
Rhodococcus opacus PD630	Lignin fermentation medium (kraft lignin) plus laccase, at 28 °C for 144 h	Simultaneous depolymerization and fermentation of lignin	Lipid production including the biodiesel precursor triacylglycerol (TGA)	Zhao et al. (2016)
Cupriavidus brasiliensis B-8	APL and NaOH treated rice straw in mineral salt medium, at 30 °C and 150 rpm	Depolymerization of lignin and *in situ* bioconversion to PHA	PHA	Si et al. (2018)
Rhodococcus jostii RHA1	APL (90%) from corn stover, supplemented with medium 9 (M9) salts at 30 °C	Depolymerization of high molecular weight lignin species and catabolization of most low molecular weight aromatics	Fatty acid methyl esters (FAME) (18:1), FAME (17) and FAME (16)	Salvachúa et al. (2015)
Pseudomonas putida KT2440 (DSM 6125) with pCAT promoter	Minimal medium in shake flasks, pine lignin and demineralized water at 30 °C	Depolymerization of lignin and production of MA from a mixture of aromatics with catechol as the central element	*Cis,cis*-muconic acid (MA)	Kohlstedt et al. (2018)
Pseudomonas putida KT2440 engineered	Base-catalyzed depolymerization (BCD) liquor from corn stover	Consume main lignin-derived monomers, and converts lignin in up to 15%	Muconic acid	Rodriguez et al. (2017)

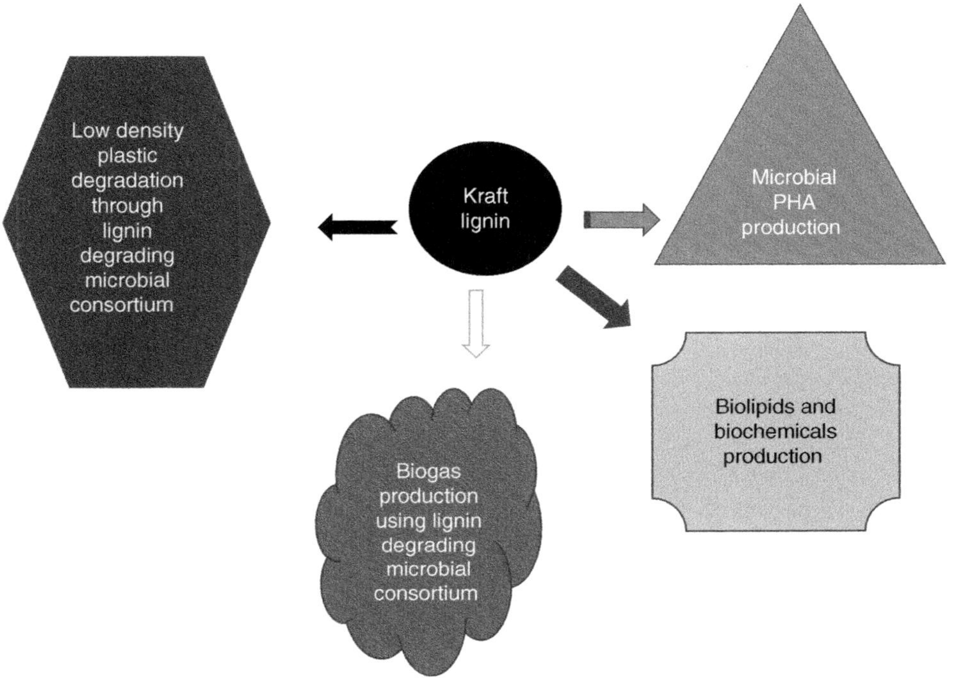

Figure 10.4 Lignin as a precursor of biobased products through lignin-based refineries. Based on Rajesh Banu et al., 2019.

and finally the purification of lignin. The biological approach presents the advantage of the simultaneous depolymerization and production of end-use products. Enzymes that break lignin are currently produced by fungi and bacteria and cover in particular peroxidase and phenoloxidase laccases. The resulting aromatic compounds of lignin depolymerization are "biologically funneled" up to a central intermediate like acetyl-CoA, which is metabolized, producing, on the way, various end-use products like *mcl*-PHA, biolipids, biogas, among others. Despite all this, many aspects to fully utilize lignin still remain to be solved.

On the other hand, the chemical transformation of lignin has been studied widely. Chemical conversion of lignin leads to the production of a myriad of valuable products such as polymers, abrasives, adhesives, adsorbents, surfactants, binders, dispersants, among others. Chem (bio) transformation of lignin has vast potential in the strengthening of a carbon neutral economy. However, the full potential of lignin is still to be unleashed at the commercial level.

Acknowledgements

A.K. Chandel is grateful to the São Paulo State Funding Agency (FAPESP) for the Grant Number 2020/12559-6 (jovem pesquisador project). A.K. Chandel is also thankful to the CNPq for the scientific productivity program (Process number: 309214/2021-1). A.E.M. Mera is thankful to the CAPES-Brazil for the doctoral fellowship.

References

Akash, B.A. (2016). Thermal depolymerization of lignin. *Int. J. Thermal Environm. Eng.* 13 (1): 17–22. https://doi.org/10.5383/ijtee.13.01.004.

Balakshin, M.Y., Capanema, E.A., Sulaeva, I. et al. (2021). New opportunities in the valorization of technical lignins, ChemSusChem. 14: 1–22. https://doi.org/10.1002/cssc.202002553.

Brunow, G. and Lundquist, K. (2010). Functional groups and bonding patterns in lignin (including the lignin-carbohydrate complexes). In: *Lignin and Lignans: Advances in Chemistry* (ed. C. Heitner, D. Dimmel and J. Schmidt), 267–299. CRC Press https://doi.org/10.1201/ebk1574444865-c7.

Chio, C., Sain, M., and Qin, W. (2019). Lignin utilization: a review of lignin depolymerization from various aspects. *Renew. Sust. Energ. Rev.* 107 (December 2018): 232–249. https://doi.org/10.1016/j.rser.2019.03.008.

de D'Almeida, M.L.O. (1988). *Celulose e Papel: Tecnologia de Fabricação do Papel*, 2e. São Paulo, IPT—Instituto de Pesquisa Tecnológica de São Paulo.

Dessbesell, L., Paleologou, M., Leitch, M. et al. (2020). Global lignin supply overview and kraft lignin potential as an alternative for petroleum-based polymers. *Renew. Sust. Energ. Rev.* 123 (February 2019): 109768. http://dx.doi.org/10.1016/j.rser.2020.109768.

Dos Santos, H.F. (2001). Análise conformacional de modelos de lignina. *Quim. Nova* 24 (4): 480–490. https://doi.org/10.1590/s0100-40422001000400009.

Du, B., Sun, Y., Liu, B. et al. (2020). From lignin-derived bio-oil to lignin-g-polyacrylonitrile nanofiber: high lignin substitution ratio and maintaining good nanofiber morphology. *Polym. Test.* 81: 106207. http://doi.org/10.1016/j.polymertesting.2019.106207.

Fang, Z., Smith, L.R., Qi, X. (eds) (2015). *Production of Biofuels and Chemicals with Ultrasound.* https://doi.org/10.1007/978-94-017-9624-8.

Gasteiger, E., Gattiker, A., Hoogland, C. et al. (2003). ExPASy: the proteomics server for in-depth protein knowledge and analysis. *Nucleic Acids Res.* 31: 3784–3788. https://doi.org/10.17226/11474.

Gutiérrez Villanueva, A., Guirola, C., Armas, A. et al. (2020). Valorización de la lignina en el concepto de biorrefinería (I). *Centro Azúcar* 47 (3): 95–105.

Higuchi, T. (1985). *Degradative Pathways of Lignin Model Compounds*, 557–578. Orlando, FL: Academic Press. ISBN: ISBN 01-234-78804 https://doi.org/10.1016/b978-0-12-347880-1.50024-6.

Hoffmann, A., Phuong, J., Porzel, A. et al. (2019). Modification of lignoboost kraft lignin from softwoods with dihydroxybenzenes. *React. Funct. Polym.* 142 (February): 112–118. http://doi.org/10.1016/j.reactfunctpolym.2019.06.011.

Howard, R.L., Abotsi, E., Van Rensburg, J. et al. (2003). Lignocellulose biotechnology: issues of bioconversion and enzyme production. *Afr. J. Biotechnol.* 2 (12): 602–619. https://doi.org/10.5897/ajb2003.000-1115.

Ikeda, T., Holtman, K., Kadla, J. et al. (2002). Studies on the effect of ball milling on lignin structure using a modified DFRC method. *J. Agric. Food Chem.* 50 (1): 129–135. https://doi.org/10.1021/jf010870f.

IUBMB (2021). *Enzyme Nomenclature.* London: International Union of Biochemistry and Molecular Biology www.qmul.ac.uk/sbcs/iubmb/enzyme.

Kohlstedt, M., Starck, S., Barton, N. et al. (2018). From lignin to nylon: cascaded chemical and biochemical conversion using metabolically engineered *Pseudomonas putida*. *Metab. Eng.* 47: 279–293. https://doi.org/10.1016/j.ymben.2018.03.003.

Leal, G.F. (2014). *Conversão catalítica de celulose utilizando catalisadores de carbeto de tungstênio suportado em carvão ativo e promovido por paládio*, vol. 110 f. Universidade de São Paulo.

Lee, J. (1997). Biological conversion of lignocellulosic biomass to ethanol. *J. Biotechnol.* 56 (1): 1–24. https://doi.org/10.1016/S0168-1656(97)00073-4.

Lee, S., Kang, M., Bae, J. et al. (2019). Bacterial valorization of lignin: strains, enzymes, conversion pathways, biosensors, and perspectives. *Front. Bioeng. Biotechnol.* 7 (September): 1–18. https://doi.org/10.3389/fbioe.2019.00209.

Lin, L., Cheng, Y., Pu, Y. et al. (2016). Systems biology-guided biodesign of consolidated lignin conversion. *Green Chem.* 18: 5536–5547. https://doi.org/10.1039/C6GC01131D.

Linger, J.G., Vardon, D., Guarnieri, M. et al. (2014). Lignin valorization through integrated biological funneling and chemical catalysis. *Proc. Natl. Acad. Sci. USA* 111 (33): 12013–12018. https://doi.org/10.1073/pnas.1410657111.

Lu, Y., Lu, Y.C., Hu, H. et al. (2017, 2017). Structural characterization of lignin and its degradation products with spectroscopic methods. *J. Spectrosc* 1–15. https://doi.org/10.1155/2017/8951658.

Northcote, D.H. (1972). Chemistry of the plant cell wall. *Annu. Rev. Plant Physiol.* 23 (1): 113–132. https://doi.org/10.1146/annurev.pp.23.060172.000553.

Rajesh Banu, J., Kavitha, S., Kannath, R. et al. (2019). A review on biopolymer production via lignin valorization. *Bioresour. Technol.* 290 (July): 121790. https://doi.org/10.1016/j.biortech.2019.121790.

Ramos, L.P. (2003). The chemistry involved in the steam treatment of lignocellulosic materials. *Quim. Nova* 26 (6): 863–871. https://doi.org/10.1590/s0100-40422003000600015.

Ramsurn, H. and Gupta, R.B. (2012). Production of biocrude from biomass by acidic subcritical water followed by alkaline supercritical water two-step liquefaction. *Energy Fuels* 26 (4): 2365–2375. https://doi.org/10.1021/ef2020414.

Rodriguez, A., Salvachúa, D., Katahira, R. et al. (2017). Base-catalyzed depolymerization of solid lignin-rich streams enables microbial conversion. ACS Sust. *Chem. Eng.* 5 (9): 8171–8180. https://doi.org/10.1021/acssuschemeng.7b01818.

Rubik, D.F. (2013). Pré-tratamento de biomassa lignocelulósica para facilitar a hidrólise enzimática visando a produção de bioetanol 2G, Master's thesis, Regional University of Blumenau, Department of Chemistry.

Salvachúa, D., Karp, E., Nimlos, C. et al. (2015). Towards lignin consolidated bioprocessing: simultaneous lignin depolymerization and product generation by bacteria. *Green Chem.* 17 (11): 4951–4967. https://doi.org/10.1039/c5gc01165e.

Shenoy, D., Pai, A., Vikas, R. et al. (2011). A study on bioethanol production from cashew apple pulp and coffee pulp waste. *Biomass Bioenergy* 35 (10): 4107–4111. https://doi.org/10.1016/j.biombioe.2011.05.016.

Si, M., Yan, X., Liu, M. et al. (2018). In-situ lignin bioconversion promotes complete carbohydrate conversion of rice straw by *Cupriavidus basilensis* B-8. *ACS Sust. Chem. Eng.* 6 (6): 7969–7978. https://doi.org/10.1021/acssuschemeng.8b01336.

Sobrinho, F.D.A.A.D.F. (2017). *Poliuretanos à Base de Lignina do endocarpo do coco seco como potencial sistema de liberação controlada de fertilizantes*. UFC, Centro de Ciências, Programa de Pós-graduação em Química, Dissertação.

Solomon, E.I., Sundaram, U.M., and Machonkin, T.E. (1996). Multicopper oxidases and oxygenases. *Chem. Rev.* 96 (7): 2563–2605. https://doi.org/10.1021/cr950046o.

Stöcker, M. (2008). Biofuels and biomass-to-liquid fuels in the biorefinery: catalytic conversion of lignocellulosic biomass using porous materials. *Angew. Chem.* (International Edition in English) 47 (48): 9200–9211. https://doi.org/10.1002/anie.200801476.

Tumuluru, J.S., Sokhansanj, S., Hess, J.R. et al. (2011). A review on biomass torrefaction process and product properties for energy applications. *Ind. Biotechnol.* 7 (5): 384–401. https://doi.org/10.1089/ind.2011.0014.

Zhao, C., Xie, S., Pu, Y. et al. (2016). Synergistic enzymatic and microbial lignin conversion. *Green Chem.* 18: 1306–1312. https://doi.org/10.1039/C5GC01955A.

Zhu, J.Y., Zhu, W., OBryan, P. et al. (2010). Ethanol production from SPORL-pretreated lodgepole pine: preliminary evaluation of mass balance and process energy efficiency. *Appl. Microbiol. Biotechnol.* 86 (5): 1355–1365. https://doi.org/10.1007/s00253-009-2408-7.

11

Downstream Processing in Lignocellulose Conversion

Current Challenges and Future Practices

Kelly J. Dussán[1,2,3], *Débora D. V. Silva*[4], *Ana F. M. Costa*[5], *Luana C. Grangeiro*[1], *and Ellen C. Giese*[6]

[1] *Department of Engineering, Physics and Mathematics, Institute of Chemistry, São Paulo State University (UNESP), Av. Prof. Francisco Degni, 55—Jardim Quitandinha, CEP 14800-060, Araraquara, SP, Brazil*
[2] *Bioenergy Research Institute (IPBEN), São Paulo State University (UNESP), Araraquara, SP, Brazil*
[3] *Center for Monitoring and Research of the Quality of Fuels, Biofuels, Crude Oil, and Derivatives, Institute of Chemistry (CEMPEQC), São Paulo State University (UNESP), Araraquara, SP, Brazil*
[4] *Department of Biochemistry and Organic Chemistry, Institute of Chemistry, São Paulo State University (UNESP), Araraquara, SP, Brazil*
[5] *Centre Nacional de la Recherche Scientifique (CNRS), Laboratory of Chemical Engineering (LGC) and Laboratory of Fundamental and Applied Heterochemistry (LHFA), Université Paul Sabatier Toulouse, Toulouse, France*
[6] *Service of Extractive Metallurgy and Bioprocesses, Centre for Mineral Technology (CETEM), Rio de Janeiro, RJ, Brazil*

11.1 Introduction

Between the years of 1973 and 2016, the industrial sector accounted for 29% of the total energy consumed in the world. In the same period, the use of fossil fuels as energy source by industries has increased 44%. While the use of renewable energy and waste sources has risen considerably between these years (128%), the use of coal has also increased (132%) and fossil fuels are still the dominating source consumed (International Energy Agency 2018). The Paris Agreement in 2015 drives global awareness to the use of renewable sources of energy and toward environmentally friendly industrial processes. The target set by the Paris Agreement was high; in it 195 countries committed to reducing greenhouse emissions by 30% by 2030. Since both fossil fuels and industrial processes account for 65% of global greenhouse gas emissions, these sectors offer great opportunity for the implementation of environmentally effective systems. It is recognized that, for nations around the world to achieve the targets set by the Paris Agreement, a shift from fossil fuels to renewable sources of energy and chemical production will be necessary. Concerning renewable sources, waste streams are interesting sources for green energy and (bio)-chemicals (Asunis et al. 2020).

In this sense, biorefineries are proposed as sustainable options to produce biobased molecules and to propose green process pathways. Biorefinery systems are described as a facility that can sustainably process renewable-based feedstock such as biomass or organic

Lignocellulose Bioconversion Through White Biotechnology, First Edition. Edited by Anuj Kumar Chandel.

waste to generate marketable products including energy, materials, and chemicals. Biorefineries are known for their adaptability and efficient use of resources as well as integration of units. In the refining of raw materials into multiple products, the design, operation, and control of the industrial plant is also dependent on multiple unit operations with different types of technology.

The different refinery processes are based on chemical reactions and physical separations, which themselves depend on a complexity of parameters such as temperature, pressure, electrical potential, catalytic activity, etc. The good functioning between these numerous processes and parameters is essential for the formation of the different final products and their commercialization. To enhance the efficiency of biorefineries, current challenges are related to the creation of methods to optimize scale-up problems, transform old factories into a smart plant, and better understand the microorganism metabolism in biological platforms.

Various concepts of biorefinery have been presented in the literature. However, the common objective between different biorefineries is the development of environmentally friendly and economically feasible commercial-scale processes. Biorefineries are based on the idea of accelerating the development of sustainable routes or value chains that start from the bioconversion of biomass feedstock until the application of biobased products in the end-markets (Dragone et al. 2020). Different research demonstrates the use of alternative sources of biomass such as lignocellulosic biomass (LB) originating from renewable natural resources to produce valuable products and bioproducts (Isikgor and Becer 2015).

The LB biorefinery has attracted great interest since LB is described as the most abundant and cheapest bio-renewable resource on earth (Belaud et al. 2019; Bilal and Iqbal 2020a). LB biorefineries were first projected and studied in the late twentieth century when nations such as the United States and Brazil started to further invest in the production of biofuels from biomass-based feedstocks such as corn and sugarcane (Marin et al. 2016). According to Bilal and Iqbal (2020a) this investment encouraged numerous research papers concerning biofuel production, which were later broadly classified into four generations.

The LB biorefinery (Figure 11.1) is based in a process that involves five steps: biomass collection, preparation of biomass such as washing and drying, pretreatment, enzymatic hydrolysis, fermentation and downstream processing (Kumar and Verma 2020).

According to Isikgor and Becer (2015), a great variety of value-added compounds can be derived from LB by using a number of treatment methods. The application of biotreatments in the conversion of LB is a promising approach to the production of bioenergy, biomolecules, and biomaterials. It is therefore relevant that sustainable techniques can be developed to obtain this vast range of bioproducts.

The LB is mainly composed of three polymers—cellulose, hemicellulose, and lignin—but also by small amounts of other components. The main challenge in the bioconversion of LB is the degradation of the lignocellulosic material. In this sense, many studies have focused on developing different approaches for the pretreatment of LB such as physical, chemical, biological, and physicochemical techniques that aim to deconstruct lignin (Asgher et al. 2013; Belaud et al. 2019; Chuetor et al. 2019; Kumar et al. 2019). Several studies discuss the valorization of lignocellulosic waste derived from domestic, food processing plants, crop residues, forest residues, sawmill residue, and agricultural residues using enzymatic treatment (Ter-Mikaelian et al. 2015; Saini et al. 2020). The reuse of recovered waste integrates

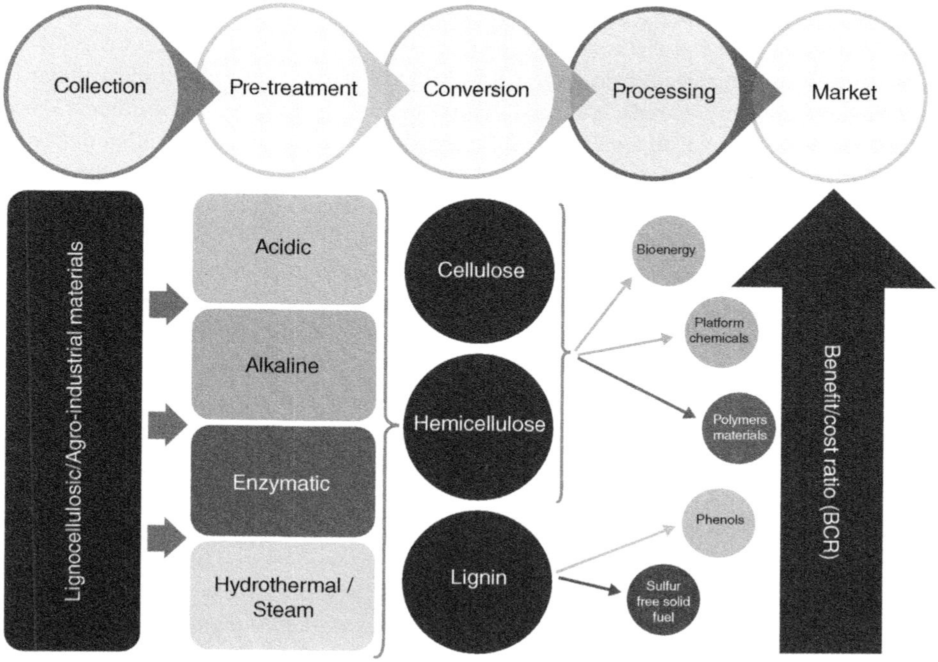

Figure 11.1 LB biorefinery concept. Reprinted from *International Journal of Biological Macromolecules*, Vol. 99/June 2017, Arevalo-Gallegos A, Ahmad Z., Asgher M., Parra-Saldivar R., Iqbal H.M.N., Lignocellulose: a sustainable material to produce value-added products with a zero waste approach—a review, pp. 308–318. Copyright 2017, with permission from Elsevier.

the concept of sustainable biorefinery and circular economy. The reprocessing of a large volume of LB promotes benefits for the environment, economy, and society.

Biorefinery processing generally involves the use of microorganisms or their enzymatic system for the bioconversion of lignocellulosic materials toward bioproducts such as bio-oil, vanillin, and pentose (C5) or hexose (C6) sugars, and biofuels such as ethanol, hydrogen, or methane (Martin et al. 2015; Bilal and Iqbal 2020a; Saini et al. 2020). It is relevant to highlight that the concept of biorefineries is linked to other areas such as sustainability, agronomy, supply chains, industrial biotechnology, geo-information, and bioinformatics. The integration of these areas helps in the building of new solutions by the development of tools, fundaments, and methodologies that support strategic decisions for future viable biorefinery (Torres et al. 2019).

11.2 Challenges and Perspectives Encompassing Circular Economy

The main challenges of modern society are to minimize dependence on fossil resources and to reduce the generation of solid and liquid waste. The current generation of solid urban waste is 2 billion tons, with an estimate that about 30% of this amount is not managed in an environmentally safe manner. Looking ahead, global waste is expected to grow to 3.4 billion tons by 2050, more than double the population growth in the same period (Kaza et al. 2018).

Compound waste corresponds to 44% green and food waste, 17% paper and cardboard, 12% plastic, 5% glass, 4% metals, 2% wood, 2% rubber and leather, and 12% others (Kaza et al. 2018). Most solid waste worldwide is destined for landfill or uncontrolled disposal causing environmental damage, damage, and financial loss. Thus, in recent years, most countries are seeking the development of sustainable management of waste by promoting increased protection of the environment and health.

Much of the green and food waste has been considered as inexhaustible sources of renewable resources. LB, whether cellulose and paper, or residues from agriculture, the food industry, forestry, as well as municipal solid waste, is currently considered as the most promising renewable resource, since it is available worldwide (Gavrilescu 2014).

Green residues or biomass residues tend to replace biomass that is especially cultivable for energy purposes, which is currently responsible for the first generation of biofuels. Lee et al. (2019a) categorized biomass residues into primary, secondary, and tertiary groups according to Figure 11.2.

Since the energy crisis of the 1970s, many countries have initiated national research, technology, and innovation programs seeking to obtain biofuels from lignocellulosic waste. However, the low prices of fossil energy have allowed it to continue to be considered as the first energy option despite the high emissions of greenhouse gases and air pollution.

It can be said that the use of fuels from fossil resources such as oil, coal, and natural gas is based on a linear flow as illustrated in Figure 11.3. The main problem related to the use

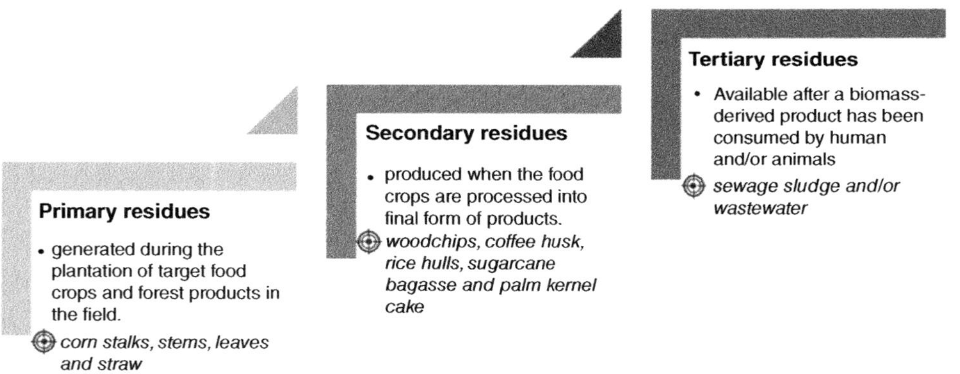

Figure 11.2 Classification of biomass residues used in the biorefineries. Based on Lee et al. (2019a).

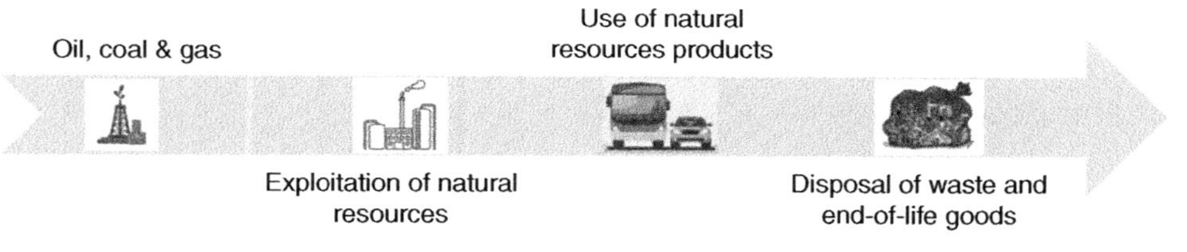

Figure 11.3 Linear economy model illustration for fossil energy.

of fossil resources is their limited supply because they are not renewable resources, which generates geopolitical and economic disputes. In the former linear economy model, fossil resources are consumed, and waste is thrown away and not converted into new raw materials. This model is outdated, and it is known that electricity generation will be unable to meet its demand in 2025–2040, requiring a major expansion of electricity generation from renewable resources in the coming years (Capellán-Pérez et al. 2014).

Only the recent discussions around sustainable development and the strong growth in the global demand for transport fuel have spurred extensive research efforts in the development of bioenergy. Biomass is a renewable resource and therefore has been considered as an alternative raw material to provide sustainable energy in the future.

In 2018, the production of first-generation bioethanol increased more than 7%, from 104 billion liters to 112 billion liters. Brazil was the third-largest producer of bioelectricity globally and the largest producer in South America; the country's capacity reached 14.7 GW and generation rose 9% to 54 TWh. Most of the bioelectricity generation was from sugarcane bagasse. The increase in global production of biofuels is estimated at 25% by 2024, a growth that may be even greater if biomass residues are considered as raw materials in the bioenergy chain. In Europe, the other globally significant market for biomethane for transport, consumption increased 13% in 2017, especially in Sweden, where methane production from food wastes is encouraged and prioritized as part of a sustainable waste reduction policy (REN21 2019).

Currently, the best vision of sustainable management for energy production is based on the vision of circular economy, which strives to reduce the generation of waste and seeks to promote the exploitation of renewable resources (Allesch and Brunner 2014). The concept of circular economy in biorefinery is of great importance as an alternative to the linear economy that has been adopted, especially for fossil fuel production and marketing practices.

The circular economy is an antagonistic approach to a linear economy in which the accumulated environmental impact is reduced concurrently with the consumption of raw materials and waste generation (Ellen MacArthur Foundation 2013). The circular economy encourages the reuse of by-products in the manufacturing industries, which seeks the use of renewable resources of biological origin and their conversion into several high-value bioproducts. In addition, the circular bioeconomy promises to mitigate the effects of climate change and provide a renewable carbon source (biomass) promoting the economic and social development of underdeveloped and developing countries through the valuation of rural areas (Ubando et al. 2019).

The circular economy models applied to biorefineries aim at the complete use of lignocellulosic residues where bioproducts are obtained in an integrated manner to the generation of bioenergy, especially biofuels, electricity, and heat (Liguori and Faraco 2016). The circular economy seeks a closed-loop biorefinery system to maximize the recovery of raw materials derived from renewable resources in permanent replacement to fossil resources (Velis 2015). It can be seen in Figure 11.4 that the approach of the circular economy concept transcends the concept of sustainable circular bioeconomy, where the wide range of renewable raw materials can generate a large portfolio of biobased products (Dahiya et al. 2018).

Biorefineries are analogous to oil refineries and integrate biomass conversion processes and technologies to produce fuels, energy, and chemicals, being composed of four pillars: input raw material; processing technologies; platforms for co-production of intermediate

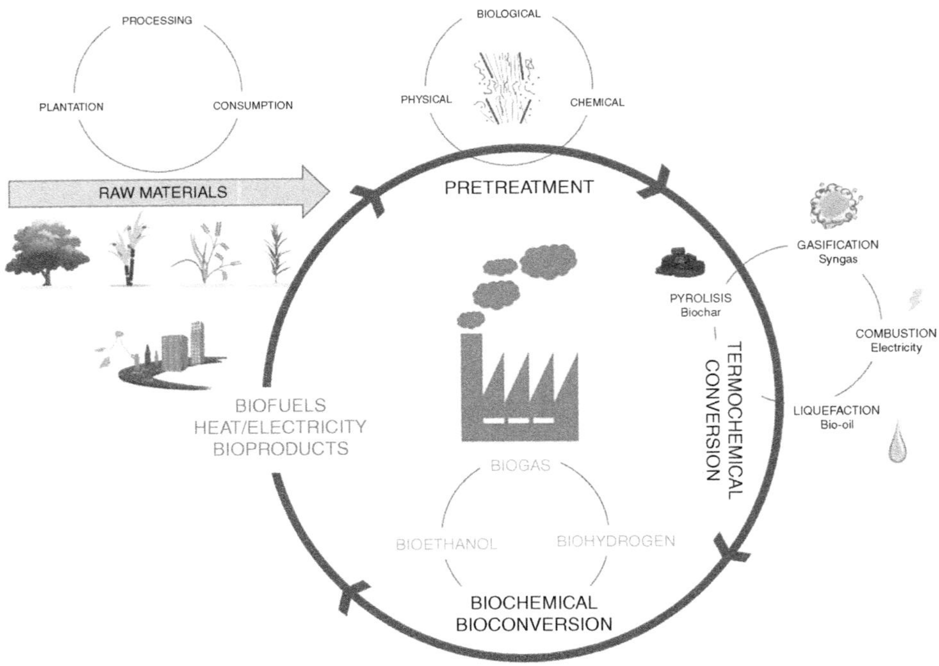

Figure 11.4 Circular economy model illustration for second-generation biofuel.

substances, and output products (Venkata Mohan et al. 2016). The use of waste as input raw material and the handling of all production waste as by-products contribute to reduce the ecological footprint and provide process within the circular economy.

Despite the imminent disruption of fossil energy generation due to high demands and a decrease in non-renewable raw materials, currently only 9% of the entire global energy is generated and supplied by biorefineries (Ahorsu et al. 2018). The challenges encountered to boost biorefineries go beyond scientific issues. It is also necessary to adopt public policies that boost the circular economy and the practice of sustainable development, which still causes some economic and political discomfort in the face of the outdated practices of making politics and global economic arrangements. Since the linear chain has been in place for a long time, it is still difficult to leave ties and realize that fossil resources must be left aside.

The main critical points regarding the good functioning of biorefineries within the concept of a circular economy ranges from the collection, transport, and handling of lignocellulosic waste to the expansion of a satisfactory scale for business profitability. The technological challenges inherent to the pretreatment processes of the raw material, to the fermentative processes, and to obtaining enzymes on a large scale have been overcome in laboratory tests, but they still fall short of exceeding the expectations of sustainable bioconversion of lignocellulosic waste so that they are competitive in the global market.

The optimization of the bioconversion of lignocellulosic residues should involve a systematic improvement in the performance of microorganisms and enzymes, with special attention to economic viability and environmental impacts. Although the research results are quite sophisticated, large-scale reproducibility is a big issue. Gaps are still found, especially regarding the technical-economic analysis of the processes.

Despite its advantages, the biological pretreatment of lignocellulosic residues by means of microorganisms is limited by some disadvantages, mainly related to the identification of ideal culture conditions such as substrate concentration and sterilization, pH, and temperature, in addition to the high variability of performance process (i.e. microorganism growth rate and moisture percentage) and the difficulty of increasing to industrial scale. Another disadvantage related to the pretreatment of lignocellulosic residues is the generation of derived compounds that end up decreasing the yield of the later stages of saccharification and fermentation (Allen et al. 2010). In the last few years, metagenomics and metabolic engineering have become a science combined with improving the performance of biorefineries by allowing direct access to microbial DNA and contributing to significantly increasing the performance of the catalysts needed in the biomass conversion into biofuel (Hess et al. 2011).

There is a consensus that integrated biorefineries will be more suitable for the bioeconomy and circular economy, using additional biobased raw materials or multiple combinations of raw materials to produce various biofuels, energy, and chemical materials. Hybrid biorefineries are a medium-term solution for reducing the cost of capital and include greater efficiency in converting biomass to final products (Nizami et al. 2017).

11.3 Improving Lignocellulose Conversion for Future Bioeconomy

Processes for production of bioproducts from lignocellulosic materials can be divided into three main stages: pretreatment of raw materials (upstream process), biotransformation, and purification (downstream process). In the first stage, plant cell wall is disrupted to facilitate the digestibility and/or availability of fermentable sugars; this can be done by chemical or biological methods (Kumar and Sharma 2017). The biotransformation stage consists in reaction of the substrates, using different types of microorganisms, which can be carried out in a reactor (fermenter), in continuous or batched regimes, producing the desired bioproducts, by-products and effluents. In this step, it is necessary to control the process variables, such as formulation of culture medium, pH, dissolved oxygen, temperature, agitation to reach a high productivity level. As soon as the bioproducts are obtained, the separation and purification steps are necessary. These steps must be planned to be unit operations that are operationally easy, economically viable, associated with obtaining a product with high purity.

Downstream is the step that aims to give greater purity to the final product; choosing the best process is necessary considering product characteristics such as its molecular size, concentration, solubility, polarity, volatility, physicochemical properties of the medium, intracellular or extracellular production (Cramer and Holstein 2011; Mores et al. 2020).

For example, after the fermentation process, the broth containing the target product also contains by-products, microorganisms, cell fragments, dead cells, salts, carbohydrates, lipids, amino acids, peptides, proteins, vitamins, organic acids, and many other molecules that are remaining ingredients (from the upstream stage) or metabolic waste products (Lee et al. 2019b). Microorganisms can produce intracellular or extracellular bioproducts, depending on the cytological and physiological characteristics of the cell. For an intracellular product (that accumulates inside the cell and is not secreted to the surrounding medium) it is necessary first to rupture the cells to release the target product, and more refined downstream techniques are used (Jozala et al. 2016; Yenkie et al. 2017; Wang et al. 2019b). Extracellular bioproducts have easier recovery and purification processes compared to intracellular ones because they are usually soluble and can be accumulated in the medium (Wang et al. 2019b).

Thus, knowledge of the biomolecular and medium characteristics are key points in choosing the downstream method, in order not to damage the product, remove impurities, and achieve maximum recovery rate of the bioproduct. It is important to highlight that there are no downstream processes of general application; there is not just one technique with efficient general application for all bioproduct separation and purification and each case has its specificities and requirements. Table 11.1 shows some examples of downstream processes for different bioproducts produced from different LB.

Most bioprocesses, especially those for production of high-value products, operate in batch mode, and smaller volumes are processed compared to continuous-mode processes (Gregory et al. 2019). Continuous bio-separation processes are used in the production of biobased products, such as organic acids and ethanol (Gregory et al. 2019; Mores et al. 2020). Choosing the downstream techniques will also depend on the operation regime of the production plant (Waites et al. 2009; Gregory et al. 2019). The downstream bioprocesses generally consist in a sequence of steps as shown in Figure 11.5.

The production routes of an ideal biorefinery, sustainable in broad approach, must consider the energetic, economic, and environmental aspects. These considerations could lead in the short term to lower energy costs and consequently higher profitability and reducing greenhouse gas emissions. Any kind of technical adjustments in biorefineries, aiming to optimize their processes with a macro view of bioeconomics, are linked to other important aspects such as government incentives, political regulations, market factors, among others (GBS 2018; IRENA 2019). Thus, applying bioeconomy concepts, development of lignocellulosic biorefineries can drive the transition toward a more sustainable economy, addressing the main challenges of food security, climate change, and scarcity of resources (Lewandowski et al. 2019).

The use of LB to produce bioproducts should be analyzed over technical, economic, environmental, and social aspects. This productive chain will achieve sustainability if all stages of production, from adaptation of raw material (upstream) to purification of target product (downstream), were planned to be interconnected (Figure 11.6), following the circular economy concept, which will contribute to a reduction of the environmental impact (Martín and Adams II 2019).

One of the biotechnology routes used in the biorefinery of LB involves the enzymatic hydrolysis of biomass. Biotechnology pretreatments are applied to alter the structure of biomass residues and expose the lignocellulosic fractions to facilitate the access of enzymes to the bulk and improve the rate and yield during the bioprocess (Belaud et al. 2019). Several methods for LB pretreatment have been studied in detail over the last 30 years and are nowadays considered as consolidated pretreatment technologies. However, for the success of lignocellulosic biorefineries, pretreatment efforts must be reduced. The adequate pretreatment route must be chosen in terms of available biomass, product quality, economy, and sustainability

Table 11.1 Downstream process from different LBs.

Agro-industrial residue	Downstream process	Bioproduct	Observation	References
Rice straw	Extraction solvent (dichloromethane, water)	Bio-oil produced from pyrolysis	The optimum extraction efficiency (47.36 wt%) was achieved with pyrolysis oil:dichloromethane:water = 10:10:3 (volume ratio)	Li et al. (2016)
Pine (*Pinus radiata*) and poplar (*Populus* sp.)	Lignin depolymerization using homogeneous (NaOH) and heterogeneous (Ni-, Cu-, or Ni-Cu-hydrotalcites) base catalysis and catalytic hydrogenolysis using Ru/C	Poly(hydroxy) aromatic compounds	Homogeneous base catalyzed depolymerization (BCD) and Ru/C hydrogenolysis combined on poplar lignin resulted in an aromatics amount of about 11 wt%	Cornejo et al. (2020)
Sugarcane straw	Supercritical CO_2 extraction assisted by ultrasound	Lipophilic molecules	The maximum global yield of the extracts (0.90 ± 0.10 g extract g^{-1} straw) was obtained at 60 °C and 800 W	Lachos-Perez et al. (2020)
Sugarcane bagasse and straw	Supercritical CO_2 extraction using ethanol as a co-solvent	Xylitol	The xylitol recovery was 40.51% at a high purity level of 99.59%	Silva et al. (2020)
Sugarcane bagasse	Nanofiltration membrane (NF polypiperazine amide)	Xylitol	The best nanofiltration conditions were 40 bar, 40 °C, and pH 5.4 with a xylitol purification factor of 3.3, protein separation factor of 8.4, and color removal of 96.0%	Alves et al. (2021)
Carob pods	Extraction with green solvent (lignin mixed in chloroform)	Polyhydroxybutyrate (PHB)	The highest yield (92.7%) and purity (91.5%) of PHB were obtained with a combination of 1% (w/v) lignin mixed in chloroform, in a closed-loop biorefinery strategy (using lignin obtained from the carob pods)	Manikandan et al. (2020)
Oil palm frond (OPF) bagasse	Forward osmosis (FO) and crystallization	Succinic acid	The fermentation broth was pretreatment by activated carbon before FO, which exhibited a concentration factor by 3.9-fold; succinic acid crystals were obtained with purity and yield of 90.52% and 67.09% respectively	Law et al. (2019)
Dried distillers' grains with solubles (DDGS)	Adsorption by anion exchange resin (Amberlite® IRA67, IRA12, and IRA67)	D-lactic acid	The purification process recovered approximately 80.4% D-lactic acid with 91.8% purity, reduced effectively the dark color of fermentation broth, removed oligosaccharides and converted sodium lactate to undissociated lactic acid	Zaini et al. (2019)

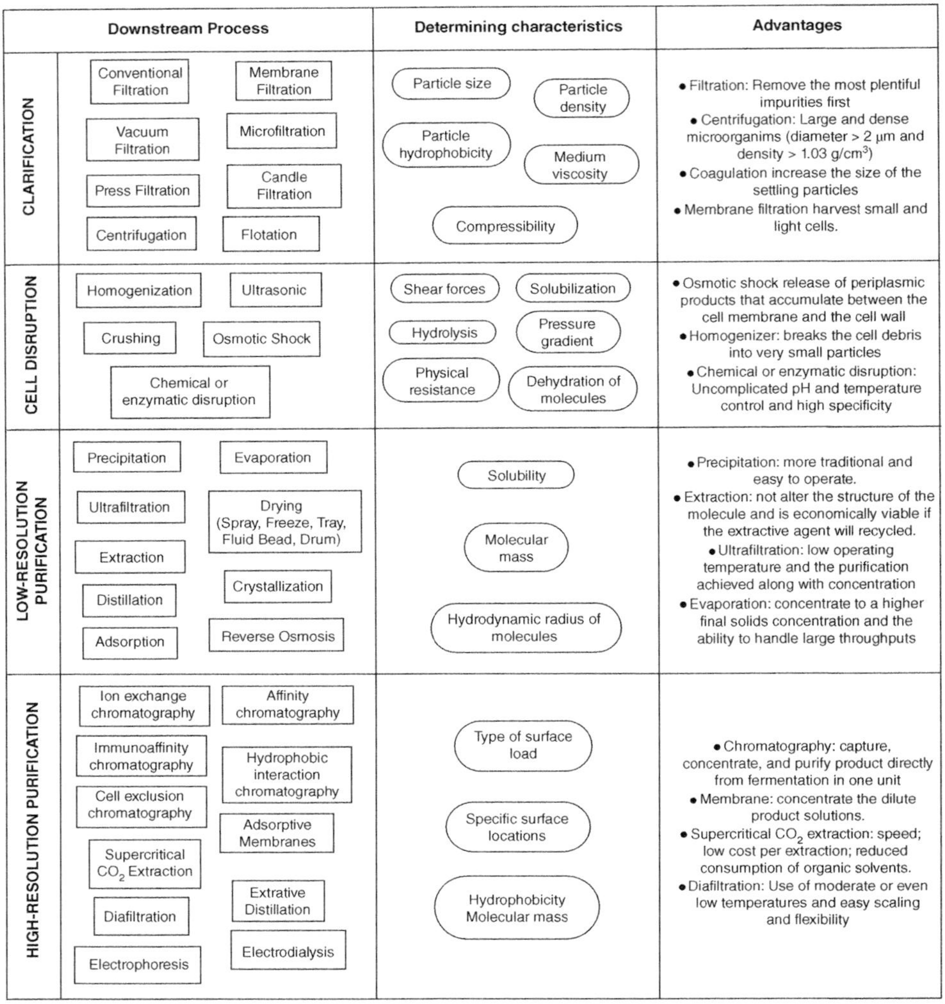

Figure 11.5 Diagram of different options of downstream processing.

(Belaud et al. 2019; Kumar and Verma 2020). Enzymatic hydrolysis is a well-established biological route for the treatment of LB; however, one of the main related challenges is the choice of an enzyme type that couples low cost and process effectiveness.

According to Olofsson et al. (2017), enzymes used in biorefineries should be regarded as a low-cost commodity if biobased products want to compete with petrochemical alternatives. Moreover, the authors declare that it is important to consider reduction of costs along the entire process of enzyme production (Figure 11.7), including the production and transportation of enzymes to the biorefinery.

The production of enzymes for biorefineries can vary based on the location and fermentation feedstock, including off-site manufacturing, on-site manufacturing, and integrated manufacturing. The best setup would allow benefits related to economy of scale and optimized

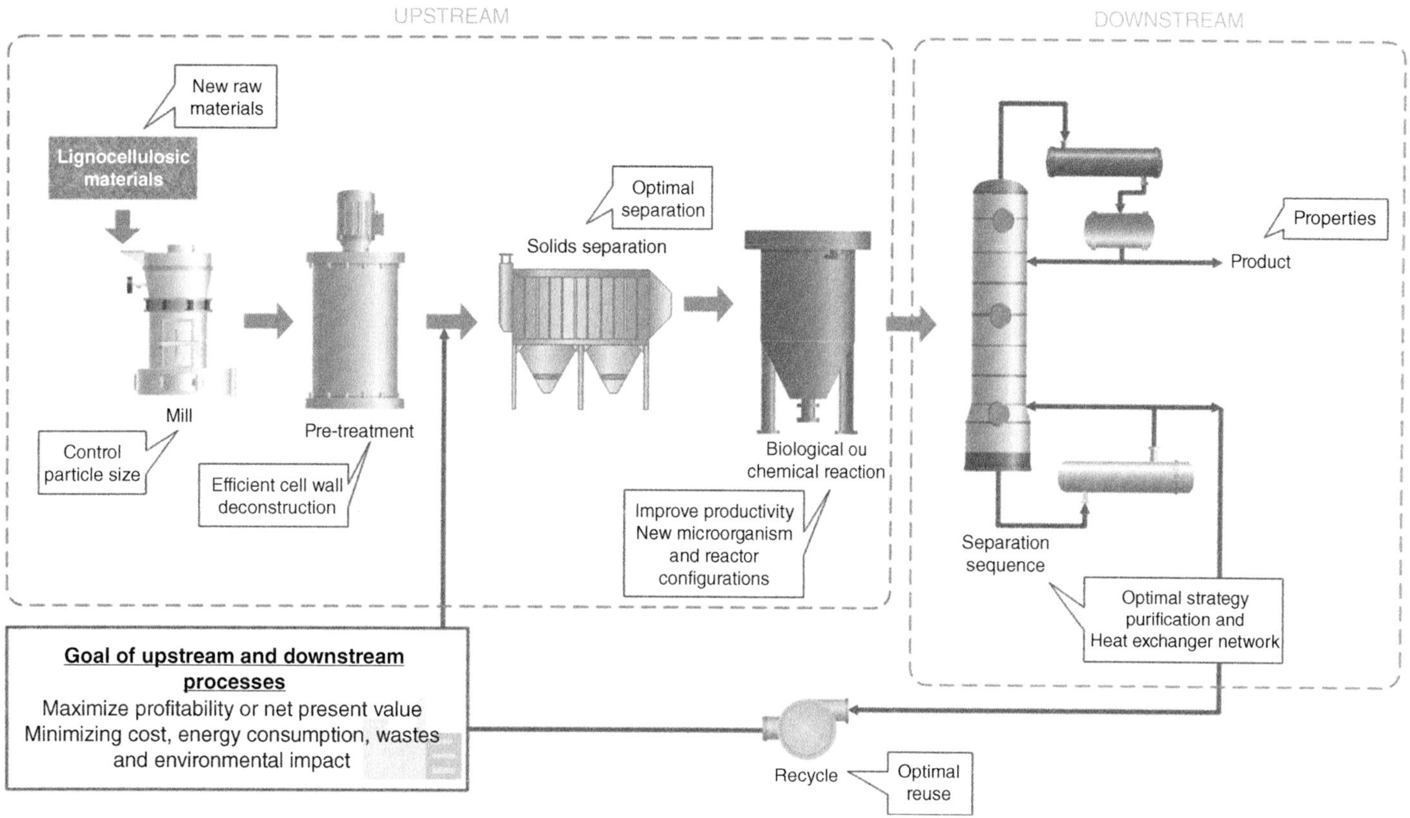

Figure 11.6 Overview of upstream and downstream steps in processes for use of lignocellulosic materials.

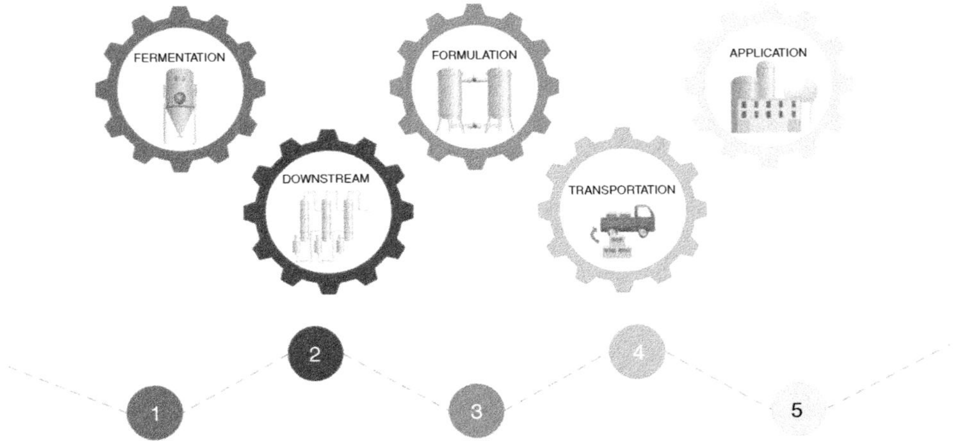

Figure 11.7 Simplified enzyme production. (1) Fermentation step to enzyme-producing strains. (2) Downstream step to recover the enzyme from the fermentation broth. (3) Formulation step of the enzyme product to ensure stability during storage and shipment. (4) Transportation step to biorefinery. (5) Application step in the enzymatic pretreatment in the biorefinery.

fermentation processes. Some authors argue that, even if off-site enzyme production is related to benefits such as process simplicity, purchasing enzymes from third parties creates risks related to supply uncertainty and costs related to formulation and transportation of the enzyme product (Humbird et al. 2011; Johnson 2016, Ellilä et al. 2017;).

Besides enzyme production costs, research is also focused on the continuous optimization of parameters related to the fermentation process in order to increase enzyme productivity (Humbird et al. 2011). Enzyme performance is an important parameter for the final production cost since the contribution of enzyme cost is linear with enzyme loading. Also, better enzyme performance can lead to the use of milder pretreatment conditions. Although considerable effort has been done to make existing commercial enzyme cocktails cheaper, enzymes are still one of the highest costs in biorefinery processes (Ellilä et al. 2017; Bentil et al. 2018).

The review of Kumar and Verma (2020) gives insight on the variety of hydrolytic enzymes produced by various microorganisms and possessing different characteristics, structures, and mechanisms. The authors argue that enzymes can act on different biomass components resulting in specific oligomers and monomers which would be transformed into biofuels and biochemicals. Despite the great enzyme variety available, the scientific community acknowledges that the application of these enzymes in biorefineries is limited. It is therefore necessary to study and better understand enzyme-based technologies in biorefinery.

According to Liguori and Faraco (2016), the ligninolytic enzymes can be used for biomass delignification as an alternative to fungal and bacterial strains. Delignification using enzymes is quicker and more productive (high rate) in comparison to microbial delignification. Moreover, the enzymatic delignification is generally carried out over wide temperature and pH ranges and does not require nutrient supplementation (Saini et al. 2020).

Today it is common to use commercial enzymes in the form of a cocktail. A study by Dragone et al. (2020) argues that typically biorefineries use standardized commercial enzymes, mostly produced by American and European companies. Also, the authors give some examples of enzyme cocktails such as Spirizyme® and Cellic®Ctec3 released by DuPont company. These enzymes operate in a complete synergy of enzymatic degradation of LB that entails a cocktail of cellulases, hemicellulases, and auxiliary proteins.

Research confirmed that the type of LB pretreatment affects the efficiency of individual enzyme components, for example the cellobiohydrolase *Tr*Cel7A from *Hypocrea jecorina* and the lytic polysaccharide monooxygenase *Ta*AA9A from *Thermoascus aurantiacus*, as a result affecting the performance of the enzyme cocktail in breaking down the feedstock (Schuerg et al. 2017). Thus, it is important to evaluate the use of industrially relevant pretreated substrates to validate the efficiency of enzyme cocktails and also a study to identify key enzyme activities in the enzyme cocktail (Chylenski et al. 2017; Østby et al. 2020).

Based on prior knowledge that enzymes are specific to their respective substrates, the enzymatic breakdown of complex biomass requires the synergistic action of different hydrolytic enzymes (Chen and Arnold 2020). Among the group of enzymes used to facilitate the hydrolysis of polymeric components are ligninolytic enzymes, cellulases, hemicellulases, lytic polysaccharide monooxygenases, pectinases, and lipases. Cellulase, hemicellulase, amylase, and pectinases have developed an important role in enzymatic hydrolysis applied in biobased biorefinery (Chylenski et al. 2017; Chen and Arnold 2020).

Saini et al. (2020) discuss the relevant development of a highly efficient enzyme cocktail to convert cellulose and hemicellulose with maximum efficiency for lignocellulose hydrolysis. It includes the synergetic action of all the enzymes to full conversion of polymeric sugars into their monomeric forms. Some data published demonstrate that the application of enzymes such as laccase and lignin peroxidase (LiP) can increase the cellulose accessibility to (hemi)cellulolytic enzymes and allows valuable phenolic compounds derived from lignin degradation to be recovered. Consequently, this combined action can be considered to develop an economically and eco-friendly biorefinery system (Chandra and Madakka 2019; Urbanek et al. 2020).

As an alternative, several non-enzymatic methods are available for LB pretreatment that generally make use of high temperatures and pressures, or pH variations by addition of acid or base. Examples of these processes include acid hydrolysis, steam explosion, or ammonia treatment (Bernardes 2011). The choice of pretreatment affects the downstream processing, the hydrolysis, and the fermentation steps. In addition, the use of certain pretreatments raises different challenges for the implementation of monitoring and controlling of the fermentation process.

Considering an economical lignocellulosic biorefinery or the process on a large scale, there is a need to reduce the cost of enzyme production once a large amount of enzyme is required for the process. It is relevant to discuss some limitations associated with the production, application, and recovery of the hydrolytic enzymes such as cost of production, stability of the enzyme in varying environmental conditions, and recovery or reuse of the enzyme. Once knowing these limiting factors, it is possible to find an approach to overcome them (Kumar and Verma 2020; Meramo-Hurtado et al. 2020). For example, some researchers suggest the idea of enzyme production from biomass such as agro-residues and the statistical optimization of their production process. This combination would result in enhanced yield at a lower cost (Robinson 2015; Bhardwaj et al. 2017). In the last decade, studies analyzed the use of agro-residues such as orange peel for pectinase production, coffee waste for lipase, rice straw for xylanase, wheat bran for cellulase, and also a mixture of rice husk, cottonseed cake, and red gram husk for lipase production (Bhardwaj et al. 2017; Kumar and Sharma 2017; Kumar and Verma 2020).

The future of the enzyme biorefinery 4.0 includes new and advanced biotechnology approaches that have been investigated and developed to enhance the quality of enzymes and their successful application in biorefinery. Among these approaches are synthetic biology, metabolic engineering, and other advanced technologies, which can be used to develop novel or synthetic microorganisms with enhanced enzymatic capabilities. In the last few years, genetic engineering and metagenomics approaches have been used to increase enzyme production and create enzymes with novel and specific properties (Adrio and Demain 2010; Ngara and Zhang 2018; Wang et al. 2019a; Kumar and Verma 2020). These approaches aim to combine enzymes from different sources and create new enzymes with maximal catalytic efficacy applied to protein engineering. Recently practical studies developed various recombinant microorganisms, for example *Abdopus aculeatus*, *Bacillus clausii*, *Bispora* sp., *Bacillus licheniformis*, *Bacillus halodurans*, and *Paenibacillius* sp. for enhanced pectinase enzyme yield with high activity and high alkali and acidic stability (Singh et al. 2019). In the same way, Bhardwaj et al. (2020) studied the expression of xylanase gene (*XynF1*) in *Escherichia coli* BL21 (DE3). A result of this work was the development of a high titer of specific xylanase activity of $1037.3\,U\,mg^{-1}$ which is 9.3-fold higher than that of the native strain *Aspergillus oryzae* LC1.

Another approach is the use of co-expression of multiple proteins on the surface of a single organism by the biorefinery industry in which the yeast cell is able to ferment the biomass but cannot produce the enzyme required for saccharification of the biomass. This approach is related to separating the steps of biomass processing and sugar conversion (separate hydrolysis and fermentation) (Taherzadeh and Karimi 2007; Tanaka and Kondo 2015; Lange 2017). On the other hand, some authors suggest reducing the number of steps during the conversion of the biomass to biofuel. For example, suggesting the saccharification and co-fermentation steps in a single bioreactor, or the delignification, saccharification, and fermentation steps in a single vessel. As result, the decrease in steps would minimize costs of the process (Lynd et al. 2005; Chandel et al. 2012).

Lots of research discuss the need to improve enzyme production for biorefinery. Due to this interest, it is necessary to develop a supply chain of enzymes at low cost with high activity, specificity, and ability for its application in biorefinery at a large scale. New and advanced approaches are essential since they can help in the discovery and improvement of enzymes such as cellulase, xylanase, laccase, and amylases that have multiple functions at the hydrolysis and fermentation steps (Dragone et al. 2020).

11.4 Industry 4.0: Advanced Technologies for the Biorefinery Platform

The term Industry 4.0 was first introduced by the German Federal Government in 2013 to refer to an industrial revolution toward the establishment of smart products and processes. The concept of Industry 4.0 has been defined as a group of key organization tools and technologies such as a cyber-physical system (CPS) and the Internet of Things (IoT) (Hermann et al. 2016). According to Reinhardt et al. (2020), the core of Industry 4.0 can be divided into two technology categories: base and front-end technologies. Base technology contains mainly four elements including IoT, big data (BDA), cloud services, and analytics. Other authors also consider CPS, information and communication technology, and BDA analysis as base technology in the Industry 4.0 (Yli-Ojanperä et al. 2019; Aceto et al. 2020). Front-end technologies have been described to encompass another four components related to smart working, smart manufacturing, smart products, and smart supply chain. However, the concept of Industry 4.0 has been adapted depending on the industry application field. Concerning the future of manufacturing industry, Ghobakhloo (2018) defends that the values of Industry 4.0 lie on the fusion between the physical and digital world of the organization (Ghobakhloo 2018). In the pharmaceutical sector, the concept of Industry 4.0 is linked to the adoption of certain processes and techniques including IoT, BDA, digitalization, and cloud computing by a company allowing it to obtain competitive advantages in regional or global markets (Reinhardt et al. 2020). A common point between the different definitions is that a transition to Industry 4.0 involves reshaping industries toward the integration of their production and systems information with communication technologies (Burmeister et al. 2015; Peruzzini and Stjepandić 2018).

The application of Industry 4.0 concepts has been especially useful to address economic opportunities (Burmeister et al. 2015) and ecological (Gisbert et al. 2014) and social issues (Schuh et al. 2014). In terms of economic development, the use of Industry 4.0 has been shown to ameliorate the creation of company value and competitiveness (Porter and Heppelmann 2014), as well as production efficiency (Hossain and Muhammad 2016). Moreover, the implementation of Industry 4.0 leads to enhanced industrial flexibility, capacity to react faster to critical events, and reduction of logistics costs (Civerchia et al. 2017). It has been recognized that one of the main challenges toward sustainable development is the integration of opportunities across electricity, heating, water, waste, transport, industry, and agriculture (Duić et al. 2020). In this sense, the application of Industry 4.0 can offer interesting opportunities from an ecological point of view improving energy efficiency (Dalenogare et al. 2018) and reducing the emission of greenhouse gases (Peukert et al. 2015) and waste (Berman 2012). The Industry 4.0 revolution toward the use of cyber-physical systems is already in progress. This evolution not only brings with it technical changes but also challenges the social organization of companies. Examples of changes in the organizational structure in the Industry 4.0 context include a new level of socio-technical interaction and inter-company network, the control of smart manufacturing, and individualized production by the employees (Kagermann et al. 2013).

The Industry 4.0 revolution resulted in promising positive impacts for industries of different sectors. For instance, in healthcare systems, the implementation of Industry 4.0 contributed to early diagnoses, continuous treatment of chronic diseases, and the reduction of costs (Aceto et al. 2020). The pharmaceutical and biopharmaceutical sector has also suffered significant operational and footprint changes over the last few years (Gautam and Pan 2016). Reinhardt et al. (2020) point out that the actual pharmaceutical model is transitioning to focus on research footprint for the development of key innovation bio-clusters and specialty products and biologics to gain access to emerging markets. In this sense, the study showed that the implementation of Industry 4.0 can help in the modernization of pharmaceutical and biopharmaceutical industries.

In the case of manufacturing organizations, the benefits of Industry 4.0 implementation are found in the optimization of decision making to satisfy customer needs, while keeping good efficiency and productivity (El-khoury et al. 2019).

As mentioned before, the biorefinery industry is in current development and represents an immense opportunity to produce biochemicals and bioenergy. Biorefinery systems are essentially adaptable and able to create strategies for the efficient valorization of biomass and wastes, while reducing the environmental footprint and forming new routes for socio-economic developments (Azapagic 2014; Brown 2018; Katakojwala and Mohan 2021). Therefore, the implementation of Industry 4.0 in biorefinery systems could be a great contribution toward new economic, environmental, and social developments.

Industry 4.0 concepts including the integration of multiple plants and terminals to allow efficient and real-time control of the process have already been adopted for well-established refinery systems such as hydrocarbons and minerals. However, for biorefineries to be able to produce more complex bioproducts using advanced processes, technological frontiers have to be overcome. Some of the technological needs include the creating of new models for the development of biological conversion and multi-scale bioprocessing industry, the development of microorganisms for the biosynthesis of different products, the design of sustainable and efficient processes. Other current challenges concerning the implementation of Industry 4.0 in the biorefinery sector are the creation of adapted models that consider metabolic flux variations with the aid of cognitive analytics, integration of metabolic modeling for online purposes, planning and scheduling in a multi-scale plant. For instance, Menezes et al. (2019) propose that artificial neural networks could be used to model the highly variable metabolic flux in the fermentation process present in first-generation biorefineries such as the ethanol and sugar industries, as well as for next generation biorefineries related to the processing of waste and more complex streams. It is also mentioned that the use of biosensing apparatus to modulate and control microbial DNA modification are expected future advances in a biorefinery 4.0 (Menezes et al. 2019).

Resuming, the term Industry 4.0 has been used to define a set of technologies that merge fields such as physics, computer science, and biology aiming at the optimizing of processes and bioprocesses (Belaud et al. 2019). The main technological concepts brought by the Industry 4.0 revolution are CPS, IoT, BDA, and internet of services (Figure 11.8). These innovative technologies have been used individually or combined to allow enterprises to make quick, assertive decisions, detect problems, avoid failures, and reduce costs in their production chain (Kagermann et al. 2013; Belaud et al. 2019).

Considering strategies to reduce costs in biorefinery platforms, recently scientific discussions put forward that the development of process biotechnology is directly linked to digital technology. Digital technology enables the creation and modeling of products used in many industrial sectors. As a consequence, digitalization or Industry 4.0 or the Industrial Internet of Things (IIoT) allows large amounts of data from the manufacturing line to be measured and analyzed and to take corrective actions (Belaud et al. 2019; Birkel et al. 2019; Ramirez-Lopez et al. 2019).

A revolution in the amount of available and potentially relevant data or advancements in data recording and storage technology is related to the concept BDA (Diebold 2003). This concept applied to biorefinery allows the cutback of feedstock and manufacturing costs by

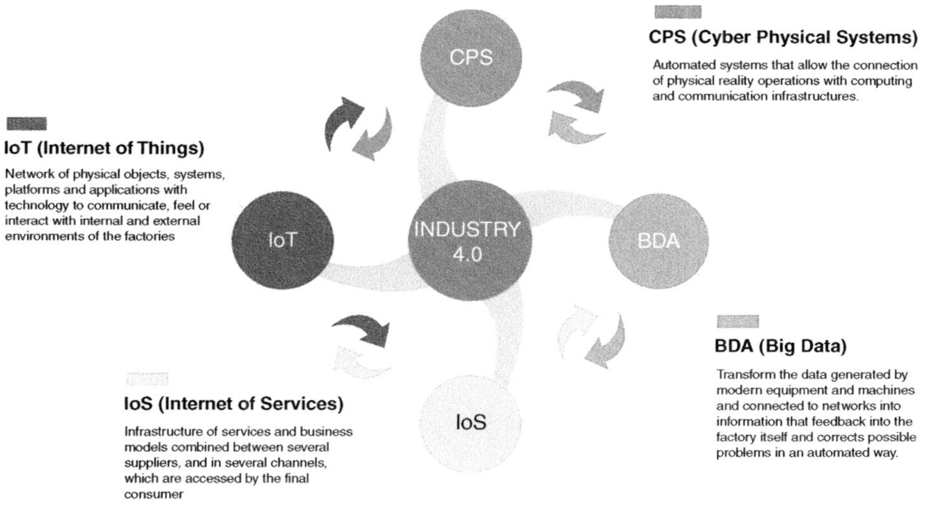

Figure 11.8 A simplified Industry 4.0 concept.

improving its processing dynamically by digitalization and virtualization of the information process, etc. Moreover, the BDA tool can be combined with other tools such as IoT and machine learning to collect, monitor, and manipulate equipment data remotely. For example, a cultivation of agricultural crops could be monitored remotely by the network at the factory. This assists in the control and optimization steps of the process to enhance productivity (Dragone et al. 2020). An example of application of BDA in biorefinery is found in the Brazilian sugar bioenergy group São Martino Mill that have developed a BDA-based Industry 4.0. This project involves automation and information technology to collect data from agricultural machines in rural areas so they can be in time processed and analyzed at the factory (Independent Group of Scientists appointed by the Secretary-General 2019; Dragone et al. 2020). Another case for study in the use of BDA integrated with artificial intelligence is given by Chr. Hansen company. This global company optimized fermentation experiments and automated sampling systems including data collection and processing. As a result, operational process is accelerated, and researchers can focus on interpretations of results and new projects instead of working mainly on data generation (Chr. Hansen 2020). BDA can also help to obtain detailed analysis of weather, soil conditions or help farmers for decision making related to aspects of planting, fertilizing, and harvesting crops. The DuPont company, for example, is a leading company in precision farming and applied the BDA to increase yields and produce less nitrogen per acre, providing the average farmer with a $22 per acre return (Chandran and Kasat 2018). In fact, there are still some challenges for BDA implementation in biorefinery, but this tool will help to virtualize experiments and provide a new set of parameters to control, update equipment, and optimize steps of the process.

A good example for integration of biorefinery and Industry 4.0 is in the algae biorefinery platform. The coupling of Industry 4.0 concepts and algae biorefineries (Figure 11.9) will allow the formation of an optimized biorefinery that produces the highest number of goods and co-products, at the same time producing the lowest amount of residues and

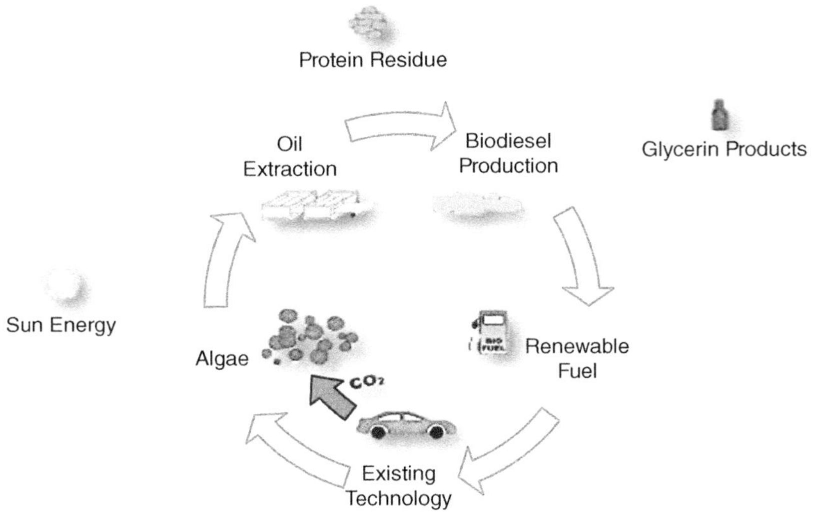

Figure 11.9 Scheme representing the cycle of algae biomass used in the production of biofuels and co-products (ElFar et al. 2021).

downstream costs assuring optimal economic returns (Léonard et al. 2010). Besides their use as biofuels, a number of products are derived from algae and microalgae biomass such as microalgae medication (trace elements and bioactive molecules), bulk proteins, carbohydrates, and lipids (Yuan et al. 2011; Leonardo et al. 2014).

Contextualizing, algae are classified as a group of photosynthetic eukaryotic organisms found in freshwater and marine habitats. The diverse algae group ranges from single cell organisms to organisms 100 ft in length. Algae have attracted research attention due to their capacity for producing food and energy. Compared to the first and second generation of biofuels that use food crops and lignocellulosic materials as biomass sources, algae can utilize wastewater and CO_2 to grow, producing high amounts of biomass and oil content. It is desirable that the algae feedstock used for bioenergy production is highly efficient and able to work in a spectrum of temperature, pH, salinity, and light intensity conditions. A fourth generation of customized and personalized algae could be a great alternative to produce cost-effective and reliable sources of energy (Milano et al. 2016; Khan et al. 2018).

Advances proposed by the implementation of Industry 4.0 in algae biorefineries would allow the real-time tracking and output control of algae by applying for instance a network of IoTs coupled with plug and play sensors (Algae Biomass Organization 2012). The use of sensors to create a digital duplicate of the algae population allows the predictions of cellular performance in real time. This forecast of cellular performance allows the algae biorefinery to anticipate the commodity demand and minimize possible wastes (McCormick et al. 2011).

The future of the algae biorefinery 4.0 includes advances in main areas such as the development of tools for alternating cultivation mode, more specific and multipurpose medium for cultivation of algae, optimization of cultivation conditions, customization of advanced reactors, and algae genetic engineering (ElFar et al. 2021). For instance, the development of hybrid photobioreactors combining air lit systems and external tubular loops have been proposed as promising tools for enhancing algae processing performance (Ungerer et al. 2012). Other tools, such as automation of biological algae libraries, which allow the simpler evaluation of novel biological designs have also been proposed as advances. The combination of the formation of clone libraries, novel strain production, modern biosensor technology, and algae phenomics analysis are examples of revolutionary developments in algae biorefinery. For instance, artificial intelligence could be applied to find the most sustainable algae strain and genetic sequence to be altered in order to form algae with more acceptable phenotype and suitable to maximize the production of useful by-products (ElFar et al. 2021).

Another strategy to boost biorefinery is the use of BDA and life-cycle assessment (LCA) analyses in an integrative approach helping to identify economic and ecological impacts in a biorefinery. This concept involves inputs and associated emissions from the bioproduction process followed by the product discharge from the manufacturing process (Bilal and Iqbal 2020b).

An application of LCA to lignocellulosic biofuels bioprocess would assist in the analysis of various environmental factors, which affect the equilibrium of natural ecosystems. It would also provide a tool for both consumers and policymakers determining the ideal environmentally friendly fuel. Also, an LCA study would identify two categories explicitly: fossil consumption/depletion and carbon and greenhouse gas discharges into the environment. Another example is the use of LCA in the ethanol production from miscanthus grown in the marginal land of Great Britain and in Germany to determine the environmental impacts of the process. The main results suggest that the type of pretreatment applied can have a strong influence on the environmental performance. The site-dependent variation in biomass composition did not have a relevant influence on the environmental performance of the analyzed pathways (Lask et al. 2019). Thus, it can be noticed that LCA is a useful tool for biorefineries and can help in the description, inventory assessment, impact analysis, and interpretation of the process (Kumar and Murthy 2012; Melis et al. 2018).

Since 2010, few studies have been published concerning climate change and BDA in the domain of agriculture. The use of BDA and LCA is presented in the work of Belaud et al. (2019), which studied the valorization of agricultural waste and by-products, applying BDA to support the management of sustainability. This study discusses the relationship between product quality and biomass pretreatment and the use of LCA to evaluate environmental aspects and potential environmental impacts throughout the life cycle of a product or process. In summary, a base was composed with relevant information from heterogeneous sources for the analysis of pretreatments for LB valorization and the BDA was used to improve sustainability management in supply chain design for the valorization of agricultural wastes and outputs of agri-food applications. Dragone et al. (2020) demonstrates that the LCA of biorefineries could also be incorporated and applied with BDA to determine the minimal distance between the harvesting of agricultural residues and the biorefinery in relation to regional costs. The incorporation of LCA and BDA in biorefinery could also be useful to predict the environmental impact based on emissions, land use, and regional labor cost.

11.5 Conclusions

The current challenge for industries is to become more sustainable. Not only by choosing unit operations that generate lower energy costs and being easy to operate, but also with the implementation of strategies that allow maximum conversion of both raw materials and by-products of the processes.

The development of a technology that allows the complete use of LB to produce bioenergy associated with bioproducts is essential for establishing a change in industrial processes, increasingly focused on environmental sustainability. In a broader sense, biorefineries that combine the concepts of bioeconomy, circular economy that encourages the reuse of by-products in the industrial network, and Industry 4.0 will become smarter in economical, ecological, and social issues. Besides, in a biorefinery applying bioeconomy and circular economy concepts, the wastes or the biomass can be monetized with a sustainability initiative and it can be boosted if combined with Industry 4.0.

Making biorefineries commercially competitive, investments are mandatory, mainly in the downstream step, one of the most important stages to bioproduct production and still the least developed sector. Improvement of downstream processes will certainly bring about a great reduction in production costs, minimize environmental impacts, and improve the state of social well-being.

References

Aceto, G., Persico, V., and Pescapé, A. (2020). Industry 4.0 and health: internet of things, big data, and cloud computing for healthcare 4.0. *J. Ind. Inf. Integr.* 18: 100129. https://doi.org/10.1016/j.jii.2020.100129.

Adrio, J.L. and Demain, A.L. (2010). Recombinant organisms for production of industrial products. *Bioeng. Bugs* 1 (2): 116–131. https://doi.org/10.4161/bbug.1.2.10484.

Ahorsu, R., Medina, F., and Constantí, M. (2018). Significance and challenges of biomass as a suitable feedstock for bioenergy and biochemical production: a review. *Energies* 11: 3366. https://doi.org/10.3390/en11123366.

Algae Biomass Organization (2012). Technical Standards Committee. Algae Industry Minimum Descriptive Language: Guidance to Evaluate Life Cycle Inputs and Outputs. 21 pp. http://algaebiomass.org/wp-content/uploads/2012/09/ABO-MDL-5.0-WEB-VERSION-Sept-2012.pdf

Allen, S.A., Clark, W., McCaffery, J. et al. (2010). Furfural induces reactive oxygen species accumulation and cellular damage in *Saccharomyces cerevisiae*. *Biotechnol. Biofuels* 3: 2. https://doi.org/10.1186/1754-6834-3-2.

Allesch, A. and Brunner, P.H. (2014). Assessment methods for solid waste management: a literature review. *Waste Manag. Res.* 32 (6): 461–473. https://doi.org/10.1177/0734242X14535653.

Alves, Y.P.C., Antunes, F.A.F., da Silva, S.S., and Forte, M.B.S. (2021). From by-to bioproducts: selection of a nanofiltration membrane for biotechnological xylitol purification and process optimization. *Food Bioprod. Process.* 125: 79–90. https://doi.org/10.1016/j.fbp.2020.10.005.

Asgher, M., Ahmad, Z., and Iqbal, H.M.N. (2013). Alkali and enzymatic delignification of sugarcane bagasse to expose cellulose polymers for saccharification and bio-ethanol production. *Ind. Crop. Prod.* 44: 488–495. https://doi.org/10.1016/j.indcrop.2012.10.005.

Asunis, F., de Gioannis, G., Dessì, P. et al. (2020). The dairy biorefinery: integrating treatment processes for cheese whey valorisation. *J. Environ. Manag.* 276: 111240. https://doi.org/10.1016/j.jenvman.2020.111240.

Azapagic, A. (2014). Sustainability considerations for integrated biorefineries. *Trends Biotechnol.* 32 (1): 1–4. https://doi.org/10.1016/j.tibtech.2013.10.009.

Belaud, J.P., Prioux, N., Vialle, C., and Sablayrolles, C. (2019). Big data for agri-food 4.0: application to sustainability management for by-products supply chain. *Comput. Ind.* 111: 41–50. https://doi.org/10.1016/j.compind.2019.06.006.

Bentil, J.A., Thygesen, A., Mensah, M. et al. (2018). Cellulase production by white-rot basidiomycetous fungi: solid-state versus submerged cultivation. *Appl. Microbiol. Biotechnol.* 102: 5827–5839. https://doi.org/10.1007/s00253-018-9072-8.

Berman, B. (2012). 3-D printing: the new industrial revolution. *Bus. Horiz.* 55 (2): 155–162. https://doi.org/10.1016/j.bushor.2011.11.003.

Bernardes, M.A.S. (2011). *Biofuel's Engineering Process Technology*. London: Intech Open eBook (PDF) ISBN: 978-953-51-4468-7. 10.5772/961.

Bhardwaj, N., Chanda, K., Kumar, B. et al. (2017). Statistical optimization of nutritional and physical parameters for xylanase production from newly isolated *Aspergillus oryzae* LC1 and its application in the hydrolysis of lignocellulosic agro-residues. *Bioresources* 12 (4): 8519–8538. https://ojs.cnr.ncsu.edu/index.php/BioRes/article/view/BioRes_12_4_8519_Bhardwaj_Statistical_Optimization_Xylanase_Production.

Bhardwaj, N., Verma, V.K., Chaturvedi, V., and Verma, P. (2020). Cloning, expression and characterization of a thermo-alkali-stable xylanase from *Aspergillus oryzae* LC1 in *Escherichia coli* BL21(DE3). *Protein Expr. Purif.* 168: 105551. https://doi.org/10.1016/j.pep.2019.105551.

Bilal, M. and Iqbal, H.M.N. (2020a). Ligninolytic enzymes mediated ligninolysis: an untapped biocatalytic potential to deconstruct lignocellulosic molecules in a sustainable manner. *Catal. Lett.* 150: 524–543. https://doi.org/10.1007/s10562-019-03096-9.

Bilal, M. and Iqbal, H.M.N. (2020b). Recent advancements in the life cycle analysis of lignocellulosic biomass. *Curr. Sust. Renew. Energy Rep.* 7: 100–107. https://doi.org/10.1007/s40518-020-00153-5.

Birkel, H.S., Veile, J.W., Müller, J.M. et al. (2019). Development of a risk framework for industry 4.0 in the context of sustainability for established manufacturers. *Sustainability* 11 (2): 384. https://doi.org/10.3390/su11020384.

Brown, T.R. (2018). Price uncertainty, policy, and the economic feasibility of cellulosic biorefineries. *Biofuels Bioprod. Biorefin.* 12 (3): 485–496. https://doi.org/10.1002/bbb.1865.

Burmeister, C., Luettgens, D., Piller, F.T. (2015). Business model innovation for Industrie 4.0: why the 'industrial internet' mandates a new perspective on innovation. Die Unternehmung, 2/2016. Available at SSRN: https://ssrn.com/abstract=2571033 or http://dx.doi.org/10.2139/ssrn.2571033.

Capellán-Pérez, I., Mediavila, M., Castro, C. et al. (2014). Fossil fuel depletion and socio-economic scenarios: an integrated approach. *Energy* 77: 641–666. https://doi.org/10.1016/j.energy.2014.09.063.

Chandel, A.K., Chandrasekhar, G., Silva, M.B., and da Silva, S.S. (2012). The realm of cellulases in biorefinery development. *Crit. Rev. Biotechnol.* 32 (3): 187–202. https://doi.org/10.3109/07388551.2011.595385.

Chandra, M.R.G.S. and Madakka, M. (2019). Comparative biochemistry and kinetics of microbial lignocellulolytic enzymes. In: *Recent Developments in Applied Microbiology and Biochemistry.* (ed. V. Buddolla), 147–159. Academic Press https://doi.org/10.1016/B978-0-12-816328-3.00011-8.

Chandran, S. and Kasat, R. (2018). Aligning data analytics and strategy in the chemical industry. In: *Aligning Business Strategies and Analytics,* Advances in Analytics and Data Science, vol. 1 (ed. M. Anandarajan and T. Harrison), 195–208. Cham: Springer https://doi.org/10.1007/978-3-319-93299-6_12.

Chen, K. and Arnold, F.H. (2020). Engineering new catalytic activities in enzymes. *Nat. Catal.* 3: 203–213. https://doi.org/10.1038/s41929-019-0385-5.

Chr. Hansen (2020). Chr. Hansen launches 2025 strategy with the ambition to create a differentiated bioscience leader to improve food and health for a sustainable future. https://www.chr-hansen.com/en/media/press-releases/2020/8/chr-hansen-launches-2025-strategy-with-the-ambition-to-create-a-differentiated-bioscience-leader. Accessed: 23 November 2020.

Chuetor, S., Champreda, V., and Laosiripojana, N. (2019). Evaluation of combined semi-humid chemo-mechanical pretreatment of lignocellulosic biomass in energy efficiency and waste generation. *Bioresour. Technol.* 292: 121966. https://doi.org/10.1016/j.biortech.2019.121966.

Chylenski, P., Petrović, D.M., Müller, G. et al. (2017). Enzymatic degradation of sulfite-pulped softwoods and the role of LPMOs. *Biotechnol. Biofuels* 10: 177. https://doi.org/10.1186/s13068-017-0862-5.

Civerchia, F., Bocchino, S., Salvadori, C. et al. (2017). Industrial internet of things monitoring solution for advanced predictive maintenance applications. *J. Ind. Inf. Integr.* 7: 4–12. https://doi.org/10.1016/j.jii.2017.02.003.

Cornejo, A., Bimbela, F., Moreira, R. et al. (2020). Production of aromatic compounds by catalytic depolymerization of technical and downstream biorefinery lignins. *Biomolecules* 10 (9): 1338. https://doi.org/10.3390/biom10091338.

Cramer, S.M. and Holstein, M.A. (2011). Downstream bioprocessing: recent advances and future promise. *Curr. Opin. Chem. Eng.* 1 (1): 27–37. https://doi.org/10.1016/j.coche.2011.08.008.

Dahiya, S., Kumar, A.N., Shanthi Sravan, J. et al. (2018). Food waste biorefinery: sustainable strategy for circular bioeconomy. *Bioresour. Technol.* 248: 2–12. https://doi.org/10.1016/j.biortech.2017.07.176.

Dalenogare, L.S., Benitez, G.B., Ayala, N.F., and Frank, A.G. (2018). The expected contribution of industry 4.0 technologies for industrial performance. *Int. J. Prod. Econ.* 204: 383–394. https://doi.org/10.1016/j.ijpe.2018.08.019.

Diebold, F.X. (2003). 'Big data' dynamic factor models for macroeconomic measurement and forecasting. In: *Advances in Economics and Econometrics: Theory and Applications, Eighth World Congress* (ed. M. Dewatripont, L. Hansen and S. Turnovsky), 115–122. Cambridge: Cambridge University Press https://doi.org/10.1017/CBO9780511610264.005.

Dragone, G., Kerssemakers, A.A.J., Driessen, J.L.S.P. et al. (2020). Innovation and strategic orientations for the development of advanced biorefineries. *Bioresour. Technol.* 302: 122847. https://doi.org/10.1016/j.biortech.2020.122847.

Duić, N., da Costa, M.M.G., Wang, Q. et al. (2020). Selected Papers from SDEWES 2018: Conferences on Sustainable Development of Energy, Water and Environment Systems.

Energy, ISBN 978-3-03921-554-6 (Hbk); ISBN 978-3-03921-555-3 (PDF). 10.3390/books978-3-03921-555-3.

ElFar, O.A., Chang, C.K., Leong, H.Y. et al. (2021). Prospects of industry 5.0 in algae: customization of production and new advance technology for clean bioenergy generation. *Energy Convers. Manag. X* 10: 100048. https://doi.org/10.1016/j.ecmx.2020.100048.

El-khoury, J., Berezovskyi, A., and Nyberg, M. (2019). An industrial evaluation of data access techniques for the interoperability of engineering software tools. *J. Ind. Inf. Integr.* 15: 58–68. https://doi.org/10.1016/j.jii.2019.04.004.

Ellen MacArthur Foundation. (2013). Towards the Circular Economy: Economic and Business Rationale for Accelerated Transition. 1530-9290. https://www.ellenmacarthurfoundation.org/assets/downloads/publications/Ellen-MacArthur-Foundation-Towards-the-Circular-Economy-vol.1.pdf

Ellilä, S., Fonseca, L., Uchima, C. et al. (2017). Development of a low-cost cellulase production process using *Trichoderma reesei* for Brazilian biorefineries. *Biotechnol. Biofuels* 10: 30. https://doi.org/10.1186/s13068-017-0717-0.

Gautam, A. and Pan, X. (2016). The changing model of big pharma: impact of key trends. *Drug Discov. Today* 21 (3): 379–384. https://doi.org/10.1016/j.drudis.2015.10.002.

Gavrilescu, M. (2014). Biomass potential for sustainable environment, biorefinery products and energy. In: *Sustainable Energy in the Built Environment— Steps Towards nZEB. Springer Proceedings in Energy.* (ed. I. Visa), 169–194. Cham: Springer https://doi.org/10.1007/978-3-319-09707-7_13.

GBS 2018—Summit (2018). Innovation in the global bioeconomy for sustainable and inclusive transformation and wellbeing. In Conference Report. Available at: https://gbs2018.com/fileadmin/gbs2018/Downloads/GBS_2018_Communique.pdf. Accessed: 17 December 2020.

Ghobakhloo, M. (2018). The future of manufacturing industry: a strategic roadmap toward industry 4.0. *J. Manuf. Technol. Manag.* 29 (6): 910–936. https://doi.org/10.1108/JMTM-02-2018-0057.

Gisbert, J.R., Palau, C., Uriarte, M. et al. (2014). Integrated system for control and monitoring industrial wireless networks for labor risk prevention. *J. Netw. Comput. Appl.* 39: 233–252. https://doi.org/10.1016/j.jnca.2013.07.014.

Gregory, F., Chalmers, J., Harrison, R.G. et al. (2019). Downstream processing: separation and purification. In: *Perry's Chemical Engineers' Handbook*, 9e (ed. D.W. Green and M.Z. Southard). New York: McGraw-Hill, Ch. 20.3 https://www.accessengineeringlibrary.com/content/book/9780071834087/toc-chapter/chapter20/section/section18.

Hermann, M., Pentek, T., Otto, B. (2016). Design principles for industrie 4.0 scenarios. In Proceedings of the Annual Hawaii International Conference on System Sciences, 49th Hawaii International Conference on System Sciences (HICSS), Koloa, HI, 3928–3937. https://doi.org/10.1109/HICSS.2016.488.

Hess, M., Sczyrba, A., Egan, R. et al. (2011). Metagenomic discovery of biomass-degrading genes and genomes from cow rumen. *Science* 331: 463–467. https://doi.org/10.1126/science.1200387.

Hossain, M.S. and Muhammad, G. (2016). Cloud-assisted industrial internet of things (IIoT)—enabled framework for health monitoring. *Comput. Netw.* 101 (4): 192–202. https://doi.org/10.1016/j.comnet.2016.01.009.

Humbird, D., Davis, R., Tao, L. et al. (2011). Process Design and Economics for Biochemical Conversion of Lignocellulosic Biomass to Ethanol: Dilute-Acid Pretreatment and Enzymatic Hydrolysis of Corn Stover. Technical Report NREL/TP-5100-47764. National Renewable Energy Laboratory. https://www.nrel.gov/docs/fy11osti/47764.pdf

Independent Group of Scientists appointed by the Secretary-General (2019). The Future is Now—Science for Achieving Sustainable Development. Global Sustainable Development Report 2019. https://sustainabledevelopment.un.org/content/documents/24797GSDR_report_2019.pdf

International Energy Agency (2018). http://www.iea.org/statistics. Accessed: 20 November 2020.

IRENA (2019). Advanced biofuels. What holds them back?, International Renewable Energy Agency, Abu Dhabi. Available: https://irena.org/-/media/Files/IRENA/Agency/Publication/2019/Nov/IRENA_Advanced-biofuels_2019.pdf. Accessed: 17 December 2020.

Isikgor, F.H. and Becer, C.R. (2015). Lignocellulosic biomass: a sustainable platform for the production of bio-based chemicals and polymers. *Polym. Chem.* 2015 (6): 4497–4559. https://doi.org/10.1039/c5py00263j.

Johnson, E. (2016). Integrated enzyme production lowers the cost of cellulosic ethanol. *Biofuels Bioprod. Biorefin.* 10 (2): 164–174. https://doi.org/10.1002/bbb.1634.

Jozala, A.F., Geraldes, D.C., Tundisi, L.L. et al. (2016). Biopharmaceuticals from microorganisms: from production to purification. *Braz. J. Microbiol.* 47: 51–63. https://doi.org/10.1016/j.bjm.2016.10.007.

Kagermann, H., Wahlster, W., Helbig, J. (2013). Recommendations for implementing the strategic initiative INDUSTRIE 4.0. Final report of the Industrie 4.0 Working Group. Acatech—National Academy of Science and Engineering. Linda Treugut, M.A. https://www.din.de/blob/76902/e8cac883f42bf28536e7e8165993f1fd/recommendations-for-implementing-industry-4-0-data.pdf

Katakojwala, R. and Mohan, S.V. (2021). A critical view on the environmental sustainability of biorefinery systems. *Curr. Opin. Green Sust. Chem.* 27: 100392. https://doi.org/10.1016/j.cogsc.2020.100392.

Kaza, S., Yao, L., Bhada-Tata, P., Van Woerden, F. (2018). What a Waste 2.0: A Global Snapshot of Solid Waste Management to 2050. Urban Development. Washington, DC: World Bank. https://openknowledge.worldbank.org/handle/10986/30317.

Khan, M.I., Shin, J.H., and Kim, J.D. (2018). The promising future of microalgae: current status, challenges, and optimization of a sustainable and renewable industry for biofuels, feed, and other products. *Microb. Cell Factories* 17 (1): 36. https://doi.org/10.1186/s12934-018-0879-x.

Kumar, D. and Murthy, G.S. (2012). Life cycle assessment of energy and GHG emissions during ethanol production from grass straws using various pretreatment processes. *Int. J. Life Cycle Assess.* 17: 388–401. https://doi.org/10.1007/s11367-011-0376-5.

Kumar, A.K. and Sharma, S. (2017). Recent updates on different methods of pretreatment of lignocellulosic feedstocks: a review. *Bioresour. Bioprocess.* 4: 7. https://doi.org/10.1186/s40643-017-0137-9.

Kumar, B. and Verma, P. (2020). Enzyme mediated multi-product process: a concept of bio-based refinery. *Ind. Crop. Prod.* 154: 112607. https://doi.org/10.1016/j.indcrop.2020.112607.

Kumar, G., Dharmaraja, J., Arvindnarayan, S. et al. (2019). A comprehensive review on thermochemical, biological, biochemical and hybrid conversion methods of bio-derived lignocellulosic molecules into renewable fuels. *Fuel* 251: 352–367. https://doi.org/10.1016/j.fuel.2019.04.049.

Lachos-Perez, D., Barrales, F.M., Martinez, J., and Maciel Filho, R. (2020). Supercritical CO_2 extraction of lipophilic molecules from sugarcane straw. *Chem. Eng. Trans.* 80: 313–318. https://doi.org/10.3303/CET2080053.

Lange, L. (2017). Fungal enzymes and yeasts for conversion of plant biomass to bioenergy and high-value products. *Microbiol. Spectr.* 5 (1): https://doi.org/10.1128/microbiolspec.FUNK-0007-2016.

Lask, J., Wagner, M., Trindade, L.M., and Lewandowski, I. (2019). Life cycle assessment of ethanol production from miscanthus: a comparison of production pathways at two European sites. *GCB Bioenergy* 11 (1): 269–288. https://doi.org/10.1111/gcbb.12551.

Law, J.Y., Mohammad, A.W., Tee, Z.K. et al. (2019). Recovery of succinic acid from fermentation broth by forward osmosis-assisted crystallization process. *J. Membr. Sci.* 583: 139–151. https://doi.org/10.1016/j.memsci.2019.04.036.

Lee, S.Y., Sankaran, R., Chew, K.W. et al. (2019a). Waste to bioenergy: a review on the recent conversion technologies. *BMC Energy* 1: 4. https://doi.org/10.1186/s42500-019-0004-7.

Lee, S.Y., Chew, K.W., and Show, P.L. (2019b). Cell separation and disruption, product recovery, and purification. In: *Essentials in Fermentation Technology.* (ed. A. Berenjian), 237–271. Cham: Springer https://doi.org/10.1007/978-3-030-16230-6_8.

Léonard, A., Rooke, J.C., Meunier, C.F. et al. (2010). Cyanobacteria immobilised in porous silica gels: exploring biocompatible synthesis routes for the development of photobioreactors. *Energy Environ. Sci.* 3: 370–377. https://doi.org/10.1039/b923859j.

Leonardo, T., Farhi, E., Boisson, A.M. et al. (2014). Determination of elemental distribution in green micro-algae using synchrotron radiation nano X-ray fluorescence (SR-nXRF) and electron microscopy techniques—subcellular localization and quantitative imaging of silver and cobalt uptake by *Coccomyxa actinabiotis*. *Metallomics* 6: 316–329. https://doi.org/10.1039/c3mt00281k.

Lewandowski, I., Bahrs, E., Dahmen, N. et al. (2019). Biobased value chains for a growing bioeconomy. *GCB Bioenergy* 11 (1): 4–8. https://doi.org/10.1111/gcbb.12578.

Li, H., Xia, S., and Ma, P. (2016). Upgrading fast pyrolysis oil: solvent–anti-solvent extraction and blending with diesel. *Energy Convers. Manag.* 110: 378–385. https://doi.org/10.1016/j.enconman.2015.11.043.

Liguori, R. and Faraco, V. (2016). Biological processes for advancing lignocellulosic waste biorefinery by advocating circular economy. *Bioresour. Technol.* 215: 13–20. https://doi.org/10.1016/j.biortech.2016.04.054.

Lynd, L.R., Van Zyl, W.H., McBride, J.E., and Laser, M. (2005). Consolidated bioprocessing of cellulosic biomass: an update. *Curr. Opin. Biotechnol.* 16 (5): 577–583. https://doi.org/10.1016/j.copbio.2005.08.009.

Manikandan, N.A., Pakshirajan, K., and Pugazhenthi, G. (2020). A closed-loop biorefinery approach for polyhydroxybutyrate (PHB) production using sugars from carob pods as the sole raw material and downstream processing using the co-product lignin. *Bioresour. Technol.* 307: 123247. https://doi.org/10.1016/j.biortech.2020.123247.

Marin, F.R., Martha, G.B., Cassman, K.G., and Grassini, P. (2016). Prospects for increasing sugarcane and bioethanol production on existing crop area in Brazil. *Bioscience* 66 (4): 307–316. https://doi.org/10.1093/biosci/biw009.

Martín, M. and Adams, T.A. II (2019). Challenges and future directions for process and product synthesis and design. *Comput. Chem. Eng.* 128: 421–436. https://doi.org/10.1016/j.compchemeng.2019.06.022.

Martin, E.W., Chester, M.V., and Vergara, S.E. (2015). Attributional and consequential life-cycle assessment in biofuels: a review of recent literature in the context of system boundaries. *Curr. Sust. Renew. Energy Rep.* 2: 82–89. https://doi.org/10.1007/s40518-015-0034-9.

McCormick, A.J., Bombelli, P., Scott, A.M. et al. (2011). Photosynthetic biofilms in pure culture harness solar energy in a mediatorless bio-photovoltaic cell (BPV) system. *Energy Environ. Sci.* 4: 4699–4709. https://doi.org/10.1039/c1ee01965a.

Melis, E., Vincis, A., and Orrù, P.F. (2018). An overview of current models and approaches to biomass supply chain design and management. *Curr. Sust. Renew. Energy Rep.* 5: 138–149. https://doi.org/10.1007/s40518-018-0108-6.

Menezes, B.C., Kelly, J.D., and Leal, A.G. (2019). Identification and design of Industry 4.0 opportunities in manufacturing: examples from mature industries to laboratory level systems. *IFAC-PapersOnLine* 52 (13): 2494–2500. https://doi.org/10.1016/j.ifacol.2019.11.581.

Meramo-Hurtado, S.I., Sanchez-Tuiran, E., Ponce-Ortega, J.M. et al. (2020). Synthesis and sustainability evaluation of a lignocellulosic multifeedstock biorefinery considering technical performance indicators. *ACS Omega* 5 (16): 9259–9275. https://doi.org/10.1021/acsomega.0c00114.

Milano, J., Ong, H.C., Masjuki, H.H. et al. (2016). Microalgae biofuels as an alternative to fossil fuel for power generation. *Renew. Sust. Energ. Rev.* 58: 180–197. https://doi.org/10.1016/j.rser.2015.12.150.

Mores, S., de Souza Vandenberghe, L.P., Júnior, A.I.M. et al. (2020). Citric acid bioproduction and downstream processing: status, opportunities, and challenges. *Bioresour. Technol.* 320 (Part B): 124426. https://doi.org/10.1016/j.biortech.2020.124426.

Ngara, T.R. and Zhang, H. (2018). Recent advances in function-based metagenomic screening. *Genomics Proteomics Bioinformatics* 16 (6): 405–415. https://doi.org/10.1016/j.gpb.2018.01.002.

Nizami, A.S., Rehan, M., Waqas, M. et al. (2017). Waste biorefineries: enabling circular economies in developing countries. *Bioresour. Technol.* 241: 1101–1117. https://doi.org/10.1016/j.biortech.2017.05.097.

Olofsson, J., Barta, Z., Börjesson, P., and Wallberg, O. (2017). Integrating enzyme fermentation in lignocellulosic ethanol production: life-cycle assessment and techno-economic analysis. *Biotechnol. Biofuels* 10: 51. https://doi.org/10.1186/s13068-017-0733-0.

Østby, H., Hansen, L.D., Horn, S.J. et al. (2020). Enzymatic processing of lignocellulosic biomass: principles, recent advances and perspectives. *J. Ind. Microbiol. Biotechnol.* 47: 623–657. https://doi.org/10.1007/s10295-020-02301-8.

Peruzzini, M. and Stjepandić, J. (2018). Editorial to the special issue "Transdisciplinary analytics in supply chain management". *J. Manage. Anal.* 5 (2): 75–80. https://doi.org/10.1080/23270012.2018.1443405.

Peukert, B., Benecke, S., Clavell, J. et al. (2015). Addressing sustainability and flexibility in manufacturing via smart modular machine tool frames to support sustainable value creation. *Procedia CIRP* 29: 514–519. https://doi.org/10.1016/j.procir.2015.02.181.

Porter, M.E. and Heppelmann, J.E. (2014). How smart, connected products are transforming competition. *Harv. Bus. Rev.* 92: 64–88. https://hbr.org/2014/11/how-smart-connected-products-are-transforming-competition.

Ramirez-Lopez, L.J., Aponte, G.P., and Garcia, A.R. (2019). Internet of things applied in healthcare based on open hardware with low-energy consumption. *Healthc. Inform. Res.* 25 (3): 230–235. https://doi.org/10.4258/hir.2019.25.3.230.

Reinhardt, I.C., Oliveira, D.J.C., and Ring, D.D.T. (2020). Current perspectives on the development of industry 4.0 in the pharmaceutical sector. *J. Ind. Inf. Integr.* 18: 100131. https://doi.org/10.1016/j.jii.2020.100131.

REN21 (2019). Renewables 2019 Global Status Report Paris: REN21 Secretariat. https://www.ren21.net/gsr-2019. ISBN 978-3-9818911-7-1.

Robinson, P.K. (2015). Enzymes: principles and biotechnological applications. *Essays Biochem.* 59: 1–41. https://doi.org/10.1042/bse0590001.

Saini, R., Osorio-Gonzalez, C.S., Hegde, K. et al. (2020). Lignocellulosic biomass-based biorefinery: an insight into commercialization and economic standout. *Curr. Sust. Renew. Energy Rep.* 7: 122–136. https://doi.org/10.1007/s40518-020-00157-1.

Schuerg, T., Gabriel, R., Baecker, N. et al. (2017). *Thermoascus aurantiacus* is an intriguing host for the industrial production of cellulases. *Curr. Biotechnol.* 6 (2): 89–97. https://doi.org/10.2174/2211550105666160520123504.

Schuh, G., Potente, T., Varandani, R. et al. (2014). Collaboration moves productivity to the next level. *Procedia CIRP* 17: 3–8. https://doi.org/10.1016/j.procir.2014.02.037.

Silva, D.D., Dussán, K.J., Idarraga, A. et al. (2020). Production and purification of xylitol by *Scheffersomyces amazonenses* via sugarcane hemicellulosic hydrolysate. *Biofuels Bioprod. Biorefin.* 14 (2): 344–356. https://doi.org/10.1002/bbb.2085.

Singh, R.S., Singh, T., and Pandey, A. (2019). Microbial enzymes—an overview. In: *Biomass, Biofuels, Biochemicals, Advances in Enzyme Technology* (ed. R.S. Singh, R.R. Singhania, A. Pandey and C. Larroche), 1–40. Elsevier. ISBN: 9780444641144 https://doi.org/10.1016/B978-0-444-64114-4.00001-7.

Taherzadeh, M.J. and Karimi, K. (2007). Enzyme-based hydrolysis processes for ethanol from lignocellulosic materials: a review. *Bioresources* 2 (4): 707–738. https://ojs.cnr.ncsu.edu/index.php/BioRes/article/view/BioRes_2_4_707_738_Taherzadeh_Karimi_EnzymeBased_Hydrol_Ethanol_Review.

Tanaka, T. and Kondo, A. (2015). Cell surface engineering of industrial microorganisms for biorefining applications. *Biotechnol. Adv.* 33 (7): 1403–1411. https://doi.org/10.1016/j.biotechadv.2015.06.002.

Ter-Mikaelian, M.T., Colombo, S.J., Lovekin, D. et al. (2015). Carbon debt repayment or carbon sequestration parity? Lessons from a forest bioenergy case study in Ontario, Canada. *GCB Bioenergy* 7 (4): 704–716. https://doi.org/10.1111/gcbb.12198.

Torres, A.F., Xu, X., Nikiforidis, C.V. et al. (2019). Exploring the treasure of plant molecules with integrated biorefineries. *Front. Plant Sci.* 10: 478. https://doi.org/10.3389/fpls.2019.00478.

Ubando, A.T., Felix, C.B., and Chen, W.H. (2019). Biorefineries in circular bioeconomy: a comprehensive review. *Bioresour. Technol.* 299: 122585. https://doi.org/10.1016/j.biortech.2019.122585.

Ungerer, J., Tao, L., Davis, M. et al. (2012). Sustained photosynthetic conversion of CO_2 to ethylene in recombinant cyanobacterium *Synechocystis* 6803. *Energy Environ. Sci.* 5: 8998–9006. https://doi.org/10.1039/c2ee22555g.

Urbanek, A.K., Mirończuk, A.M., García-Martín, A. et al. (2020). Biochemical properties and biotechnological applications of microbial enzymes involved in the degradation of polyester-type plastics. *Biochim. Biophys. Acta Protein Proteomics* 1868 (2): 140315. https://doi.org/10.1016/j.bbapap.2019.140315.

Velis, C.A. (2015). Circular economy and global secondary material supply chains. *Waste Manag. Res.* 33 (5): 389–391. https://doi.org/10.1177/0734242X15587641.

Venkata Mohan, S., Nikhil, G.N., Chiranjeevi, P. et al. (2016). Waste biorefinery models towards sustainable circular bioeconomy: critical review and future perspectives. *Bioresour. Technol.* 215: 2–12. https://doi.org/10.1016/j.biortech.2016.03.130.

Waites, M.J., Morgan, N.L., Rockey, J.S., and Higton, G. (2009). *Industrial Microbiology: An Introduction*. Wiley-Blackwell. ISBN: 978-0-632-05307-0.

Wang, J., Zhu, L., Wang, J. et al. (2019a). Enzymatic synthesis of functional structured lipids from glycerol and naturally phenolic antioxidants. In: *Glycerine Production and Transformation—an Innovative Platform for Sustainable Biorefinery and Energy*. (ed. M. Frediani, M. Bartoli and L. Rosi). London: IntechOpen https://doi.org/10.5772/intechopen.85913.

Wang, Y., Ling, C., Chen, Y. et al. (2019b). Microbial engineering for easy downstream processing. *Biotechnol. Adv.* 37 (6): 107365. https://doi.org/10.1016/j.biotechadv.2019.03.004.

Yenkie, K.M., Wu, W., and Maravelias, C.T. (2017). Synthesis and analysis of separation networks for the recovery of intracellular chemicals generated from microbial-based conversions. *Biotechnol. Biofuels* 10 (1): 119. https://doi.org/10.5772/intechopen.85913.

Yli-Ojanperä, M., Sierla, S., Papakonstantinou, N., and Vyatkin, V. (2019). Adapting an agile manufacturing concept to the reference architecture model industry 4.0: a survey and case study. *J. Ind. Inf. Integr.* 15: 147–160. https://doi.org/10.1016/j.jii.2018.12.002.

Yuan, X., Shi, X., Zhang, D. et al. (2011). Biogas production and microcystin biodegradation in anaerobic digestion of blue algae. *Energy Environ. Sci.* 4: 1511–1515. https://doi.org/10.1039/c0ee00452a.

Zaini, N.A.M., Chatzifragkou, A., Tverezovskiy, V., and Charalampopoulos, D. (2019). Purification and polymerisation of microbial d-lactic acid from DDGS hydrolysates fermentation. *Biochem. Eng. J.* 150: 107265. https://doi.org/10.1016/j.bej.2019.107265.

12

Scale-up Process Challenges in Lignocellulosic Biomass Conversion and Possible Solutions to Overcome the Hurdles

Henrique M. Baudel[1,2], *Danielle Matias Rodrigues*[1,3], *Eduardo Diebold*[1], *and Anuj Kumar Chandel*[4]

[1] *America Biomass Technologies, Piracicaba, 13.419-190, Brazil*
[2] *Department of Chemical Engineering, Federal University of Pernambuco, Recife, 50100-100, Brazil*
[3] *Post-graduation Program, Bioenergy, University of Campinas (UNICAMP), Campinas, 13083-970, Brazil*
[4] *Department of Biotechnology, Engineering School of Lorena (EEL), University of São Paulo, Lorena, 12.602-810, Brazil*

12.1 Introduction

In a typical biorefinery, lignocellulosic feedstocks are fractionated and processed into a myriad of intermediates and final products such as biofuels and chemicals (Chandel et al. 2020; De Jong et al. 2020). Different lignocellulosic biomass feedstocks can be processed individually or simultaneously by using a common integrated platform. Lignocellulosic sugar-based biorefineries often encompass both chemical and biochemical routes in integrated production systems (IPSs) for biofuels and chemicals of industrial interest, as shown in Figure 12.1.

Over the last few decades, several initiatives have been implemented worldwide aiming to develop biorefinery configurations at different plant scales based on the technology readiness level (TRL). Table 12.1 presents some important examples of biomass-based operations from various countries.

In this scenario, scaling-up from bench to pilot, from pilot to integrated pilot and from integrated pilot to flagship (demonstration) plants has been pointed out as one of the most challenging efforts to effectively implement and consolidate biorefineries at commercial scale from economic and environmental feasibility standpoints. Table 12.2 shows important initiatives on biomass conversion from Europe highlighting different TRLs, status, and production/processing capacities.

Efficient lignocellulosic conversions in biorefinery configurations at demo (flagship) and commercial scale require relevant mechanical, chemical, and biochemical issues to be identified and managed in order to overcome costly hurdles prior to operating. High recovery and selective conversion of lignocellulosic sugars into target products, effective integration and reduced process downtime levels can be pointed out as major scale-up challenges being faced by the diverse flagships and commercial-scale biorefineries worldwide. High

Lignocellulose Bioconversion Through White Biotechnology, First Edition. Edited by Anuj Kumar Chandel.

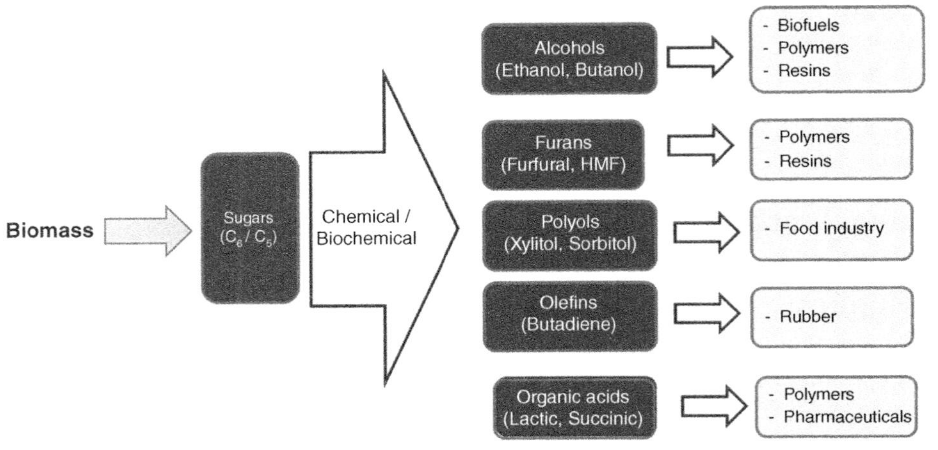

Figure 12.1 Basic description of biomass to products via sugar-based platforms. *Source:* adapted from www.americabio.com.br.

Table 12.1 Lignocellulosic based biorefineries worldwide funded by companies.

Company	Country	Target products/technology	Remarks
ORLEN Poludnie	POL	Cellulosic ethanol from agro residues	Flagship plant
UPM	GER	Chemicals and fuels from wood	Greenlighted. €500 millions 25 kton year^{-1} 2G ethanol
Bharat	IND	Biofuels from rice straws and corn stalks	Greenlighted. 7.5×10^6 gal year^{-1}
Croatian O&G (INA)	CRO	Cellulosic ethanol from miscanthus and wheat straw	300 kton year^{-1} wastes and crops Sisak plant. 55 kton year^{-1} 2G ethanol
Graanul Invest	STO	Biomaterials, sugars and lignin products from wood wastes	Flagship plant
PRAJ	IND	2G ethanol from sugarcane bagasse	Pilot plant (TBC)
GranBio	BRA	2G ethanol from sugarcane straw, cellulose products	Expansion to commercial
Raízen	BRA	2G ethanol from sugarcane bagasse	Expansion to commercial
SCA Biorefinery	SWE	2G ethanol, lignin, and biomaterials from wood wastes	Expansion to commercial Co-located at Östram pulp mill
Versalis	ITA	2G ethanol, PHAs and polymers from lignocellulosic biomasses	Crescentino flagship plant
MSF	AUS	Sugarcane and agave products for biofuels, aviation fuels and biomaterials	55 000 m^3 ethanol Queensland. Greenlighted
GeoSynFuels	USA	C_5 fermentation technology	Flagship WBE plant
Stora Enso	SWE	Biofuels and products from wood	Flagship plant. Expansion forecast by 2024

Table 12.1 (Continued)

Company	Country	Target products/technology	Remarks
POET Biorefining	USA	2G ethanol production from corn stover	Commercial
Nova Pangaea Technologies	UK	C_5/C_6 sugars, lignin char REFNOVA (thermolysis) technology	License to JS Group (Pakistan) Commercial (25 kton SCB per day)
Avantium	HOL	FDCA, MEG and PEF from industrial sugars. Dawn, YXY and Ray technologies	Expansion of flagship plant Lignin for biobased asphalt
Annelotech	USA	BTX for renewable chemicals and fuels	Flagship plant, expansion to 500 tons per day pine
Sweetwater Energy	USA	Biomass sugars, lignin and cellulose. SunBurst technology	License to Graanul flagship plant (Estonia) wood wastes
Chempolis	FIN	2G ethanol, furfural, acetic acid. FormicoBio technology	NSF (India) plant €160 million plant. 60×10^6 L 2G ethanol
Arbiom	USA	Protein and amino acids from wood	19 kton furfural, 11 kton acetic acid. Pilot plant for trials

Abbreviations: PHA, polyhydroxyalkanoate; SCB, sugarcane bagasse.

Table 12.2 Biomass-to-ethanol biorefineries: European initiatives at different technology readiness level (TRL), status and production/processing capacities.

Company	Plant/technology	Country	TRL	Start-up year	Status (2020)	Capacity (ton year^{-1})
Borregaard	ChemCell Ethanol	Norway	9	1938	Operational	16
Dömsjoe Fabriker		Sweden	8	1940	Operational	19
St1	Cellulonix	Finland	6–7	2017	Operational	8
Chempolis	Biorefining plant	Finland	6–7	2008	Operational	5
Clariant	Sunliquid	Germany	6–7	2012	Operational	1
IFP	Futurol	France	6–7	2016	Operational	350
SEKAB	Biorefinery Demo	Sweden	8	2004	Operational	160
Borregaard	Biorefinery Demo	Norway	6–7	2012	Operational	110
Clariant	Sunliquid	Romania	8	2021	Construction	50
AustroCel		Austria	8	2020	Construction	30
Enviral	Sunliquid	Slovakia	9		Planned	50
ORLEN Poludnie	Sunliquid	Poland	9		Planned	25
INA	Sunliquid	Croatia	8		Planned	55
St1	Cellulonix Kajaani	Finland	8	2024	Planned	40
St1	Cellulonix	Finland	8	2024	Planned	40

(*Continued*)

Table 12.2 (Continued)

Company	Plant/technology	Country	TRL	Start-up year	Status (2020)	Capacity (ton year^{-1})
St1	Cellulonix Follum	Finland	8	2024	Planned	40
Versalis	Crescentino	Italy	8	2020	Planned	40
Sainc Energy	Cordoba	Spain	8	2020	Planned	25
Kanteleen Volma	SEKAB CelluApp	Finland	6–7	2021	Planned	65

Borregaard, Dömsjoe and AustroCell use spent pulping liquor as feedstock.

feedstock recalcitrance, biomass clogging in the feeding devices, accumulation of biomass in curves and pipelines, erosion and corrosion of equipment, devices and valves emerge as frequent issues that increase operational and capital costs and process downtime, thus reducing overall process reliability (Chandel et al. 2019). The adoption of pseudo-continuous (synchronized batch) configurations coupled with preconditioning of the biomass feedstock using sprinkling or water washing and stripping, the use of catalysts and promoters to reduce thermal-mechanical severities and the proper design of curves and pipelines have emerged as effective alternatives to minimize problems related to downtime as well as costly problems related to the erosion and corrosion of metallic parts used in equipment and lines. Scaling-up lignocellulosic conversion processes from bench and pilot plants to flagship (demo) plants might be very challenging and risky, particularly when sequential chemical and biochemical routes are integrated in biorefinery configurations. Scalable products like lactic acid, cellulosic ethanol and mycoproteins are getting attraction in European countries (Table 12.3).

There are several challenges associated with scale-up of biomass processing. Reasons include lack of understanding of feedstock characteristics, handling of the feedstock, use of conditions or configurations successfully employed at bench scale (feedstock sampling and preparation, low solid loadings) but not replicable at pilot (single or integrated) or at flagship scale, and inadequate modeling among other aspects such as non-systemic approach coupled with insufficient testing at integrated pilot scale prior to replication at demo/flagship scale. Integration is crucial to overcome the inaccuracy of mathematical models to describe complex chemical

Table 12.3 Lignocellulosic based flagships funded by BBI-JU (Europe).

Company/project	Country	Products	Fund (€ million)
Agrichem Whey (County Tipperary)	Ireland	Lactic acid (building block for PLA), minerals and fertilizers from dairy processing side streams	22.0
Plenitude (Ghent)	Belgium	Mycoproteins and bioethanol from cereal crops	17.0
Lignoflag	Romania	2G ethanol from wheat and barley straw	25.0
Bioskoh	Slovakia	2G ethanol, bioethylene oxide from wastes/crops	21.6

BBI JU, Bio-based Industry Joint Undertaking: a €3.7 billion partnership between the European Union and the Bio-based Industry Consortium. PLA, polylactic acid.

reactions and biochemical pathways. Hence, the design and operation of integrated pilot and demonstration plants consist of the most reliable resources to generate the information and data required to properly select the main and ancillary equipment that will be acquired and used at a commercial scale. Improper equipment selection, particularly with regard to the processing of heterogeneous and crude feedstocks, results in costly downtime and operational expenditure (OPEX) related issues such as inadequate loading of the feedstock, low mixture performance, mechanical and heat transfer problems, low efficiency of separation devices, as well as relevant capital expenditure (CAPEX) related issues (e.g. corrosion and erosion) that considerably reduce the lifetime of the main equipment (e.g. pretreatment reactors) as well as of pipelines and ancillary devices such as valves and connections (Warner 2015).

The present work describes the relevant information associated with the challenges, problem-solving issues, and opportunities evidenced at scale up from bench to pilot (single and integrated) and flagship plant of biomass-to-products routes mostly related to long-run (500–1200 h) campaigns performed in Brazil, the United States and European geographies for the production of 2G bioethanol and chemicals from lignocellulosic feedstocks such as sugarcane bagasse/straw, eucalyptus, bamboo, elephant grass, and rice shells using steam pretreatment. It is noteworthy that such initiatives involve the participation of multifunctional teams from technology providers, engineering and biotechnology companies, and R&D institutions (e.g. universities and technology centers) working in close collaborations over the last 20 years.

12.2 Lignocellulosic Conversion Processes and Engineering: Challenges and Possible Solutions

Basically, the lignocellulose to bioproducts consists of chemical and biochemical routes, followed by product purification and recovery of by-products, and finally process integration, including utilities and useful recyclable wastes, as shown in Figure 12.2. From the process point of view, major steps consist of lignocellulose fractionation into holocellulose (cellulose and hemicelluloses) and lignin, saccharification of holocellulose into convertible sugars, and lastly chemical or biochemical conversion of sugars into biofuels and chemicals. Other important steps include raw/intermediate materials collecting, handling, conditioning (which might include pretreatment) and auxiliary operations such as separations/detoxification/purification, product purification and recovery, integration of streams and waste management, including valorization treatment and disposal (Virmond 2013). From a production point of view, engineering, design and operational standpoints, the aspects that significantly influence the techno-economic and environmental feasibility of the biorefinery configuration as an IPS are mostly related to process, integration, operational capability, and waste management. In this context, proper equipment selection, process design, use of cost-effective, safe, and environmentally friendly chemicals and additives, as well as productive and robust microorganisms are mandatory.

Cost-effective biomass fractionation and conversion of sugars into biofuels and biochemicals using physicochemical, enzymatic and biochemical routes requires adequate understanding on how to effectively combine feedstock characteristics, process engineering fundamentals, equipment selection, and operational issues. In particular, proper biomass

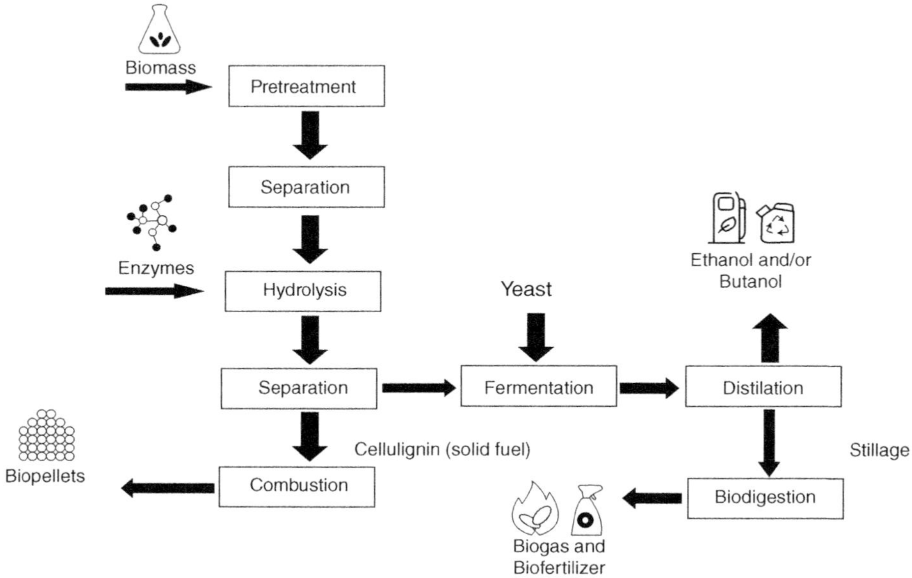

Figure 12.2 Typical flow diagram of a biorefinery for the production of biofuels and other products from lignocellulosic biomasses. Separate hydrolysis and fermentation (SHF), with clear mash and cellulignin recovery. *Source:* adapted from Baudel and Diebold (2018).

conditioning (handling and pretreatment) is a crucial step to overcome biomass recalcitrance aiming to produce holocellulosic sugars for production of biofuels, biochemicals, and other products of industrial interest via chemical technology and biotechnology platforms.

The buffering capacity of lignocellulosic biomass feedstocks significantly reduces the decarboxylation of the hemicelluloses during the pretreatment, and thus decreases hemicellulose removal. Consequently, pretreated materials tend to present lower fiber reactivity, leading to lower conversions of the holocellulose into fermentable sugars after enzymatic hydrolysis. Usually, the procedure to estimate the chemical (non-structural) recalcitrance (buffering capacity) of a certain lignocellulosic feedstock consists of immersion of the biomass in a solution of H_2SO_4 using a proportion of 0.2 g H_2SO_4 (g dry biomass)$^{-1}$ and 10% of the biomass. After 60 min of immersion, the pH of the mixture is measured and the material is autoclaved at 120 °C for 60 min. After cooling the mixture, a pH measurement is carried out. The primary recalcitrance R_1 (atmospheric acidification at room temperature) and secondary R_2 (hot acidification under pressure) are determined using the following equations:

$$R_1 = 100 * \left(\mathrm{pH_i} - \mathrm{pH_f}\right)^{-1}$$

$$R_2 = \left(\mathrm{H_f^+} - \mathrm{H_i^+}\right)^{-1}$$

As shown in Table 12.4, the aqueous mixture produced from the *Eucalyptus globulus* showed a greater recalcitrance (buffering capacity) in relation to *Eucalyptus grandis*,

Table 12.4 Primary and secondary recalcitrance from different feedstocks.

	R_1 (primary)			R_2 (secondary)		
Biomass	**Native**	**Pre-washed**	**Removal (%)**	**Native**	**Pre-washed**	**Removal (%)**
Eucalyptus globulus (14 years)	22.5	19.4	13.8	46.5	44.1	5.2
Eucalyptus grandis (14 years)	21.3	18.5	13.1	32.1	30.8	4.0
Sugarcane bagasse (crushers)	18.4	16.7	9.2	24.7	22.9	7.3
Sugarcane bagasse (diffuser)	16.1	15.5	3.7	24.3	22.6	7.0
Sugarcane straw (dry cleaning)	20.9	17.2	17.7	26.7	24.3	9.0
Elephant grass (milled)	19.9	16.7	16.1	27.6	25.9	6.2
Corn stover (milled)	18.5	15.9	14.1	21.1	19.0	10.0
Bamboo fines (micronizer)	21.8	17.9	17.9	25.4	24.1	5.1
Rice shells	23.7	19.5	17.7	33.3	32.3	3.0

Pretreatment conditions: 0.18 g H_2SO_4 (g dry biomass)$^{-1}$; 10 g dry biomass (100 g water)$^{-1}$; 120 °C; 60 min.

evidenced by the lower pH and [H^+] variations in relation to white (sample acidified at room temperature and atmospheric pressure), as well as in relation to the higher value of the secondary recalcitrance (R_2), arising from the most severe conditions used during the hydrothermal process conducted for the pretreatment in the autoclave.

Overcoming biomass recalcitrance followed by enzymatic hydrolysis of structural polymeric carbohydrates with cost-efficient liberation of fermentable sugars is one of the most challenging technical and economic barriers to the success of biorefineries (Zhang et al. 2009; Chandel et al. 2019). Pretreatment is certainly crucial pathway for the effective production and conversion of lignocellulose sugars, being thus the important key to producing low-cost cellulosic ethanol (Yang and Wyman 2008). Pretreatment of highly recalcitrant lignocellulosic materials is essential to achieve high sugar yields from enzymatic hydrolysis for the biological conversions. Nevertheless, pretreatment costs might represent above 25% of the total operational and capital costs (OPEX and CAPEX) associated with the production of a single product such as cellulosic ethanol. Basically, pretreatment strongly influences practically all downstream operations such as enzymatic hydrolysis, biological conversions via fermentation, as well as auxiliary operations such as separations (e.g. filtration, centrifugation), solid processing, product recovery, and conditioning of by-products and wastes (Baudel 2006; Yang and Wyman 2008).

Various pretreatments involving biological, chemical, physical, and hydrothermal approaches have been investigated over the last decades; however, a minor part of these can be considered as scalable alternatives in terms of operational, techno-economic viability, and environmental aspects. In this scenario, pretreatments such as using dilute acid, liquid hot water, steam and wet (hydrothermal) explosion, wet alkaline oxidation and ammonia fiber expansion (AFEX) have emerged as promising alternatives (Bhatia et al. 2019; Chandel et al. 2019). Such pretreatments present different combinations of mechanical and chemical severities, as shown in Figure 12.3, as well as process and operational related attributes that impact overall technical, economic, operational, and environmental feasibilities.

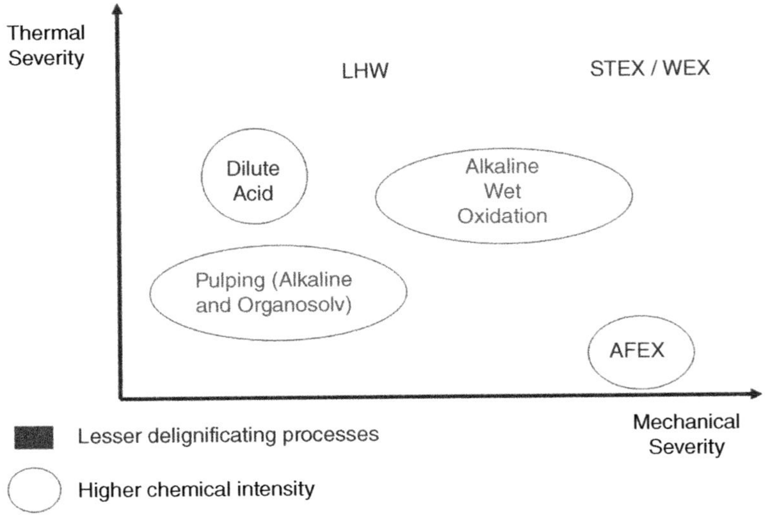

Figure 12.3 Thermal, chemical, and mechanical severity levels of different physicochemical pretreatments. *Source:* adapted from Baudel (2009).

Although a number of pretreatment methods have been proposed and investigated on the laboratory scale and even pilot plant scale during the past several decades, many of the data in the literature have been obtained and reported under different standards, leading to miscommunication issues among different players such as academic researchers, technology providers, and potential industrial investors (Zhang et al. 2009; Chandel et al. 2018). In order to circumvent such issues, the adoption of an efficiency index criterion (as shown in Table 12.5) emerged as a tool for comparing, selection, and decision-making processes among different pretreatment options, considering the specific needs and requirements of the industry.

Table 12.5 Process and operational parameters between different pretreatments.

Parameter	DA	LHW	WEX	STEX	WAO	AFEX
Ethanol yield	75	80	90	100	75	95
Complexity/SHE issues	100	95	90	85	85	80
Solid loading	80	80	90	100	85	85
Chemicals	95	100	100	100	85	80
Water (usage/cost)	90	90	95	80	85	100
Energy (usage/cost)	90	90	90	80	90	100
Equipment/installations	100	100	90	85	85	80
Lignin processing	100	100	100	100	90	90
Efficiency index	98	98	100	99	92	96

Abbreviations: AFEX, ammonia fiber expansion; DA, dilute acid; LHW, liquid hot water; SHE, Safety Health Environment; STEX, WEX, steam and wet (hydrothermal) explosion; WAO, wet alkaline oxidation. *Source:* Adapted from Baudel (2009).

12.2.1 Steam Pretreatment: Issues and Potential Problems

Steam explosion (STEX®) is recognized as one of the most cost-effective pretreatment processes for lignocellulosic biomasses such as hardwoods and agricultural residues, but it is less effective for softwoods (Sun and Cheng 2002). However, there are some important process related issues and limitations such as degradation of sugars with generation of compounds that may be inhibitory to microorganisms for the fermentation reactions. Also, difficulties are related to complex chemical reaction kinetics due to the presence of consecutive, parallel, and side reactions (e.g. dehydration and oxidation of xylose into furfural and organic acids) and non-selective fractionation of lignin–carbohydrate complex. The use of aggressive chemicals such as H_2SO_4, H_3PO_4, and SO_2 as catalysts might result in problems associated with the corrosion of equipment, devices, and pipelines, as well as with the emission of acidic vapors into the atmosphere. Because of the formation of degradation products that are inhibitory to microbial growth, enzymatic hydrolysis, and fermentation, steam pretreated biomasses need to be conditioned in order to remove the inhibitory compounds along with water-soluble hemicelluloses (Sun and Cheng 2002).

Other hurdles can be related to the transportation of the feedstock and subsequent loading of the biomass in the pretreatment reactor, formation of impurities and inert materials inside the reactor, as well as mechanical problems related to the occlusion or erosion of pipes, reactor lines, connections, and ancillary devices (e.g. valves). However, mostly problems that can negatively affect the efficiency of the pretreatment (and thus the downstream operation steps) can be overcome by proper management of attributes of the biomass feedstock such as the origin and chemical/structural characteristics, and by adopting proper process engineering related techniques, such as the use of moderate catalysts and promoters, equipment design and configuration of the IPS, as well as the incorporation of auxiliary operations such as post-conditioning of pretreated material, for example.

As described in Figure 12.4, biomass fractions with fairly similar holocellulose and lignin contents that underwent the same steam pretreatment might result in pulps with different fiber reactivities. Hence, it is important to have in mind the intrinsic variability and reliability of the steam pretreatment due to small heterogeneities of the feedstock composition. The adoption of proper ranges of process severity levels combined with the use of selective catalysts tend to reduce such differences.

Sugarcane bagasse originating from diffusers (extractors that remove sucrose-rich juices from sugarcane using pressurized water) tends to be less recalcitrant for the removal of holocelluloses (mostly hemicelluloses) comparably to the bagasse fractions generated from mill/crushers, notably if rigid cane varieties were steam pretreated in the absence of catalysts, as shown in Figure 12.5. The removal of hemicelluloses (until a certain level) tends to enhance fiber reactivity of the steam pretreated biomasses, as shown in Figure 12.6. Again, the adoption of suitable severities combined with the use of low to severe catalysts tends to significantly reduce such differences thus enhancing fiber reactivity.

Our experimental findings confirmed the densification of the feedstock after stripping with saturated steam. This stripped biomass can be loaded in the pretreatment reactor with a higher solid amount. Stripping of biomass feedstock such as sugarcane bagasse and straw, rice shells and bamboo fines, can facilitate the higher biomass loadings by 10%–15%. Stripping the feedstock prior to the steam pretreatment facilitates in removal of inert compounds and impurities, in turn increasing the density of the biomass. Stripping of the biomass during the

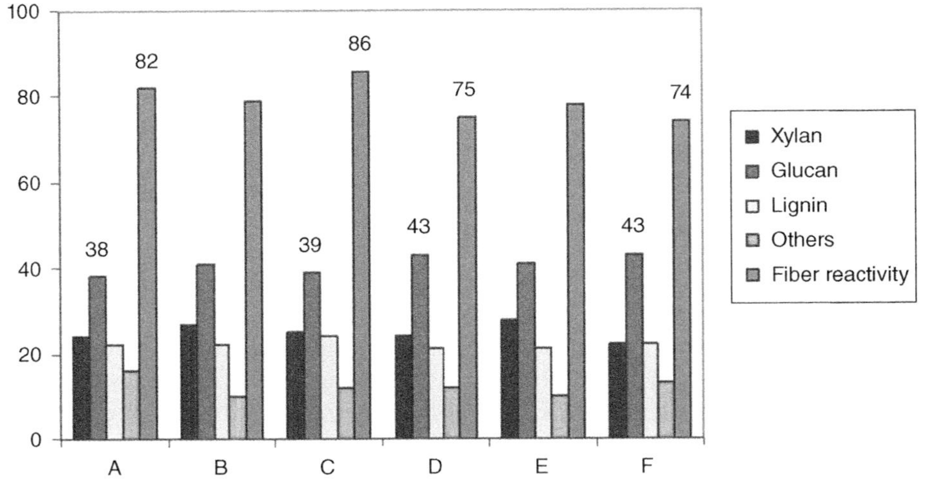

Figure 12.4 Fiber reactivity (FR) of steam-pretreated pulps originating from sugarcane bagasse fractions collected at the same mill but in different harvesting months. STEX: 200 °C, 5 min, no catalyst. 24 h enzymatic hydrolysis, 2% WIS, pH 4.8–5.2, 50 °C, 15 FPU/g (FPU, filter paper unit). FR values are expressed in conversion (%) of cellulose presented in the pretreated bagasse. Glucan, xylan, and lignin contents are expressed in g $(100\,g)^{-1}$. The *x*-axis refers to the type of the sample. The *y*-axis refers to the contents of xylan, glucan, lignin, and others (expressed in g (100 g dry raw bagasse)$^{-1}$) and to the fiber reactivity (expressed in percentage cellulose conversion).

pre-heating step using saturated steam facilitates in removal of air entrapped in the feedstock cell wall. Stripping may facilitate the loading of higher biomass concentration in pretreatment. The removal of impurities decreases the buffering capacity (chemical recalcitrance) and coupled with the removal of the "inert air-seal" and volatile contaminants from the loaded biomass results in more efficient decarboxylation and subsequent removal of hemicelluloses, thus increasing fiber reactivity of the pretreated material as well as enhancing sugar (C_6) recovery and fermentability of the hydrolysates, as shown in Table 12.6.

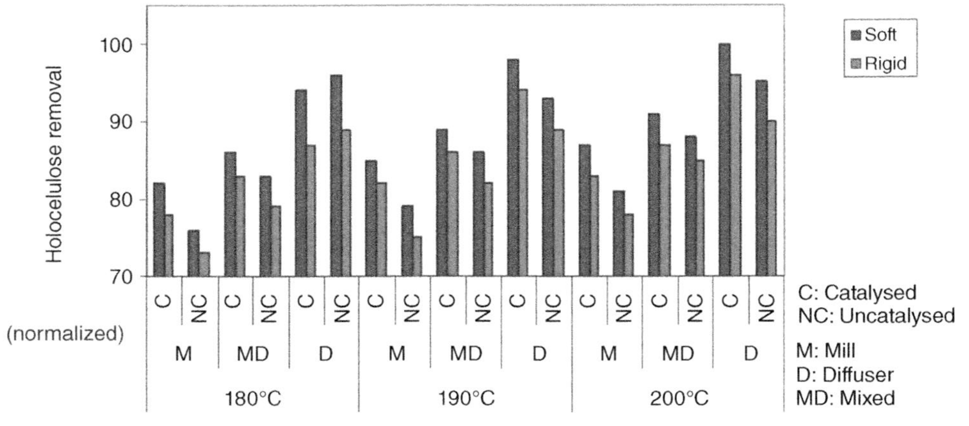

Figure 12.5 Holocellulose removal of steam pretreated sugarcane bagasse for different feedstock origins and processing, for catalyzed and uncatalyzed processes. STEX conditions: 205 °C, 10 min, 20 bar steam, with stripping. Catalyst CO_2 (1.0 g (100 g DM)$^{-1}$)/H_2SO_4 (0.2 g (100 g DM)$^{-1}$) (DM, dry matter; normalized values).

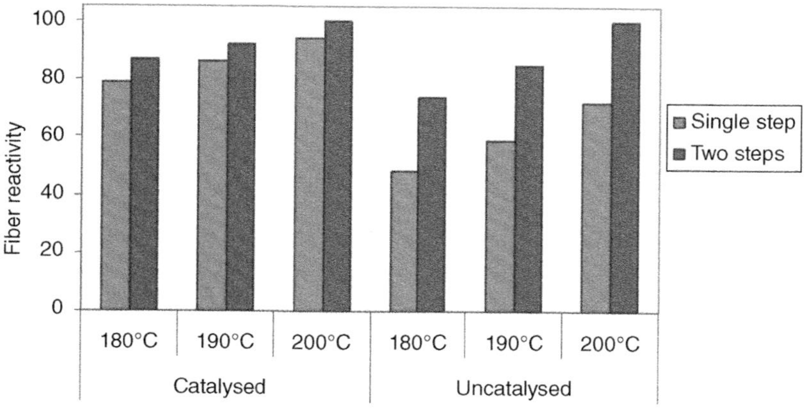

Figure 12.6 Effect of catalyst on steam pretreatment of water-sprinkled eucalyptus bark. 205 °C, 10 min, 20 bar steam, with stripping. Catalyst CO_2 (1.0 g (100 g DM)$^{-1}$)/H_2SO_4 (0.2 g (100 g DM)$^{-1}$). Fiber reactivity determined by enzymatic hydrolysis. Enzyme usage 4 mg protein (g cellulose)$^{-1}$, 12% WIS, 50 °C, pH 4.8–5.2, 48 h, Cellic Ctec3. Normalized values shown.

Table 12.6 Comparison of sprinkling and stripping of lignocellulosic biomasses.

Feedstock	SPK	STP	HC removal (%)	Fiber reactivity (%)	C_6 recovery (%)	C_6 fermentability (g EtOH/g C_6)
Eucalyptus chips	NO	NO	87.2	66.6	62.1	0.42
	NO	YES	88.9	70.5	65.2	0.42
	YES	NO	92.4	72.3	66.4	0.43
	YES	YES	96.5	76.4	72.1	0.45
Bamboo fines	NO	NO	88.1	67.4	62.4	0.42
	NO	YES	92.4	71.3	65.3	0.42
	YES	NO	91.0	73.3	67.1	0.43
	YES	YES	97.2	76.9	72.5	0.44
Sugarcane bagasse	NO	NO	88.3	67.1	62.2	0.42
	NO	YES	89.6	71.4	66.3	0.42
	YES	NO	94.5	74.0	69.0	0.45
	YES	YES	97.5	78.5	74.8	0.45
Sugarcane straw	NO	NO	86.4	67.3	62.2	0.41
	NO	YES	89.0	71.1	65.9	0.41
	YES	NO	92.1	74.7	69.4	0.44
	YES	YES	97.0	77.9	75.3	0.45
Rice shells	NO	NO	85.0	66.4	63.2	0.39
	NO	YES	87.7	69.4	67.1	0.39
	YES	NO	90.4	72.7	68.5	0.42
	YES	YES	95.3	75.6	74.0	0.42

One-step uncatalyzed steam pretreatment: 205 °C (17.5 bar), 8 min.
Enzymatic hydrolysis: 12% WIS, 50 °C, pH 4.8–5.2, 48 min, 3 mg protein (g cellulose)$^{-1}$ (Cellic Ctec3), fed-batch mode.
Ethanolic fermentation: Mauri Y-904 yeast, 1.0% (w/w), 32 °C, 24 h.
Abbreviations: HC, hemicelluloses; SPK, sprinkling; STP, stripping (5.0 bar, 2 min).

Proper combination of thermal severity coupled with the use of mild catalysts in a two-step configuration might enhance pretreatment performance in terms of the fiber reactivity and enzymatic digestibility of the steam pretreated eucalyptus bark, as shown in Figure 12.7.

Lignocellulose conditioning of the steam pretreated material using hot water (90 °C) washing might effectively remove impurities and contaminants, preventing problems such as lignin recondensation and redeposition (coating) into the fibers, as well as issues related to the inhibition of the enzymes and fermenting yeasts, for example. Nevertheless, this technique requires proper integration of streams and reuse via incorporation into other process and production steps in order to be cost effective and environmentally feasible. Mechanical conditioning techniques such as extrusion refining of the steam pretreated biomass are effective at bench and non-integrated pilot plants, but difficult to scale-up to demo, flagship, and industrial plants due to the different (and not reproducible at scale) dimension ratio between the mechanical devices of the "bigger" equipment (e.g. extruder screw, refiner discs, mill spheres) and granulometry of the pretreated biomass (particularly for pretreated sugarcane bagasse and straw) compared to tailored bench-scale versus standard commercial size equipment. As an example, 500-h runs of mechanical treatment (Century extruder) performed at pilot scale evidenced significant improvement for steam pretreated woody wastes such eucalyptus bark/chips and bamboo chips, but not at same level for sugarcane bagasse and cane straw. Figure 12.8 describes an integrated process for the production of 2G ethanol and bio-lignin from sugarcane bagasse and eucalyptus wastes using mechanical conditioning (extrusion) of steam pretreated biomasses.

Producing convertible sugars via holocellulose saccharification using enzymatic hydrolysis might require exploring some unconventional approaches such as pre-liquefaction of the pretreated material coupled with the adoption of a fed-batch profile for enzymatic

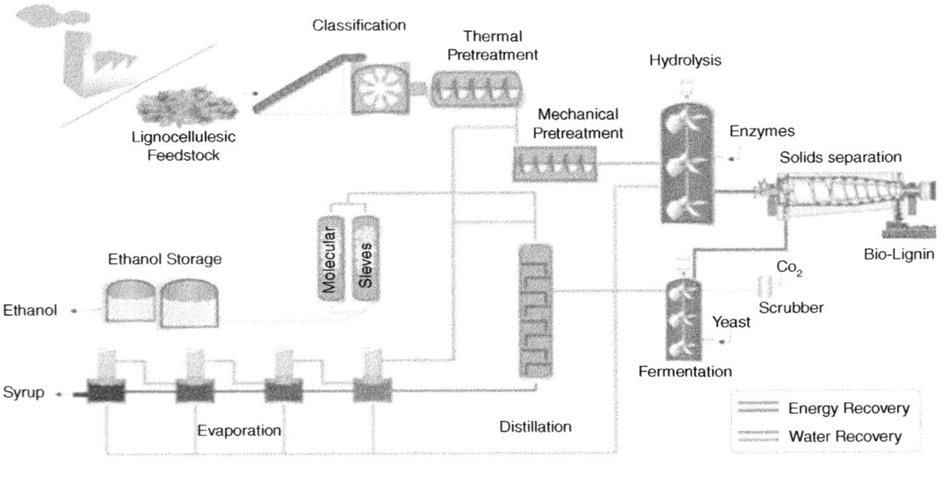

Figure 12.7 Integrated process for production of 2G ethanol and bio-lignin from sugarcane bagasse and eucalyptus wastes using mechanical conditioning (extrusion) of steam pretreated biomasses. Western Biomass Energy (WBE) flagship biorefinery plant (Wyoming, USA). *Source:* adapted from Baudel et al. (2014).

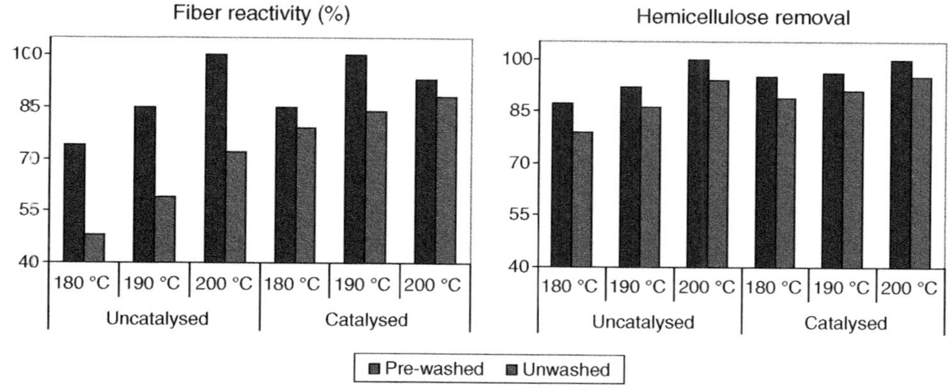

Figure 12.8 Correlation between hemicellulose removal and fiber reactivity of the steam pretreated sugarcane washed and unwashed straw. Sprinkle-washed (blue) and unwashed (red). Steam treatment: 205 °C, 10 min, 20 bar steam, with stripping. Catalyst CO_2 (1.0 g (100 g DM)$^{-1}$)/ H_2SO_4 (0.2 g (100 g DM)$^{-1}$). Fiber reactivity determined by enzymatic hydrolysis. Enzyme usage 4 mg protein (g cellulose)$^{-1}$, 12% WIS, 50 °C, pH 4.8–5.2, 48 h, Cellic Ctec3. Normalized values shown.

hydrolysis. In principle, previous liquefaction using a horizontal reactor that promotes a "cascading effect" of the pretreated material due to the circular motion coupled with the use of lower-cost endoglucanases tends to produce less viscous slurries mostly constituted by cellulose-derived oligomers which can be easily transferred to the enzymatic hydrolysis reactors by gravity or pumping. Separately, the fed-batch addition of the slurry allows simultaneous operations using proper liquid-to-fiber, water-insoluble solids (WIS) contents and enzyme-to-cellulose ratios at fairly lower instantaneous viscosity of the reaction medium. From the operational standpoint, important positives related to better control of pH and temperature, lower energy consumption and higher biomass saccharification yield (ton sugar m^{-3} h^{-1}) can be obtained with the use of combined liquefaction fed-batch profiles, as shown in Table 12.7.

Table 12.7 Influence of liquefaction and feeding mode of steam pretreated sugarcane straw.

	Direct liquefaction		Pre-liquefaction	
	Batch	**Fed batch**	**Batch**	**Fed batch**
WIS (%)	18	20	20	22
Process time	72	60	48	48
Time	100	90	88	88
Enzyme usage	100	84	84	81
Cost factor	100	96	92	87

One-step catalyzed steam treatment. 205 °C, 10 min, 2 min stripping. Sugarcane straw.
Enzyme usage (normalized) to reach 82% conversion of cellulose. Cellic Ctec3, 48–52 °C, pH 4.8–5.2.
Cost factor (normalized): CF = [(CAPEX + PEX)/CAPEX]/ton C_6.

Converting biomass sugars using chemical and biochemical routes requires full integration with existing facilities and downstream process steps. Interesting examples consist of the use of stillage from ethanol purification in enzymatic hydrolysis to dilute and acidify alkaline steam pretreated biomasses, use of CO_2 recovered from fermentation using scrubbers as catalyst for steam pretreatment and for cleaning-up of hydrogenation catalysts, use of cane molasses (high BRIX) as a source of nutrients, and concentrated sugars properly mixed with dilute enzymatic hydrolysates, for example.

Scaling up and integrating processes for efficient lignocellulosic conversions in biorefinery configurations require several mechanical, chemical, and biochemical issues to be managed in order to overcome the hurdles prior to operating at commercial scale. Basically, the approach of increasing the production scale for converting biomass into chemicals of interest requires reaching efficient conversion of the holocellulose into fermentable sugars, maintaining or improving the efficiency of conversion, selectivity and yields of such sugars into products of interest (particularly ethanol from lignocellulose), proper transportation of biomass between unit operations (as solids, liquids, or suspensions), with moderate viscosity, minimizing water and energy consumption in the various unit operations, and achieving the longest possible time of operation (minimum downtime) in accordance with the engineering design standards.

The recovery of pentoses is an important part of the process of reducing the unit cost of the final product, since their production, in principle, does not require enzymes. Additionally, the pentose-rich liquid streams can be efficiently used for production of ethanol, furfural, xylitol, and biogas, or even as dilution (reducing water usage) of molasses. The Renner-AmericaBio process of steam pretreatment of lignocellulosic and starchy biomasses (Patent Request WIPO WO 2017/168207, Baudel and Diebold 2017) addresses potentially interesting ways to simultaneously optimize pentose recovery and minimize the generation of inhibitors that inhibit the subsequent enzymatic hydrolysis and ethanolic fermentation steps. Such process adopts the configuration of a blow tank specially designed to recover pentoses and remove furans and organic acids via stripping, rendering a material with high fiber reactivity and cellulose accessibility. Hence, the blow tank configuration allows the main functions of fiber exposure and pentose hydrolysis to be performed in the STEX process. Being closed and connected to a gas scrubber/condenser, the blow tank avoids the release of sugars to the atmosphere, and deactivates contaminants and inhibitors generated by the side reactions that take place inside the steam pretreatment reactor.

As mentioned by Silveira et al. (2018), numerous challenges are associated with the scale-up of biomass pretreatment in continuous STEX operations for hours or even days. Among these challenges are the continuous biomass feeding in a pretreatment reactor and smooth recovery of pretreated slurry, the holding-up of the desired steam pressure inside the STEX reactor without biomass leakage, and minimization of evaporation of sugars in order to obtain near-quantitative mass balance. Operational issues such as mass transportation discontinuity and equipment wearing have been responsible for most downtime issues faced by integrated pilot and flagship plants of 2G ethanol (cellulosic ethanol). Such problems have been mostly attributed to the erosion of the metallic structure of the pretreatment reactors, valves, connections, and pipelines by the solid biomass itself as well as by impurities that are brought into the process (e.g. extraneous soil material from sugarcane). As mentioned earlier, the preconditioning of the biomass feedstock using techniques such as sprinkling or water washing, stripping the feedstock with saturated steam prior to the reaction itself, and use of catalysts and promoters to reduce thermal-mechanical severities has emerged as effective alternatives to minimize problems related to the erosion and corrosion of metallic parts used in equipment and lines.

By reapplying the concepts adopted by major chemical and biomass processing industries (e.g. pulp and paper), the technologies already put into practice on an industrial scale have opted for continuous processes to convert heterogeneous feedstocks into ethanol and chemicals, following the approach used by medium or large production capacities within the biofuels industry. However, this approach has shown some important negative issues faced by the integrated pilot and flagship plants associated with the high downtime levels evidenced by the continuous process configurations. These problems are significantly magnified for plants that process withing the range of 50 000–100 000 tons dry biomass per year, with major incidences of biomass clogged or stuck in the feeding devices, accumulation of biomass in curves and pipelines, as well as friction wear in equipment and valves, for example. In some cases, such problems cause serious interruptions of regular operations, as well as leading to downtime and process reliability problems related to the need to restart the whole system.

The above-mentioned disadvantages can be minimized or even eliminated by using a pseudo-continuous (synchronized batch) process strategy. Pseudo-continuous processing allows the transport of the wet solid biomass to the liquefaction prior to enzymatic hydrolysis. The transport is done by means of rubber mats and/or screw conveyors. The volume of intermediate stocks to regulate flows tends to be minimal, being no greater than 1 h of biomass consumption between operations and start-ups. Additionally, pseudo-continuous configurations facilitate the maintenance of production, at least partially, of the biorefinery and avoids the need to carry out a general start-up of the plant after solving a problem. The increase in CAPEX, even including redundancies between equipment, is practically negligible because the initial system becomes much simpler to operate using cheaper main and ancillary equipment. This approach is particularly suitable for biorefineries with processing capacity within 200–700 tons biomass per day. For capacities between 700 and 1100 tons per day the continuous or pseudo-continuous systems are fairly equivalent. Continuous systems are more suitable for capacities of biomass processing greater than 1100 tons biomass per day.

It is important to mention that the decision criteria also depend on the country/region where the biorefinery will be installed. The vast majority of biorefineries in Europe process approximately 200–700 tons biomass per day because of the availability and logistics of the feedstock coupled with the discouragement of cultivating "energy crops." Such a scenario is completely different within geographies such as the United States and Brazil, where the average processing capacity might range from 500 to 1100 tons biomass per day (Diebold and Baudel 2020). Table 12.8 shows some relevant differences between continuous and pseudo-continuous configurations of steam pretreatment for sugarcane bagasse, cane straw, and eucalyptus wastes in terms of process, cost and operational aspects, metrics and parameters.

The use of rubber belts, metallic grids or screw conveyors transport the material without any major problems, even at moisture contents varying from 35% to 75%. With regard to the incidence of wear of pipelines and equipment, many projects on a commercial scale need to manage unexpected higher abrasion capacity from minerals and impurities originating from the soil and from the feedstock itself. An alternative to circumvent such problems

Table 12.8 Comparison of continuous and pseudo-continuous biomass processing configurations. Diebold and Baudel, 2020.

	Sugarcane bagasse/straw		Eucalyptus wastes	
Parameters	**Continuous**	**Pseudo-continuous**	**Continuous**	**Pseudo-continuous**
Downtime (%)	38	19	16	11
Build-up/occlusion	High	Low	Medium	Very low
Recovery	70	72	72	72
Degradation	5%	3%	2%	2%
Fiber reactivity	82%	82%	82%	82%
Enzyme usage	94	86	100	90
Cost factor	100	92	92	89

One-step catalyzed steam treatment. 205 °C, 10 min, 2 min stripping.
Enzyme usage (normalized) to reach 82% conversion of cellulose. Cellic Ctec3.
Fed-batch mode, 48–52 °C, pH 4.8–5.2, 48 min.
Cost factor (normalized): CF = [(CAPEX + OPEX)/CAPEX]/ton C_6 (includes the incidence of build-up/occlusion due to the presence of impurities inside the pretreatment reactor).

might consist of using longer curves and collecting/conditioning vessels, particularly with solid wet biomasses with moisture contents greater than 60%. An illustrative example is the STEX reactor/blow tank set. The initial project foresaw the bottom outlet of the reactor fitted with a 90° open curve and 20-m-long pipeline connected to the blow tank that collects and spreads the jet of biomass, steam, and condensed water along the side wall of the equipment. The friction of the biomass along the path and on the blow tank wall itself is important to increase the exposure of the fibers, but a similar effect can be obtained if we position the reactor above a median position and make an exit with a 45° curve placing an additional 45° curve close to the blow tank inlet. Inside the blow tank a steel sacrifice plate alleviates the impact of the high-speed wet steam treated biomass promoting a "peeling effect" that "refines" the material, significantly increasing fiber reactivity and accessibility, while maintaining the integrity of the blow tank walls and ancillary devices at minimum repairing cost.

12.3 Ethanol from Eucalyptus Wastes

The pretreatment by STEX in a pilot plant (1000 L ethanol day^{-1}) was performed using a STEX reactor (AmericaBio, Brazil) equipped with a 5 m^3 expansion tank ("blow tank"). The fresh biomass was weighed, transported in an elevator, and loaded into the reactor. Subsequently, saturated steam (20 kgf cm^{-2}) was injected into the reactor in order to perform a stripping through the biomass for 5 min. After that, the purge valve was closed and the mixture was heated to 200 °C (internal pressure 17.5 kgf cm^{-2}) and stayed for 10 min cooking time. The STEX reactor was discharged (sudden decompression) by opening the quick discharge valve, and the material was expelled from the reactor and dispersed,

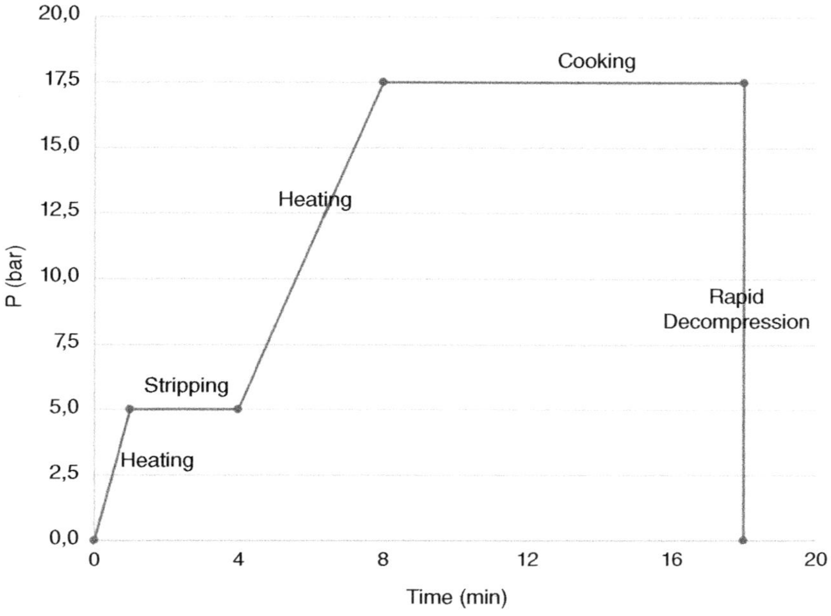

Figure 12.9 Thermodynamic profile of steam pretreatment of eucalyptus wastes.

expanded, and collected in the blow tank. The exhausting vapors and gases were condensed and collected in a scrubber/condenser equipment. The thermodynamic profile used in the pretreatment is shown in Figure 12.9.

The pretreated biomass (BPT1) was subjected to a hot water (90 °C) extraction and then impregnated with an amphoteric nitrogen-based catalyst in order to promote selective changes to the lignocellulose structure that enhance both fiber reactivity and cellulose accessibility to the hydrolytic enzymes (cellulases) used in the subsequent saccharification step. The impregnated biomass was then subjected to another steam treatment (T2) conducted under similar conditions, producing the pretreated biomass (BPT2). The liquids produced during the pretreatment were mixed and pumped to the pre-liquefaction reactor, in order to maximize the sugar recovery. No detoxification of the liquids was necessary at this stage due to the presence of nitrogen-based species that inactivate the inhibitors of the fermentation step such as furfural, hydroxymethylfurfural, and organic acids.

The lignocellulose pulp was pre-liquefied in 500-L STIRFLEX horizontal hydrolyzate reactors were designed, ced by AmericaBio, a company owned by the authors Eduardo Diebold and Henrique Baudel. Using ABT-420C (endoglucanases) at 50 °C, pH 4.5–5.5, for 40 min. The slurries were then pumped and batch fed into a 5000 m^3 vertical hydrolyzer (ABT5K) for the subsequent "finishing hydrolysis" step. The material was treated with commercial enzymes (Cellic Ctec3) provided by Novozymes at a dosage of 3 mg protein enzyme (3 mgP enzyme) (g cellulose)$^{-1}$, 50 °C, pH 4.8–5.2 (without use of any buffer), for 48 h reaction time, with absolute WIS content of 28% and relative WIS of circa 15%. The hydrolyzed slurry was then filtered (filter drum), the solid fraction (cellulignin cake) was

Table 12.9 Process and production metrics of 2G ethanol and biomethane from eucalyptus wastes.

Metric	Value
Pentose recovery (%)	88.5
Pentose degradation (%)	4.0–5.3
Hexose recovery (%)	80.0–82.0
Ethanol yield (L (ton dry biomass)$^{-1}$)	218.8
Biogas yield (m^3 methane (m^3 vinasse)$^{-1}$)	12–15

Vinasse ratio: approximately 12 L vinasse (L ethanol)$^{-1}$.

resuspended in water and the liquid was recycled and reused in the subsequent liquefaction, to recover sugars and desorbed enzymes. The "clean" cellulignin was hot-air dried and compacted into energy briquettes (solid fuel). The liquid fraction (C_6-rich enzymatic hydrolysate) was concentrated under vacuum to 180 g L^{-1} of (monomeric) glucose and then detoxified with urea/active carbon for subsequent ethanol production via conventional fermentation processes employed in Brazilian sugar/ethanol mills. It is important to remember that the unconverted pentoses were transferred to the stillage (80 kg chemical oxygen demand (COD) m^{-3}), being converted into biomethane via conventional anaerobic digestion. Tables 12.9 and 12.10 present the metrics and the most frequent issues associated with the integrated production of 2G ethanol and biogas from eucalyptus wastes, as well as their respective possible causes and solutions.

Table 12.10 Most frequent issues and solutions for 2G ethanol from eucalyptus wastes.

Main issues	Possible cause(s)	Solution(s)
Biomass handling and transportation Clogging of pipelines and valves Erosion and corrosion of equipment, valves, curves, and pipelines	Contamination with impurities from soil Slurries with high content of solids or higher particle size Friction of abrasive or corrosive pretreated biomass	Washing/water extraction Dilution or suspension of fibers Reduction or elimination of acidic catalysts Use of open curves and sacrifice plates in critical regions of the line
Steam pretreatment Non-uniform biomass cooking and differences in aspect and composition of the pretreated biomass Low hemicellulose removal Low recovery of pentoses Generation of inhibitors and contaminants	Moisture content of the feedstock below 30% Inadequate severity Insufficient stripping High feedstock recalcitrance and heterogeneity	Biomass hydration prior to the steam treatment Adjust process severity (heating profile, stripping, and reaction time) Use of catalysts

Table 12.10 (Continued)

Main issues	Possible cause(s)	Solution(s)
Enzymatic hydrolysis Presence of organic acids Heterogeneous temperature and pH profiles	Microbial contamination Inadequate stirring Higher viscosity of the slurry	CIP (cleaning in place) and sterilization Reduce WIS content Use of surfactants
Ethanolic fermentation Low fermentability of the hydrolysates Different temperature and pH levels inside the medium Low selectivity	Presence of inhibitors in the hydrolysate Inadequate stirring Higher viscosity of the slurry Microbial contamination	Detoxification with urea and active carbon Sanitization and CIP

12.4 Ethanol and Xylitol Production from Sprinkled Sugarcane Straw

The pretreatment by STEX in a pilot plant (1000 L ethanol per day) was performed using a STEX reactor (AmericaBio, Brazil) equipped with a 5 m^3 expansion tank ("blow tank"). The fresh biomass was weighed, transported in an elevator, and loaded into the reactor. Subsequently, saturated steam (20 kgf cm^{-2}) was injected into the reactor in order to perform a stripping through the biomass for 5 min. After that, the purge valve was closed and the mixture was heated to 180 °C (internal pressure 17.5 kgf cm^{-2}) and kept for 15 min cooking time. The STEX reactor was discharged (sudden decompression) by opening the quick discharge valve, and the material was expelled from the reactor and dispersed, expanded, and collected in the blow tank. The exhausting vapors and gases were condensed and collected in a scrubber/condenser equipment.

The pretreated biomass (BPT1) was subjected to a water extraction and a hydrothermal treatment was then carried out (180 °C, 30 min, 12% consistency) using sulfuric acid and a Lewis acid as homogeneous catalyst, to selectively remove the pentoses present in the hemicelluloses of greater chemical recalcitrance, as well as to promote selective structural modifications on the lignin–carbohydrate complex to enhance both fiber reactivity and cellulose accessibility to the hydrolytic enzymes (cellulases) used in the subsequent saccharification step. After the hydrothermal treatment, the material was filtered (press filter) producing a solid fraction (crude pulp) and a pentose-rich liquid fraction (hydrolysate).

The crude pulp (60%–65% cellulose content) was then pre-liquefied in three 500-L STIRFLEX horizontal hydrolyzers (AmericaBio, Brazil) using ABT-420C (endoglucanases) at 50 °C, pH 4.5–5.5, for 40 min. The slurries were pumped (positive displacement) and batch fed (four cycles) into a 5000 m^3 vertical hydrolyzer (ABT5K) for the subsequent "finishing hydrolysis" step. The material was treated with commercial enzymes (Cellic

Ctec3, Novozymes, USA) at 3 mgP enzyme (g cellulose)$^{-1}$, 50 °C, pH 4.8–5.2 (without use of any buffer), for 48 h reaction time, with absolute WIS content of 28% and relative WIS of circa 15%. The hydrolyzed slurry was then filtered (filter drum), the solid fraction (cellulignin cake) was resuspended in water, and the liquid was recycled and reused in the subsequent liquefaction, to recover sugars and desorbed enzymes. The "clean" cellulignin was hot-air dried and compacted into energy briquettes (solid fuel).

The liquid fraction (enzymatic hydrolysate) was mixed with sugarcane molasses (BRIX 80) in order to adjust the concentration of total fermentable sugars to approximately BRIX 20 (approximately 200 g L^{-1}), for subsequent ethanol production via conventional fermentation processes employed in sugar/ethanol mills.

The pentose-containing liquids produced by the steam and hydrothermal treatments were mixed and subsequently treated with $CaCO_3$ to remove sulfur (thus preventing catalyst deactivation), acidified with CO_2 (recovered from the ethanolic fermentation), stripped with saturated steam and then vacuum concentrated to 300 g L^{-1} of monomeric xylose. The "xylose syrup" was hydrogenated under batch-fed mode at 120 °C/20 bar during 120 min using Ni/Ru catalyst (0.05 g catalyst (g xylose)$^{-1}$) in a 500-L steel reactor (AmericaBio, Brazil). The catalyst was regenerated by flushing CO_2 (recovered from ethanolic fermentation) followed by an ultrasonic treatment after three consecutive operational runs and reincorporated as "make-up" to the subsequent set of three batches. Figure 12.10 shows the profile of conversion of xylose into xylitol via catalytic hydrogenation (Ni/Ru) of xylose-rich liquid streams generated from sequential steam pretreatment. Tables 12.11 and 12.12 show the metrics and the most frequent issues associated with the integrated production of xylitol and 2G ethanol from sugarcane straw, as well as their respective possible causes and solutions.

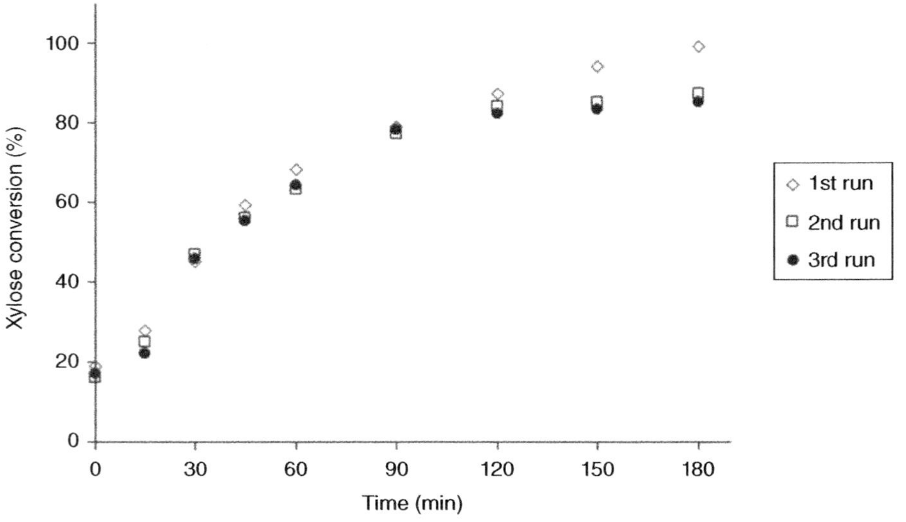

Figure 12.10 Selective conversion of xylose into xylitol via catalytic hydrogenation of a mixture of xylose-rich liquid streams generated from sequential steam pretreatment.

Table 12.11 Process and production metrics for xylitol and 2G ethanol from sugarcane straw.

Metric	Value1
Pentose recovery (%)	92.5
Pentose degradation (%)	3.0–3.5
Hexose recovery (%)	82.0–84.0
Ethanol yield (L (ton dry biomass)$^{-1}$)	208.5
Xylose conversion (%)	97.4
Selectivity xylitol (%)	99.5
Xylitol yield (kg (ton dry biomass)$^{-1}$)	184.0

Stillage ratio: approximately 12 L stillage (L ethanol)$^{-1}$.

Table 12.12 Most frequent issues and solutions for xylitol and 2G ethanol production from sugarcane straw.

Main issues	Possible cause(s)	Solution(s)
Low removal of hemicelluloses	1) Feedstock recalcitrance 2) Buffering capacity	Removal of extractives prior to the steam treatment using two-step extraction with hot water and anhydrous ethanol
Low selectivity into xylitol	1) Higher oligo/mono ratio of xylose 2) Oxidation of pentose sugars 3) High ratio of catalyst/sugars	1) Acidification by bubbling CO_2 2) Removal of oxygen from the headspace by injection of CO_2 prior to the addition of H_2 3) Addition of catalyst by pulse
Catalyst deactivation	1) Contamination with sulfur 2) Coke laydown by phenolic compounds generated by fractionation of lignin	1) Treatment with $CaCO_3$ 2) Treatment with active charcoal

12.5 Conclusions and Remarks

Scaling up and integrating processes for efficient lignocellulosic conversions in biorefinery configurations require a myriad of mechanical, chemical, and biochemical issues to be managed in order to overcome hurdles prior to operating at commercial scale. High downtime levels evidenced by the continuous process configurations have emerged as a major problem faced by the integrated pilot and flagship plants. Biomass clogging or sticking in the feeding devices, accumulation of biomass in pipelines, and friction wear in equipment and valves are the most frequent issues that increase downtime, thus reducing overall process reliability. Overcoming biomass recalcitrance followed by batch fed or continuous enzymatic hydrolysis of structural polymeric carbohydrates with cost-efficient liberation of fermentable sugars is one of the most challenging technical and economic barriers to the commercial success of biorefineries. The adoption of pseudo-continuous (synchronized batch) configurations coupled with preconditioning of the biomass feedstock using sprinkling or water washing and stripping, use of catalysts and promoters to reduce thermal-mechanical severities, and proper design of curves and pipelines have emerged as effective alternatives to minimize problems related to downtime as well as costly problems related to the erosion and corrosion of metallic parts used in equipment and lines.

Acknowledgment

AKC gratefully acknowledges the financial support from CNPq for scientific productivity program (Process number: 309214/2021-1).

References

Baudel, H.M. (2006). *Produção de Etanol a Partir de Bagaço de Cana-de-Açúcar: Pré-Tratamento e Hidrólise*. IPT—Instituto de Pesquisas Tecnológicas. III WORKSHOP PPPP—HIDRÓLISE. Oral presentation. São Paulo, Brazil. http://www.apta.sp.gov.br/cana

Baudel, H.M. (2009). Suitable processes for pretreatment and enzymatic hydrolysis of sugarcane bagasse. *FAPESP. BIOEN Workshop on Processes for Ethanol Production—Sugarcane Processing and Engineering*. Oral presentation. São Paulo, Brazil.

Baudel, H.M. and Diebold, E. (2017). Pretreatment process for lignocellulose biomass for producing carbohydrates and other byproducts. Brazil Patent Request. WO2017/168207 A1.

Baudel, H.M. and Diebold, E.F. (2018). Novel saccharification of lignocellulosic biomasses using "whole broth" enzymes: towards on-site manufacturing (OSM) of enzymes for production of 2G ethanol from eucalyptus wastes in biorefinery configurations. 3rd World Biotechnology Congress. São Paulo, Brazil.

Baudel, H.M., Aguiar, R.S., Baudel, P.H.S., and Abreu, C.A.M. (2014). Small-scale biorefineries: towards the biofuel and bioenergy sustainability at local basis. *RCN Conference on Pan American Biofuel and Bioenergy Sustainability*. AICHe. Oral presentation. Recife, Brazil.

Bhatia, L., Garlapati, V.K., and Chandel, A.K. (2019). Scalable technologies for lignocellulosic biomass processing into cellulosic ethanol. In: *Horizons in Bioprocess Engineering*, 73–90. Switzerland, Springer International Publishing.

Chandel, A.K., Garlapati, V.K., Singh, A.K. et al. (2018). The path forward for lignocellulose biorefineries: bottlenecks, solutions, and perspectives on commercialization. *Bioresour. Technol.* 264: 370–381.

Chandel, A.K., Albarelli, J.Q., dos Santos, D.T. et al. (2019). Comparative analysis of key technologies for cellulosic ethanol production from Brazilian sugarcane bagasse at the commercial-scale. *Biofuels Bioprod. Biorefin.* 13: 994–1014.

Chandel, A.K., Garlapati, V.K., Kumar, S.P.J. et al. (2020). The role of renewable chemicals and biofuels in building a bioeconomy. *Biofuels Bioprod. Biorefin.* 14: 830–844.

De Jong, E., Stichnothe, H., Bell, G., and Jørgensen, H. (2020). Bio-Based Chemicals: A 2020 Update. Report. Published by IEA Bioenergy. ISBN 978-1-910154-69-4.

Diebold, E.F. and Baudel, H.M. (2020). Technical and Economic Evaluation of Eucalyptus and Sugarcane-based Biorefineries: a comparison between continuous and pseudo-continuous configurations for steam pretreatment of eucalyptus (bark, roots and chips) and sugarcane (bagasse and straw) lignocellulosic wastes. Renner-Herrman Report Nr. 04-2019/20, Porto Alegre, Brazil.

Silveira, M.H.L., Chandel, A.K., Vanelli, B.A. et al. (2018). Production of hemicellulosic sugars from sugarcane bagasse via steam explosion employing industrially feasible conditions: pilot scale study. *Bioresour. Technol. Rep.* 03: 138–146.

Sun, Y. and Cheng, J. (2002). Hydrolysis of lignocellulosic materials for ethanol production: a review. *Bioresour. Technol.* 83: 1–11.

Virmond, D. (2013). Valorization of agroindustrial solid residues and residues from biofuel production chains by thermochemical conversion: a review, citing Brazil as a case study. https://www.scielo.br/scielo.php?script=sci_arttext&pid=S0104-66322013000200001

Warner, M. (2015). What makes scale-up of industrial biotechnology so difficult? https://www.linkedin.com/pulse/what-makes-scale-up-industrial-biotechnology-so-difficult-mark-warner.

Yang, B. and Wyman, C.E. (2008). Pretreatment: the key to unlocking low-cost cellulosic ethanol. *Biofuels Bioprod. Biorefin.* 2, 1): 26–40.

Zhang, Y.-H.P., Berson, E., Sarkanen, S., and Dale, B.E. (2009). Sessions 3 and 8: Pretreatment and biomass recalcitrance. *Appl. Biochem. Biotechnol.* 153 (1–3)): 80–83.

13

Techno-economic Analysis of Bioconversion of Woody Biomass to Ethanol

Deepak Kumar[1], *Anuj Kumar Chandel*[2], *and Lakhveer Singh*[3]

[1] Department of Chemical Engineering, SUNY College of Environmental Science and Forestry, Syracuse, NY, 13210, USA
[2] Department of Biotechnology, University of São Paulo (USP), Lorena, São Paulo, Brazil
[3] Department of Environmental Science, SRM University-AP, Amaravati, Andhra Pradesh, 522502, India

13.1 Introduction

Bioethanol, produced from the fermentation of sugars derived from starchy or lignocellulosic feedstocks, is considered the most promising renewable liquid transportation fuel. Ethanol can be used as a transportation fuel in the existing infrastructure and modern spark-ignited combustion engines after blending in low amounts with gasoline (up to 20%), for example E10, a mixture of 10% ethanol and 90% gasoline by volume. Other high amount blends such as E85 (a mixture of 85% ethanol and 15% of gasoline) can be used in modified vehicles, also known as "flexible-fuel vehicles" (Wingren et al. 2003; Fornell and Berntsson 2012). Despite low energy density (about two-thirds of gasoline), the high octane number of the ethanol blend (106–110 vs. 91–96 for gasoline) and high oxygen provides several benefits of use in vehicles compared to gasoline (Kumar and Singh 2019), and make an ethanol blend a very promising transportation fuel.

Currently, most of the bioethanol is produced using food crops. The United States is the global leader in bioethanol production (15.8 billion gallons in 2019; 54.4% of the world's production) and uses corn as the main feedstock (more than 90% ethanol production). Brazil is the second-biggest producer of bioethanol (8.6 billion gallons), mostly using sugarcane as the feedstock (RFA 2021). However, ethanol production from food crops like corn or sugarcane, commonly referred to as the first-generation biofuel, is criticized for various reasons, including food versus fuel priority, intensive agricultural inputs, land and freshwater use, and capacity limitations (Shafiei et al. 2011; Kumar and Murthy 2012; Olofsson et al. 2017). Alternatively, the use of lignocellulosic biomass, such as agricultural residues, grasses, forestry products and wastes, municipal solid wastes, and energy crops, can address some of these concerns related to the first-generation biofuel. Lignocellulosic feedstocks are non-food/feed resources low in economic value and provide relatively superior environmental

Lignocellulose Bioconversion Through White Biotechnology, First Edition. Edited by Anuj Kumar Chandel.

benefits during ethanol production compared to starchy crops (Spatari et al. 2005; Kumar and Murthy 2012; Wang et al. 2012a; Therasme et al. 2021).

Extensive research efforts have been made in the last few decades to identify potential lignocellulosic feedstocks and to develop technology to efficiently convert these feedstocks into ethanol at a low cost. Although most of the research and development attention has been focused mainly on agricultural residues (e.g. corn stover, wheat straw, rice straw, etc.) and dedicated energy crops (e.g. switchgrass, Miscanthus), the woody biomass (hardwoods and softwoods) and waste products from the forest industry remain high potential and an important feedstock for cellulosic ethanol production. In the United States only, about 370 million tons (dry weight) of woody biomass (about 30% of the total biomass available) is estimated to be available annually for biofuel production (Zhu and Pan 2010). Similarly, by-products such as sawdust are produced in large quantities in forest product industries and can be efficiently converted to biofuels. About 2.3 million dry tons of sawdust is produced every year in Canada (Roy and Dutta 2013). Woody biomass is also available in significantly large quantities in many other regions of the world and is believed to play a critical part in the biomass supply chain in the future bioeconomy (Zhu and Pan 2010).

Despite the biomass availability and advantages of woody biomass and other lignocellulosic ethanol, ethanol production from cellulosic biomass on a commercial scale is still not very successful because of technical bottlenecks and high processing costs. Conversion of structural carbohydrates like cellulose and hemicellulose to sugar monomers is relatively very difficult compared to storage polymer like starch (in the case of corn ethanol) and requires relatively much higher cost and energy. Although the lignocellulosic feedstocks are of relatively low economic value, however, still feedstock cost is the maximum in total raw material costs in most cases. A variety of technologies have been developed and investigated on various woody biomass to improve the process yields. Although several of these technologies look attractive on the laboratory scale, the success of these technologies depends on economic viability and energy use on commercial-scale applications. The material and energy balances of all unit processes included in the bioconversion pathway are required to determine the product and by-product yields, capital investments, raw material amounts, and their costs, etc., and to translate these parameters into overall technical and economic feasibility. As most of these technologies are only in the conceptual phase, process simulation models are used to conduct techno-economic analysis (TEA) and determine the technical and economic feasibility of the processes on a commercial scale.

This chapter summarizes the findings from the techno-economic studies on woody biomass to ethanol conversion and discusses how the TEA determines the hot-spots (energy or cost-intensive unit processes or process inputs) and provides a direction for future research to determine the most sustainable technology approach.

13.2 Techno-economic Analysis

TEA is an effective and commonly used approach to determine the technological, energetic, and economic feasibility of an industrial process, which provides critical information about the commercial-scale viability of the process and major bottlenecks that need to be investigated to further improvise the process. The TEA has become an integral part of the research in both academia and industry. In addition to providing estimates of process economics, TEA helps in determining critical process parameters and inputs, and their effect on the overall functionality and profitability of a process. For every engineering system or industrial process, there are several possible technologies and raw material choices. For example, in the case of woody biomass to ethanol, there are several choices of pretreatment processes (physical, chemical, physicochemical, etc.), hydrolysis (acid vs. enzymatic), lignin valorization, and many other processes. Similarly, various types of woody feedstocks can be used to produce ethanol using the same process. However, the process yields, raw material inputs, and energy use would vary with the change in any of the parameters discussed above. Most of the time, there is a tradeoff between process yields (amount of final product produced), economics (capital investment and production cost), and energy use among different choices. The TEA acts as a process evaluation tool to conduct this comprehensive analysis and help in understanding this tradeoff for a specific technology (Shah et al. 2016). The analysis even becomes critical for the process in the early stage of development as the TEA can help in process optimization based on product output and economics constraints by integrating it with the experiments (Figure 13.1). A comprehensive TEA provides sufficient details of technical evaluation and economic performance of a process that it can be used to determine the most feasible alternative among different choices. The technical details include information on equipment capacity requirements, utilities (heating and cooling agents), maintenance, equipment installation, skill requirement, etc. (Towler and Sinnott 2012; Shah et al. 2016). The economic evaluations provide details on several direct and indirect costs associated with the capital investment, working capital, operational costs, and revenues from product and co-product sales. Depending on the desired accuracy levels of process economics, the TEA can be performed at multiple levels: order of magnitude estimate (±10%–50% accuracy), sturdy estimate (±30% accuracy),

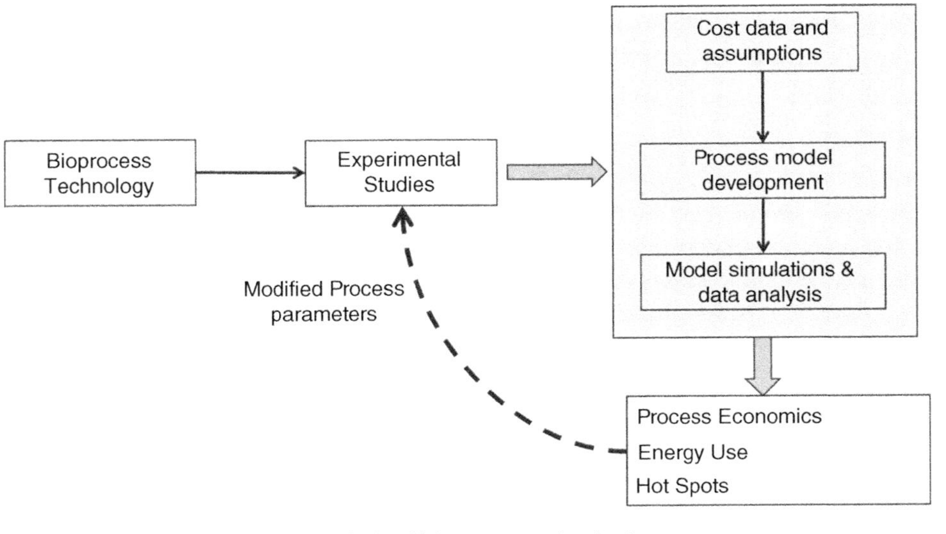

Figure 13.1 Techno-economic analysis of bioprocess technologies.

preliminary estimate (±20% accuracy), definitive estimate (±10% accuracy), and detailed estimate (±5% accuracy) (Murthy 2017). Although the accuracy levels are higher in the detailed estimates, the expense of the analysis is also significantly higher (Murthy 2017).

The development of a detailed process flow model is usually a first step in the TEA of a process. The model includes all necessary equipment and the technical and operational details of all unit operations of the process. For example, Figure 13.2 illustrates a process model of lignocellulosic biomass (grass straw) to ethanol, developed in SuperPro Designer. For biorefineries, the process flow model can be developed using various process modeling software including both commercial software, such as Aspen Plus, SuperPro Designer, and open access platforms, such as BioSTEAM (Cortes-Peña et al. 2020). The process model allows a detailed material and energy balance to be conducted and the equipment design, the capacities, and associated purchase and operational cost to be determined. The operational data (e.g. conversion efficiencies in various process operating conditions) required as an input to the model can be obtained from experimental studies, mathematical models, pilot-scale tests, and commercial facilities. The equipment cost associated with the capacity calculated from the model simulation can be obtained from various sources, including vendor quotations, trade journals, past data from similar equipment (literature), and built-in cost models in the simulation software (Murthy 2017). In addition to the direct equipment purchase cost, several other costs, such as installation, piping, electrical facilities, insulation, and auxiliary facilities, are taken into account while estimating the total capital investment (Kumar and Murthy 2011; Shah et al. 2016; Murthy 2017; Somavat et al. 2018). These additional costs are dependent on the industry and can be estimated using various multiplication factors obtained from the previous reports or chemical engineering process modeling textbooks. The operational costs are calculated in terms of raw material costs, utility costs, facility dependent costs (maintenance and depreciation), labor costs, quality control, waste disposal costs, etc. A detailed cash flow analysis is conducted using all these cost numbers along with the revenues to calculate the profitability. As the scope of this chapter is to discuss the TEA of woody biomass to ethanol, only a brief discussion on the TEA methodology is provided here. The readers can refer to some specific book chapters (Shah et al. 2016; Murthy 2017) that provide the basics and details of the TEA methodology.

13.3 Bioconversion of Woody Biomass to Ethanol

Lignocellulosic biomass is a complex matrix of three biopolymers: cellulose (20%–50%), hemicellulose (15%–35%), and lignin (5%–30%) (Table 13.1) (Kumar and Murthy 2013). For bioethanol production, the carbohydrate fractions are hydrolyzed to sugar monomers that are further fermented to ethanol (Figure 13.3). The pure ethanol is recovered from the fermentation broth. Conversion of sugar polymers to monomers (hydrolysis) and fermentation of sugar monomers can be performed in a single step, known as simultaneous saccharification and fermentation (SSF). The SSF process provides several advantages over the separate hydrolysis and fermentation process and results in high ethanol yields (Olofsson et al. 2008; Kumar and Singh 2019). Although these steps are similar to the starch-based ethanol production process, unlike corn or any starch-rich feedstock the

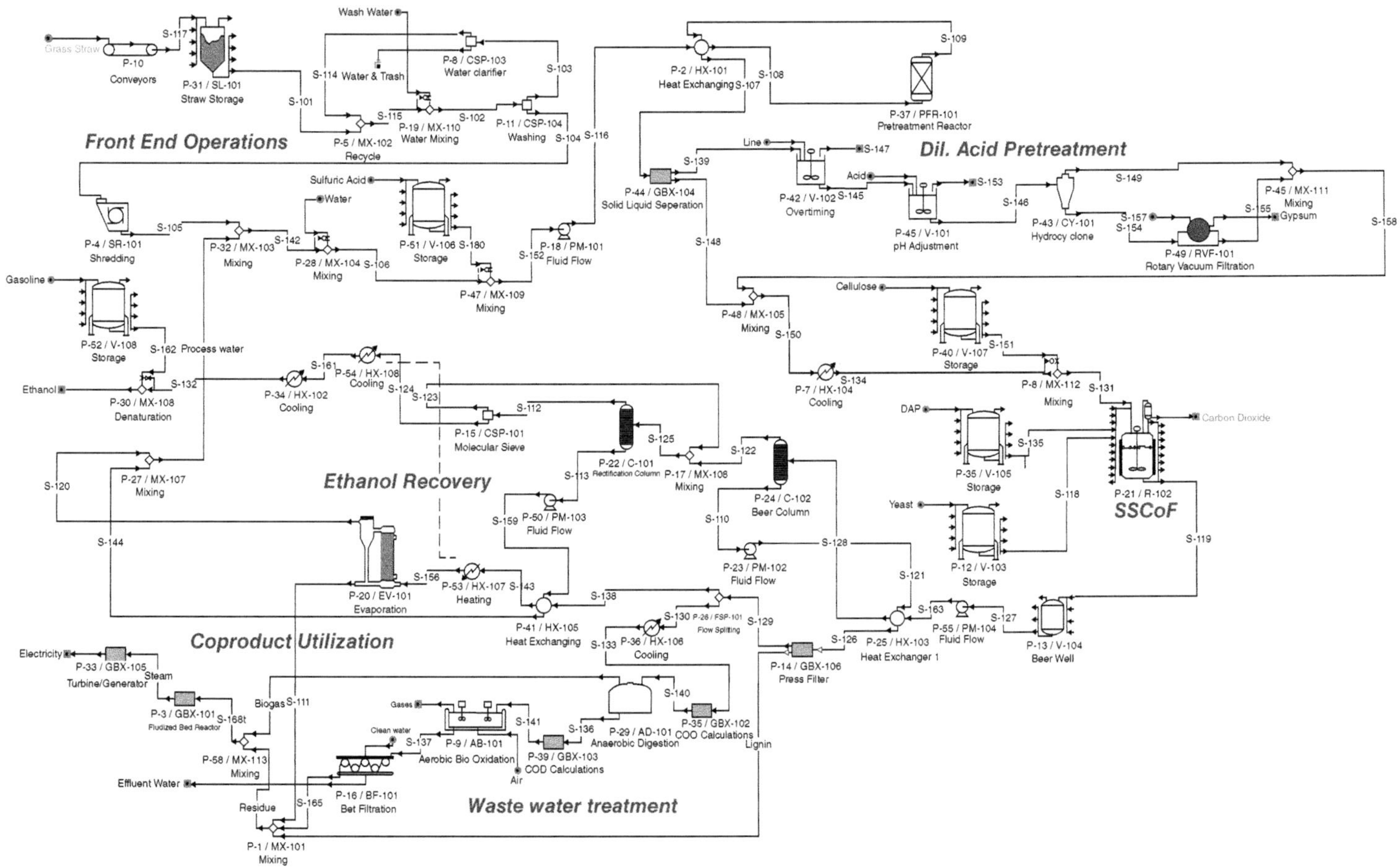

Figure 13.2 Simple flowsheet of a process model of the lignocellulosic ethanol production process (using dilute acid pretreatment) developed in SuperPro Designer. *Source:* Adapted from Kumar and Murthy, 2011 / Springer Nature / CC BY-2.0.

Table 13.1 Composition of woody biomass.

Biomass	Cellulose (%)	Hemicellulose (%)	Lignin (%)	Comments	References
Craft mill pulp	33	18	39	–	Fornell and Berntsson (2012)
Lodgepole pine	42.5	20.8	27.9	Mannans are (55.8%) major fraction of hemicellulose	Zhu and Pan (2010)
Ponderosa pine	41.7	22.8	26.9	Mannans are (47.4%) major fraction of hemicellulose	Zhu and Pan (2010)
Loblolly pine	45	21.8	28	Mannans are (50.5%) major fraction of hemicellulose	Zhu and Pan (2010)
Douglas fir	44	21.2	32	Mannans are (51.8%) major fraction of hemicellulose	Zhu and Pan (2010)
Red maple	46	22.5	24	Xylans are (84.4%) major fraction of hemicellulose	Zhu and Pan (2010)
Yellow poplar	42.1	19	23.3	Xylans are (79.5%) major fraction of hemicellulose	Zhu and Pan (2010)
Pine tree wastes	20.7–26.7	–	34.2–38.7%	Higher xylan percentage in needles and higher mannan in the branches	Salehian and Karimi (2013)
Willow sawdust	35.6	21.5	28.7		Alexandropoulou et al. (2017)
Birch	34.7	31.2	18.7		Raghavendran et al. (2018)
Spruce	37.6	27.4	32.6		Raghavendran et al. (2018)
Spruce	43.2	21.3	28.3	Mannans are (54%) major fraction of hemicellulose	Zhu and Pan (2010)
Spruce	41.6	25.2	29.5		Janzon et al. (2014)
Spruce bark	34.1	15.8	33.6	High amount of extractives (18%)	Janzon et al. (2014)
Spruce bark	20.6	30.1	34.4		Rasi et al. (2019)
Pine bark	24.9	31.1	35.6		Rasi et al. (2019)

accessibility to carbohydrates and their hydrolysis to sugar monomers is relatively difficult in the case of lignocellulosic feedstocks like woody biomass. Cellulose, hemicellulose, and lignin form a complex matrix and result in a highly recalcitrant structure. Lignin acts as the glue that provides a physical barrier and makes the cellulose impervious to enzymes. Hence, a pretreatment step is necessary to break this complex matrix of carbohydrates and lignin, to decrease the cellulose crystallinity, and subsequently to increase the access of hydrolytic enzymes for efficient hydrolysis. Compared to agricultural residues, woody biomass has relatively higher lignin content, and further the lignin composition (ratio of

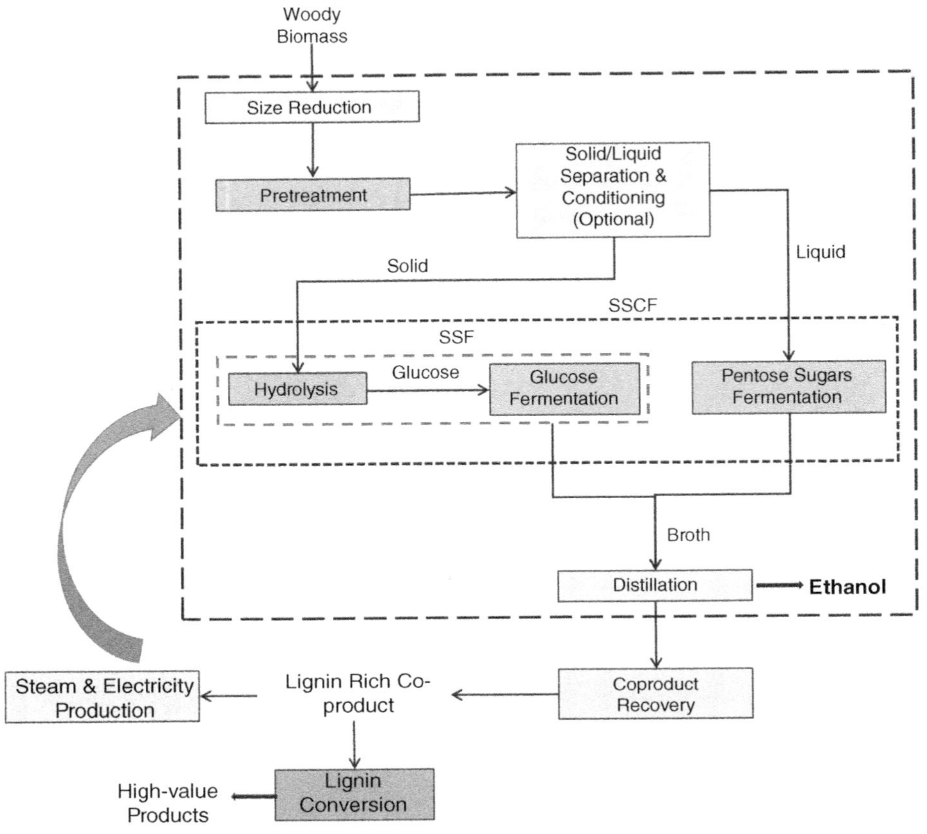

Figure 13.3 Bioconversion process of ethanol production from woody biomass.

different phenylpropane monomers) varies among different types of woods (softwood vs. hardwood) that affect the process efficiency (Zhu and Pan 2010). Over the last two decades, various pretreatment strategies including physical (e.g. mechanical comminution), chemical (e.g. dilute acid, alkali, ionic liquid, organosolv, etc.), physicochemical (e.g. steam explosion), and biological (e.g. using white rot fungi) have been developed and investigated over various lignocellulosic biomass to enhance the hydrolysis efficiency (da Costa Sousa et al. 2009; Galbe and Wallberg 2019; Sankaran et al. 2020). Dilute acid pretreatment, which includes heating the biomass at high temperature and pressure in a dilute acid solution, is the extensively investigated pretreatment approach. Although the process is effective on a wide range of feedstocks, it suffers from the limitations of sugar degradation, inhibitory compound formation, and the need for corrosion-resistant reactors. Alkali pretreatment (high pH) results in delignification, cellulose fibril swelling, and improves cellulose accessibility. Similarly, most of the organosolv pretreatment processes use various organic solvents to delignify or fractionate the biomass to remove lignin and enhance the hydrolysis of cellulose (Zhao et al. 2009; Amiri and Karimi 2015). The use of ionic liquids has also been extensively investigated as a pretreatment approach that addresses several limitations of traditional organic solvents (Hyvärinen et al. 2011; Miyafuji 2015). Hot-water pretreatment is another approach that avoids the use of any chemical and hydrotreats biomass at high temperatures and relatively long durations (Mosier 2013; Juneja et al. 2020). However, the hydrolysis yields from only hot-water pretreated biomass are usually lower than acid pretreated biomass (Kim et al. 2016b). The yield can be improved by performing hot water pretreatment in combination with another pretreatment (e.g. sequential hot water and disk refining) in a sequential manner (Kim et al. 2016a). This approach has been found highly effective on various lignocellulosic feedstocks, and the disk refining of hot-water pretreated biomass could reduce the milling energy by more than 90% compared to the untreated biomass (Zhu et al. 2010b; Cheng et al. 2019). Several review papers have discussed the mechanisms and details of these pretreatment approaches and their effectiveness on woody biomass (da Costa Sousa et al. 2009; Alvira et al. 2010; Zhu et al. 2010a; Bhatia et al. 2020). Several techno-economic studies have concluded that pretreatment is the most critical step in the woody biomass-based biorefineries and choice of pretreatment affects the overall process feasibility. The details are discussed in the next section of this chapter.

After pretreatment, the structural carbohydrates are hydrolyzed to sugar monomers that can be subsequently or simultaneously fermented to ethanol. The cellulose to glucose hydrolysis can be achieved through chemical (acid hydrolysis) or biological routes (enzymatic hydrolysis). Although acid hydrolysis is a relatively fast process, enzymatic hydrolysis is preferred over acid hydrolysis for bioethanol production due to the advantages of lower energy use, no sugar degradation, no corrosion of equipment, and neutralization requirement, and no disposal cost. However, the enzymatic hydrolysis of cellulose is a complex process and requires a synergistic action of multiple enzymes, mainly endoglucanases (EC3.2.1.4), exoglucanases, also known as cellobiohydrolase (EC3.2.1.91), and β-glucosidase (EC3.2.1.21) (Kumar and Murthy 2016). All these enzymes act in a specific action pattern (e.g. endoglucanases act randomly on the surface but cellobiohydrolase acts processively from the chain ends) and result in an effective synergism for efficient hydrolysis of cellulose (Kumar and Murthy 2016). A detailed discussion of the enzymatic hydrolysis process, enzyme synergism, and challenges can be found in numerous recent review papers and book chapters (Van Dyk and Pletschke 2012; Wang et al. 2012b; Kumar and Murthy 2016). The sugars obtained in the hydrolysate are fermented to ethanol using microorganisms such as *Saccharomyces cerevisiae*, *Escherichia coli*, *Zymomonas mobilis*, *Klebsiella oxytoca*, etc.. *S. cerevisiae* is the most common industrial strain used for ethanol production; however, it is not effective in fermenting pentose sugars obtained from hemicellulose hydrolysis. Various approaches, including the development of pentose fermenting microorganisms, have been investigated to improve the conversion of pentose sugars to ethanol and enhance the overall process yield (Radhika et al. 2011). The lignin-rich fraction obtained after ethanol recovery can be used to produce in-house steam and electricity or can be processed to produce biochemicals and bioproducts using a biorefinery approach. In recent years, the lignin utilization approaches have been investigated extensively, and it has been concluded that lignin utilization can improve the economic viability of the process (Yoo et al. 2020). Lignin can be recovered at the front-end also using selective pretreatment technologies and can be further processed to produce high-value bioproducts (Si et al. 2017; Yoo et al. 2020; Yoo and Ragauskas 2021).

13.4 Techno-economic Analysis of Woody Biomass to Ethanol

Several studies have investigated the techno-economic feasibility of using different forest residues and wood-based products to produce ethanol through developing process simulation models (Table 13.2). In addition to total capital investment and unit production cost, TEA helps in identifying major cost centers in the process and most critical raw materials. Most of the studies on woody biomass ethanol have concluded that the energy use and cost associated with biomass pretreatment and the cost of feedstock are major critical factors determining the overall economic feasibility of the process (Sassner et al. 2008; Shafiei et al. 2011, 2014). In many cases, the pretreatment process is responsible for one-third of the total production costs in a bioethanol production plant (da Silva et al. 2019). Dilute acid and steam explosion are the two most thoroughly investigated methods for lignocellulosic conversions to ethanol and are believed to be ready for commercial-scale use. However, the use of these conventional pretreatments can lead to sugar degradation and the production of inhibitory compounds. Therefore, the use of a suitable pretreatment method that avoids these inhibitory compound formations is of vital importance (Kabir et al. 2015). Organosolv pretreatments, such as using *N*-methylmorpholine-*N*-oxide (NMMO), have been getting attention over the last few years due to mild pretreatment conditions and high sugar recoveries without any inhibitory compound formation (Shafiei et al. 2014; Zhou et al. 2018; da Silva et al. 2019).

Shafiei et al. (2014) compared the economic feasibility of ethanol production from pinewood using steam explosion or NMMO pretreatment. The ethanol plants were modeled to process 100 000 tons pinewood annually. The capital investment for plants using NMMO and steam explosion pretreatment was estimated at €44 million (equivalent to $53.9 million in 2014) and €40.5 million (equivalent to $58.2 million in 2014) respectively. Ethanol production cost was found to be €0.64 ($0.85) and €0.53 ($0.71) per liter of ethanol using NMMO and steam explosion pretreatment, both lower (18%–39% lower) than gasoline prices in that year. Although the efficiency is better with the NMMO pretreatment process (1.4 times ethanol yields compared to steam explosion), higher capital and higher raw material expenses (due to expensive solvent) lead to higher unit production costs. The purchase price of NMMO, feedstock, and enzymes were the most critical factors. An increase of 50% in the price of feedstock resulted in an 8% and 13% increase in unit production cost for NMMO and steam explosion pretreatment respectively. The change was lower in the case of the NMMO process because the contribution of feedstock cost in overall raw material cost was relatively higher due to the high cost of solvent. In the case of processes like steam explosion, the raw material is a major contributor to total raw material costs and the most important parameter in affecting the cost of ethanol (Sassner et al. 2008). Similar observations were made by Fornell and Berntsson (2012) for the alkali pretreatment process.

Sassner et al. (2008) investigated the economic feasibility of ethanol production from spruce wood and Salix in Sweden using SO_2-catalyzed steam pretreatment followed by SSF. The ethanol yields were estimated at 239 and 292 L ethanol per dry metric ton of Salix and spruce, respectively, resulting in 47.8 and 58.4 million L ethanol production annually. Total capital investments for the ethanol plants were found to be SEK1087 million

Table 13.2 Summary of results from techno-economic studies on ethanol production from woody biomass.

Material	Process	Annual production capacity	Biomass cost (\$ ton^{-1})	Capital cost	Production cost	References
Sawdust (from sawmills)	Biochemical route, lime pretreatment and SSF	20 000 kL	80	\$38 million	\$1.04 L^{-1}	Roy and Dutta (2013)
Sawdust (from forest residues)	Biochemical route, lime pretreatment and SSF	20 000 kL	60	\$38 million	\$0.98 L^{-1}	Roy and Dutta (2013)
Spruce wood	Biochemical route, NMMO pretreatment; non-isothermal simultaneous saccharification and fermentation (NSSF); 200 000 tons $year^{-1}$ biomass processing	38 million L	68	€58.3 million (\$81 million)	\$0.58 L^{-1}	Shafiei et al. (2011)
Spruce wood	Biochemical route, SO_2-catalyzed steam pretreatment and SSF; 200 000 tons $year^{-1}$ biomass processing	58.4 million L	75.4	SEK 1025 million (\$146 million)	\$0.63 L^{-1}	Sassner et al. (2008)
Salix (hardwood)	Biochemical route, SO_2-catalyzed steam pretreatment and SSF	47.8 million L	79.2	SEK 1087 million (\$155 million)	\$0.78 L^{-1}	Sassner et al. (2008)
Pinewood	Biochemical route, NMMO and steam explosion pretreatment; biochemical route; 100 000 ton $year^{-1}$ biomass processing	30.01 milllion L (NMMO) 22.13 million L (steam explosion)	84	€44 million (\$58.5 million) €40.5 million (\$53.9 million)	\$0.85 L^{-1} (NMMO) \$0.71 L^{-1} (steam explosion)	Shafiei et al. (2014)
Pulp from softwood-based kraft mill	Kraft mill converted to ethanol plant; alkali (NaOH) pretreatment; 1800 dry ton per day wood processing	130 million L	104.25 (wood)	282[a]	\$0.88 L^{-1}	Fornell and Berntsson (2012)
Spruce wood chips	Biochemical route, chemical (2% sulfur dioxide) assisted steam explosion pretreatment; 200 000 dry tons $year^{-1}$ biomass processing		60	\$127 million (base case) \$121.8 to \$130.2 million for other scenarios	\$0.591 L^{-1} (base case) \$0.546 to \$0.589 (for other scenarios)	Wingren et al. (2008)

[a] \$232 million for the existing pulp mill and \$50 million for the ethanol process.

(\$155 million) and SEK1025 million (\$146 million) for Salix and spruce, respectively. Due to high ethanol yields and lower feedstock costs (\$75.4 ton^{-1} vs. \$79.2 ton^{-1}) for spruce, the ethanol production cost from spruce was found to be lower than that from Salix (\$0.63 L^{-1} vs. \$0.78 L^{-1}).

Shafiei et al. (2011) also investigated the economic feasibility of ethanol production from spruce wood, the most abundant wood species in Sweden, using NMMO pretreatment and the non-isothermal simultaneous saccharification and fermentation process. As the capital investment for a plant does not increase linearly with its capacity, the effect of plant size on economic feasibility was also investigated by performing simulation for five plant sizes between 50 000 to 800 000 tons of biomass processing per year. For the base case with a processing capacity of 200 000 tons of spruce wood per year, the capital cost was estimated at €58.3 million (equivalent to \$81 million, in 2010). Larger plants were found to be more profitable with less unit production cost, but they require huge capital investments and a large supply of feedstock. The capital cost associated with the hydrolysis and fermentation section was maximum among all sections and contributed about 28% of total capital investment. The ethanol production cost for the base case was estimated at €0.44 (L ethanol)$^{-1}$ (equivalent to \$0.58 (L ethanol)$^{-1}$). This ethanol production cost was lower than the production cost (\$0.63 L^{-1}) estimated by Sassner et al. (2008) for the same feedstock and similar plant capacity but using the steam explosion pretreatment process. This could be mainly due to the lower feedstock cost assumed in this study (\$68 vs. \$75.4). Lower feedstock cost results in overall low raw material costs and decreases the unit production cost. Capital investment per unit liter (\$2.13) was also found to be lower compared to the value (\$2.5 (L ethanol)$^{-1}$) estimated by Sassner et al. (2008), which would decrease the facility-dependent costs and annual operating costs, ultimately reducing the unit production cost. The capital cost could be lower in this case due to the different pretreatment processes and lower ethanol yields, which reduce the costs associated with the equipment required for ethanol recovery.

Wingren et al. (2008) investigated the economic feasibility of ethanol production from spruce chips and also evaluated the feasibility of various downstream energy-saving alternative process configurations, such as multiple-effect evaporation and integration of evaporation and distillation, use of mechanical vapor recompression, and replacement of evaporators with an anaerobic digester, etc. The base case consisting of ethanol recovery using two strippers and a rectifier column resulted in the ethanol production cost of SEK4.14 (\$0.591) per liter of ethanol. Feedstock accounted for 34% of the cost of ethanol. The cost for alternate scenarios ranged between SEK3.82 (\$0.546) and SEK4.12 (\$0.589), with the minimum for the process using a mechanical vapor recompression system for evaporation, mainly due to high income from the co-products.

Fornell and Berntsson (2012) investigated the economic feasibility of ethanol production from kraft pulp from softwood in a biorefinery integrated with a kraft mill. The model was developed that includes modification of the pulp mill to an ethanol plant by repurposing the pulping line into an ethanol line. The plant capacity was 1800 metric tons of dry wood processing per day. Wood was pretreated with alkali (NaOH), a process commonly used in the pulping process (soda pulping) but not popular in the ethanol production process. Ethanol production cost was estimated at €0.635 L^{-1} (\$0.88 L^{-1}), relatively higher compared to ethanol production from other woody biomass (Sassner et al. 2008; Wingren et al. 2008; Shafiei et al. 2011, 2014). This was mainly observed because of the significantly higher capital cost of the refinery ($282 million), which results in the large annual facility-dependent costs and ultimately high unit production costs. Other than facility-dependent costs, feedstock cost was another major contributor to ethanol production cost. The production cost of ethanol from the conversion of sawdust ($0.98 L^{-1}–$1.04 L^{-1}), estimated from the process simulation model developed by Roy and Dutta (2013), was also higher than ethanol production costs estimated from wood chips. Feedstock cost ($0.33 L^{-1}–0.44 L^{-1} ethanol) was the major contributor to ethanol production cost. At this production cost, bioethanol is not economically competitive with fossil fuels. The authors concluded that biofuel production from sawdust in Canada is only possible with the introduction of incentives from programs like Feed-in Tariff (FiT) technology.

Although the costs of enzymes used in the ethanol production process have decreased significantly over the last decade, they still contribute a significant fraction to the annual operating costs (Olofsson et al. 2017). One potential approach to reduce this cost and increase the economic feasibility of the ethanol process is on-site enzyme production by growing microorganisms using a part of the biomass. Although the use of integrated production (using part of the feedstock for growing microorganisms) reduces the fraction of biomass available for ethanol conversion, use of the whole fermentation broth potentially improves the hydrolysis and fermentation yields due to the effect of mycelium-bound enzymes (Olofsson et al. 2017). From the process simulations of ethanol production from spruce, Olofsson et al. (2017) found a minimum selling price of €0.581 L^{-1} ($0.639 L^{-1}) ethanol for the off-site enzyme production case. In the case of in-house enzyme production, the minimum ethanol selling price was as low as €0.568 L^{-1} ($0.625 L^{-1}) ethanol.

13.5 Integrated TEA and life cycle assessment (LCA)

Although the scope of this chapter is TEA, for sustainable renewable fuel, along with economic feasibility, the fuel should provide benefits in terms of environmental factors (energy use, emissions, etc.). This section briefly discusses the LCA results of some studies on woody biomass to ethanol conversion.

Similar to the process economics, the energy use and other environmental impact results are dependent on several factors, including the source and type of feedstocks, conversion process, and assumptions used in the studies. Pretreatment and ethanol recovery are usually observed as two major hotspots in terms of energy consumption (Kumar and Murthy 2011; Shafiei et al. 2011; Roy and Dutta 2013). Shafiei et al. (2011) reported that the pretreatment process alone causes more than 80% of total energy use in the cellulosic ethanol process.

Roy and Dutta (2013) conducted an LCA of ethanol production from sawdust through the biochemical conversions route using lime pretreatment, followed by SSF. Sawdust was assumed to be obtained either from sawmills or from forest residues. Total energy consumption in the collection, transportation, and sawdust preparation in the case of sawdust from sawmills was only 0.11 MJ L^{-1} compared to 1.2 MJ L^{-1} ethanol in the case of sawdust from forest products. This was observed because, in the case of sawmills, sawdust was assumed to be readily available at the gate and required no collection and preparation. Net fossil energy use in the process was estimated at 12.29–13.37 MJ L^{-1} of ethanol, resulting in net energy ratios (ratio of total energy output to fossil energy input) of 1.59–1.73. Sandilands et al. (2009) also reported that ethanol production from forest residues and willow is highly energy efficient in New Zealand, with net energy ratios of 3.67 and 4.79 from forest residue and willow, respectively. Both of these studies used a biochemical pathway for ethanol production. Daystar et al. (2015) conducted an LCA of ethanol production from several forest woody biomass (loblolly pine, eucalyptus, unmanaged hardwoods, forest residues) using the thermochemical platform, and found a 95%–97% reduction in fossil fuel consumption with production and use of ethanol compared to petroleum-based fuel. It was also observed that these energy savings were higher than that of ethanol production from switchgrass, a dedicated energy crop. This was observed because crops like switchgrass require relatively more soil preparation and agricultural inputs.

According to the study of Roy and Dutta (2013), the pretreatment process was observed as the main energy-intensive process followed by the distillation process to recover ethanol. The same was the case for greenhouse gas (GHG) emissions. The pretreatment process followed by the distillation process contributed a maximum in terms of GHG emission production and resulted in an overall 0.75–0.94 kg CO_2 eq L^{-1} ethanol emissions during its life cycle. The emissions in all cases were lower than that of gasoline (1.5 kg CO_2 eq L^{-1} ethanol equivalent) in Canada. Daystar et al. (2015) reported 73% lower GHG emissions from production and use of ethanol from forest residues compared to gasoline in the southern United States. In addition to lower GHG emissions, ethanol production from forest-based feedstocks resulted in lower acidification, carcinogenic, non-carcinogenic, respiratory effects, eutrophication, and ecotoxicity compared to gasoline production and use (Daystar et al. 2015).

Khoshnevisan et al. (2016) conducted a detailed LCA of ethanol production from pinewood and compared the results between two pretreatment processes: NMMO and steam explosion. It was observed that ethanol production using the steam explosion pretreatment process was less environmentally friendly compared to the NMMO pretreatment process, causing more impact on the damage category of human health (17.4 vs. 11.9 disability-adjusted life years). Due to the relatively milder pretreatment conditions required for NMMO pretreatment, the process was found to be more environmentally friendly. Out of total impact, the distillation process alone was found to be responsible for about 48% impact. The pretreatment process was the second major contributor to the impact on the damage category of human health. Similarly, ethanol production using the steam explosion process produced 34.6% higher GHG emissions compared to using the NMMO pretreatment process, during its life cycle. This was mainly observed because of less steam used in the NMMO pretreatment process. Due to lower carbohydrate content and relatively low solid loadings in the case of the lignocellulosic ethanol process, the fermentation broth contains a low ethanol concentration (usually <5% v/v), which requires significant energy and resources to recover pure ethanol (Sassner et al. 2008; Wingren et al. 2008; Kumar and Murthy 2011; Kumar 2014). Due to high sugar recovery, the ethanol production in the case of NMMO pretreatment was more than the steam explosion process, which results in high ethanol concentration in the fermentation broth and relatively low energy required to recover pure ethanol in the distillation process.

One potential approach to reduce distillation energy required is the use of high gravity conditions, which could allow the use of a high percentage of solids in the process during fermentation and potentially lead to high ethanol titers at the end of fermentation. According to Janssen et al. (2016), the amount of water that needs to be removed during distillation is reduced to 10 L (L ethanol)$^{-1}$ at 30% solid loadings compared to 25 L (L ethanol)$^{-1}$ at 8% solid loadings (conventional process), with a subsequent decrease in energy use. Interestingly, it was observed that the decrease in energy use and water use does not guarantee the reduction in other environmental impact categories, such as global warming potential. Among several different process configurations investigated by Janssen et al. (2016), 12 process configurations showed a reduction in energy use during downstream processing compared to the base case configuration (8% solids); however, only six configurations had a lower global warming potential than the base case configuration. In all of the process configurations, enzyme production and use were a significant contributor to all environmental impact categories, an observation reported in several studies (Spatari et al. 2010; Kumar and Murthy 2012; Olofsson et al. 2017). Olofsson et al. (2017) conducted an LCA of ethanol production from spruce chips and compared the energy use and GHG emissions between the ethanol production process with integrated enzyme production versus off-site enzyme production. It was concluded that although fossil energy use and process economics were not significantly different in the two cases, the GHG emissions were significantly lower in the integrated case compared to off-site enzyme production (22.2–24.4 g CO_2 eq. MJ^{-1} ethanol vs. 50.4 g CO_2 eq. MJ^{-1}).

13.6 Conclusions

Woody biomass is an abundant and high-potential feedstock for carbon-negative cellulosic ethanol production in the world. Similar to any other lignocellulose biomass, woody biomass conversion into ethanol also needs to be pretreated to unlock the biomass assessing for cellulase action into lignocellulosic sugar production. The hydrolysis of pretreated woody biomass can be coupled with ethanol fermentation by a suitable microorganism growing on pentose and hexose sugars. Process consolidation is an important step in the overall economization of bioethanol production. Process parameters like effective pretreatment and hydrolysis at high solid loadings without the requirement of washing followed by the fermentation of high sugar mixture solution can provide higher ethanol yield and productivities using the recombinant ethanologen. TEA of second-generation ethanol production from woody biomass plays a pivotal role in realizing the most suitable process parameters and their range, and in turn the process economization of woody biomass refineries. Each fraction of biomass is essential to be taken into consideration. The lignin of woody biomass needs to be converted into valuable products for the overall economic viability of biorefinery. TEA and LCA determine the hot spots (energy or cost-intensive unit processes or process inputs) and provide a direction for future research for the valorization of woody biomass into renewable fuels and low carbon chemicals.

References

Alexandropoulou, M., Antonopoulou, G., Fragkou, E. et al. (2017). Fungal pretreatment of willow sawdust and its combination with alkaline treatment for enhancing biogas production. *J. Environ. Manag.* 203: 704–713.

Alvira, P., Tomás-Pejó, E., Ballesteros, M., and Negro, M.J. (2010). Pretreatment technologies for an efficient bioethanol production process based on enzymatic hydrolysis: a review. *Bioresour. Technol.* 101 (13): 4851–4861.

Amiri, H. and Karimi, K. (2015). Improvement of acetone, butanol, and ethanol production from woody biomass using organosolv pretreatment. *Bioprocess Biosyst. Eng.* 38 (10): 1959–1972.

Bhatia, S.K., Jagtap, S.S., Bedekar, A.A. et al. (2020). Recent developments in pretreatment technologies on lignocellulosic biomass: effect of key parameters, technological improvements, and challenges. *Bioresour. Technol.* 300: 122724.

Cheng, M.-H., Dien, B.S., Lee, D., and Singh, V. (2019). Sugar production from bioenergy sorghum by using pilot scale continuous hydrothermal pretreatment combined with disk refining. *Bioresour. Technol.* 289: 121663.

Cortes-Peña, Y., Kumar, D., Singh, V., and Guest, J.S. (2020). BioSTEAM: a fast and flexible platform for the design, simulation, and techno-economic analysis of biorefineries under uncertainty. *ACS Sust. Chem. Eng.* 8 (8): 3302–3310.

da Costa Sousa, L., Chundawat, S.P., Balan, V., and Dale, B.E. (2009). Cradle-to-grave' assessment of existing lignocellulose pretreatment technologies. *Curr. Opin. Biotechnol.* 20 (3): 339–347.

Daystar, J., Reeb, C., Gonzalez, R. et al. (2015). Environmental life cycle impacts of cellulosic ethanol in the Southern U.S. produced from loblolly pine, eucalyptus, unmanaged hardwoods, forest residues, and switchgrass using a thermochemical conversion pathway. *Fuel Process Technol* 138: 164–174.

Fornell, R. and Berntsson, T. (2012). Process integration study of a kraft pulp mill converted to an ethanol production plant—part A: Potential for heat integration of thermal separation units. *Appl. Therm. Eng.* 35: 81–90.

Galbe, M. and Wallberg, O. (2019). Pretreatment for biorefineries: a review of common methods for efficient utilisation of lignocellulosic materials. *Biotechnol. Biofuels* 12 (1): 1–26.

Hyvärinen, S., Damlin, P., Gräsvik, J. et al. (2011). Ionic liquid fractionation of woody biomass for fermentable monosaccharides. *Cellul. Chem. Technol.* 45 (7–8): 483–486.

Janssen, M., Xiros, C., and Tillman, A.-M. (2016). Life cycle impacts of ethanol production from spruce wood chips under high-gravity conditions. *Biotechnology for Biofuels* 9 (53): 1–19.

Janzon, R., Schütt, F., Oldenburg, S. et al. (2014). Steam pretreatment of spruce forest residues: optimal conditions for biogas production and enzymatic hydrolysis. *Carbohydr. Polym.* 100: 202–210.

Juneja, A., Kumar, D., Singh, V.K., and Singh, V. (2020). Chemical free two-step hydrothermal pretreatment to improve sugar yields from energy cane. *Energies* 13 (21): 5805.

Kabir, M.M., Rajendran, K., Taherzadeh, M.J., and Sárvári Horváth, I. (2015). Experimental and economical evaluation of bioconversion of forest residues to biogas using organosolv pretreatment. *Bioresour. Technol.* 178: 201–208.

Khoshnevisan, B., Shafiei, M., Rajaeifar, M.A., and Tabatabaei, M. (2016). Biogas and bioethanol production from pinewood pre-treated with steam explosion and N-methylmorpholine-N-oxide (NMMO): a comparative life cycle assessment approach. *Energy* 114: 935–950.

Kim, S.M., Dien, B.S., and Singh, V. (2016a). Promise of combined hydrothermal/chemical and mechanical refining for pretreatment of woody and herbaceous biomass. *Biotechnol. Biofuels* 9 (1): 97.

Kim, S.M., Dien, B.S., Tumbleson, M. et al. (2016b). Improvement of sugar yields from corn stover using sequential hot water pretreatment and disk milling. *Bioresour. Technol.* 216: 706–713.

Kumar, D. (2014) Biochemical conversion of lignocellulosic biomass to ethanol: experimental, enzymatic hydrolysis modeling, techno-economic and life cycle assessment studies. Ph.D. Dissertation.Oregon State University.

Kumar, D. and Murthy, G.S. (2011). Impact of pretreatment and downstream processing technologies on economics and energy in cellulosic ethanol production. *Biotechnol. Biofuels* 4 (1): 27.

Kumar, D. and Murthy, G.S. (2012). Life cycle assessment of energy and GHG emissions during ethanol production from grass straws using various pretreatment processes. *Int. J. Life Cycle Assess.* 17 (4): 388–401.

Kumar, D. and Murthy, G.S. (2013). Stochastic molecular model of enzymatic hydrolysis of cellulose for ethanol production. *Biotechnol. Biofuels* 6 (1): 1–20.

Kumar, D. and Murthy, G.S. (2016). Enzymatic hydrolysis of cellulose for ethanol production: fundamentals, optimal enzyme ratio, and hydrolysis modeling. In: *New and Future Developments in Microbial Biotechnology and Bioengineering*, V.K. Gupta (ed.) Philadelphia, PA 65–80. Elsevier.

Kumar, D. and Singh, V. (2019). Bioethanol production from corn. In: *Corn*, Sergio O. Philadelphia, PA Serna-Saldivar(ed.) 615–632. Elsevier.

Miyafuji, H. (2015). Application of ionic liquids for effective use of woody biomass. *J. Wood Sci.* 61 (4): 343–350.

Mosier, N.S. (2013). Fundamentals of aqueous pretreatment of biomass. In: *Aqueous Pretreatment of Plant Biomass for Biological and Chemical Conversion to Fuels and Chemicals*, Wyman CE (ed) Wiley, Weinheim 129–140.

Murthy, G.S. (2017). Techno-economic assessment. In: *Bioenergy: Principles and Applications*, vol. 1 (ed. Y. Li and S.K. Khanal), Boston, Massachusetts 507–520. Wiley.

Olofsson, K., Bertilsson, M., and Lidén, G. (2008). A short review on SSF—an interesting process option for ethanol production from lignocellulosic feedstocks. *Biotechnol. Biofuels* 1 (1): 1–14.

Olofsson, J., Barta, Z., Börjesson, P., and Wallberg, O. (2017). Integrating enzyme fermentation in lignocellulosic ethanol production: life-cycle assessment and techno-economic analysis. *Biotechnol. Biofuels* 10 (1): 1–14.

Radhika, K., Ravinder, R., and Ravindra, P. (2011). Bioconversion of pentose sugars into ethanol: a review and future directions. *Biotechnol. Mol. Biol. Rev.* 6 (1): 8–20.

Raghavendran, V., Nitsos, C., Matsakas, L. et al. (2018). A comparative study of the enzymatic hydrolysis of batch organosolv-pretreated birch and spruce biomass. *AMB Express* 8 (1): 1–8.

Rasi, S., Kilpeläinen, P., Rasa, K. et al. (2019). Cascade processing of softwood bark with hot water extraction, pyrolysis and anaerobic digestion. *Bioresour. Technol.* 292: 121893.

RFA (2021). *Annual Fuel Ethanol Production*. Washington, DC.: Renewable Fuel Association.

Roy, P. and Dutta, A. (2013). Life cycle assessment of ethanol derived from sawdust. *Bioresour. Technol.* 150: 407–411.

Salehian, P. and Karimi, K. (2013). Alkali pretreatment for improvement of biogas and ethanol production from different waste parts of pine tree. *Ind. Eng. Chem. Res.* 52 (2): 972–978.

Sandilands, J., Kellenberger, D., Nicholas, I., and Nielsen, P. (2009). Life cycle assessment of wood pellets and bioethanol from wood residues and willow. *N. Z. J. For. Sci.* 53: 25–33.

Sankaran, R., Cruz, R.A.P., Pakalapati, H. et al. (2020). Recent advances in the pretreatment of microalgal and lignocellulosic biomass: a comprehensive review. *Bioresour. Technol.* 298: 122476.

Sassner, P., Galbe, M., and Zacchi, G. (2008). Techno-economic evaluation of bioethanol production from three different lignocellulose materials. *Biomass and Bioenergy* 32: 422–430.

Shafiei, M., Karimi, K., and Taherzadeh, M.J. (2011). Techno-economical study of ethanol and biogas from spruce wood by NMMO-pretreatment and rapid fermentation and digestion. *Bioresour. Technol.* 102 (17): 7879–7886.

Shafiei, M., Karimi, K., Zilouei, H., and Taherzadeh, M.J. (2014). *Economic impact of NMMO pretreatment on ethanol and biogas production from pinewood*, 320254. Biomed Res. Int.

Shah, A., Baral, N., and Manandhar, A. (2016). Technoeconomic analysis and life cycle assessment of bioenergy systems. In: *Advances in Bioenergy*, vol. 1 (ed. Y. Li and X. Ge), 189–247. MA, USA: Elsevier.

Si, X., Lu, F., Chen, J. et al. (2017). A strategy for generating high-quality cellulose and lignin simultaneously from woody biomass. *Green Chem.* 19 (20): 4849–4857.

da Silva, A.R.G., Giuliano, A., Errico, M. et al. (2019). Economic value and environmental impact analysis of lignocellulosic ethanol production: assessment of different pretreatment processes. *Clean Technol. Environ. Policy* 21 (3): 637–654.

Somavat, P., Kumar, D., and Singh, V. (2018). Techno-economic feasibility analysis of blue and purple corn processing for anthocyanin extraction and ethanol production using modified dry grind process. *Ind. Crop. Prod.* 115: 78–87.

Spatari, S., Zhang, Y., and MacLean, H.L. (2005). Life cycle assessment of switchgrass- and corn stover-derived ethanol-fueled automobiles. *Environ. Sci. Technol.* 39 (24): 9750–9758.

Spatari, S., Bagley, D.M., and MacLean, H.L. (2010). Life cycle evaluation of emerging lignocellulosic ethanol conversion technologies. *Biores Technol* 101: 654–667.

Therasme, O., Volk, T.A., Eisenbies, M.H. et al. (2021). Life cycle greenhouse gas emissions of ethanol produced via fermentation of sugars derived from shrub willow (*Salix* ssp.) hot water extraction in the Northeast United States. *Biotechnol. Biofuels* 14 (1): 1–15.

Towler, G. and Sinnott, R. (2012). *Chemical Engineering Design: Principles, Practice and Economics of Plant and Process Design*. Kidlington, Oxford, UK: Elsevier.

Van Dyk, J. and Pletschke, B. (2012). A review of lignocellulose bioconversion using enzymatic hydrolysis and synergistic cooperation between enzymes—factors affecting enzymes, conversion and synergy. *Biotechnol. Adv.* 30 (6): 1458–1480.

Wang, M., Han, J., Dunn, J.B. et al. (2012a). Well-to-wheels energy use and greenhouse gas emissions of ethanol from corn, sugarcane and cellulosic biomass for US use. *Environ. Res. Lett.* 7 (4): 045905.

Wang, M., Li, Z., Fang, X. et al. (2012b). Cellulolytic enzyme production and enzymatic hydrolysis for second-generation bioethanol production. In: *Biotechnology in China III: Biofuels and Bioenergy* (ed. F.-W. Bai, C.-G. Liu, H. Huang and G.T. Tsao), 1–24. Berlin, Heidelberg.

Wingren, A., Galbe, M., and Zacchi, G. (2003). Techno-economic evaluation of producing ethanol from softwood: comparison of SSF and SHF and identification of bottlenecks. *Biotechnol. Prog.* 19 (4): 1109–1117.

Wingren, A., Galbe, M., and Zacchi, G. (2008). Energy considerations for a SSF-based softwood ethanol plant. *Bioresour. Technol* 99: 2121–2131.

Yoo, C.G. and Ragauskas, A.J. (2021). Opportunities and challenges of lignin utilization. In: *Lignin Utilization Strategies: From Processing to Applications* (ed. C.G. Yoo and A. Ragauskas), 1–12. Washington, DC: ACS Publications.

Yoo, C.G., Meng, X., Pu, Y., and Ragauskas, A.J. (2020). The critical role of lignin in lignocellulosic biomass conversion and recent pretreatment strategies: a comprehensive review. *Bioresour. Technol.* 301: 122784.

Zhao, X., Cheng, K., and Liu, D. (2009). Organosolv pretreatment of lignocellulosic biomass for enzymatic hydrolysis. *Appl. Microbiol. Biotechnol.* 82 (5): 815–827.

Zhou, Z., Lei, F., Li, P., and Jiang, J. (2018). Lignocellulosic biomass to biofuels and biochemicals: a comprehensive review with a focus on ethanol organosolv pretreatment technology. *Biotechnol. Bioeng.* 115: 2683–2702.

Zhu, J. and Pan, X. (2010). Woody biomass pretreatment for cellulosic ethanol production: technology and energy consumption evaluation. *Bioresour. Technol.* 101 (13): 4992–5002.

Zhu, J.Y., Pan, X., and Zalesny, R.S. (2010a). Pretreatment of woody biomass for biofuel production: energy efficiency, technologies, and recalcitrance. *Appl. Microbiol. Biotechnol.* 87 (3): 847–857.

Zhu, W., Zhu, J., Gleisner, R., and Pan, X. (2010b). On energy consumption for size-reduction and yields from subsequent enzymatic saccharification of pretreated lodgepole pine. *Bioresour. Technol.* 101 (8): 2782–2792.

14

Environmental Indicators, Life Cycle Analysis and Ecological Perspective on Biomass Conversion

Andreza A. Longati[1], *Ediane S. Alves*[2], *Simone C. Myoshi*[2], *Andrew M. Elias*[2,3], *Felipe F. Furlan*[2,4], *Everson A. Miranda*[1], *and Roberto C. Giordano*[2,4]

[1] *Department of Materials and Bioprocess Engineering, School of Chemical Engineering, University of Campinas, Campinas, SP, 13083-852, Brazil*
[2] *Chemical Engineering Graduate Program, Universidade Federal de São Carlos (UFSCar), Rodovia Washington Luís (SP-310), km 235, São Carlos, SP, 13565-905, Brazil*
[3] *EBRAPA Instrumentation, Brazilian Agricultural Research Corporation – EMBRAPA, São Carlos, SP, 13560-970, Brazil*
[4] *Department of Chemical Engineering, Universidade Federal de São Carlos (UFSCar), Rodovia Washington Luís (SP-310), km 235, São Carlos, SP, 13565-905, Brazil*

14.1 Introduction

14.1.1 The Role of Biomass in a Sustainable Economy

It is almost a scientific unanimity that greenhouse gases (GHGs) emitted by human activities are the primary driver of global warming (Giordano et al. 2020; Nyambuu and Semmler 2020). Natural resources expenditure has been consistently growing during the past century, due to the increase in human population and the demand for goods. The perspective that this trend continues in the future has to be taken into account in the assessment of climate change caused by mankind in the future (Höök and Tang 2013). Within this context, a change in the organization of human society toward a low-carbon production matrix most certainly will have to occur (Giordano et al. 2020; Nyambuu and Semmler 2020). Circular bioeconomy plays an important role for a better and integrated organization of the use of natural resources, since it offers new solutions to enable the transition from the traditional linear economy to a low carbon one (Kokkinos et al. 2020).

With the goal of dealing with the climate crisis, the 21st Conference of Parties (COP21), organized by the United Nations Framework Convention on Climate Change (UNFCCC), established the Paris Agreement. This deal aims to limit the effects of global warming to a maximum of 2 °C above the pre-industrial average by the end of the century (Klein et al. 2019). To reach this goal, net zero GHG emissions must be attained in the latter half of the twenty-first century, and net negative emissions thereafter (Rissman et al. 2020).

A variety of technologies have been developed aimed at a rapid and cost-effective reduction of energy and chemicals consumption, and GHG emissions, for a broad range of

Lignocellulose Bioconversion Through White Biotechnology, First Edition. Edited by Anuj Kumar Chandel.

industrial fields. However, the development of technologies is only a part of the picture. Policy interventions are essential to develop and improve sustainable technologies, making investment in cleaner industrial processes more attractive. According to Rissman et al. (2020), public policies can spread innovations through international supply chains, improving companies in countries that lack strong policies of their own.

Under the Paris Agreement, each country has individual targets for CO_2 mitigation, to reach the global goals (Su et al. 2015). Examples of public programs for low-carbon products include the Buy Clean California Act (BlueGreen Alliance 2017), Japan's Act on Promoting Green Purchasing (Tanabe 2016), Ujala program for efficient lighting in India (Ujala 2020), Brazilian National Biofuel Policy (RenovaBio) (Klein et al. 2019), German National Biomass Action Plan, Draft Russian Energy Strategy till 2035, and Draft Position Paper on South African Biofuels Regulatory Framework (Saravanan et al. 2020).

The generation of energy contributes almost 70% of the GHG emissions (Höök and Tang 2013; Thornley and Adams 2018). Renewable energy from biomass plays an important role in this scenario, since this feedstock can be converted into several useful forms of energy using different processes (Figure 14.1) (Gent et al. 2017). Biomass refers to the mass of living organisms, including plants, animals, and microorganisms, or, from a biochemical

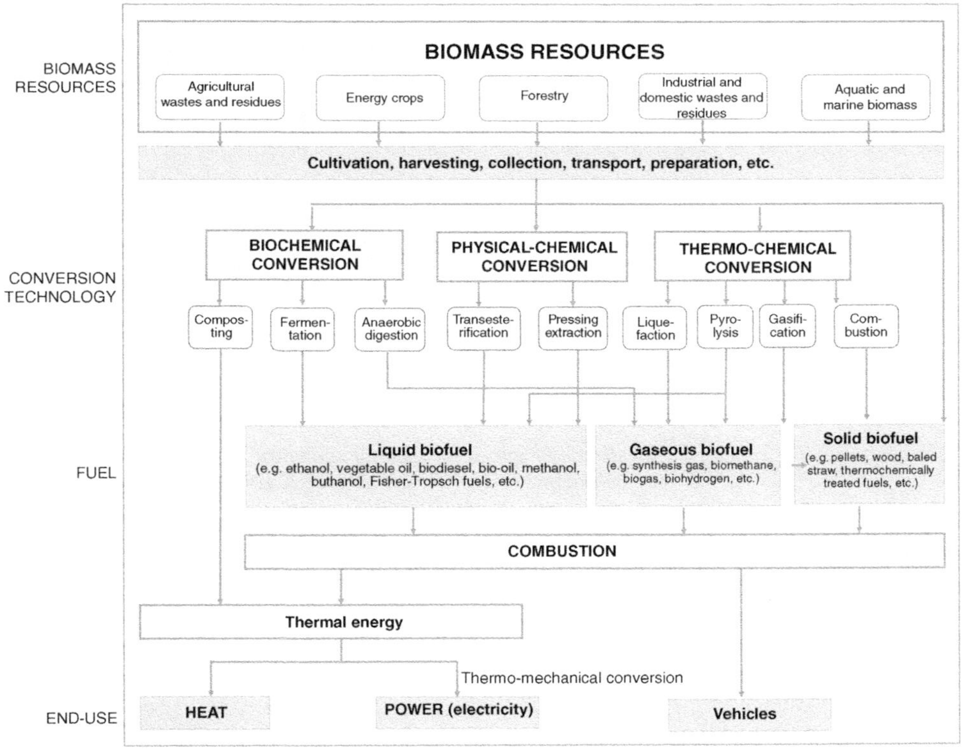

Figure 14.1 Schematic representation of biomass to bioenergy conversion pathways. *Source:* Adapted from Thrän et al. (2015) under the terms of the Creative Commons Attribution 4.0 International License (http://creativecommons.org/licenses/by/4.0).

perspective, cellulose, lignin, sugars, fats, and proteins (Houghton 2008). Biomass includes agricultural wastes (such as straw, husk, cobs, stalks, leaves, bark, fruits, and cutting vines), energy crops, forestry, industrial and domestic waste and residues, and aquatic and marine biomass (Balat et al. 2009). Multistage processes are required to convert biomass into a useful energy source (such as cultivation, harvesting, storage, transportation, extraction, hydrolysis, saccharification, etc.) (Adams et al. 2018). Conversion of biomass to energy is based on different principles and concepts. It is undertaken using three main process technologies—biochemical, physicochemical and thermochemical—to produce fuels, (solid, liquid, and gaseous) used in vehicles and to generate heat and power (Thrän et al. 2015).

Until recently, molecules for many different applications (chemicals, solvents, building blocks, pharmaceutical industry, food industry, etc.) have been derived from petroleum. This non-renewable feedstock provides carbon (together with hydrogen and other elements) for these molecules (Giordano et al. 2020). Therefore, to complete the transition from the traditional linear economy to a low carbon one, it is also necessary to replace fossil oil by biomass as the source for these molecules. According to Bender et al. (Bender et al. 2018), the production of chemicals from biomass can cover a broad range of structural complexity, from high-value jet fuels to monomers for high-performance polymers, and to pharmaceutical intermediates. It should be stressed that bio-kerosene is the only sustainable alternative to replace fossil jet fuel. With this biofuel the aviation sector could make possible the goal of zero-emissions flight. Table 14.1 is a list of some examples of chemicals produced from biomass.

Table 14.1 Examples of chemicals produced from biomass.

Chemical/molecule	Biomass source	Route conversion	References
Ammonia	Sugarcane bagasse	Thermo-chemical (gasification)	Flórez-Orrego et al. (2019)
Anhydrosugars (levoglucosan and levoglucosenone)	Cellulose	Thermo-chemical (pyrolysis)	Dai et al. (2020)
Anti-oxidants	Corn stover	Enzymatic hydrolysis	Guo et al. (2018)
Biosurfactant	Molasses (agro-industrial by-product)	Biochemical (bioreaction)	Makkar and Cameotra (1997)
Biosurfactant (rhamnolipid)	Molasses (agro-industrial by-product)	Biochemical	Rashedi et al. (2005)
Biosurfactant (sophorolipid)	Sugarcane bagasse	Biochemical	Marcelino et al. (2019)
Biosurfactant (xylolipids)	Rice straw	Biochemical	Panjiar et al. (2020)
Carotenoids (b-carotene, torulene and torularhodin)	Eucalyptus	Biochemical	Lopes et al. (2020)
Carotenoids (carotene)	Wheat straw hydrolysates	Biochemical	Liu et al. (2020)

Table 14.1 (Continued)

Chemical/molecule	Biomass source	Route conversion	References
Carotenoids (g-carotene, b-carotene, torulene and torularhodin)	Sugarcane bagasse	Biochemical	Pinheiro et al. (2020)
Cellulase	Corn cob, sugarcane bagasse, and sugarcane straw	Biochemical	Vieira et al. (2021)
Citric acid	Sugarcane (bagasse and harvesting residues)	Biochemical	Özüdoğru et al. (2019)
Hydrocarbons (benzene, toluene, xylene, naphthalene, aldehydes, furans, acetones, acetic acid, aromatics, oleafins, phenols and alkanes, aliphatic hydrocarbons)	Pine wood, furan, rice stalk, beech wood, poplar wood, napier grass, wood sawdust	Thermo-chemical (pyrolisis)	Dai et al. (2020)
Glycerine	Palm kernel oil	Chemical (transesterification of oils)	Ghannadzadeh and Tarighaleslami (2020)
Glucaric acid	Corn stover	Biochemical	Thaore et al. (2020)
Glutamic acid	Sugarcane (bagasse and harvesting residues)	Biochemical	Özüdoğru et al. (2019)
Lactic acid	Cassava biomass	Biochemical	Sharma et al. (2020)
Oxygenates (phenolic compounds to produce phenolic resins, adhesives, synthetic drugs, dyes, and fuel additives)	Corn stover-based activated carbon	Thermo-chemical (pyrolisis)	Dai et al. (2020)
Sunscreen	Corn cob and pines	Enzymatic hydrolysis	Qian et al. (2015)
Vanillin	Wood	Sulfite process	Fache et al. (2016)
Xylitol	Corn cob	Biochemical	Dasgupta et al. (2021)
Xylitol	Sugarcane (bagasse and harvesting residues)	Biochemical	Özüdoğru et al. (2019)
Xylitol	Corn residues (corn stover, husk, and cob)	Biochemical	Irmak et al. (2017)
Xylitol	Rice straw	Biochemical	Singh et al. (2021)
Xylitol	Bamboo	Biochemical	Song et al. (2020)

The diversity in the chemical composition of biomass can affect the conversion technologies employed for production of biofuels and high-value products (Gent et al. 2017). Several factors must be considered when selecting a conversion process of biomass that should be adopted as an industrial process: the type, availability, quantity, and characteristics of the biomass feedstock, the desired form of the energy or end-use requirements, environmental standards, governmental policies, economic conditions, and project-specific factors (Adams et al. 2018). A key feature to determine the most appropriate route for processing the biomass is the classification and availability of the biomass (Gent et al. 2017).

A fully integrated biorefinery arises as an option to replace the complex oil refinery petrochemical industry. This facility converts biomass into a portfolio of primary and secondary products (either biofuels or biochemicals) (Bender et al. 2018; Longati et al. 2020). Consequently, the products of an integrated biorefinery could meet future demands for energy, chemicals, and materials. It is important to stress that, even though the technologies used in the petroleum-centered economy may differ from those needed for a bio-based chemical economy, technologies are in constant development to overcome the bottlenecks of the biomass conversion processes. Shifting from an established economy, based on petroleum, to another one founded on renewable sources implies important technical and financial risks to manufacturers (Bender et al. 2018). Therefore, the path toward a low-carbon production matrix must focus on both the economic viability and the environmental sustainability of the new biomass conversion pathways.

14.2 Life Cycle Assessment (LCA)

As previously mentioned, the Intergovernmental Panel on Climate Change (IPCC) special report warns of the need to curb climate change caused by human activity (IPCC and Report 2018). The warming generated by the emission of GHG is estimated to raise the global temperature by at least 1.5 °C over the next 30 years (Raftery et al. 2017). The climate crisis will affect, and even menace, the survival of all species on the planet. Therefore, using technological routes that have less environmental impact is fundamental to mitigating anthropogenic actions in nature (IPCC and Report 2018). In this context, LCA is a method that analyzes environmental impacts generated by a product or service during all the stages of a product's (or service's) life.

The importance of LCA can be illustrated in a practical example. Biofuels have always been seen as a more sustainable option compared to fossil fuels; and they are, indeed. But common sense would say that the CO_2 emitted from the combustion of biofuels has a "neutral" character on climate change, because it would be part of the biogenic carbon cycle (Reijnders and Huijbregts 2008). However, when analyzing the process "from the cradle," it is observed that the cultivation and transportation of the raw material are responsible for most of the environmental impacts associated with the production of biofuel. This fact derives, above all, from cultivation of the raw material, with a substantial use of petroleum-based products such as diesel for the machinery, chemical fertilizers, and pesticides (Tonini et al. 2016; Longati et al. 2017). The LCA, provided it is correctly applied, can help to indicate stages of the process or service that effectively contribute to environmental impacts.

From this new perspective, new strategies can be adopted to make production more sustainable in the field (Souza and Seabra 2013).

The rules and requirements associated with the LCA are defined by international standards ISO 14040 (2006) and ISO 14044 (2006). They focus on principles and structure (ISO 14040a) and requirements and guidelines (ISO 14044b). LCA is structured in four stages: (i) definition of objectives and scope; (ii) inventory analysis; (iii) impact assessment; and (iv) interpretation of results.

LCA can follow an attributional or consequential approach. Attributional analysis aims at assessing the environmental impacts directly associated with the consumption or production of the product or service. On the other hand, consequential analysis is appropriate when trying to identify the environmental consequences of consumption or production of the product or service. In other words, it is not limited to the impacts of activities occurring within a product's value chain, but considers activities that are affected at a global level (Schrijvers et al. 2020). More details of the two approaches are beyond the scope of this text, but it should be stressed that, in general, the attributional approach is the most used to quantify the emissions of a process, especially in the chemical engineering field.

In order to quantify the environmental impacts of a product or service, LCA makes assumptions which in fact simplify the reality. This fact implies limitations on the conclusions of the analyses. It should be stressed that whole method is in continuous improvement, however.

In the stage of defining objectives and scope, the limits and simplifications of the LCA are defined. That is, a description and delimitation of the system to be analyzed, referenced to a functional unit, are done. This provides a quantified reference and guarantees the equivalence of the systems (Peters 2016).

Defining the purpose and scope of the LCA means describing the reason for carrying out the life cycle analysis, and which questions should be answered. So, the intended use of the results (application) should be clearly stated, and the expected stakeholders and target audience must be clearly defined (Curran 2017). At this stage, the indicators for the LCA and their allocation to the products/services under analysis are defined (Curran 2017).

In multifunctional systems, it is necessary to distribute the environmental impacts among all products or services generated by the analyzed system. Multifunctionality is very common in the biomass processing sector. For example, in the production of bioethanol from sugarcane, ethanol and electricity are co-products of the system, eventually alongside with sugar and other co-products of the biorefinery. Therefore, it is necessary to divide the environmental burden between all co-products, or co-functions (Moretti et al. 2020a). The ISO 14044 (2006) standard provides a hierarchy for the methods that can be used.

The first solution that this standard recommends for multifunctionality is subdivision: by dividing the process into two or more subprocesses, the impacts can be attributed to each product or service. If subdivision is not possible, system expansion should be used, which consists of expanding the product system to include the functions of the co-products within the system's boundaries.

In many cases common in the industrial field, however, it is not feasible to obtain functional equivalence by dividing the system into subprocesses. Likewise, isolating the primary function of a process from the secondary functions by expanding the system frequently is not a feasible approach. This is the case when the system, for instance a biorefinery, is composed of highly integrated processes, through mass and energy flows (Bjørn et al. 2018).

In such cases, the only feasible approach for the problem is allocation. In this case, environmental impacts are divided between co-products according to a pre-defined basis for partitioning among the co-products, which can be mass, energy, price, etc. (ISO 14044). ISO 14044 provides a systematic approach to solve the problem of multifunctionality, but there is a wide variety of allocation procedures in the literature following this logic (Schrijvers et al. 2020). The application of different allocation methodologies leads to different LCA results (Murali and Shastri 2019). For example, when using an economic base allocation for sugarcane bagasse, the impacts may be less than when mass allocation is used, since the economic value of the bagasse produced during the cultivation of sugarcane is usually lower than the products from sugarcane juice (Murali and Shastri 2019). Therefore, the LCA result for different products or processes should not be compared if different allocation approaches are adopted. So, the selection of how to allocate environmental impacts is decisive for the result of the LCA.

The next step, life cycle inventory (LCI), consists of the collection of information related to the life cycle of the product that passes through the boundaries of the system, in the form of incoming and outgoing streams of matter and energy (Guinée et al. 2002). This phase is considered the most complex in an LCA study, and is the one that requires the longest working time from the analyst (Finnveden et al. 2009). To facilitate the construction of the LCI, databases have been developed such as GaBi professional, Probas, US Life Cycle Inventory Database (NREL), European reference Life Cycle Database, International Reference Life Cycle Database, and Ecoinvent, a world leader with a long history of data collection, covering more than 4000 processes (Herrmann and Moltesen 2015).

Usually, databases are integrated in the LCA software (Finnveden et al. 2009). When comparing a random sample of 100 unit processes in software with different databases (SimaPro and GaBi), it was observed that most of the LCA results were similar. However, the databases may differ considerably. Some of these differences are so large that they can influence the conclusions drawn from the LCA study (Herrmann and Moltesen 2015). Therefore, the choice of the database, which clearly can influence the results, can become strategic depending on the motivation (Kalverkamp et al. 2020). The divergences in the calculated impact values, using different databases, are generated by the divergence in the inflows and outflows of the inventory implemented in the database (Kalverkamp et al. 2020). Therefore, before choosing a database, careful verification is essential, including the sources.

In the third phase of the construction of the LCA, life cycle impact assessment (LCIA), potential environmental impacts are quantitatively and qualitatively assessed using the results of the second phase (LCI). The LCIA converts the results of the inventory to a selected group of impacts, such as greenhouse effect, mortality, ozone layer destruction, eutrophication, photochemical ozone formation, toxicity, area occupation, among others (Rosenbaum et al. 2018). Environmental impact indicators are generally divided into midpoint or endpoint. Midpoint indicators have this denomination because they are located between the process and the effect generated in the environment. In other words, they are the direct emission of the process, such as the emission of GHGs, acidification, eutrophication, etc. (Guinée et al. 2002).

Endpoint indicators consider the entire mechanistic of environmental impacts, up to their final point, that is, they refer to a specific damage related to human health, natural environment, or natural resources (Mendes et al. 2015).

There are many available LCIA methods and indicators. Thus, the task of selecting which ones to use is not straightforward, requiring the analyst to understand the main characteristics of these methods (Rosenbaum et al. 2018).

Table 14.2 presents a survey LCIA applied to the analysis of biomass conversion processes. There is no prevalence of one specific method in the literature. ReCiPe, ILCD and CML are frequent in the analysis of the conversion of biomass.

Table 14.2 Some publications on biomass conversion LCA.

References	IPCC	ILCD	ReCiPe	TRACI	CML
Den Boer et al. (2020)		●			
Murali and Shastri (2019)			●	●	
Moreno et al. (2020)			●		
Chary et al. (2018)			●		●
Dyjakon et al. (2019)		●			
Kylili et al. (2016)		●			
Rodríguez et al. (2018)					●
Chen et al. (2016)			●	●	
Gaglio et al. (2019)					●
Bello et al. (2019)			●		
Parascanu et al. (2018)			●		
Dasgupta et al. (2021)	●				●
Moretti et al. (2020b)		●			
Ghannadzadeh and Tarighaleslami (2020)			●		
Scherer et al. (2017)					●
Zang et al. (2020)					●

Notes: IPCC is the Intergovernmental Panel on Climate Change. Its methodology is used to estimate greenhouse gas emissions and removals.
ILCD is the International Reference Life Cycle Data system and provides a common basis for life cycle studies, building on and further specifying the broad ISO standards.
ReCiPe can be understood as a fusion of two methodologies, taking the midpoint indicators from CML and the endpoint indicators from Eco Indicator.
TRACI, a Tool for the Reduction and Assessment of Chemical and Other Environmental Impacts, was developed by the US Environmental Protection Agency. The impact categories were developed specifically for US conditions.
CML is an impact assessment method that restricts quantitative modeling to early stages in the cause–effect chain to limit uncertainties. Results are grouped in midpoint categories.

14.3 New Brazilian National Biofuel Policy (RenovaBio): A Case Study for Sugarcane Distilleries

RenovaBio is a Brazilian biofuel policy, established in 2017 and starting in 2020, that aims at establishing a framework for the commercialization of decarbonization credits linked to the carbon intensity of several biofuels. The RenovaBio policy defines a method focusing on the accountability of the carbon credits, through a calculation spreadsheet called RenovaCalc (Matsuura et al. 2018).

RenovaCalc uses the following premises for the LCA: attributional method, climate change impact category, GWP100 characterization factor, and cradle-to-wheel scope. It uses a reference database to analyze the impacts, for example IPCC (2006) for emissions of fuels after combustion and EcoInvent 3.1 (Wernet et al. 2016) for raw materials inputs. All the residues are considered to have a zero emission.

In order to use RenovaCalc, it is important to specify the emissions inventory of the inputs and outputs of the process. All the biofuel production process is considered: the agricultural and industrial stages. The calculation of the impacts is quite simple. It sums the multiplication of the input's consumption by the correspondent life cycle emission factor. To this value, all the emissions produced during the process are added. The total sum is the total quantity of CO_2 equivalent emitted, as shown in Eqn (14.1).

$$\text{total emitted} = \begin{array}{l} \sum_i \text{input}_i\,\text{consumption} \cdot \text{life cycle emission factor}_i + \\ \sum_j \text{qty emission}_j\,\text{emission factor}_j \end{array} \tag{14.1}$$

Since the manufacturing of biofuels is, in general, a multiproduct process that many times includes the production of bioelectricity, RenovaCalc uses an energy-based allocation method. The quantity emitted for each product is obtained by a weighted average of the energy content plus the transportation emission, as shown in Eqn (14.2).

$$\text{emitted for product}_j = \text{total emitted} \cdot \frac{\text{output}_j\,\text{production} \cdot \text{output}_j\,\text{pci}}{\sum_i \text{output}_i\text{production} \cdot \text{output}_i\,\text{pci}} + \text{transp.emission}_j \tag{14.2}$$

Since the scope of RenovaCalc is cradle-to-wheel, the methodology also considers the emissions caused by the burning of the fuel. The carbon intensity (CI) is the total amount of gCO_2 eq per MJ emitted since the cradle to the wheel, measured in gCO_2 eq per MJ and calculated as shown in Eqn (14.3).

$$\text{CI} = \text{emitted for product } j \text{ in MJ} + \text{emitted in burning in MJ} \tag{14.3}$$

The energy environmental efficiency grade (EEEG) is obtained by comparing the carbon intensity of the biofuel with an equivalent fossil fuel's CI, as in Eqn (14.4).

$$\text{EEEG} = \text{CI} - \text{emissions per MJ of a fossil combustible of reference} \tag{14.4}$$

Following the RenovaBio framework, an example of a 1G ethanol production process is presented in Tables 14.3–14.5. As mentioned, the RenovaBio accounts for the agricultural stage, the industrial stage, transportation and burning emissions. The agricultural stage for production of 1 ton of sugarcane is presented in Table 14.3, based on Hassuani et al. (2005), Seabra et al. (2011), Cavalett et al. (2012), Matsuura et al. (2018), and Potrich et al. (2020).

Table 14.3 shows that the total sum of inputs and emissions is 38 544.1 gCO_2 eq. per ton of sugarcane. Adding the transportation of the sugarcane to the distilleries, in a radius of 32.2 km, increases it to a total of 41 347.7 gCO_2 eq per ton.

Table 14.3 Mass flow of inputs and outputs and g CO_2 eq produced per ton of sugarcane.

Description	Unit	Amount	g CO_2 eq
Inputs			
Fertilizers			
Limestone ($CaCO_3$)	kg	5.183[a]	190.919
Gypsum	kg	2.790[b]	7.893
Nitrogen in fertilizer			
Ammonia	kg	0.109[a]	214.957
Urea	kg	1.146[c]	3680.426
Ammonium nitrate	kg	0.287[a]	2365.058
MAP	kg	0.008[a]	25.350
Phosphorus in fertilizer (P_2O_5)			
Acidulated phosphates	kg	0.227[a]	536.487
MAP	kg	0.022[a]	37.252
Potash in fertilizer	kg	0.980[a]	446.104
Stillage	L	570.00[a]	0.000
Filtercake	kg	31.00[a]	0.000
Ashes and shoot	kg	14.00[a]	0.000
Unburned trash	kg	70.00[d]	0.000
Agro-chemicals			
Herbicide, Roundup Ready, as glyphosate	kg	2.789E-03[b]	32.093
Herbicide, Nortox, as 2,4D	kg	6.015E-04[b]	3.057
Insecticide/herbicide, others	kg	1.396E-02[b]	149.97
Fuel consumption			
Diesel (B0)	kg	2.440[a]	1358.094
Biodiesel (B100)	kg	0.220[a,e]	217.116
Emissions			
N_2O synthetic fertilizers	kg	0.032	8553.266
N_2O organic mineral fertilizers	Kg	0.020	5361.288
N_2O crop residues	kg	0.013	3402.381
CO_2 eq limestone and gypsum	kg	2.471	2470.563
CO_2 eq urea	kg	1.788	1788.267
CO_2 eq diesel burn	kg	7.700	7699.908

(Continued)

Table 14.3 (Continued)

Description	Unit	Amount	g CO_2 eq
CO_2 eq biodiesel burn	kg	0.004	3.630
Total agricultural phase	g CO_2 eq ton^{-1}		**38 544.1**
Ethanol with the transport phase			
Distance	km	32.3	2803.64
Transported sugarcane	kg	1000	**41 347.7**

Abbreviation: MAP, monoammonium phosphate.
[a] Seabra et al. (2011).
[b] Matsuura et al. (2018).
[c] Cavalett et al. (2012).
[d] 50% of the trash remains on field—140 kg trash ton^{-1} sugarcane (Hassuani et al. 2005).
[e] Calculated based on 8% of biodiesel on diesel.

Similar calculations are applied to the industrial stage. An LCA of the ethanol production with an annexed plant of sugar based on Cavalett et al. (2012) and Matsuura et al. (2018) is shown in Table 14.4.

Table 14.4 LCA based on GWP100, for ethanol production (distillery annexed to a sugar plant).

Description	Units	Amount	g CO_2 eq
Inputs			
Sugarcane	ton	1.00	41 347.7
Water	ton	1.50[a]	13.20
Phosphoric acid	kg	0.18[a]	267.03
Flocculants	kg	0.00[a]	5.50
Antibiotic	kg	0.00[a]	0.64
Trash	kg (dry basis)	70.00[b]	0.00
Bagasse	kg (dry basis)	138.00[a]	0.00
Lime (CaO)	kg	1.01[c]	971.35
Sulfuric acid	g	358.00[c]	39.15
Lube oil	kg	0.01[c]	15.16
Zeolites	g	16.40[c]	82.17
Emissions to air			
CO_2 eq from fermentation	kg	41.40[a]	0.00
CO_2 eq from bagasse burn	kg	138.00[a]	858.64
CO_2 eq from trash burn	kg	70.00[a]	435.54
Total emissions	g CO_2 ton^{-1} sugarcane		**44 036.08**

[a] Scherer et al. (2017).
[b] 50% of the trash remains on field—140 kg straw ton^{-1} of sugarcane (Cavalett et al. 2012).
[c] Wernet et al. (2016).

Table 14.5 Energy allocation of products and co-products of distilleries annexed to plants of sugar.

Description	Units	Amount	g CO_2 eq
Outputs—energy allocation			
Anhydrous ethanol	kg	40.40	19 182.58
Sugar	kg	50.90	13 845.36
Electricity	kW h	182.00	11 008.13
Stillage	kg	558.00	
Filter cake	kg	10.30	
Ashes and shoot	kg	8.90	
Transport			
Road distance	km	700.00[a]	2454.70
Ethanol with the transport phase			
Transported anhydrous ethanol	kg	40.40	21 637.29

[a] Average logistics distance for ethanol in road model.
Source: RenovaCalc.

As the ethanol production in an annexed plant of sugar yields ethanol, sugar, and bioelectricity, it is necessary to allocate the emissions to these products. The RenovaCalc applies energy allocation, that is, it shares the emissions according to the energetic content of each product. Table 14.5 shows the energy allocation of the example.

After the energy allocation shown in Table 14.5, transport emissions are added, obtaining a GWP100 of 40.40 gCO_2 eq kg^{-1} of anhydrous ethanol produced. This represents 535.58 gCO_2 eq kg^{-1} or 18.95 gCO_2 eq MJ^{-1}. As the anhydrous ethanol generates 0.44 gCO_2 eq MJ^{-1} in the burning, the carbon intensity of the anhydrous ethanol under the premises considered is 19.39 gCO_2 eq MJ^{-1}. Compared with gasoline (carbon intensity of 87.4 gCO_2 eq MJ^{-1}), the EEEG in this case is 68.01 gCO_2 eq MJ^{-1}. This means that the emission of 68.01 gCO_2 eq MJ^{-1} is avoided by using ethanol in substitution for gasoline.

14.4 Process Systems Engineering Tools for Biomass LCA

LCA is a holistic analysis, since it simultaneously considers the entire life cycle of a process or product and multiple environmental impacts (Ögmundarson et al. 2020).

However, although avoiding these problems, LCA creates a complex system which is not simple to analyze in full. In the case of the application of LCA to complex biomass processing systems (biorefineries), other challenges are added. These are defining a set of alternative scenarios to be assessed from a large number of conversion pathways, methodological choices for dealing with multifunctionality, and interdependence between subsystems, among others (Vadenbo et al. 2017). Moreover, it is necessary to consider simultaneously the process and product design, and biomass availability and distribution (Calvo-Serrano et al. 2019). The use of process systems engineering (PSE) tools may be helpful in this task.

PSE can consider a great number of feasible system configurations simultaneously. Thus, PSE tools may provide means to capture the consequences of multifunctional product systems, and the subsystems' relations (Vadenbo et al. 2017). Therefore, PSE makes possible the identification of opportunities and bottlenecks, and may provide valuable information for both governmental and private decision-making. From the pool of tools that PSE provides, which may be applied to LCA problems, three deserve closer attention: global sensitivity analysis (GSA), uncertainty analysis (UA), and process optimization.

GSA is a method for determining how the uncertainty in the output of a model can be attributed to its inputs. Through GSA, one can find the inputs that might be considered non-influential regarding the chosen output. These can be fixed for further analysis, simplifying the model. GSA can also be used to find the most influential inputs, which might be prioritized in an effort to decrease the uncertainty in the model output, thus contributing for a more sound decision (Saltelli et al. 2007). For example, Azadi et al. (2014) applied GSA in conjunction with LCA to a biodiesel production process from algae. Through GSA, the authors found the dominant input parameters affecting the metrics chosen, GHGs and non-renewable energy consumption. They concluded that the sustainability of the process relies mainly on using a low carbon solution for reducing the water content of the feedstock such as solar drying. GSA was able to identify the five most influential inputs among the 23 inputs initially considered, depending on the process. Similarly, Carpio et al. (2020) applied GSA on first- and second-generation bioethanol from a sugarcane biorefinery considering the effect of RenovaBio's decarbonization credits on the economic feasibility of the process. GSA showed that a set of only three variables was influential (from the original set of 10 variables). The authors also concluded that the price of decarbonization credits did not change the relative relevance of the analyzed parameters, for the prices considered.

Uncertainties are naturally embedded in LCA due to constraints (technical, conceptual, etc.) in the verification, validation, and confirmation of results (Igos et al. 2019). Therefore, an adequate analysis of these uncertainties, their sources and types, helps to increase transparency and reliability of the results, preventing misguided decisions (Lima et al. 2020). UA may be performed using a Monte Carlo approach. As an example of UA, Zhao et al. (2020) evaluated the environmental impact potential of producing bioethanol from corn stover. The authors assessed seven different technological configurations and assessed global warming potential. An UA was conducted to prioritize technologies for further development based on their probabilities of promoting savings regarding the environmental impacts. The analysis shows that steam explosion and ammonia-based biomass pretreatments have the highest probabilities of promoting savings regarding the environmental impacts: 100% and 40% of non-toxic and toxic impacts, respectively.

Finally, LCA can also be combined with optimization techniques to obtain optimal environmental results (minimum impacts). Since economic metrics are usually included in these analyses, multi-objective optimization problems are defined. One option is to treat the problem using one single objective function, by combining both economic and environmental indicators in a single metric. That was the approach proposed by Ögmundarson et al. (2020) which presents a framework that combines environmental and economic performance in a single score. The strategy is to use a LCA monetization to enable the combination of several indices into one. Vadenbo et al. (2017), on the other hand, used a fuzzy interval formulation ranging from marginally acceptable (based on the current process) to fully acceptable results (based on a mono-objective optimization of the single metric) for each environmental metric. The fuzzy interval was used as inequality constraints, granting that any feasible solution would present an environmental impact lower than or equal to the current process. Another option is to use multi-objective techniques to find the locus of optimal solutions which represent tradeoffs between all indicators (Carpio et al. 2021). Calvo-Serrano et al. (2019) presents a framework for optimizing the environmental impact and economic feasibility of a network of biomass to fuel, chemicals, and electricity. The authors applied the methodology to the whole European Union production network, showing that as much as 1.81 Gtons CO_2 eq could be reduced annually by exploiting biomass resources. Finally, Yue et al. (2016) present a framework for performing the life cycle optimization which not only intends to evaluate the environmental impact but also compares different alternatives and identifies the best decisions. The framework considers both direct and indirect emissions, which may be underestimated by LCA.

14.5 Retro Techno-economic Environmental Analysis

The conventional approach of LCA and techno-economic analysis (TEA) demands the simulation of a process with predetermined operational conditions. Process variables are calculated with the aid of process simulators. With these data, the economic performance and environmental impacts are assessed, usually in a different software environment.

This methodology is intrinsically iterative. However, instead of asking about the environmental performance or economic feasibility of a given process, one could ask "what process conditions would turn this process feasible and with environmental impacts below a target value?" Process conditions should be understood as ranges of process variables, ranges of specifications, and design constraints (Giordano et al. 2020). For each solution provided by the process simulator, the TEA and LCA would be undertaken, providing one value for each economic and environmental metric. The procedure would go on until achieving a feasible operational condition.

To overcome this exhausting procedure, Furlan et al. (2016) proposed a novel approach, the retro-TEA (RTEA), that turns this problem upside down. Later, Elias et al. (2021) expanded the RTEA incorporating the LCA, originating the retro techno-economic environmental analysis (RTEEA). In this methodology, rather than iteratively searching for a process setup that would comply with economic or environmental metrics, the authors explicitly included in the simulation the desired specification, such as net present value equal to zero or the global warming potential equal to a desired target. This was just another equation that had to be satisfied, for the solver to converge (Giordano et al. 2020). Of course, the equations that compute the environmental and economic performances must be solved simultaneously with the models of the process units (Longati et al. 2018). In this case, equation-oriented process simulators are demanded (Furlan et al. 2016). To keep the problem well-posed, that is, zero degrees of freedom, another equation or specification has to be suppressed, selected by a sensitivity analysis. Using RTEEA, a problem like "what conversion is demanded in the saccharification reactor for the 2G ethanol process to be feasible" can be answered straightly, without the need of an iterative approach.

The RTEEA methodology draws windows of operation (Woodley and Titchener-Hooker 1996) whereby the process would be technically, economically, and environmentally viable. For each metric (environmental and economic indices), a plot, named an "isometric curve," is created (Elias et al. 2021). It provides threshold values of the main process variables. These values become targets to be pursued by R&D teams, in order to seek feasible operational conditions taking into account the overall process (Elias et al. 2018). The technique can be applied for multidimensional problems, working simultaneously with multiple variables (n) (Giordano et al. 2020). In this case, $n - 1$ variables are specified and the last one is calculated by the process simulator in order to meet the economic or the environmental metric.

The RTEEA methodology is composed of four steps: construction of a base case, incorporation of TEA and LCA analysis into the process simulation, selection of key variables through GSA, and delimitation of the feasible space (Elias et al. 2021). After modeling the desired process, economic and environmental metrics are included as extra variables and equations. So, these will be solved by the simulator, together with the process units' models (energy and mass balances, thermodynamic and physical properties, etc.) (Elias et al. 2021).

Key variables have a significant impact on the unconditional variance of the model's output (economic and environmental metrics). In the retro analysis, the variable that has the largest share in the metric variance is calculated, as mentioned above. This main variable is selected through GSA. However, the dimension of the input variables' space can be very large. To reduce this burden, a prescreening of the most important variables, based on their influence on the selected metric, is required before applying RTEEA. Saltelli et al. (2007) recommend the elementary effects method to perform the factor fixation, that is, discarding non-influential factors by fixing them at a predetermined value. After that, a more precise method (and more computationally costly) can list the process variables according to their influence on the unconditional metrics variance, that is, a variable prioritization is performed. Variance-based GSA is undoubtedly the most common approach in the literature for factor prioritization (Saltelli et al. 2007).

In order to facilitate visualization for the analyst, windows of operation are constructed in three-dimensional diagrams. The axes of the diagram are the main variables of the process, selected in the factor prioritization step. For each metric, a curve is plotted according to the desired specification, such as global warming potential equal to the target value. These isometric curves delimit the feasible region. However, each metric (economic and environmental) can be influenced by a different group of variables. This is especially true when multiple outputs are analyzed simultaneously. In this situation, to facilitate the graphical visualization it is recommended to group the metrics, using some clustering technique, according to the results of the GSA in the factor prioritization mode. More details about the RTEEA methodology can be found in Elias et al. (2021).

Acknowledgements

The authors would like to thank FAPESP (São Paulo State Research Foundation, Brazil), for the financial support (grant #2019/15851-2, #2017/13993-9, #2016/10636-8, #2020/15450-5, and #2014/21252-0) and CNPq (National Council for Scientific and Technological

Development, Brazil). This study was financed in part by the Coordenação de Aperfeiçoamento de Pessoal de Nível Superior—Brasil (CAPES)—Finance Code 001.

References

Adams, P., Bridgwater, T., Lea-Langton, A. et al. (2018). Biomass conversion technologies. In:*Greenhouse Gases Balances of Bioenergy Systems* (ed. P.,.T.,.P.(e. Adams), 107–139. Cambridge, MA: Elsevier Academic Press.

Azadi, P., Brownbridge, G., Mosbach, S. et al. (2014). The carbon footprint and non-renewable energy demand of algae-derived biodiesel. *Appl. Energy* 113: 1632–1644.

Balat, M., Balat, M., Kırtay, E., and Balat, H. (2009). Main routes for the thermo-conversion of biomass into fuels and chemicals. Part 1: Pyrolysis systems. *Energy Convers. Manag.* 50 (12): 3147–3157.

Bello, S., Salim, I., Méndez-Trelles, P. et al. (2019). Environmental sustainability assessment of HMF and FDCA production from lignocellulosic biomass through life cycle assessment (LCA). *Holzforschung* 73 (1): 105–115.

Bender, T.A., Dabrowski, J.A., and Gagné, M.R. (2018). Homogeneous catalysis for the production of low-volume, high-value chemicals from biomass. *Nat. Rev. Chem.* 2: 35–46.

Bjørn, A., Owsianiak, M., Laurent, A. et al. (2018). Scope definition. In: *Life Cycle Assessment* (ed. M. Hauschild, R. Rosenbaum and S. Olsen), 75–116. Cham: Springer.

BlueGreen Alliance (2017). Buy Clean California Act clamps down on imported carbon emissions. https://www.bluegreenalliance.org/the-latest/buy-clean-california-act-clamps-down-on-imported-carbon-emissions/ (accessed 3 December 2020).

Boer, J.D., Dyjakon, A., Boer, E.D. et al. (2020). Life-cycle assessment of the use of peach pruning residues for electricity generation. *Energies* 13 (11): 1–16.

Calvo-Serrano, R., Guo, M., Pozo, C. et al. (2019). Biomass conversion into fuels, chemicals, or electricity? A network-based life cycle optimization approach applied to the European Union. *ACS Sust. Chem. Eng.* 7 (12): 10570–10582.

Carpio, R.R., Miyoshi, S.C., Elias, A.M. et al. (2020). Economics of climate change: a sensitivity analysis study applied to integrated first- and second-generation ethanol biorefinery. In: *Computer Aided Chemical Engineering*, vol. 48, 1705–1710. Elsevier.

Carpio, R.R., Miyoshi, S.C., Elias, A.M. et al. (2021). Multi-objective optimization of a 1G–2G biorefinery: a tool towards economic and environmental viability. *J. Clean. Prod.* 284: 125431.

Cavalett, O., Junqueira, T.L., Dias, M.O.S. et al. (2012). Environmental and economic assessment of sugarcane first generation biorefineries in Brazil. *Clean Techn. Environ. Policy* 14 (3): 399–410.

Chary, K., Aubin, J., Guindé, L. et al. (2018). Cultivating biomass locally or importing it? LCA of biomass provision scenarios for cleaner electricity production in a small tropical island. *Biomass Bioenergy* 110 (December 2017): 1–12.

Chen, L., Pelton, R.E.O., and Smith, T.M. (2016). Comparative life cycle assessment of fossil and bio-based polyethylene terephthalate (PET) bottles. *J. Clean. Prod.* 137 (November 2016): 667–676.

Curran, M. (2017). *Goal and Scope Definition in Life Cycle Assessment*. Dordrecht: Springer Netherlands.

Dai, L., Wang, Y., Liu, Y. et al. (2020). A review on selective production of value-added chemicals via catalytic pyrolysis of lignocellulosic biomass. *Sci. Total Environ.* 749: 142386.

Dasgupta, D., Sidana, A., Ghosh, P. et al. (2021). Energy and life cycle impact assessment for xylitol production from corncob. *J. Clean. Prod.* 278: 123217.

Dyjakon, A., den Boer, J., Szumny, A., and den Boer, E. (2019). Local energy use of biomass from apple orchards—an LCA study. *Sustainability* 11 (6): 1604.

Elias, A.M., Furlan, F.F., Ribeiro, M.P.A., and Roberto, C. (2018). Retro-techno-economic-environmental analysis (RTEEA) from the cradle : a new approach for process development. In: *Computer Aided Chemical Engineering*, vol. 43, 1541–1546. Elsevier.

Elias, A.M., Longati, A.A., de Campos Giordano, R., and Furlan, F.F. (2021). Retro-techno-economic-environmental analysis improves the operation efficiency of 1G–2G bioethanol and bioelectricity facilities. *Appl. Energy* 282 (PA): 116133.

Fache, M., Boutevin, B., and Caillol, S. (2016). Vanillin production from lignin and its use as a renewable chemical. *ACS Sust. Chem. Eng.* 4 (1): 35–46.

Finnveden, G., Hauschild, M.Z., Ekvall, T. et al. (2009). Recent developments in life cycle assessment. *J. Environ. Manag.* 91 (1): 1–21.

Flórez-Orrego, D., Maréchal, F., and de Oliveira Junior, S. (2019). Comparative energy and economic assessment of fossil and biomass-based routes for ammonia production. *Energy Convers. Manag.* 194: 22–36.

Furlan, F.F., Costa, C.B.B., Secchi, A.R. et al. (2016). Retro-techno-economic analysis: using (bio) process systems engineering tools to attain process target values. *Ind. Eng. Chem. Res.* 55 (37): 9865–9872.

Gaglio, M., Tarburini, E., Lucchesi, F. et al. (2019). Life cycle assessment of maize-germ oil production and the use of bioenergy to mitigate environmental impacts: a gate-to-gate case study. *Resources* 8: 60.

Gent, S., Twedt, M., Gerometta, C., and Almberg, E. (2017). Introduction to thermochemical conversion processes. In: *Theoretical and Applied Aspects of Biomass Torrefaction* (ed. S. Gent, M. Twedt, C. Gerometta and E. Almberg), 1–16. Elsevier Butterworth-Heinemann.

Ghannadzadeh, A. and Tarighaleslami, A.H. (2020). Environmental life cycle assessment of glycerine production: energy transition from natural gas to biomass. *Sust. Energy Technol. Assess.* 42: 100775.

Giordano, R.C., Elias, A.M., and Furlan, F.F. (2020). Virtual biorefineries: computational tools for bridging the gap toward a low carbon economy. In: *Recent Advances in Bioconversion of Lignocellulose to Biofuels and Value Added Chemicals within the Biorefinery Concept* (ed. E.X.F. Filho, E.A. Ximenes, L.R.S. Moreira and C.S. Farinas), 233–258. Elsevier.

Guinée, J.B., Gorrée, M., Heijungs, R. et al. (2002). *Handbook on Life Cycle Assessment. Operational Guide to the ISO Standards*. New York: Kluwer Academic Publishers.

Guo, M., Jin, T., Nghiem, N.P. et al. (2018). Assessment of antioxidant and antimicrobial properties of lignin from corn stover residue pretreated with low-moisture anhydrous ammonia and enzymatic hydrolysis process. *Appl. Biochem. Biotechnol.* 184 (1): 350–365.

Hassuani, S.J., Leal, M.R.L.V., and Macedo, I.C. (2005). *Biomass Power Generation: Sugar Cane Bagasse and Trash* (ed. A.V. Bridgwater). Piracicaba, Blackwell Science.

Herrmann, I.T. and Moltesen, A. (2015). Does it matter which life cycle assessment (LCA) tool you choose?—A comparative assessment of SimaPro and GaBi. *J. Clean. Prod.* 86: 163–169.

Höök, M. and Tang, X. (2013). Depletion of fossil fuels and anthropogenic climate change—a review. *Energy Policy* 52: 797–809.

Houghton, R.A. (2008). Biomass. In: *Encyclopedia of Ecology* (ed. S.E. Jørgensen and B.D. Fath), 448–453. Elsevier.

Igos, E., Benetto, E., Meyer, R. et al. (2019). How to treat uncertainties in life cycle assessment studies? *Int. J. Life Cycle Assess.* 24 (4): 794–807.

IPCC (2006). IPCC Guidelines for National Greenhouse Gas Inventories: v. 2 Energy. https://www.ipcc-nggip.iges.or.jp/public/2006gl/vol2.html (accessed March 2022).

IPCC and Report (2018). *Global warming of 1.5°C—IPCC*. https://www.ipcc.ch/sr15/ (accessed December 2020).

Irmak, S., Canisag, H., Vokoun, C., and Meryemoglu, B. (2017). Xylitol production from lignocellulosics: are corn biomass residues good candidates? *Biocatal. Agric. Biotechnol.* 11: 220–223.

ISO 14040 (2006). Environmental management—life cycle assessment—principles and framework. Technical Committee ISO/TC 207/SC 5.

ISO 14044 (2006). Environmental management—life cycle assessment—requirements and guidelines. Technical Committee ISO/TC 207/SC 5.

Kalverkamp, M., Helmers, E., and Pehlken, A. (2020). Impacts of life cycle inventory databases on life cycle assessments: a review by means of a drivetrain case study. *J. Clean. Prod.* 269: 121329.

Klein, B.C., Chagas, M.F., Watanabe, M.D.B. et al. (2019). Low carbon biofuels and the new Brazilian National Biofuel Policy (RenovaBio): a case study for sugarcane mills and integrated sugarcane-microalgae biorefineries. *Renew. Sust. Energ. Rev.* 115: 109365.

Kokkinos, K., Karayannis, V., and Moustakas, K. (2020). Circular bio-economy via energy transition supported by fuzzy cognitive map modeling towards sustainable low-carbon environment. *Sci. Total Environ.* 721: 137754.

Kylili, A., Christoforou, E., and Fokaides, P.A. (2016). Environmental evaluation of biomass pelleting using life cycle assessment. *Biomass Bioenergy* 84: 107–117.

Lima, R.S., de Azevedo Caldeira-Pires, A., and Cardoso, A.N. (2020). Uncertainty analysis in life cycle assessments applied to biorefineries systems: a critical review of the literature. *Process Integr. Optim. Sust.* 4 (1): 1–13.

Liu, Z., Feist, A.M., Dragone, G., and Mussatto, S.I. (2020). Lipid and carotenoid production from wheat straw hydrolysates by different oleaginous yeasts. *J. Clean. Prod.* 249: 119308.

Longati, A.A., Cavalett, O., and Cruz, J.G. (2017). Life cycle assessment of vinasse biogas production in sugarcane biorefineries. In: *Computer Aided Chemical Engineering*, vol. 40. Paris: Elsevier Masson SAS.

Longati, A.A., Lino, A.R.A., Giordano, R.C. et al. (2018). Defining research and development process targets through retro-techno-economic analysis: the sugarcane biorefinery case. *Bioresour. Technol.* 263 (February): 1–9.

Longati, A.A., Batista, G., and da Cruz, A.J.G. (2020). Brazilian integrated sugarcane-soybean biorefinery: trends and opportunities. *Curr. Opin. Green Sust. Chem.* 26: 100400.

Lopes, H.J.S., Bonturi, N., and Miranda, E.A. (2020). Rhodotorula toruloides single cell oil production using *Eucalyptus urograndis* hemicellulose hydrolysate as a carbon source. *Energies* 13 (4).

Makkar, R.S. and Cameotra, S.S. (1997). Utilization of molasses for biosurfactant production by two Bacillus strains at thermophilic conditions. *J. Am. Oil Chem. Soc.* 74 (7): 887–889.

Marcelino, P.R.F., Peres, G.F.D., Terán-hilares, R. et al. (2019). Biosurfactants production by yeasts using sugarcane bagasse hemicellulosic hydrolysate as new sustainable alternative for lignocellulosic biorefineries. *Ind. Crop. Prod.* 129: 212–223.

Matsuura, M.I.S.F., Scachetti, M.T., Chagas, M.F. et al. (2018). RenovaCalc: Método e ferramenta para a contabilidade da Intensidade de Carbono de Biocombustíveis no Programa RenovaBio, Technical report from government https://www.gov.br/anp/pt-br/assuntos/consultas-e-audiencias-publicas/consulta-e-audiencia-publica/2018/arquivos-consultas-e-audiencias-publicas-2018/cap-10-2018/cp10-2018_nota-tecnica-renova-calc.pdf (accessed December 2020).

Mendes, N.C., Bueno, C., and Ometto, A.R. (2015). Avaliação de Impacto do Ciclo de Vida: revisão dos principais métodos. *Production* 26 (1): 160–175.

Moreno, J., Iglesias, J., Blanco, J. et al. (2020). Life-cycle sustainability of biomass-derived sorbitol: proposing technological alternatives for improving the environmental profile of a bio-refinery platform molecule. *J. Clean. Prod.* 250: 119568.

Moretti, C., Corona, B., Edwards, R. et al. (2020a). Reviewing ISO compliant multifunctionality practices in environmental life cycle modeling. *Energies* 13 (14): 1–24.

Moretti, C., Corona, B., Rühlin, V. et al. (2020b). Combining biomass gasification and solid oxide fuel cell for heat and power generation: an early-stage life cycle assessment. *Energies* 13 (11): 2773.

Murali, G. and Shastri, Y. (2019). Life-cycle assessment-based comparison of different lignocellulosic ethanol production routes. *Biofuels* (13, 2): 1–11.

Nyambuu, U. and Semmler, W. (2020). Climate change and the transition to a low carbon economy—carbon targets and the carbon budget. *Econ. Model.* 84 (May 2018): 367–376.

Ögmundarson, Ó., Sukumara, S., Herrgård, M.J., and Fantke, P. (2020). Combining environmental and economic performance for bioprocess optimization. *Trends Biotechnol.* 38 (11): 1203–1214.

Özüdoğru, H.M.R., Nieder-Heitmann, M., Haigh, K.F., and Görgens, J.F. (2019). Techno-economic analysis of product biorefineries utilizing sugarcane lignocelluloses: xylitol, citric acid and glutamic acid scenarios annexed to sugar mills with electricity co-production. *Ind. Crop. Prod.* 133: 259–268.

Panjiar, N., Mattam, A.J., Jose, S. et al. (2020). Valorization of xylose-rich hydrolysate from rice straw, an agroresidue, through biosurfactant production by the soil bacterium *Serratia nematodiphila*. *Sci. Total Environ.* 729: 138933.

Parascanu, M.M., Puig Gamero, M., Sánchez, P. et al. (2018). Life cycle assessment of olive pomace valorisation through pyrolysis. *Renew. Energy* 122: 589–601.

Peters, K. (2016). Methodological issues in life cycle assessment for remanufactured products: a critical review of existing studies and an illustrative case study. *J. Clean. Prod.* 126: 21–37.

Pinheiro, M.J., Bonturi, N., Belouah, I. et al. (2020). Xylose metabolism and the effect of oxidative stress on lipid and carotenoid production in *Rhodotorula toruloides*: insights for future biorefinery. *Front. Bioeng. Biotechnol.* 8: 1008. https://doi.org/10.3389/fbioe.2020.01008.

Potrich, E., Miyoshi, S.C., Machado, P.F.S. et al. (2020). Replacing hexane by ethanol for soybean oil extraction: modeling, simulation, and techno-economic-environmental analysis. *J. Clean. Prod.* 244: 118660.

Qian, Y., Qiu, X., and Zhu, S. (2015). Lignin: a nature-inspired sun blocker for broad-spectrum sunscreens. *Green Chem.* 17 (1): 320–324.

Raftery, A.E., Zimmer, A., Frierson, D.M.W. et al. (2017). Less than 2 °C warming by 2100 unlikely. *Nat. Clim. Chang.* 7 (9): 637–641.

Rashedi, H., Mazaheri Assadi, M., Bonakdarpour, B., and Jamshidi, E. (2005). Environmental importance of rhamnolipid production from molasses as a carbon source. *Int. J. Environ. Sci. Technol.* 2 (1): 59–62.

Reijnders, L. and Huijbregts, M.A.J. (2008). Biogenic greenhouse gas emissions linked to the life cycles of biodiesel derived from European rapeseed and Brazilian soybeans. *J. Clean. Prod.* 16 (18): 1943–1948.

Rissman, J., Bataille, C., Masanet, E. et al. (2020). Technologies and policies to decarbonize global industry: review and assessment of mitigation drivers through 2070. *Appl. Energy* 266: 114848.

Rodríguez, R., Espada, J.J., Moreno, J. et al. (2018). Environmental analysis of spirulina cultivation and biogas production using experimental and simulation approach. *Renew. Energy* 129: 724–732.

Rosenbaum, R.K., Hauschild, M.Z., Boulay, A. et al. (2018). Life cycle impact assessment. In:*Life Cycle Assessment: Theory and Practice* (ed. M. Hauschild, R. Rosenbaum and S. Olsen), 167–270. Springer International Publishing https://doi.org/10.1007/978-3-319-56475-3_10.

Saltelli, A., Ratto, M., Andres, T. et al. (2007). *Global Sensitivity Analysis. The Primer*. Chichester, UK: Wiley.

Saravanan, A.P., Pugazhendhi, A., and Mathimani, T. (2020). A comprehensive assessment of biofuel policies in the BRICS nations: implementation, blending target and gaps. *Fuel* 272: 117635.

Scherer, M.D., de Oliveira, A.C., Filho, F.J.C.M. et al. (2017). Environmental study of producing microalgal biomass and bioremediation of cattle manure effluents by microalgae cultivation. *Clean Techn. Environ. Policy* 19 (6): 1745–1759.

Schrijvers, D., Loubet, P., and Sonnemann, G. (2020). Archetypes of goal and scope definitions for consistent allocation in LCA. *Sustainability* 12: 2–3.

Seabra, J.E.A., Macedo, I.C., Chum, H.L. et al. (2011). Life cycle assessment of Brazilian sugarcane products: GHG emissions and energy use. *Biofuels, Bioprod. Biorefin.* 5: 519–532.

Sharma, A., Pranaw, K., Singh, S. et al. (2020). Efficient two-step lactic acid production from cassava biomass using thermostable enzyme cocktail and lactic acid bacteria: insights from hydrolysis optimization and proteomics analysis. *3 Biotech* 10 (9): 409.

Singh, S., Kaur, D., Yadav, S.K., and Krishania, M. (2021). Process scale-up of an efficient acid-catalyzed steam pretreatment of rice straw for xylitol production by *C. tropicalis* MTCC 6192. *Bioresour. Technol.* 320: 124422.

Song, Y., Gyo Lee, Y., Jin Cho, E., and Bae, H.-J. (2020). Production of xylose, xylulose, xylitol, and bioethanol from waste bamboo using hydrogen peroxide–acetic acid pretreatment. *Fuel* 278: 118247.

Souza, S.P. and Seabra, J.E.A. (2013). Environmental benefits of the integrated production of ethanol and biodiesel. *Appl. Energy* 102: 5–12.

Su, Y., Zhang, P., and Su, Y. (2015). An overview of biofuels policies and industrialization in the major biofuel producing countries. *Renew. Sust. Energ. Rev.* 50: 991–1003.

Tanabe, N. (2016). Toward a sustainable society and economy—green purchasing network celebrating 20th anniversary. *JFS Newsletter,* p. 165. https://www.japanfs.org/en/news/archives/news_id035576.html. (accessed February 2021).

Thaore, V.B., Armstrong, R.D., Hutchings, G.J. et al. (2020). Sustainable production of glucaric acid from corn stover via glucose oxidation: an assessment of homogeneous and heterogeneous catalytic oxidation production routes. *Chem. Eng. Res. Des.* 153: 337–349.

Thornley, P. and Adams, P. (2018). *Greenhouse Gas Balances of Bioenergy Systems*. London: Elsevier.

Thrän, D., Dotzauer, M., Lenz, V. et al. (2015). Flexible bioenergy supply for balancing fluctuating renewables in the heat and power sector—a review of technologies and concepts. *Energy. Sust. Soc.* 5 (1): 35.

Tonini, D., Hamelin, L., and Astrup, T.F. (2016). Environmental implications of the use of agro-industrial residues for biorefineries: application of a deterministic model for indirect land-use changes. *GCB Bioenergy* 8 (4): 690–706.

Ujala (2020). National Ujala Dashboard. http://ujala.gov.in/ (accessed December 2020).

Vadenbo, C., Tonini, D., and Astrup, T.F. (2017). Environmental multiobjective optimization of the use of biomass resources for energy. *Environ. Sci. Technol.* 51 (6): 3575–3583.

Vieira, M.M., Kadoguchi, E., Segato, F. et al. (2021). Production of cellulases by *Aureobasidium pullulans* LB83: optimization, characterization, and hydrolytic potential for the production of cellulosic sugars. *Prep. Biochem. Biotechnol.* 51: 153–163.

Wernet, G., Bauer, C., Steubing, B. et al. (2016). The ecoinvent database version 3 (part I): overview and methodology. *Int. J. Life Cycle Assess.* 21 (9): 1218–1230.

Woodley, J.M. and Titchener-Hooker, N.J. (1996). The use of windows of operation as a bioprocess design tool. *Bioprocess Biosyst. Eng.* 14: 263–268.

Yue, D., Pandya, S., and You, F. (2016). Integrating hybrid life cycle assessment with multiobjective optimization: a modeling framework. *Environ. Sci. Technol.* 50 (3): 1501–1509.

Zang, G., Zhang, J., Jia, J. et al. (2020). Life cycle assessment of power-generation systems based on biomass integrated gasification combined cycles. *Renew. Energy* 149: 336–346.

Zhao, Y., Damgaard, A., Liu, S. et al. (2020). Bioethanol from corn stover—integrated environmental impacts of alternative biotechnologies. *Resour. Conserv. Recycl.* 155: 104652.

15

Green Consumerism and Role in Uplifting Lignocellulose Bioeconomy

BS Dhanya

Department of Biotechnology, Udaya School of Engineering, Kanyakumari, 629204, TN, India

15.1 Introduction

It is predicted that the increase in human population will reach about 8 billion at world level in the year 2030 from the current 7.2 billion (United States Census Bureau 2015). There is an upsurge in per capita energy consumption with increase in birth rate. For life sustenance, in addition to the basic amenities, an energy source is also required. The world energy scenario involves the utilization of fossil fuels such as gasoline, natural gas, crude oil, or coal that together make a market value of $1.5 trillion (Shafiee and Topal 2009). It is said that the fossil fuels are depleting in a fast mode with a time-scale of 37, 35, and 107 years for gas, oil, and coal respectively (Shafiee and Topal 2009). The upsurge in population growth directly influences the use of energy resources which is analyzed to be about 11% per annum in 2030 (International Energy Agency 2007). In 2008 the energy consumption was about 533 EJ which was found to be increased to 653 EJ in 2020 and an energy consumption of 812 EJ is foreseeable by 2030 (Wu et al. 2010). Of this energy consumption, 8%, 84%, and 6% is from renewable sources, fossil fuel and nuclear sources respectively (Shafiee and Topal 2009; Wu et al. 2010). In India, the demand for fossil fuel is increasing at an alarming rate of 3.9% yearly (International Energy Agency 2008). In 2008, global utilization of petroleum daily was about 85.7 million barrels per day which escalated to 97.6 million per day, and it may tend to increase to 112.2 million barrels daily in 2035 (US Energy Information Administration 2011). One of the most promising alternatives for fuels from fossils are the biobased fuels that have the ability to mitigate the effects of climate change. Since biofuels are generated from renewable feedstocks, supply shortages will not be a factor to be considered. With futuristic advancements in technology, the probability of minimization of production cost is high (Nanda et al. 2015).

Even though rapid industrialization through increased use of fossil fuel is a positive sign, adverse effects such as the generation of greenhouse gases (GHGs), acid rain, global warming, and climate change prevail. Also, the price rise for current transportation fuel has not reduced its demand worldwide. Focused factors such as domestic energy security, global

Lignocellulose Bioconversion Through White Biotechnology, First Edition. Edited by Anuj Kumar Chandel.

warming, exploitation of fossil fuel resource, upsurge in fuel price, increased energy demand, and high GHG emissions have accelerated the need for renewable sources of energy (Nanda et al. 2014a). Existing energy resources such as solar, nuclear, wind, or geothermal cannot be used in generating transportation fuel. Alternatively, waste plant used as biomass which is highly rich in organic carbon can be used in producing transportation fuels. Economic and environmental advantages are high with regular use in bioenergy resources. Research in biofuels is increasing among national bodies to promote environmental and agricultural sustainability. Additionally, policy makers invest in increasing the bioenergy usage in an efficient way, contributing to economic development, employment generation, and income (United States Census Bureau 2015). For the achievement of Millennium Development Goals 2015, the bioenergy sector is included on the agenda for sustainability (Kartha and Larson 2000). For well-established economic growth, lignocellulosic waste biomass is an effective energy source.

15.2 Lignocellulosic Biomass and its Contribution in Bioeconomy

15.2.1 Lignocellulosic Biomass

Globally, lignocellulosic biomass consumption comprises 50% of the total dry weight of 11.4 billion tons (Pg year^{-1}) lignocellulosic biomass (Figure 15.1) (Van Dam et al. 2019). Plant residues that are not edible constitute lignocellulosic materials that can be obtained from agricultural areas, forests, or savannahs. They also include agricultural residues such as sugarcane bagasse, rice husk, corn stover, wheat straw, and rice hulls. In addition to

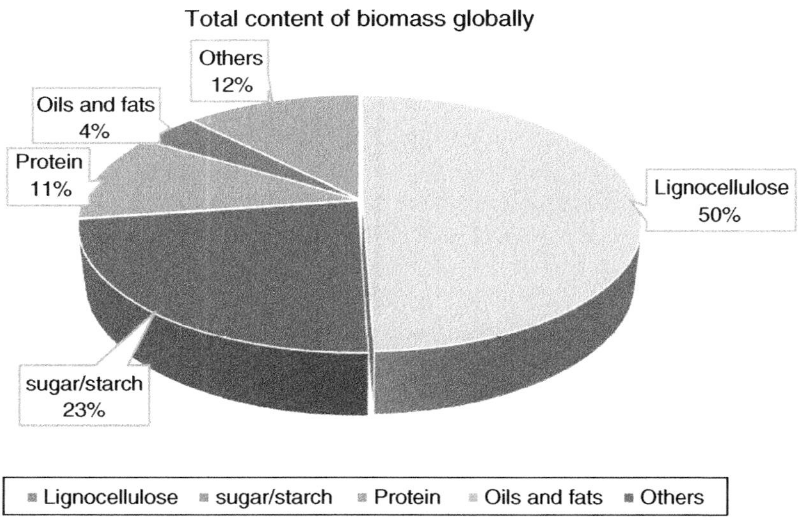

Figure 15.1 Total content of biomass globally. Based on Van Dam et al., 2019.

agricultural residues, sawmill residues, municipal paper waste, wood residues, and paper mill refuse also constitute lignocellulosic feedstocks. The abundant availability of lignocellulosic feedstocks supports the production of biofuels (Nanda et al. 2013). The components in lignocellulosic biomass that aid in biofuel generation constitute hemicellulose, cellulose, lignin, etc. The percentage of hemicellulose, cellulose, and lignin is 20%–40%, 30%–60%, and 15%–25% respectively. Components in lignocellulosic materials, the conversion of biomass to biofuel technology, different operating parameters, and the feedstock origin are some of the factors that favor the potentiality of biofuel generation from lignocellulosic biomass (Vassilev et al. 2012).

Lignocellulosic biomass is used as the most widely available resource by ancestors due to its ease of availability, ease of usage, and its renewable nature. 2G feedstock lignocellulosic materials are supramolecular structures that are capable of producing biobased energy, chemicals, and materials (Figure 15.2). The development of bioconversion technologies will significantly increase futuristic energy generation using forest residues, energy crops, and agricultural residues to about 500 EJ annually by 2050 (Binder et al. 2007). Basic conversion technologies are through biochemical or thermochemical modes (Figure 15.3). The pyrolysis process and liquefaction are included in the thermochemical conversion process. Bio-oil can be produced using the thermochemical conversion route. Acid hydrolysis and

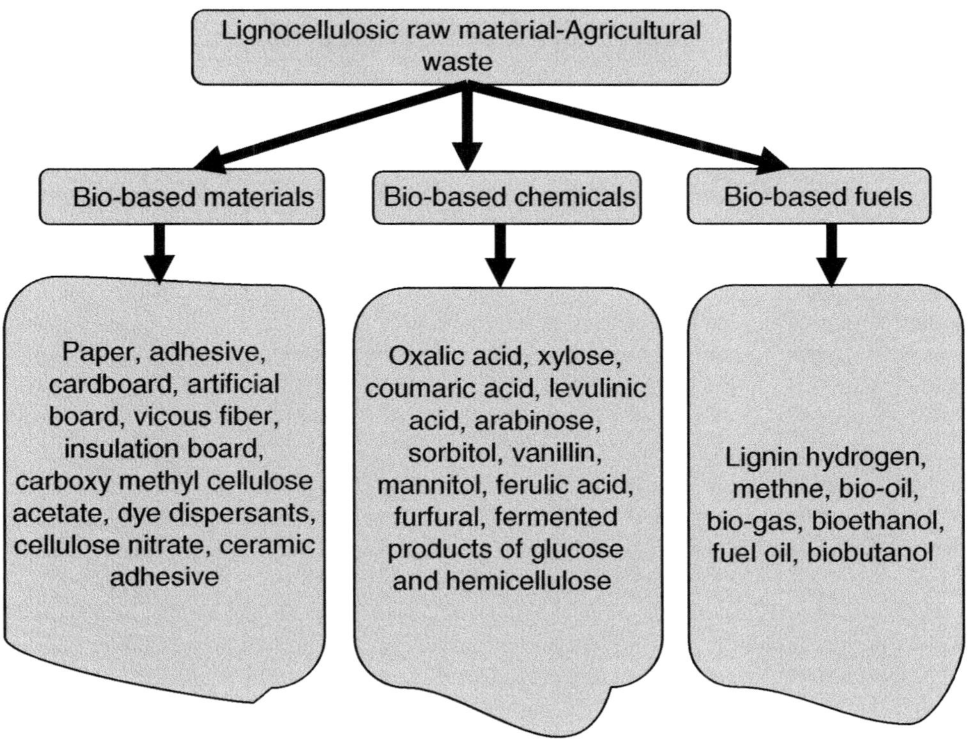

Figure 15.2 Lignocellulosic products.

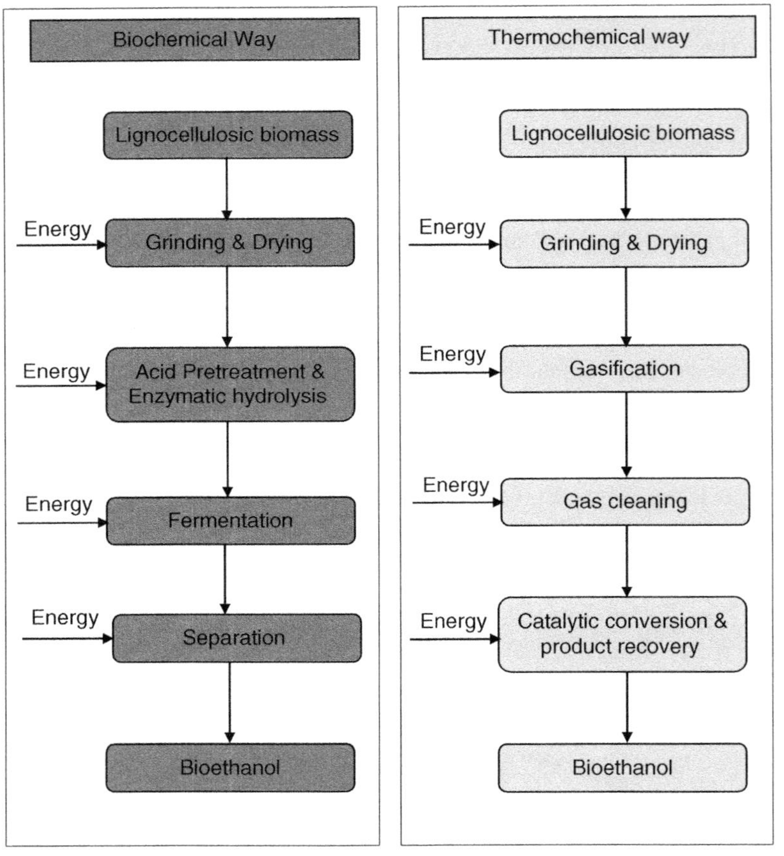

Figure 15.3 Bioethanol production from lignocellulosic biomass.

subsequent fermentation for fuel generation are included in the biochemical conversion route (Nanda et al. 2014b). Using lignocellulosic biomass, bioethanol, biochar, solvents, aromatic compounds, bio-oil, butanol, organic acids, phenols, etc. can be produced (Nanda et al. 2015).

Total consumption of energy in the form of liquid fuels is about 40%. Currently, bioethanol and biodiesel are produced and marketed economically with their feasible use in blending forms. Bioethanol and biodiesel can be replaced with biobutanol in the fuel market (Green 2011). The compatibility of biobutanol with other transportation fuels is high. Also, the performance of biobutanol is higher than bioethanol. The production potential and energy density is more for biobutanol (Kumar and Gayen 2011).

In future, there is wide scope of generating renewable fuels and biobased products using lignocellulosic biomass. The US Department of Energy predicted that about 5% electricity, 20% transportation fuels, and 25% chemicals could be replaced with lignocellulosic biomass products by 2030, where the raw material consumption will be about 1×10^9 tons. It is estimated by Royal Shell company that lignocellulosic biomass will be used to develop 30% of fuels and chemicals globally (Chen 2015).

15.2.2 Life Cycle Assessment (LCA) of Lignocellulosic Biomass

LCA is an important reliable tool that reveals the predictable analysis of the life cycle of biofuel on the ecosystem, focused on three economic platforms such as industry, agriculture, and energy. The required lignocellulosic materials, their energy flow, and end waste disposal are analyzed in an LCA study, thereby minimizing the greenhouse gas emissions (GHGs) (Milazzo et al. 2013). LCA is crucial in determining the risks and also the viable opportunities related to lignocellulosic biorefining methods. The LCA of lignocellulosic feedstock, analyzing its advantages and disadvantages, is pivotal for proper planning of its sustainable usage. In particular, LCA includes the technologies involved in biofuel production, biomass production options, its conversion process, co-product allocation, and environmental impacts. The LCA of lignocellulosic biomass is a prerequisite for identifying the socio-ecological impacts over production, energy, and its associated emissions (Singh et al. 2010). It analyses the effects and consequences associated with the production of lignocellulosic products for the ecosystem (Guinee et al. 2011). Energy consumption and the emissions generated during production, distribution, transportation, and consumption are calculated during a LCA study (Sills et al. 2013). The LCA study depends on the chosen system boundary, allocation methods, production parameters (heating rate, fermentation time, temperature, and enzyme loading), and functional units (Wiloso et al. 2012). It predicts that about 60% GHG emissions can be reduced by using a lignocellulosic biorefinery in comparison with the fossil fuel biorefinery (Singh et al. 2010). The importance of LCA lies in addressing environmentally sustainable criteria such as land usage pattern alteration, optimizing the process involved in minimizing the emissions of GHG and energy consumption, treating waste water, water usage for biomass production process and biorefining, supply chain of feed stocks, increasing the efficiency of machines used, involving the adaptation of genetically altered organisms for maximum fermentation, saving energy by combining heat and power, etc. (Nanda et al. 2015).

Applying LCA analysis depends on the precursors available in a particular geographical location (Wiloso et al. 2012). In addition to this, carbon storage should also be considered during the LCA study (Gnansounou 2010). The principal components considered during LCA analysis comprise energy usage (Borrion et al. 2012), GHG emissions (Borrion et al. 2012), chemicals, enzymes, and nutrients used during the conversion process (Nanda et al. 2015), pretreatment processing strategies (Nanda et al. 2015), ecological issues (Borrion et al. 2012), fertilizer and pesticide usage (Borrion et al. 2012), by-products formed during the energy production (Nanda et al. 2014a), and the precursor availability (Gnansounou 2010).

Most countries are utilizing bioethanol as a substitute for gasoline. The emission of particulate matter and GHG is much less when ethanol, a fuel with 35% oxygen, is subjected to combustion. With use of the blended E85 form (85% ethanol + 15% gasoline), the consumption of gasoline can be minimized by 32%. Countries such as the United States (E85 and E10), Sweden (E85 and E5), China (E10), Columbia (E10), Thailand (E10), India (E5), Canada (E85 and E10), Brazil (E25 and E20), Australia (E10), Peru (E10) and Paraguay (E7) have come forward with ethanol–gasoline programs (Balat 2011). Bioethanol derived from lignocellulosic biomass is referred to as second-generation biofuel. 2G ethanol has lower GHG emission compared to 1G ethanol (Wang et al. 2007). Lignocellulosic biomass that has been utilized for bioethanol production includes forest residues and wastes from agriculture and energy providing crops. Poplar, corn stover, bagasse, wood residues, oil palm residues, municipal solid wastes, mustard refuse, alfalfa stem, straw, flax shives, hemp herd, recycled paper, etc. are used as biomass for bioethanol production (Wiloso et al. 2012).

15.3 Lignocellulosic Bioeconomy and its Sustainability in the World

Bioeconomy is defined as an economy contributing sustainability with production of green products such as bioenergy, feed, bioproducts, food, etc. utilizing renewable resources (Strategy 2012). It is also defined as biological resource utilization with innovative principles and knowledge related production; with all aspects leading to a sustainable product generation (Global Bioeconomy Summit 2015). It constitutes both agriculture and food related industries where the agriculture products are processed as commodities for serving the economy globally (Peerlings 1993). It includes different sectors such as forestry, aquaculture, paper, building, agriculture, fishery, food, pharmaceuticals, etc. Advanced studies using nanotechnological approaches are pivotal for a well-refined bioeconomy. It is reported that by 2025 the world biotech market will rise to $7271 billion (Grand View Research Inc. 2018). From the beginning of year 2015, the bioeconomy agenda has been propagated and pursued with the involvement of both public and private stakeholders.

Evaluation of environmental consequences of biofuels and the source production is very important during the start of the biofuel industry. Moreover, the motto of sustainability should be achieved through entering into a bioeconomy business (Ewing and Msangi 2009). In 2009, the European Union came out with a directive setting the achievement of 20% renewable energy consumption by the year 2020 (Gnansounou 2010) with sustainability parameters such as achieving 10% production of biofuel and its import and 60% reduction in GHG emission.

Unutilized lignocellulosic feedstocks constitute about 40 million tonnes (Sanderson 2011) and converting this feedstock to biofuel is socio-economically feasible and viable. Such types of feedstocks contribute to the stimulation of rural and local economies pertaining to increased production of biofuel. Increasing the use of lignocellulosic materials can minimize the competition and demand for agricultural commodities (Biswas et al. 2011). Production of lignocellulosic feedstocks positively influences restoration of the environment by conversion of barren land into vegetative land, protecting the land from water shedding, rainwater harvesting, habitat creation, restoration of degraded soils, etc. During the conversion of lignocellulosic biomass to biofuel, the by-product biochar is synthesized, which is used on the land again for producing feedstocks (Nanda et al. 2014a). Biochar, which contains rich mineral content, is remarkable for soil fertility improvement (Mohanty et al. 2013) and neutralization of soils that are acidic (Azargohar et al. 2013, 2014). Carbon sequestration is high in terms of using biochar (Laird 2008), favoring the improvement of the lignocellulosic bioeconomy.

Government level institutions have been framed worldwide as a part of uplifting the biomass related bioeconomy. In 2018, the Global Bioeconomy Summit was held at Berlin with the motto of innovation, growth, and sustainable development. The whole world was categorized into different regions such as Eastern Africa, Latin America and Caribbean, America, Central and Eastern Europe, Northern Europe, and Asia and Pacific region. The Bioresource Innovation Network for Eastern Africa Development (Eastern Africa—Ethiopia, Rwanda, Uganda, Burundi, Kenya, Tanzania), United Nations Latin America and Caribbean (Latin America and Caribbean—Brazil, Paraguay, Argentina, Uruguay), National Economic Council White House Office of Science and Technology Policy (America—United States), Biomass Research and Development Board (America—United States), Department of Energy (America—Canada), Canadian Council of Foreign Ministers (America—Canada), European Structural Investment Funds (Central and Eastern Europe), Central Eastern European Initiative for Knowledge-based Agriculture (Central and Eastern Europe), Aquaculture and Forestry in the Bioeconomy (Central and Eastern Europe), Biobased Industries Joint Undertaking (Central and Eastern Europe), Biobased Industries Consortium (Central and Eastern Europe), Nordic Bioeconomy Strategy (Northern Europe), Baltic Sea Region Bioeconomy Council, Japan Bioindustry Association (Asia and Pacific—Japan), Association of Biotechnology-Led Enterprises (Asia and Pacific—India), Queensland Department of State Development (Asia and Pacific—Australia), Central Committee of China's Communist Party, National People's Congress and Chinese State Council (Asia and Pacific—China), Ministry for Primary Industries supported by the Minister of Science and Innovation and the Prime Minister's Chief Science Advisor and Pracharath Collaboration (Asia and Pacific—New Zealand), Ministry of Industry and the PTT Global Chemicals Public Company Limited are the institutions that reinforce the bioeconomy development (Asia and Pacific—Thailand) (Global Bioeconomy Summit 2018; Samdin and Kamaruddin 2019).

15.3.1 Lignocellulose Bioeconomy in Malaysia

The year 2013 marks the starting of Malaysia's journey toward bioeconomy. The government established BiotechCorp in the year 2005 serves as a steep inclination toward biobased investment. BiotechCorp paves a platform for other biotech companies to indulge in the energy sector. The Bioeconomy Transformation Program (BTP) is introduced with widened goals towards attaining a full bioeconomy world. Its main focus is on long-term sustainable development of society thereby minimizing the carbon footprint and maximizing the value chain. BTP acts as a knot in increasing the emergence of bioeconomy in Malaysia and also provides a way for all private sectors to increase business opportunities. It forms a bridge between government agencies for profiteering investments in agricultural and biotech sectors. There are about 20 projects initially and 13 new projects additionally. The gross national income is expected to increase with 16 300 job creations. There is a surge in the bioeconomy of Malaysia with 10 entry point projects in the field of agriculture, biomedical, and bioindustrial sectors. The synthesis of biomaterials, biochemical, high-value foods, drugs, etc. are on the list of entry projects. The bioeconomy influences the society in multiple ways such as improvisation in rural income, promoting long-term sustainable environment, improving the state of health by developing biosimilars of low cost, etc. Among the 10 projects, five projects are based on the bio-industrial sector—lignocellulosic bioeconomy. A biomass (woodchip/rice husk) power generation plant is planned at Naka, Kedah. This is a setup of Majunaka Eco Energy Sdn Bhd (Samdin and Kamaruddin 2019).

15.3.2 Lignocellulose Bioeconomy in Japan

About 66% of the area in Japan constitutes forest cover which is about $251 \times 103\,km^2$. Wood chips are important resources for lignocellulose and its total production is about 2800×10^4 tons per year globally. The traditional use of lignocellulose includes wood processing and lumber industry and the paper and pulp industry. Lumber production is about $2224 \times 10^4\,m^3$ per year. Cross-laminated timber is of interest in the Japan bioeconomy with a future targeted production of $50 \times 104\,m^3$ cross-laminated timber per year by 2024. In the paper and pulp industry, about 477×10^4 tons lumber per year was utilized in 2016 and about 1417×10^4 tons and 1157×10^4 tons of paper and board was produced during 2016. Research is being done on developing solid fuel products from lignocellulose. Wood chips are utilized as solid fuel and their production was about 12×10^4 tons per year in 2015. The cost of wood pellets was about \$275–\$642 per ton. One of the developed products using solid fuel is torrefaction pellets that can increase the energy density thereby decreasing the transportation costs. Large-scale production of this technology is executed by Nippon Paper Industries Co. Ltd. In addition to this product, biocokes are produced from solid fuels. Biocokes produced from lignocellulose are hard and are not breakable during transportation. The liquid fuels bioethanol and biomethanol have been produced in Japan through bio or thermal processing routes. In 1990, the pilot-scale generation of biomethanol from lignocellulose was begun. Increase in the use of renewable source for electricity generation prevailed after 2012 with enforcement of the Act on Special Measures Concerning Procurement of Renewable Electric Energy by Operators of Electric Utilities. The purchasing price was fixed by the tenders as per the revised price fixed by the government. It is found that the purchase price for renewable electricity from wood, lumber, building material waste, and biomass containing waste is about 22 cents kW/h (capacity $<$10 000 kW), 29 cents kW/h (capacity $<$2000 kW), 12 cents kW/h, and 16 cents kW/h respectively. Chemicals synthesized from lignocellulose include rosin and turpentine. Fat molecules from pine trees are used for manufacturing rosin, which finds application as a sizing agent. Pine chips are used as raw material for extracting turpentine which acts as a solvent (Hirata 2019).

Research and development in Japan is focused on synthesizing nanocellulose, which is of nanoscale material produced from lignocellulose. Nanocellulose can be classified into cellulose nanofiber, lingo-cellulose nanofiber, and carbon nanocrystal, based on the manufacturing process. Cellulose nanofiber is produced more in Europe, Asia, and North America and carbon nanocrystals are produced more in Israel and North America. Nanocellulose has specific properties such as high tensile strength, low weight, transparency, high surface area holding functional groups, etc. These specific properties influence the usage of nanocellulose in developing packaging materials, thickeners, stabilizers, composite materials, electronic materials, filter materials, etc. There are about 10 companies that manufacture nanocellulose in Japan. \$460 million is allocated by the Japanese government for undergoing research in nanocellulose. Investments have been done in the area of generating novel products for automobile parts, housing items, electric appliances, etc. (Hirata 2019).

Moreover, in an agreement with Kyoto University, the Ministry of Economy, Trade, and Industry is putting forward the intention to convert lignocellulose to cellulose nanofiber reinforced resin which finds application in automobile parts, kitchen appliances, and other homely items. The allocated budget in financial year 2018 for cellulose nanofiber reinforced resin was $36 million per year. The replacement of metal used in automobile parts with cellulose nanofiber is the key motto in addition to minimizing the emission of GHGs (Hirata 2019).

Nippon Paper constructed a new plant for producing cellulose nanofiber reinforced resin with a production capacity of 10 tons per year (Hirata 2019).

15.3.3 Lignocellulose Bioeconomy in European Countries

Most of the European countries put less effort into increasing the supply chain of the bioeconomy. Using smart strategies, the gaps in activities related to favoring the lignocellulose bioeconomy are analyzed. About 98.6% of European countries are involved in innovative research work for enhancing the bioeconomy with alteration in denominations. Changes include green growth, innovative production methods, eco-innovation, low carbon, sustainable agriculture, green chemistry, and the circular economy. Variation exists in terms of the different modes of biobased industries and the production process. One of the approaches that is being noted in European countries is the view of bioeconomy in the way of value chain analysis. In total, this includes the chain from supplying biomass for processing to production of biobased products. Biofuel, foods, bioenergy, and beverages are the focused approaches. Overall ranking in innovation is done among the European countries and about 5% (Finland, Austria, Central Jutland, Germany, Sweden, England, Flemish region, southern Netherlands) are classified under a high bioeconomy maturity index. About 23% and 36% of countries come under a middle and low maturity index respectively. Sustainable innovation capacity is predicted to be high in the European Union. Stakeholders include sectoral associations, academic institutions, governments, business intermediaries, technological institutions, and public administrations. Gathering stakeholders for an efficient bioeconomy is pivotal and it is possible with clustering that integrates the people, government, and research for unique bioeconomy sector development (Samdin and Kamaruddin 2019).

15.4 Green Consumerism and its Upsurge in the Lignocellulosic Bioeconomy

The demand for green products is on the rise due to their pro-environment benefits. The necessity for green products starts back in the 1970s. Global initiatives are in constant growth showing the path of sustainability. The United Nations Intergovernmental Panel on Climate Change, the Kyoto Protocol and the Paris Climate Agreement are examples of global initiatives. Green consumers make choices on green products and green consumerism is based on some characteristics such as products with less environmental impact, products with a biodegradable nature, and post-use requirement. The pivotal role of green consumerism comprises mitigating the emissions, minimizing the waste, consuming healthy foods that are good for both humans and the environment, and maximizing the energy production. The Global Web Index reports the choice of green products by different aged consumers. The report stresses the role of consumers in the age group 22–35 as active members in green consumerism (Malyan and Duhan 2018). In the year 2016, about 17×10^3 tons of rosin was produced but rosin consumption was about 69×10^3 tons (Hirata 2019). The responsibility for increasing green consumerism is vested in the ideal green product manufacturers which in turn indicate the vital role of government bodies and multiple stakeholders (Malyan and Duhan 2018).

According to a report, 23.1×10^6 tons nanocellulose is demanded globally per year in various industries such as textiles, paper, paint, resin, water treatment, pharmaceutical, cement, cosmetics, etc. The developed product cellulose nanofiber is used as a thickener in gel-ink pens where TEMPO-oxidized cellulose nanofiber increases the thixotropy, thereby favoring the ink flowing smoothly without the necessity for applying pressure. Since cellulose nanofiber has a large surface area and additional functional groups, silver ions are added that exhibit an antibacterial effect with increased deodorization power and it is efficiently applied in diaper production. As an energy saver, cellulose nanofiber is used as disposable cleaning paper. Carboxymethylated cellulose nanofiber is used in the cosmetic industry as a fragrance gel, hand cream, etc. due to its increased water holding and viscosity control properties. Other applications of cellulose nanofiber include its use as powder in paper making, as additives in food, etc. (Hirata 2019).

The main reason for an upsurge in the lignocellulosic bioeconomy is the wide scope in green consumerism. In addition to this, government subsidies, eco-friendly competitive advantage, and corporate social responsibility also favor the development of the lignocellulose bioeconomy (Malyan and Duhan 2018).

15.4.1 Wide Scope in Green Consumerism

Numerous reports have described awareness of green product consumption and issues arising with the use of chemical products in the environment. A large part of marketing can be targeted with this wide green consumerism approach by customers. The increase in green consumerism increases the demand for a particular green product and thereby manufacturers come out with new solutions to increase the supply with a cost reduction motive. Consequently, this will be a double-sided benefit for both consumers and manufacturers. Manufacturers will achieve profit in the marketing economy and consumers will receive product safety and satisfaction. Figure 15.4 depicts the role of green consumerism in uplifting the lignocellulose bioeconomy (Malyan and Duhan 2018).

15.4.2 Government Subsidies

The exploitation of natural resources and its frequency in depletion creates a major concern among all governments to safeguard our available resources. In particular, in minimizing the environmental issues, governments stress that manufacturers need to develop green products. In India, the Karnataka government banned the use of plastic bags and also urged manufacturers to stop plastic bag production. The government encouraged the manufacturers to produce user-friendly paper bags. On achieving the production of more eco-friendly products, subsidies were provided by the government (Malyan and Duhan 2018).

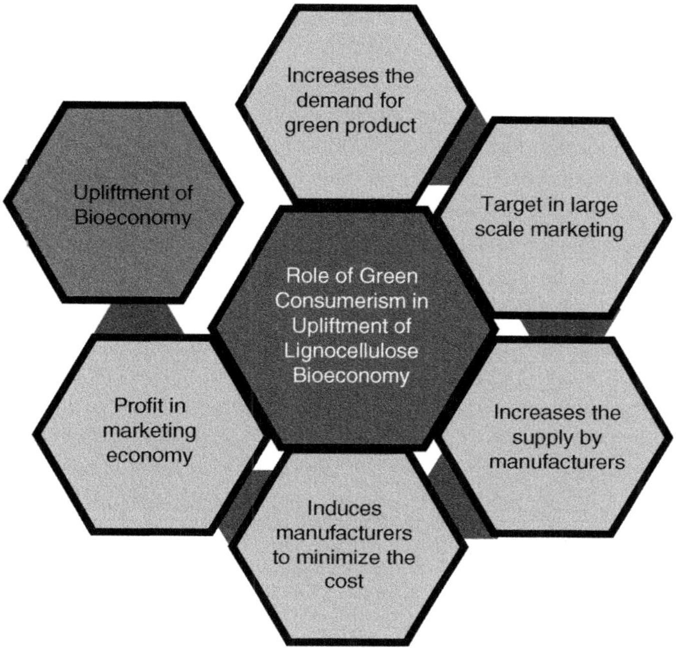

Figure 15.4 Role of green consumerism in the development of lignocellulose bioeconomy.

15.4.3 Eco-friendly Competitive Advantage

Manufacturers have to develop more novel eco-friendly technologies which in turn aid in green product generation. An increase in green product servicing to humankind will encourage consumers to adapt to the new products and will mitigate the negative effects caused to the environment by using chemical products (Malyan and Duhan 2018).

15.4.4 Corporate Social Responsibility

Effective implementation of eco-friendly solutions by manufacturers ideally reflects the concern society and the environment. Green marketing and green consumerism is vested as a social responsibility of corporate firms. This is used as a marketing tool to increase the economy with profit benefits. Both consumers and manufacturers will be satisfied with the sustenance of green consumerism policies (Malyan and Duhan 2018).

15.5 Challenges in Green Consumerism

Even though technologies are upgraded, there exists the problem of 100% green consumerism of lignocellulosic products. For example, in Japan, the developed torrefaction pellet is not within the reach of green consumers (Hirata 2019). Two sectors that gain importance for developing the lignocellulosic bioeconomy are the biotechnology and agriculture sectors.

It is predicted that 75% of the contribution toward bioeconomy will be from the biotechnology and agriculture sectors (OECD 2009). The main goal of the lignocellulosic biorefinery is the maximum exploitation of the resources available without any wastage. Generally, generating green products using lignocellulosic resources and bringing them to the global economy is a timely necessity. Efficient use of lignocellulosic biomass is possible with an effective biorefinery. Currently, there exist many challenges in the production of bioproducts. Achieving a better biorefinery technology is possible with the aid of developing bioconversion technologies and integrating eco-industrial chain and production process engineering (Chen and Wang 2008). Multilateral development and the involvement of a multidisciplinary approach is an essential aspect for maximizing the development of lignocellulosic biomass exploitation. Some of the challenges in a lignocellulosic biorefinery includes poor performance of the developed product, increased production cost, limited access to the industrial products, difficulties in large-scale production, undeveloped refining technologies, lag in effective equipment, poor technology integration, and high cost at small-scale production (Figure 15.5) (Chen 2015). The performance of the product developed seems to be a negative correlator with the existing petro-products, the reason being the variation in properties.

Also, the cost associated with the utilization of lignocellulosic biomass is high. Ethanol produced from lignocellulosic material costs about \$3–\$6 L^{-1}. In Japan, the production of bioethanol, biomethanol, and biodiesel from lignocellulose is implemented on a pilot scale but it fails to run in the competitive market due to the influence of high cost. Most of the technologies developed are acceptable for small-scale generation and are not yet commercialized.

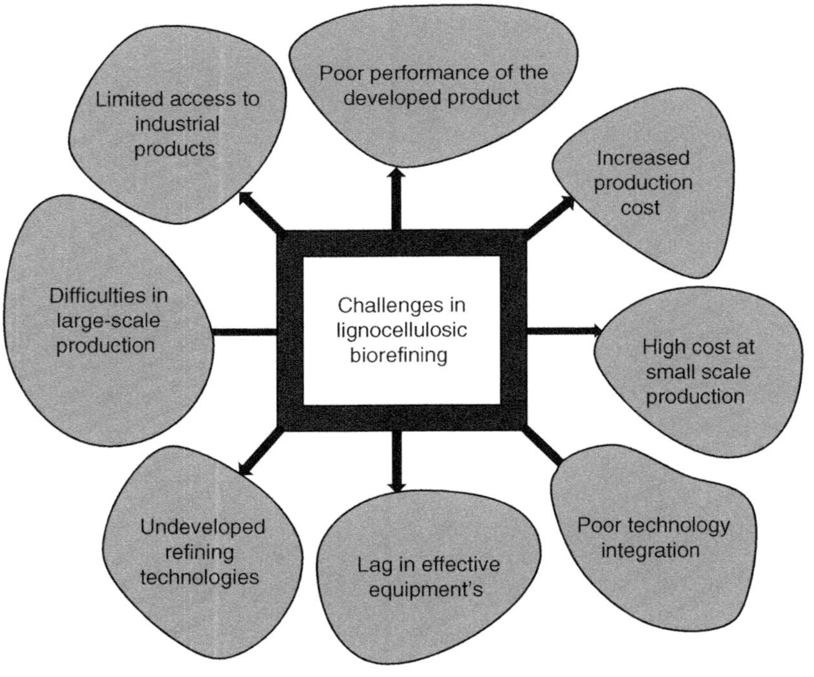

Figure 15.5 Challenges in lignocellulosic biorefining. Based on Chen, H. 2015.

Also, the fuel cost for biomass electricity generation is 70% higher than the generation cost of power from biomass which indicates that the economic feasibility is challenged (Hirata 2019).

Complete development of industrialized product is limited and it sets a drawback for developing lignocellulosic biorefinery. In addition to this, large-scale production is lagging due to capital costs and most of the technologies are immature in handling multiple parametric tasks. The relationship existing between sustainability and product acceptance by consumers is complex. All these challenges have to be accomplished for a highly effective development of lignocellulosic bioeconomy (Chen 2015).

15.6 Future Prospects

Most of the green products generated from lignocellulose are of high cost. For example, in Japan the cost of cellulose nanofiber is \$180–\$460 kg^{-1} which is expensive (Hirata 2019). Cellulose nanofiber can be used as a filler in plastics but, considering the cost, it is not automated in the market economy. So, it is necessary to cut down the production cost by improvisation in the production process. Some of the processes are difficult to scale up and hence getting a logistical method with cost viability is an added advantage for effective green consumerism and its development in the lignocellulose bioeconomy. Launching an integrated lignocellulose biorefinery process is required for a bioeconomy uplift. Multiple biomolecules are present in the lignocellulose and so it is important to utilize all these components. Moreover, the waste generated from the lignocellulose process must be used in some way and should be converted to useful products. Tremendous research and development is required for finding cost-effective novel separation technologies.

15.7 Conclusion

Advanced technologies will aid in innovating economically feasible bioconversion methods, making lignocellulosic feedstocks an effective source in mitigating climate change. Recently, great importance has been given for bioenergy from lignocellulosic feedstocks. A focused understanding on the availability of lignocellulosic feedstocks, conversion technologies, along with assessment of the lifecycle is important for development of the lignocellulose bioeconomy. A smaller carbon footprint, abundancy in resources, and flexibility in production are some of the promising features of lignocellulosic biomaterials, which overcome the criticisms of first-generation feedstocks such as maximum energy input, less yield, food competition, less energy balance, etc. The maximum lignocellulosic material supply is predicted to be contributed from North and South America. Support in public policies will aid in the introduction of lignocellulosic biorefineries in Europe. A regular monitoring of bioenergy strategy and exact planning is required to improve the use of lignocellulosic feedstocks in Africa. Technological advancements in bioconversion methodologies and feedstock usage could pave the way for lignocellulosic bioeconomy development in Asia. Moreover, green consumerism has a wide scope in the improvement in the lignocellulose bioeconomy. The easy move of consumers toward green products indirectly influences the manufacturers to increase the supply, thereby maximizing the profiteering marketing economy. Even though green consumerism has its own advantages in developing the lignocellulose bioeconomy, it faces multiple challenges such as lag in implementation of effective government policies, minimal importance given to the research and development of the novel technology and product generation, less implementation of biorefining processes, etc. Overcoming all these challenges will lead to the development of the lignocellulose bioeconomy without any hindrance in the future.

References

Azargohar, R., Nanda, S., Rao, B.V.S.K., and Dalai, A.K. (2013). Slow pyrolysis of deoiled canola meal: product yields and characterization. *Energy Fuel* 27 (9): 5268–5279.

Azargohar, R., Nanda, S., Kozinski, J.A. et al. (2014). Effects of temperature on the physicochemical characteristics of fast pyrolysis bio-chars derived from Canadian waste biomass. *Fuel* 125: 90–100.

Balat, M. (2011). Production of bioethanol from lignocellulosic materials via the biochemical pathway: a review. *Energy Convers. Manag.* 52 (2): 858–875.

Binder, T.P., Benemann, J.R., Cavalieri, R.P. et al. (2007). Topic paper #8: Biomass. In Working Document of the NPC Global Oil and Gas Study. http://downloadcenter.connectlive.com/events/npc071807/pdf-downloads/Study_Topic_Papers/8-STG-Biomass.pdf.

Biswas, W.K., Barton, L., and Carter, D. (2011). Biodiesel production in a semiarid environment: a life cycle assessment approach. *Environ. Sci. Technol.* 45 (7): 3069–3074.

Borrion, A.L., McManus, M.C., and Hammond, G.P. (2012). Environmental life cycle assessment of lignocellulosic conversion to ethanol: a review. *Renew. Sust. Energy Rev.* 16: 4638–4650.

Chen, H. (2015). Future perspectives for lignocellulose biorefinery engineering. In: *Lignocellulose Biorefinery Engineering:Principles and Applications*, 247–251. Cambridge, MA: Woodhead Publishing.

Chen, H.Z. and Wang, L. (2008). Research progress on key process and integrated eco-industrial chains of biobased products—proposal of biobased product process engineering. *Chin. J. Process. Eng.* 8 (4): 676–681.

Ewing, M. and Msangi, S. (2009). Biofuels production in developing countries: assessing trade-offs in welfare and food security. *Environ. Sci. Pol.* 12 (4): 520–528.

Global Bioeconomy Summit (2015). *Making Bioeconomy Work for Sustainable Development.* Berlin, Germany: Communique.

Global Bioeconomy Summit (2018). Innovation in the global bioeconomy for sustainable and inclusive transformation and wellbeing, Communique: Berlin, Germany.

Gnansounou, E. (2010). Production and use of lignocellulosic bioethanol in Europe: current situation and perspectives. *Bioresour. Technol.* 101 (13): 4842–4850.

Grand View Research Inc. (2018). Biotechnology market worth USD 7271 billion. Available from: https://www.grandviewresearch.com/press-release/global-biotechnology-market Accessed 25.05.18

Green, E.M. (2011). Fermentative production of butanol—the industrial perspective. *Curr. Opin. Biotechnol.* 22 (3): 337–343.

Guinee, J.B., Heijungs, R., Hupps, G. et al. (2011). Life cycle assessment: past, present, and future. *Environ. Sci. Technol.* 45: 90–96.

Hirata, S. (2019). Lignocellulose-based bioeconomy in Japan. In: *Lignocellulose for Future Bioeconomy* (ed. H. Ariffin, S.M. Sapuan and M.A. Hassan), 243–264. Elsevier.

International Energy Agency (2007). *World energy outlook 2007: China and India insights.* Paris: IEA.

International Energy Agency (2008). *World energy outlook 2008.* Paris: IEA.

Kartha, S. Larson, E.D.(2000). Bioenergy Primer: Modernised biomass energy for sustainable development, United Nations Development Programme, New York.

Kumar, M. and Gayen, K. (2011). Developments in biobutanol production: new insights. *Appl. Energy* 88 (6): 1999–2012.

Laird, D.A. (2008). The charcoal vision: a win–win–win scenario for simultaneously producing bioenergy, permanently sequestering carbon, while improving soil and water quality. *Agron. J.* 100 (1): 178–181.

Malyan, R.S. and Duhan, P. (2018). *Green Consumerism: Perspectives, Sustainability, and Behavior*, 1e. New York, Apple Academic Press.

Milazzo, M.F., Spina, F., Primerano, P., and Bart, J.C.J. (2013). Soy biodiesel pathways: global prospects. *Renew. Sust. Energ. Rev.* 26: 579–624.

Mohanty, P., Nanda, S., Pant, K.K. et al. (2013). Evaluation of the physiochemical development of biochars obtained from pyrolysis of wheat straw, timothy grass and pinewood: effects of heating rate. *J. Anal. Appl. Pyrolysis* 104: 485–493.

Nanda, S., Mohanty, P., Pant, K.K. et al. (2013). Characterization of North American lignocellulosic biomass and biochars in terms of their candidacy for alternate renewable fuels. *Bioenergy Res.* 6 (2): 663–677.

Nanda, S., Mohammad, J., Reddy, S.N. et al. (2014a). Pathways of lignocellulosic biomass conversion to renewable fuels. *Biomass Conv. Bioref.* 4 (2): 157–191.

Nanda, S., Azargohar, R., Kozinski, J.A., and Dalai, A.K. (2014b). Characteristic studies on the pyrolysis products from hydrolyzed Canadian lignocellulosic feedstocks. *Bioenergy Res.* 7 (1): 174–191.

Nanda, S., Azargohar, R., Dalai, A.K., and Kozinski, J.A. (2015). An assessment on the sustainability of lignocellulosic biomass for biorefining. *Renew. Sust. Energ. Rev.* 50: 925–941.

OECD (2009). The Bioeconomy to 2030: Designing a Policy Agenda, Paris: Organisation for Economic Co-operation and Development.

Peerlings, J. (1993). An Applied General Equilibrium Model for Dutch Agribusiness Policy Analysis, Dissertation, Agriculture Economics and Rural Policy, Peerlings 254.

Samdin, Z. and Kamaruddin, N. (2019). The future of lignocellulose in Malaysian bioeconomy. In: *Lignocellulose for Future Bioeconomy* (ed. H. Ariffin, S.M. Sapuan and M.A. Hassan), 301–331. Elsevier.

Sanderson, K. (2011). Lignocellulose: a chewy problem. *Nature* 474 (7352): S12–S14.

Shafiee, S. and Topal, E. (2009). When will fossil fuel reserves be diminished? *Energy Policy* 37 (1): 181–189.

Sills, D.L., Paramita, V., Franke, M.J. et al. (2013). Quantitative uncertainty analysis of life cycle assessment for algal biofuel production. *Environ. Sci. Technol.* 47: 687–694.

Singh, A., Pant, D., Korres, N.E. et al. (2010). Key issues in life cycle assessment of ethanol production from lignocellulosic biomass: challenges and perspectives. *Bioresour. Technol.* 101: 5003–5012.

Strategy (2012). *Innovating for Sustainable Growth: A Bioeconomy for Europe*. Brussels, Belgium: European Commission.

United States Census Bureau (2015). US and world population clock (http://www.census.gov/popclock) [accessed on April 28, 2015].

United States Energy Information Administration (2011). *International Energy Outlook*. Washington, DC: USEIA.

Van Dam, J.E., Meesters, K., Elbersen, H.W. et al. (2019). Bioresources and biocommodities for the global bioeconomy. In: *Lignocellulose for Future Bioeconomy* (ed. H. Ariffin, S.M. Sapuan and M.A. Hassan), 287–300. Elsevier.

Vassilev, S.V., Baxter, D., Andersen, L.K. et al. (2012). An overview of the organic and inorganic phase composition of biomass. *Fuel* 94: 1–33.

Wang, M., Wu, M., and Huo, H. (2007). Life-cycle energy and greenhouse gas emission impacts of different corn ethanol plant types. *Environ. Res. Lett.* 2 (2): 024001.

Wiloso, E.I., Heijungs, R., and de Snoo, G.R. (2012). LCA of second generation bioethanol: a review and some issues to be resolved for good LCA practice. *Renew. Sust. Energy Rev.* 16: 5295–5308.

Wu, X., McLaren, J., Madl, R., and Wang, D. (2010). Biofuels from lignocellulosic biomass. In: *Sustainable Biotechnology* (ed. O.V. Singh and S.P. Harvey), 19–41. Dordrecht: Springer.

16

Going Green

Achieving the Circular Economy with Sustainable Biorefineries, Process Scale-Up, and Fermentation Optimization

Sreenivas R. Ravella[1], *David N. Bryant*[1], *Phil J. Hobbs*[2], *Ana Winters*[1], *David J. Warren- Walker*[1], *and Joe Gallagher*[1]

[1] *Institute of Biological Environmental and Rural Sciences (IBERS), Aberystwyth University, Aberystwyth, UK*
[2] *Anaerobic Analytics, Okehampton, UK*

16.1 Introduction

The European Union (EU) aims to be climate neutral by 2050, with an ambitious green deal action plan and objectives related to bioenergy and biorefining including (i) clean, affordable, and secure energy, (ii) an industrial strategy for a clean and circular economy, (iii) farm to fork strategy, and (iv) EU methane strategy (EU Green Deal). To deliver the UK net zero carbon and EU climate neutral strategies, advances in biorefining (biomass supply chain, conversion, and scalable technologies), waste utilization, and waste refining technologies are needed. Further to extract maximum value in a circular economy requires identifying strategic advances that will enable delivery (Scarlat et al. 2015; Pérez-Camacho and Curry 2018; Zetterholm et al. 2020). Moreover, to deliver these ambitious targets, pilot and demonstration-scale processing plants are critical to understanding and developing the technology required. These developments and challenges are addressed in several reports (IB 2025 2009; OECD 2009, 2011; Anonymous 2011; European Commission 2011a, 2011b; Star-COLIBRI 2011; European Commission 2012; IEA BIOENERGY Task42 BIOREFINING 2014; Besi and McCormick 2015; BIO-TIC 2015; National Research Council 2015; Hodgson et al. 2016b; Nattrass et al. 2016; Meyer 2017; Banks et al. 2018; Committee on Climate Change 2018; UK Bioeconomy Strategy 2018; De Buck et al. 2020) and highlight how pilot facilities bridge the gap between developing laboratory-scale technologies and demonstrating them at commercial scale. EU projects, such as EU AGRO BIOGAS and BEACON (http://beaconwales.org), have helped advance the renewable and biorefining sectors while working with national and international companies. Here, we discuss three important objectives for a large-scale biorefining plant build:

1) Identifying process changes due to scaling up.
2) Advances in process optimization, data interpretation, and control at pilot-scale.
3) Understanding the relevance of the microbial population profile associated with better performance.

Lignocellulose Bioconversion Through White Biotechnology, First Edition. Edited by Anuj Kumar Chandel.

Most small companies cannot afford to scale up until the technology/process has been demonstrated at a pilot scale. Pilot facilities are expensive and require substantial additional costs to employ skilled staff and to operate. Much of the focus has been on microbial performance which is important for the conversion of waste feedstocks into greener products. However, to implement effective fermentation requires feedback from monitoring relevant parameters. Effective or optimum control can be achieved with adaptive and intelligent data interpretation. There are numerous software options available.

Microbial processes are responsible for fermentation products but are often driven by feedstock inputs and physical parameters. Fermentation mostly behaves differently on scaling up and needs to be robustly characterized before progressing to demonstration trials. Building pilot or demonstration facilities reduces the risk of economic or product process failure. Pilot-scale development enables realistic optimization, saves time, and can reduce the risk of expensive investment cost and delay. Increasingly, new technology approaches are necessary and need to be focused on the integration of engineering and biology to enable fermentation optimization. Such approaches have been identified as the Fourth Industrial Revolution of Industry 4.0 (Chen et al. 2017; Kaya et al. 2020).

Over a decade we have built and operated small-scale biogas reactors and a two-stage $2\,m^3$ pilot plant for biogas production. Twelve commercial biogas plants were benchmarked for performance in Europe as part of an EU FP6 program—the EU AGRO BIOGAS project on European biogas initiative to improve the yield of agricultural biogas plants (Hobbs et al. 2010). One of the industrial partners on this project, Vogelsang developed several pretreatment technologies including Quick Mix, BioCrack, CC-Mix solid matter feeder, X Ripper and Disruptor. Further we have worked on the Climate KIC biogas program (Sustainable Biogas Process Optimization and Implementation Strategies [Biogas2Market] and Canadian Commonwealth Scholarships in Water and Energy on biogas). Currently we are working with companies including Coomb Farm, Alltech (BEACON projects on enzymes), Fre-energy and Membranology (BIOREVIEW project on volatile fatty acids [VFAs]) for biogas biorefining.

The European project BIOCORE that investigated lignocellulosic biomass (LCB) for a spectrum of products through an organosolv pretreatment process found that cellulose and lignin could be successfully extracted, but a formic/acetic acid pretreatment resulted in a hemicellulose fraction that was unfermentable without significant detoxification (O'Donohue 2014). Furthermore, it is apparent that while designing and modeling biorefinery processes environmental sustainability assessment needs to be considered (Rettenmaier et al. 2014). High impact delivery programs such as the Biobased Industries Joint Undertaking (BBI JU) have assisted the development and delivery of new biorefining technologies (Sierra et al. 2021) actively promoting and demonstrating new biobased technologies and products (Hodgson et al. 2016b). Recently, the BBI JU funded 11 flagship projects in the EU for biorefining to support the EU bioeconomy (www.bbi.europa.eu).

16.2 Sustainable Biorefineries and Supply Chain Aspects

Feedstock availability and supply chain configurations are important to start any biorefining or bioenergy plants. A commercial biorefinery generally requires at least 150 000 tons of feedstock per annum and the consequences of competing demand for substantial volumes

Table 16.1 List of biomass, agri-waste and other waste streams exploded and processed to yield sugars, lignin, and hydrolysates at Aberystwyth University's BEACON Biorefining Centre.

Corn stover	Pine	Bran	Sawmill shavings
Willow	Birch	Oat hulls	Bracken
Wheat straw	Silver birch	Corn cobs and corn fiber	Gorse
Sugarcane bagasse	Vine clippings	Grass fiber	Heather
Miscanthus	Waste wood fines	Brewers spent grain	Chalk scrub
Paper sludge	Mixed waste wood	Seaweed	Rhododendron

of feedstock can often be overlooked (Schröder et al. 2019). The availability of biomass for example at local level (Charlton et al. 2009; Pitcairn et al. 2017), national level (Martínez-Pérez et al. 2007; USDA 2011; Akgul et al. 2012; Morais and Bogel-Lukasik 2013; Alexander et al. 2015; Purdy et al. 2017; Moll et al. 2020), and regional level (Prochnow et al. 2009; Meyer et al. 2018a) needs to be assessed. Zaimes et al. (2015) identified numerous issues, mostly associated with retrospective outcomes of feedstock choices. These include ignoring economic and ecological outcomes resulting in a sub-optimal solution. Policy-based targets, investments, and installation design have an impact on sustainability (Markard et al. 2016). The supply chain needs to be assessed, but more on a case-by-case basis. The use of fermentation waste for energy is necessary for sustainable biorefineries and the green economy. A sustainable biorefinery depends on the variety of feedstock and adapting the process for optimization. Table 16.1 lists sources of different feedstocks tested at BEACON pilot-scale facility, most of which require pretreatment before being amenable to extraction of bioproducts or energy.

BEACON (http://beaconwales.org) has a breadth of expertise in numerous areas of biorefining and supports transition to a low carbon economy; the research that BEACON has undertaken with hundreds of companies is helping to foster the transition to a low carbon society, presenting significant environmental and business impacts. BEACON received the European Commission's 2014 RegioStars Award for Sustainable Growth.

Numerous pretreatment techniques have been covered in recent reviews (Hendriks and Zeeman 2008; Harmsen et al. 2010; Baruah et al. 2018) which highlight new and more sustainable processes that lead to no waste and CO_2 neutral targets. A good example of an integrated biorefinery is defined by Borregaard (Rødsrud et al. 2012). Rødsrud et al. (2012) state that a biorefinery should maximize the value of products from the biomass. Borregaard continually changes product lines and advances biorefining of all three major components of biomass (cellulose, hemicellulose, and lignin).

To make biorefineries more profitable in a changing market, they need to produce a range of products, often from different feedstock, and waste stream treatment is necessary to achieve a net-zero circular economy (Anaerobic Digestion and Bioresources Association, https://adbioresources.org/tag/net-zero). This approach requires solving some fundamental and applied research challenges with regard to material or waste utilization that can include secondary refining. More processes or technologies need to be tested at pilot level to become mature technologies (Figure 16.1).

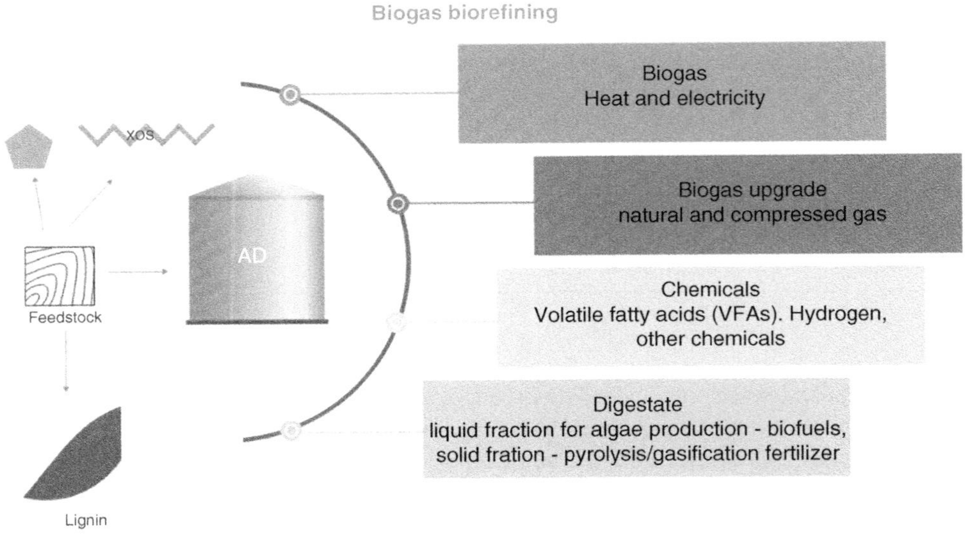

Figure 16.1 Biogas biorefining model.

16.3 Pretreatment of Biomass Using Pilot-Scale Steam Explosion Rigs

Here we focus on steam explosion (SE) as an example that produces several products from biomass and demonstrates how control optimization improves performance. SE protocols have been identified as a topic for further optimization; also, residual biomass from the process can be used to produce biogas by anaerobic digestion (AD). When considering the AD process, there are a range of influencing factors, including pretreatment, feedstock blending, and mixing that can all improve the overall efficiency and yield of biogas. This section highlights the importance of pretreatment, process development, and optimization.

16.3.1 Steam Explosion (SE) of Miscanthus and Methane Production from Miscanthus as an Example

Pretreatment is important, especially when using crop biomass because of recalcitrance to hydrolysis (Williams et al. 2017). At the Aberystwyth University (AU) BEACON Biorefining Centre, staff have operated a pilot-scale SE since 2013 investigating multiple biomass feedstocks for downstream processing. One feedstock investigated intensively is Miscanthus, which shows variation in chemical composition as well as within accessions grown at AU (Hodgson et al. 2011). To release sugars from Miscanthus, a relatively severe pretreatment is required. We have demonstrated that SE releases sugars and causes the breakdown of some lignin components in Miscanthus for further processing to xylose, xylitol (Walker et al. 2018), xylo-oligosaccharides, and soluble lignin (Bhatia et al. 2020). We also identified the factors influencing SE using design of experimental (DOE) concepts such as Taguchi methods (Walker et al. 2018).

Our preliminary results have shown improvement of biogas production from steam exploded Miscanthus compared to non-steam exploded Miscanthus (Ravella et al. 2019). SEs were performed with 400 g of Miscanthus stem samples at 12 bar for 3 min retention time and 15 bar flash. Our studies demonstrated increased methane production (20%) from biomass following pretreatment with SE using the Automatic Methane Potential Test System II (AMPTS II, Bioprocess Control). The AMPTS was also used for process optimization methods.

SE is historically a scalable technology and in the past few years has been operated by some commercial biogas plants. One site treats around 7500 tons of wheat straw per year (www.futurebiogas.com). There is more scope for this technology to be employed throughout the AD industry (Rodriguez et al. 2017).

The economy of SE pretreatment for biogas production was studied showing that the production cost of biogas could be lowered by 36% (Shafiei et al. 2013). This is possible because steam exploded material yields more biogas than untreated material per cubic meter of digester. SE also improves the efficiency of the AD process in terms of mixing, pumping, and microbial accession to the steam exploded fiber. Valmet (https://www.valmet.com) developed pilot-, demonstration-, and commercial-scale SE systems for biomass hydrolysis and biorefining. BioGasol ApS developed a scalable BioGasol's Carbofrac® pretreatment system based on SE. Their trials demonstrated that the pretreatment increased biogas production by about 40% from manure fibers and wheat straw. In addition, the Carbofrac pretreatment also helps to separate C5 sugar fractions. We are currently working with Bangor University on continuous steam treatment methods. Cambi (www.cambi.com) developed a robust thermal hydrolysis process for sludge management. In the UK Nova Pangaea Technologies have pretreatment capability for producing C5 sugar streams (www.novapangaea.com).

Inbicon A/S (a subsidiary of DONG Energy A/S) also developed a one-step hydrothermal pretreatment at 180–200 °C at demonstration scale to produce 5.4×10^6 L of cellulosic ethanol, 11 400 metric tonnes of lignin pellets, and 13 900 metric tonnes of C5 molasses (http://www.inbicon.com).

The above research demonstrates that SE pretreatment helps to separate hemicellulose and lignin from lignocellulosic material which has a commercial value. The global market for xylitol is more than 125 000 tons per annum (Ravella et al. 2012) and the xylitol market is expected to reach $1.37 billion by 2025 (https://www.bccresearch.com). Recently Sweet Appeal™ Natural Products announced the first commercial sale of xylitol produced through a fermentation process (https://sweet-appeal.com; https://zuchem.dcc.site/news). Another xylan-derived product, xylo-oligosaccharide is projected to achieve a global market of $144.9 million by 2026 (www.marketwatch.com). The lignin market size exceeded $730 million in 2019 with an estimated compound annual growth rate of 5.7% between 2020 and 2026 (http://www.gminsights.com/industry-analysis/lignin-market).

16.3.2 Heat Requirement of Biorefineries

We consider the resources required and especially the use of heat which can be critical to the economic viability of the plant also for SE. Heat can be generated from treating the biomass residues in an ancillary biogas plant. If we assess the commercial viability, we must consider the throughput and cost of the appropriate plant size. If, for example, the plant produces about 75 m^3 of biogas per tonne from the residue then this can generate about 180 kW h of electricity and about 250 kW h of heat. The biogas plant will require about half the heat for fermentation, leaving about 125 kW for the biorefinery. The electrical requirement of the plant is also important and if not required a smaller power unit can be used with a biogas burner to generate a greater proportion of heat, for example 100 kW h of electricity and 350 kW h of heat (about 225 kW h for the biorefinery) from 1 ton of residue (40 kW h of heat raise the temperature of 1 ton of water from 15 to 50 °C). Once the plant is operating successfully, optimizing performance is the biggest challenge. There is a wealth of known statistical and process control approaches available but there are differences for feedstock type and even for the same biomass crop these may require changing process parameters.

16.4 Taguchi Methodology for Process Optimization

Taguchi methodology is a DOE concept for robust process optimization. We have been using this DOE for more than a decade and optimized several processes in fermentations and pretreatment (Rao et al. 2008). The Taguchi method, a statistical approach, is a powerful technique for process optimization. It uses orthogonal arrays based on a matrix design and reduces the experiments required to optimize a process. This significantly reduces R&D time and costs where there are limited fermentation or pilot facilities. The Taguchi method has established popularity with engineering companies. We have applied this technique for optimizing pilot-scale SE pretreatment (Walker et al. 2018), germination of Miscanthus seed (Awty-Carroll et al. 2020), pyrolysis process optimization (Hodgson et al. 2016a), and biogas production (Adu-Gyamfi et al. 2012).

To achieve process optimization, it is necessary to follow through process development from laboratory to commercial scale. Any small change at commercial scale can have a very large impact, for example optimizing a large-scale fermentation process through comparative statistics resulted in a major improvement in enzyme production (Glynn 2017). A large-scale fermentation process (150 000 L capacity bioreactor) for production of an aldolase related enzyme by a bacterial species was optimized based on reviewing several years of process variables. By achieving an increase of 2 U mL^{-1} from 20 to 22 U mL^{-1} resulted in a total yield improvement of around 2.5×10^8 units of enzyme activity from each production batch (Glynn 2017). To reduce the 10% variation in the production process several trials were performed at a pilot plant and a yield improvement estimated to be $>2.5 \times 10^8$ U mL^{-1} batch was achieved at large-scale production in 2017 (Glynn 2017).

16.5 Process Automation

16.5.1 Automation

The key objective of the pilot plant is to bridge the scaling up gap between laboratory and commercial scale. The differences not only of the fermentation process but also of the technology and the changes in design required can have a more complex interactive influence.

We observed significant differences in fermentation by product type and conversion efficiencies when working at different scales. Also, the differences between laboratory batch processes and continuous processes are so significant that the risk of going straight to a production scale is too high. Generally, for fermentation, any volume of approximately 10 m^3 and above can be considered representative of the large scale. Significant importance is associated with the design and engineering which can be site, resource, and financially constricted. Here we are concentrating on the fermentation aspects and waste treatment that add to product and circular economy outcomes.

Fermentation in the pilot plant must be automated for dynamic or continuous representation of what may happen at the industrial scale. Investigation of the effects of mixing and viscosity are important for the substrate and microbes to produce the desired product. Suspended solids can coagulate and cause flotation and sedimentation and result in blockages within the pipework.

Clearly fermentation can only be monitored effectively in real time for optimal efficiency. Important parameters can be monitored by sensors both directly and indirectly. Indirect sensor monitoring uses sensor results and derives the important parameter to produce a soft-sensor response.

While there are numerous theoretical interpretations of what occurs in fermentation processes the process chemist must deal with the observations available. Fermentation is often not ideal and can be prone to delays of physical processes as well as microbial lag. Such nonlinear processes require good understanding; for example a localized optimum can prevent the best optimization from being achieved.

Interpretation of the sensor response or online monitoring data is a significant topic with statistical and artificial intelligence (AI) methods of evaluation. Interpretation ranges from the now accepted standard deviation evaluation method where straying from the straight and narrow production target is a cause for changing parameters to return to normal operations. AI methods increase the complexity of interpretation but enable better outcomes of fermentation processes that are often nonlinear and can lead to more better outcomes being achieved. Building expertise to adopt and implement automation for biopharmaceutical processes is particularly challenging (Goldrick et al. 2019).

Below are some of the latest advances that highlight topics for further investigation. Biorefining or biological fermentations are not well automated with little use of advanced process control especially compared to oil refining or other large-scale industrial processes (Goldrick et al. 2019). A simulation process IndPenSim was developed using the historical batch records of a 100 000 L penicillin fermentation process utilizing a high-yielding industrial strain of *Penicillium chrysogenum*. The simulation was used to predict input variables for biomass and penicillin production. The control depended on application of a process analytical technology (PAT) analyzer. The calibration and prediction models were used to control the feeding rate of the fermentation process (Goldrick et al. 2015). The process was monitored using real-time Raman spectroscopy which measured the control process parameters and resulted in 20% more penicillin per annum (Goldrick et al. 2014, 2019).

There are more readily available tools in recent years, for example a comprehensive automation control system was built from a mathematical toolbox. Fermentation was successfully tested under laboratory conditions (Kuprijanov et al. 2012). In another study commercial yogurt fermentation (part of Industry 4.0) was automatically controlled by a

LabVIEW software. The application was created in the form of a SCADA system (Simon et al. 2018). Real-time monitoring of biomass hydrolysis (Bryant et al. 2011) and fermentation processes (Ward et al. 2008a, 2008b, 2011a, 2011b; Ward 2016) with sensors is also important in process optimization and automation.

16.5.1.1 Mobile Phone and Real-time Control

There has been a significant increase in control technologies; this is demonstrated by the processing power of the modern mobile phone. We highlight the possibility because much of the real-time information does not directly measure new bioprocesses and needs to be derived or inferred. We have developed an app that can be used to receive inputs to upload to the cloud as well as process within the phone to identify immediate changes to the loading rate (Dr Phil Hobbs, Sylvarum Services Ltd). Further cloud storage means an overview can be achieved with a longer review of the data. Uploading data on the spot and remotely is a significant advantage for busy plant operators and is more likely to be recorded and recorded accurately.

16.5.1.2 BrewMonitor® System

Precision Fermentation offers a real-time monitoring solution that enhances control over the fermentation process for producing brewed beverages and other fermented products. Precision Fermentation's flagship product, the BrewMonitor System (www.precisionfermentation.com), helps ensure complete product consistency, increases manufacturing efficiency, and enables greater business profitability. Together, the solution's sensor-array device and web-based software suite (patent pending) continuously live-streams data from an active fermentation, offering complete fermentation process visibility and control that replaces trial-and-error measurements and enables excellent product and business outcomes.

Precision Fermentation Inc. was founded in 2017 and is based in the Raleigh-Durham, North Carolina, area of the United States. Its leadership is a forward-thinking group of brewing science and business professionals with deep experience in craft brewing, cell-cycle biology, bio-computational science, food and beverage business, and leading-edge web application development.

The BrewMonitor Sensor Hub attaches easily, cleans easily and live-streams critical data to web-enabled devices:

- dissolved oxygen
- pH
- gravity
- pressure
- fluid temperature
- ambient temperature
- conductivity

16.5.2 Process Optimization and Artificial Intelligence

There has been recent progress in control techniques capable of handling nonlinear fermentation processes. These improvements target accurate modeling and optimization.

There are numerous "non" AI techniques that include response surface methodology to enforce methods such as factorial design, central composite design for ethanol production (El-Gendy et al. 2013), and the Taguchi methodology (Rao et al. 2008). The AI techniques include neural network design for hydrogen production (Mu et al. 2006), fuzzy logic (Horiuchi 2002; Demirel et al. 2009; Vasickaninova et al. 2017), and genetic algorithms in which the latter has outperformed a conventional proportional–integral–derivative (PID) controlled process. Interestingly some AI techniques use the input of an experienced operator, so we do not have to start from "the ground floor." The performance can then be further improved or refined with the fuzzy logic inference system. We have identified the use of the Mamdani and Sugeno approaches as effective for good process control (Vasickaninova et al. 2017; www.mathworks.com/help/fuzzy/types-of-fuzzy-inference-systems.html). The fuzzy system should initially be supplied with the full range of parameters from which to learn; otherwise the fuzzy systems will provide strange results outside the supplied learning range. The fuzzy logic system can perform well, and these techniques are growing in popularity for nonlinear processes. However, robustness, as for other process control methods, must be assured. For example, a fuzzy logic system used to control the alkalinity of a fermentation enabled the maximum biogas output from a cellulose feedstock. The controller maintained the alkalinity and pH within the desired stability range to enable optimized performance (Anastasios 2013). Recently Carlsberg used AI for "beer fingerprinting" using Microsoft's AI solutions and sensors from iNano at Aarhus University.

National Instruments (NI) LabVIEW (http://www.ni.com/labview) offers a graphical programming approach that helps you to operate, acquire data from and control a fermenter (Ward et al. 2008b). The graphic representation simplifies control for automation, customization, and optimization. LabVIEW can be used to operate pilot plant operations remotely without any manual intervention to achieve automation of a biogas pilot plant. Here we give an example of our pilot plant operations in automated mode for more than 3 months and compared with manual operations. Figure 16.2 shows LabVIEW hardware

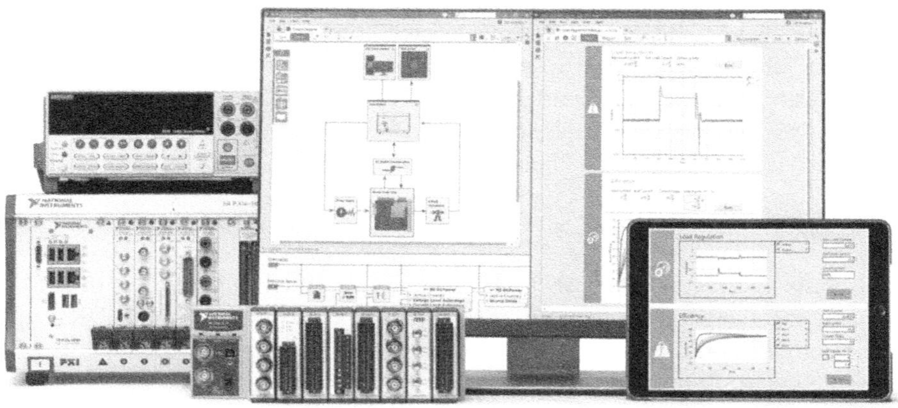

Figure 16.2 LabVIEW seamlessly integrates thousands of measurement and control devices through NI hardware.

that can be used for automation. Further fuzzy-logic-based algorithms can help with remote control once connected to sensors and motors.

16.5.3 Biogas Pilot Plant

Biogas production occurs through four recognized fermentation processes from long chain natural polymers to smaller molecules that result in the production of methane and carbon dioxide. The resulting biogas also contains trace components that include (hydrogen and other) sulfides, ammonia, and water vapor.

Biogas production is an anaerobic process occurring below −330 mV redox at a pH of between 6.7 and just over 8. Addition of feedstock should be limited by the concentration of the resulting VFAs. More effective measurements are the acid to base ratio, with an optimum level for most plants in the ratio 0.3–0.4. The methanogens which are of the Archaea taxa are temperature sensitive, requiring a range of ±1 °C during optimum biogas production. The operating temperature can be in the mesophilic (35–44 °C) or the thermophilic range (50–60 °C). Mesophilic digester populations are largely influenced by the feedstock used, whereas the thermophilic digester microbial populations are influenced by the higher temperature, having greater enzyme and microbial activity.

Methanogens can produce methane from VFAs as well as hydrogen-based sources. However, they are sensitive to other by-products such as hydrogen sulfide and ammonia, and undissociated small fatty acids that permeate the cell wall to have a toxic effect. Methanogens require macro-nutrients, nitrogen, carbon, sulfur in appropriate ratios. Further methanogens require trace elements, nickel, cobalt, iron, selenium, and numerous others.

The use of a pilot plant is common in the biorefining sector; in this section we are limiting the discussion to pilot plants configured for biogas production and the respective fermentation processes and products. We built a biogas pilot plant with a two-stage process consisting of a 1 m^3 hydrolysis tank and a 1.5 m^3 methanogenesis tank (EU funded AGRO BIOGAS project); the process control was performed by a LabVIEW application. It was a collaborative project with industrial partners including Hugo Vogelsang and General Electric along with commercial biogas plants located in Europe (Amon et al. 2009). The pilot plant has inbuilt flexibility for research purposes to develop sensor and automatic remote early warning systems. Two tanks (T1 and T2) are installed with multiple probes (Figure 16.3). This pilot plant was built not only to develop early warning systems but also to optimize biogas production from different feedstocks.

16.5.4 Sensors

Both hydrolysis and methanogenesis tanks have probes fitted to measure pH, redox, conductivity, and temperature. Partech Instruments (St Austell, Cornwall, UK) supplied the electrodes to monitor pH, redox, conductivity, and temperature for the soft-sensor development. Data from the probes were downloaded into a file via LabVIEW commands.

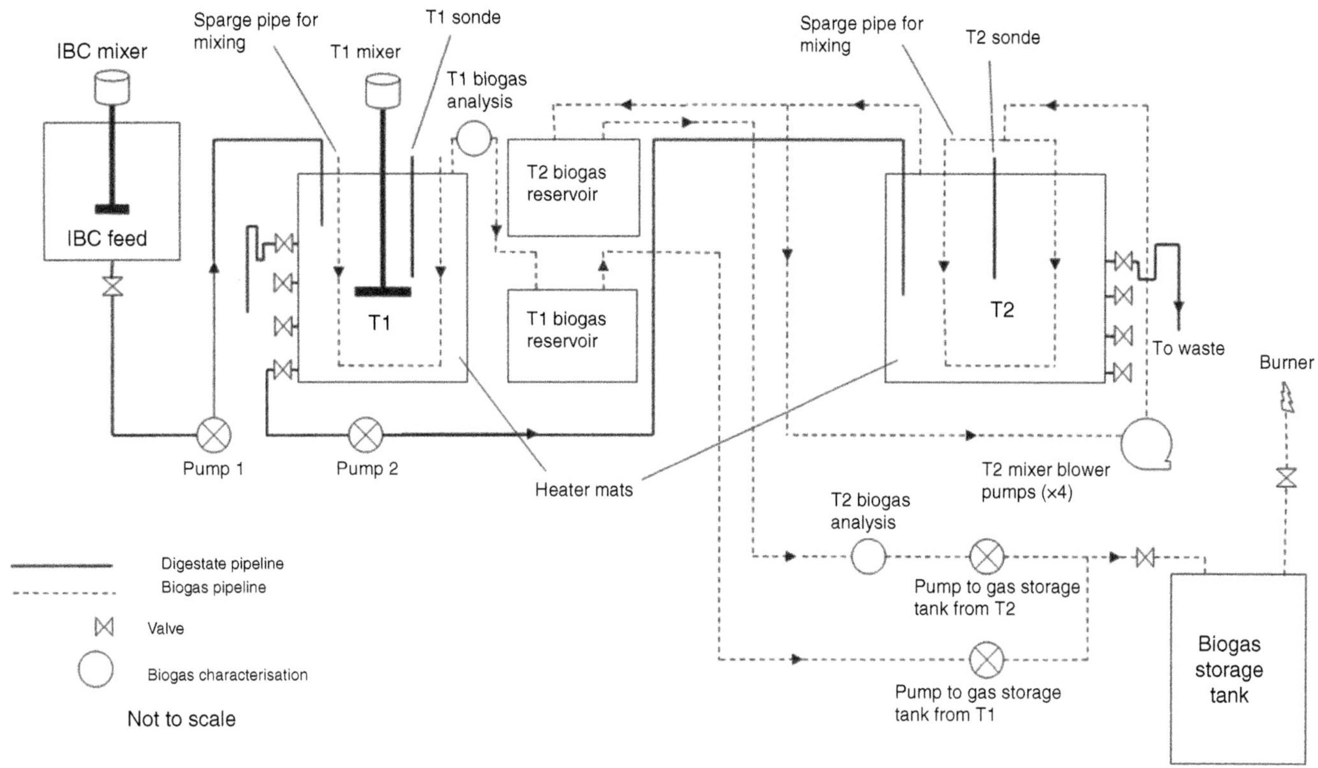

Figure 16.3 Biogas pilot plant process schematic.

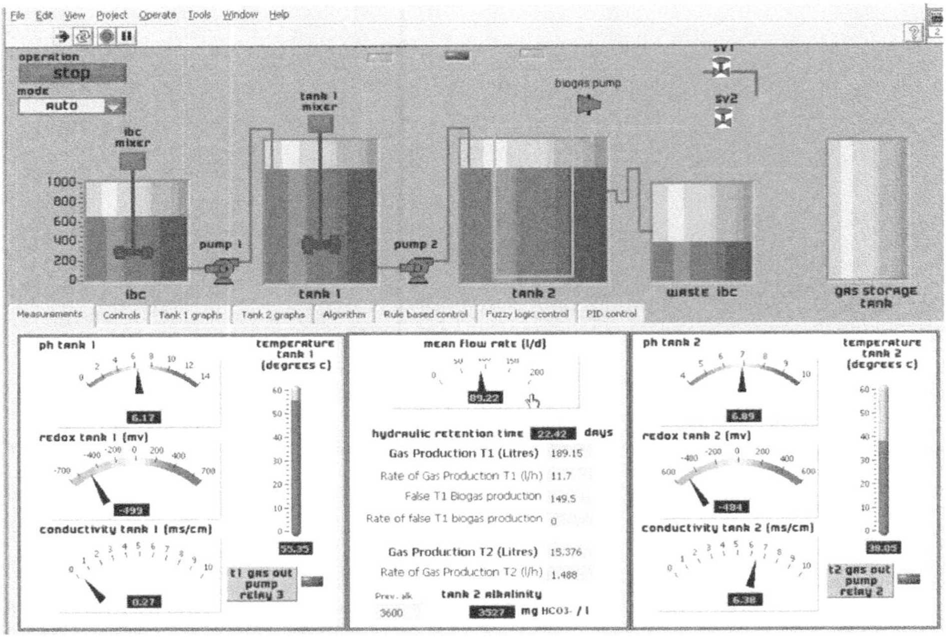

Figure 16.4 Schematic of the process control interface for the pilot plant showing status and sensor responses.

16.5.5 Process Control Configuration with LabVIEW and NI Data Acquisition (DAQ) Devices

16.5.5.1 Connect Sensors and Signals to a DAQ Device

The pilot plant was monitored and controlled using NI software; this enabled remote monitoring and control. The LabVIEW acquires information through an NI DAQ device, displays, and saves real-time data from the temperature, pH, conductivity, and redox probes installed in each of the tanks (Figure 16.4). The software also controls mixing and pumping events. A soft sensor (rule-based control system/automated mode) was programmed in LabVIEW to predict the value of the alkalinity (HCO_3^-) from real-time values of pH, redox potential, and conductivity. Process control of the organic loading rate (OLR) was adjusted depending on the real-time measurement of the alkalinity and the current feeding rate of the digester. As the predicted alkalinity increased the OLR was increased according to a rule-based structure.

16.5.6 Rule-based Control Structure

When the pilot plant was controlled in automated mode, the real-time predicted alkalinity was compared to the alkalinity set points (Figure 16.5). During this experiment the OLR was adjusted automatically by a rule-based automatic loading regime. For each of the set

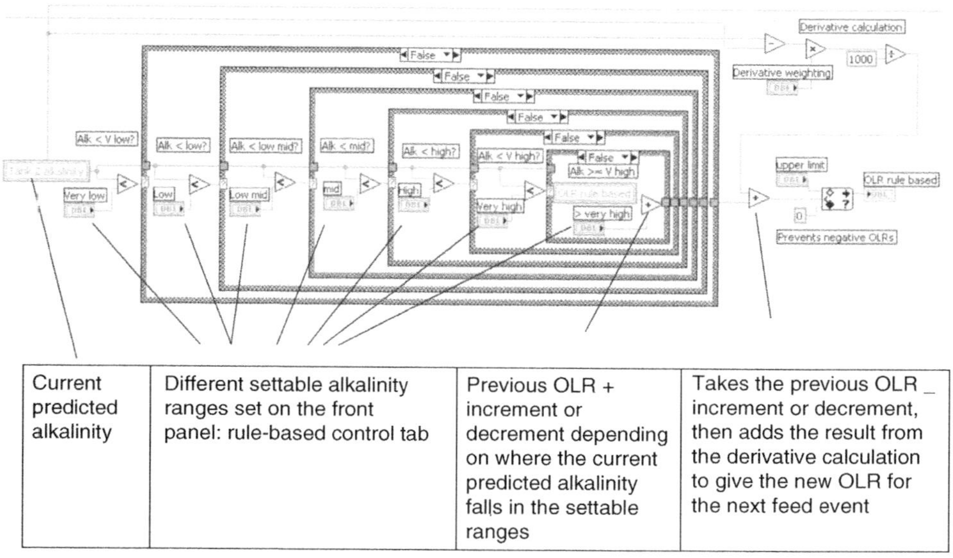

Figure 16.5 Part of the LabVIEW block diagram code explaining how the OLR is calculated based on the current and previous predicted alkalinity. This algorithm calculates the difference in magnitude between the current and previous alkalinity, together with comparing the current predicted alkalinity with designated set points and the previous OLR.

points above or below the midpoint, the OLR at the next feed event was changed by the value of the OLR increase/decrease respectively. Essentially, if the current predicted alkalinity is greater than the midpoint, the OLR will be increased at the next feed event. If the current predicted alkalinity is below the midpoint, the organic loading is decreased at the next feed event. The further away from the midpoint the current predicted alkalinity is, the greater the increase or decrease in OLR.

16.5.7 Pilot Plant Data

The LabVIEW VI was programmed to save data hourly in a new data file each day. Data saved for both tanks include temperature, pH, redox potential, conductivity, biogas volume tally, and hourly rate of biogas production.

16.5.8 LabVIEW Application for Laboratory-scale, Pilot-scale and Industrial Fermentations

LabVIEW can be used to control small laboratory-scale (Gregory and Thornhill 1992; Rocha and Ferreira 2001; Mkondweni et al. 2005; Wolf et al. 2012), pilot-scale (Huang et al. 2005, 2012; Wang et al. 2020), and industrial-scale fermentations (Simon et al. 2018). Figures 16.6 and 16.7 show the pilot-scale 350 L fermenter at BEACON biorefining facility, run by using LabVIEW applications.

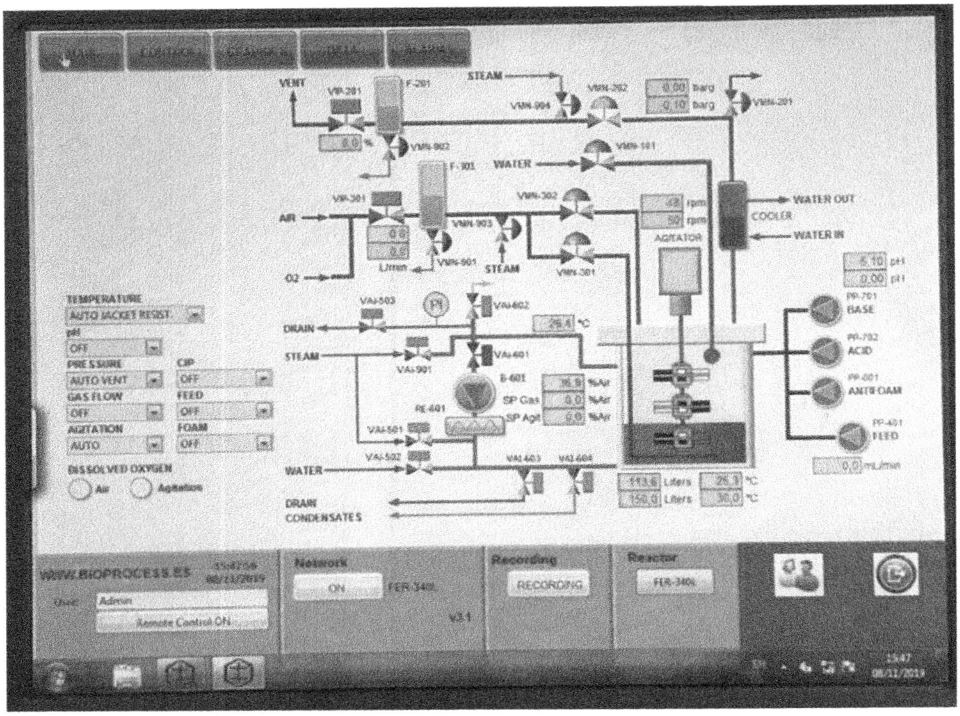

Figure 16.6 PID control of 350 L fermenter at BEACON biorefining facility, the data acquisition and controlling developed by using LabVIEW software.

16.5.8.1 LabVIEW Datalogging and Supervisory Control Module

The NI LabVIEW Datalogging and Supervisory Control (DSC) Module helps to develop HMI/SCADA or high-channel-count datalogging applications. The LabVIEW DSC Module is add-on software for the LabVIEW applications.

16.5.9 Advantages of LabVIEW in Automation and Monitoring Commercial Plants

NI developed several hardware products to install in big industries; for example Duke Energy modernized their power generation plant using NI hardware and software tools. Duke Energy implemented a wireless condition and monitoring system using NI waveform generating sensors mounted on their 30 sites. These sensors can be mounted anywhere, indoors and outdoors, and can be mounted on hard to reach sites. By using this technology Duke Energy replaced 60 000 data collection sites with remote data collection and monitoring. This saves technicians time of traveling to different remote plant sites and hard to reach installations. The early warning systems (NI InsightCM™ software) allows diagnosis of problems before traveling to the plant. This saves valuable time and resources for routine maintenance of the plant. NI hardware can allow connection to thousands of industrial assets, such as gas and steam turbines, water pumps, and power generators. For more

Figure 16.7 350 L fermenter at BEACON biorefining facility, the complex engineering system controlled (automation for mixing, pumping, sterilization, and managing alarms) by using LabVIEW applications.

examples and case studies see the NI website (NI.com, Energy Research User Handbook; NI.com, Hardware Integration with NI LabVIEW).

16.6 Microbial Adaptation, Evolution, and Diversity for Process Optimization

There are three important aspects that would target biomass conversion to both wanted and abundant bioproducts. These are a better understanding of how microbial communities (i) adapt and (ii) evolve to the feedstock presented and (iii) remain diverse during the fermentation process. Even after several years of studying biogas microbial communities

(Jackson et al. 1999; Schlüter et al. 2008; Jünemann et al. 2017; Theuerl et al. 2019a, 2019b; Zhang et al. 2019), understanding of the molecular and ecological basis of community-level functions is poor. Sierocinski et al. (2018) investigated the causal link between methane productivity and diversity by manipulating cultures by dilution. Dilution led to the loss of the rarer species from the communities. A positive relationship between methane production and the number of taxa was found, suggesting that rare species play an important role. A positive relationship between species richness and methane production emerged following ecological selection imposed by the laboratory conditions. There was little functional redundancy, so any loss of diversity both natural and resulting from changes in propagation conditions during AD would reduce methane production.

The degradation or hydrolysis of substrates in a fermenter is complex because of the large number and diversity of species present; the seeming functional redundancy among some of them; and their specific structural associations. It is believed that the spatial structure in the form of heterogeneous biofilms and granules can enhance the metabolic interactions in the digesters (Winkler et al. 2011; Lu et al. 2013). The functional differences in the biofilms and the location of the granules within the reactors was studied (Winkler et al. 2011); similarly functions of the microbial communities within the layers of anaerobic granules were also studied (Gagliano et al. 2017, 2020), methanogenic archaea found in the interior of the granule and fermentative bacteria on the granule surface.

To understand the microbiology of biogas we have performed several studies based on phospholipid fatty acids (Hobbs et al. 2008), extracellular polysaccharides, and genomics of a commercial biogas reactor (Arunasri et al. 2010a, 2010b, 2015; Ravella et al. 2010b, 2011). Novel microbes were discovered (Ravella et al. 2010a). Further studies was focused on the evolution of organisms associated with methanogenic communities during the initiation of biogas fermentation with two substrate feeding strategies using the 454 sequencing tool (Ravella et al. 2011).

A study on microbial communities from a commercial mesophilic biogas plant from the UK was performed (Arunasri et al. 2015). Two methanogenic-specific 16S rRNA gene clone libraries (MT1 and MT2) and two bacterial 16S rRNA gene clone libraries (HW1 and HW2) were constructed using DNA isolated from two samples collected from two separate tanks of commercial biogas plant. In the Euryarchaeota-based libraries both samples revealed clones belonging to *Methanomicrobiales* and *Methanosarcinales* of the Euryarchaeota phylum. The majority of the archaeal clone sequences belonged to *Methanoculleus bourgensis;* this also revealed that hydrogenotrophic driven methanogens are abundant in the biogas reactors. The complete gene sequence of *M. bourgensis* has been identified as the hydrogenotrophic microbe (Manzoor et al. 2016). 16S rRNA gene-based bacterial libraries also revealed the ubiquitous presence of Firmicutes (*Clostridium straminisolvens* were abundant in sample 1, *Clostridium thermocellum* predominated in sample 2) and also Proteobacteria phyla. Cellulolytic bacterial clone sequences were abundant in both samples (Arunasri et al. 2015). Fluorescent in situ hybridization results also revealed the presence of *Methanomicrobiales* and *Methanosarcinales* as dominant groups (Arunasri et al. 2015).

Another study to optimize biogas production demonstrated (Hobbs et al. 2008) clear differentiation between the microbial community structure in a biogas system that supports a higher microbial density utilizing twice the input material and produces double the biogas output. Fungal populations are higher in the hydrolysis stage than the methanogenesis stage (Hobbs et al. 2008). There was a higher density of microorganisms for the system with double the feedstock. There are fewer organisms in the mobile phase for the methanogenesis tank with the microbial support. The microbial support also aids process stability producing fewer perturbations during microbial cycling. This study identifies how we can improve biogas yield and reduce capital costs by minimizing the volume of the methanogenesis digester (Hobbs et al. 2008; Adu-Gyamfi et al. 2012). The above studies have highlighted the array of the microbial community in biogas reactors; further studies are required to understand the functionality of the community using proteomics tools to understand their role in the hydrolysis of feedstock and community evolution.

The evolution of organisms associated with methanogenic communities during the initiation of a pilot-scale biogas fermentation with two substrate feeding strategies was studied using the 454 sequencing tool (Ravella et al. 2011). One of our studies on thermophilic methanogenic community adaptability was investigated by changing the feedstock. The experiment demonstrated that changing feedstock allows microorganisms to evolve to more readily anaerobically digest each of the feedstock types even after a considerable absence. This research was done to understand how a thermophilic AD process is influenced by low and high C:N ratios. The objectives of the study were threefold: (i) to gain insight into the extent that the C:N ratio and nutrients affect biogas production; (ii) to understand the taxonomic composition and methane-producing microbial community in thermophilic biogas reactors, and (iii) to evaluate the effectiveness of feed in several biogas reactors on the stability of the process and the evolution of the microbial community. We have identified prominent thermotolerant fungal communities that survived high temperatures and were useful for the hydrolysis of cellulosic material, xylose utilization, polysaccharide production, and biofilm formation. The results also showed that Clostridial groups were abundant, almost double in phase 1 reactors (high C:N ratio 25) than phase 2 reactors (low C:N ratio 14) (changing the feed based on C:N ratio). *C. thermocellum*, *Clostridium cellulolyticum*, *Clostridium lentocellum*, *Geobacillus stearothermophilus*, and *Acetivibrio cellulyticus* of Ruminococcaceae were dominant in the hydrolysis state of the process. More thermophilic bacteria belonging to Thermotogae phylum evolved in phase 2 reactors than phase 1 reactors. A starch utilizing, thermophilic bacterium *Petrotoga mobilis* of the phylum Thermotogae (95% of) was dominant in phase 2 reactors. Within phase 2 reactors *P. mobilis* doubled in number where two reactors were pulsed with more starch, that is, a high C:N ratio compared to a low C:N ratio, whereas acetoclastic methanogens were more abundant in low C:N ratio reactors.

16.6.1 Microbiology of Volatile Fatty Acids (VFAs) Production in AD

VFAs are important intermediates in biogas and biorefined production. VFAs also have commercial applications, highlighting their importance in advanced AD refineries (Bhatia and Yang 2017; Atasoy et al. 2018; Wainaina et al. 2019; Aktij et al. 2020). Measurement of VFAs and analyzing VFAs producing and degrading microbes gives a robust understanding of the AD process. Production can be inhibited with high concentrations of ammonia. In some experiments VFA concentrations had an effect on methanogenic communities where acetoclastic methanogens Methanosaeta were completely replaced by Methanosarcina; further in this study, ammonia was demonstrated to have an effect on VFA dynamics, the propionic acid degrading genus Syntrophobacter being replaced by yet unknown propionic acid degraders (Bonk et al. 2018). Advancements in AD microbiology will help us to explore and isolate novel potential producers of VFAs. For example, in our recent metagenomic studies (using Miscanthus as a feedstock), we found the presence of two highly efficient acetoclastic electroactive bacteria *Geobacter sulfurreducens* and *Desulfuromonas acetexigens* for acetate utilization (Figure 16.8). *G. sulfurreducens* is currently considered the most important bacterium in AD biorefining. Metagenomics is also helpful as a novel tool and advances in bioinformatics help us to discover VFA biosynthesis routes since the pathways for VFAs are still not well understood. Butyric, acetic, and propionic acids are the most important VFAs in AD digesters; other VFAs such as isovaleric, valeric, and caproic acids also accumulate in the digester prior to biogas production. Accumulation of butyric acid is regarded as a signal of higher hydrogen partial pressure in the digesters. VFA profiles will change based on feedstock and operational fermentation conditions (Bonk et al. 2018). Promoting key acidogenic bacteria populations is a key factor for maximizing VFA production in the AD digesters, and reduction of the methanogenic activity also plays a key role in maximizing VFA accumulation. In our metagenomic analysis Acetobacteraceae bacterial species such as *Gluconobacter oxydans*, *Granulibacter bethesdensis*, *Acetobacter pasteurianus*, *Acidiphilium cryptum* dominate acetic acid producing strains present in the microbial community (Figure 16.9). We have identified key VFA producing species including Ruminococcus sp. (Figure 16.10). Ten genera from Acetobacteraceae make up the acetic acid bacteria. Acetate is produced largely by *Bifidobacteria* and *Lactobacilli*, but *Ruminococcus* spp. also produce acetate (Figure 16.11). We have identified butyrate-producing *Faecalibacterium prausnitzii*, *Eubacterium rectale*, *Eubacterium hallii, and Eubacterium cellulosolvens* (Figure 16.12), and VFA oxidizing species *Syntrophomonas wolfei*, *Syntrophus aciditrophicus*, *Syntrophothermus lipocalidus*, and *Syntrophobacter*

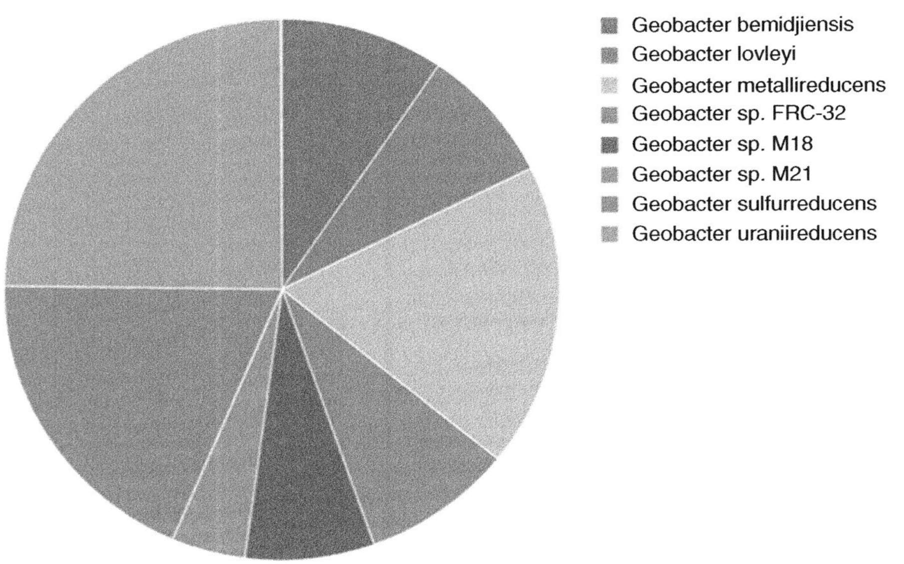

Figure 16.8 Acetate utilizing bacterial species in the biogas reactor.

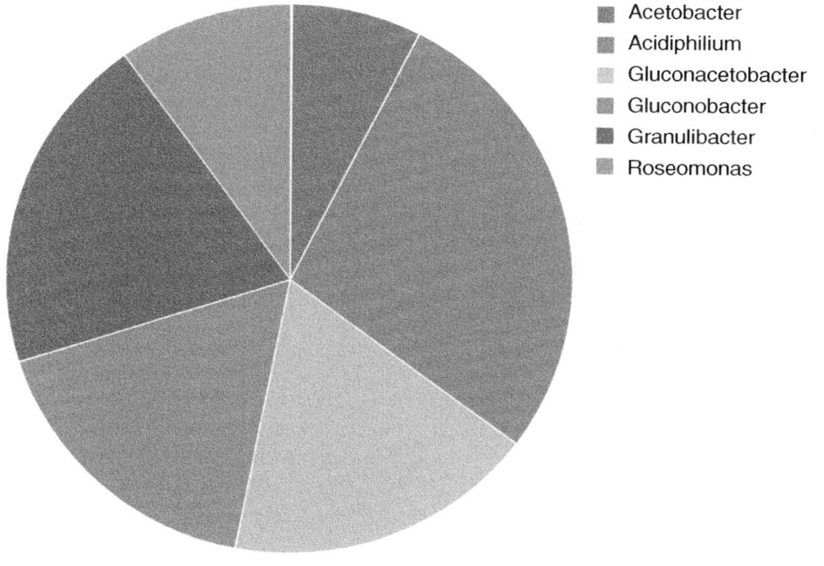

Figure 16.9 Acetic acid producing strains in the biogas reactor.

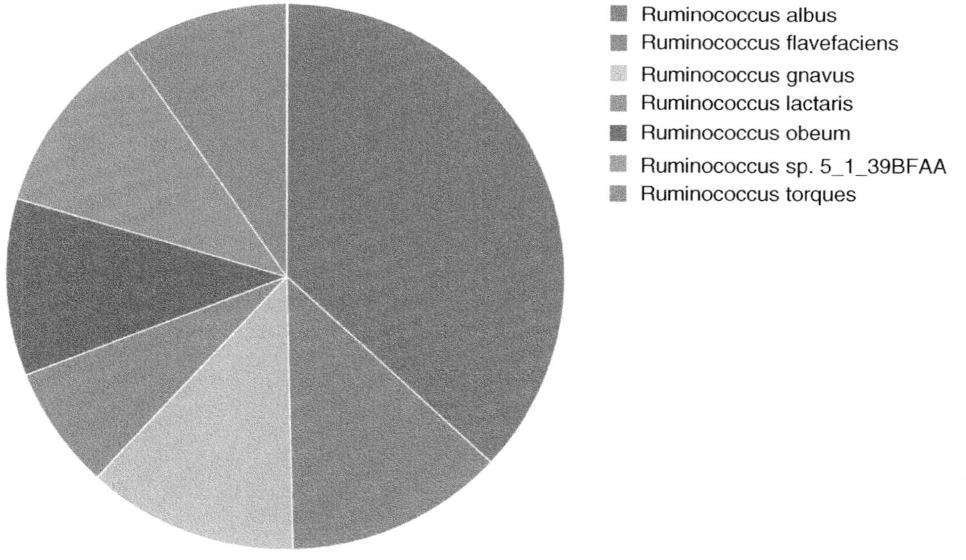

Figure 16.10 Key VFA producing species in the biogas reactor.

fumaroxidans, a species of syntrophic propionate-degrading sulfate-reducing bacterium within the microbial community. The important VFA degrading microbe genomes have been fully sequenced (Plugge et al. 2012). Hydrogenotrophic methanogens were more resistant than acetoclastic methanogens to high VFAs and ammonium concentrations. We also found that acetoclastic methanogens *Methanosarcina barkeri*, *Methanosarcina*

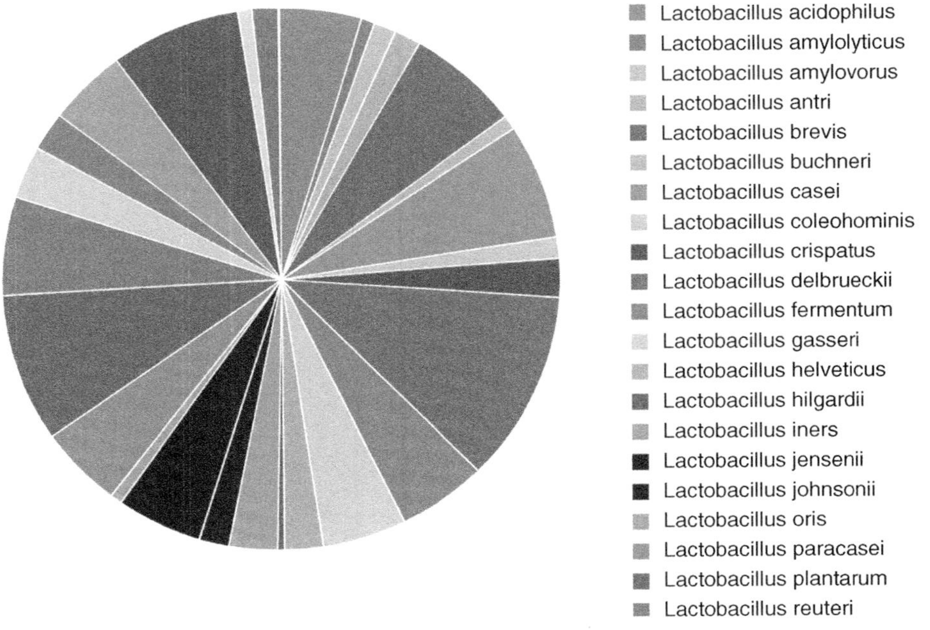

Figure 16.11 Acetate is producing strains in the biogas reactor.

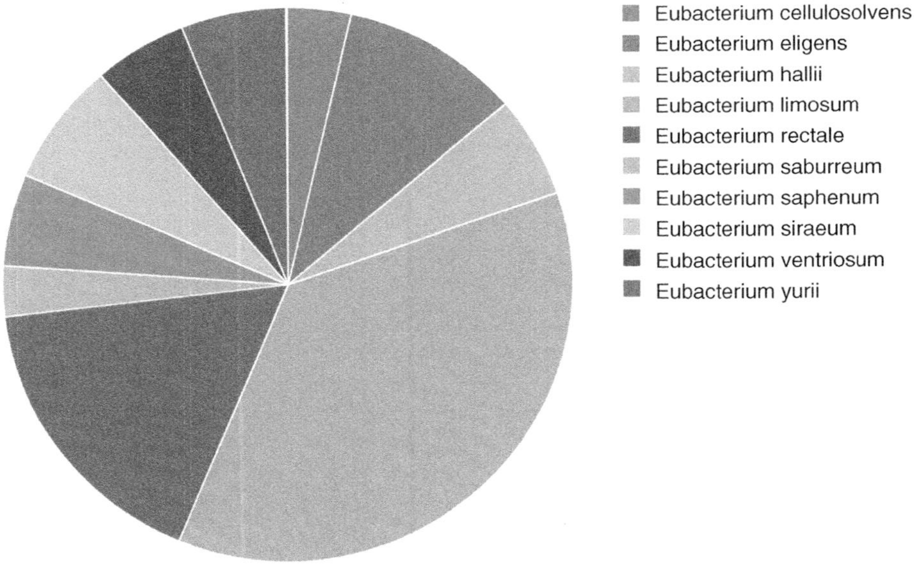

Figure 16.12 Butyrate producing bacterial species in the biogas reactor.

acetivorans, and *Methanosarcina mazei* (Figure 16.13) belonging to methanosarcinales were dominant in the methanogenic community in our AD samples. *M. barkeri* is the most common species of the genus Methanosarcina. *M. barkeri* is able to grow in a variety of different substrates including methanol, acetate, methylamines. *M. acetivorans* is a

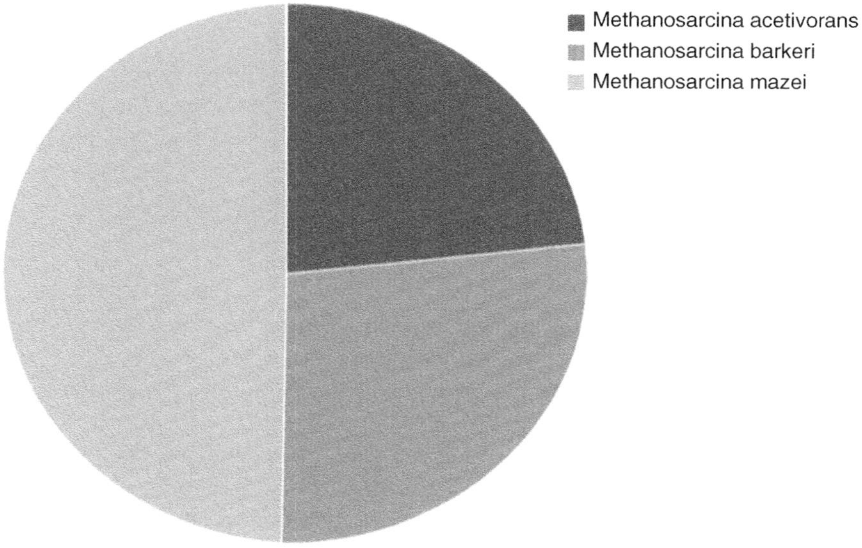

Figure 16.13 Acetoclastic methanogens in the biogas reactor.

versatile methanogen capable of forming multicellular colonies (Ferry 2020). Acetoclastic methanogens are most susceptible to ammonia and VFAs.

Studying microbial communities using advanced sequencing techniques (Jünemann et al. 2017; Beauchemin et al. 2020; Lim et al. 2020) and analytical tools (Keegan et al. 2016; Meyer et al. 2018b; Zhang et al. 2019) will help us to understand responses to feed and to optimize fermentation conditions. Studying microbial communities will enable targeting of maximum VFAs and biogas production. Microbiology of AD is complex but important for biorefining and circular economy advances.

In our current BIOREVIEW (http://www.bioreviewproject.org) project we are working on refining AD samples for pilot-level VFA production from biogas digesters.

16.7 Final Remarks and Conclusions

Despite the considerable number of studies monitoring small-scale through to commercial biogas reactors there are still many unknowns that limit process efficiency. Many of the studies show the need to optimize these technologies as discussed above. Biogas studies are focused on hydrolysis in the current literature. Further SE is of interest because there are several products extractable after the process that can be considered relevant for the next generation of bioproducts.

The biogas industry has been on the bow wave of using biomass for society. The biogas incentives have virtually disappeared in the UK and profits are minimal at 200–500 kW plants. Biorefineries similarly receive little or no incentives. Hydrolysis is the common link between biogas production and bioproducts for society. Some of the biogas plant control technology and engineering approaches are transferable and can greatly benefit the development of biorefineries. This is because cheaper and more effective engineering has been developed and is available.

The development of commercial biorefineries is driven by the available or possible technology and in the long run the economics of the process. Such economics can drive the need to maximize products and utilize the residues for energy production. Much of the technology is new or in its infancy. There are various means to reduce the biomass to smaller molecules or a usable form, ranging from physical to biological processes.

16.7.1 Main Conclusions

- To deliver sustainable biorefineries biomass supply chains need to be considered to utilize all the waste streams to generate value-added products.
- To develop new technologies pilot-scale systems are paramount to progress and achieve financial success.
- More biorefining demonstration plants need to be available for companies to work collaborative programs, for example centers like BEACON and AU Aber Innovation.
- Integrated biorefining models need be developed for multi process and product development.
- SE can provide access to biorefinery products from cellulosic biomass sources.
- Advance technologies like wireless technologies, software, hardware, process optimization statistical methods and sensor-based technologies are necessary for biorefining.
- Studies of the fermentation microbial communities are in their infancy.
- The focus should be on the dynamic nature of the population and how we can influence change or indeed stability to provide an improved output.
- Advanced molecular tools, bioinformatics, advanced computing systems are used to understand microbiological science and aid processes.
- Application of molecular techniques is crucial to understand microbial communities in response to altering process parameters.
- AD can be used as a final step for waste streams in a circular economy as profits have recently increased with electricity and biomethane increasing in value e.g. kWh is now worth about 20 to 24p.

Acknowledgements

We thank WEFO for BEACON pilot-scale biorefining center and BEACON project (80851), BBSRC and INNOVATE UK for Newton Bhabha Fund—BIOREVIEW project (BB/S011994/1), Interreg (EU) for CCC Hub project and EU for EU AGRO BIOGAS project.

References

Adu-Gyamfi, N., Ravella, S.R., and Hobbs, P.J. (2012). Optimizing anaerobic digestion by selection of the immobilizing surface for enhanced methane production. *Bioresour. Technol.* 120: 248–255. https://doi.org/10.1016/j.biortech.2012.06.042.

Akgul, O., Shah, N., and Papageorgiou, L.G. (2012). Economic optimisation of a UK advanced biofuel supply chain. *Biomass Bioenergy* 41: 57–72. https://doi.org/10.1016/j.biombioe.2012.01.040.

Aktij, S.A., Zirehpour, A., Mollahosseini, A. et al. (2020). Feasibility of membrane processes for the recovery and purification of bio-based volatile fatty acids: a comprehensive review. *J. Ind. Eng. Chem.* 81: 24–40.

Alexander, P., Moran, D., and Rounsevell, M.D.A. (2015). Evaluating potential policies for the UK perennial energy crop market to achieve carbon abatement and deliver a source of low carbon electricity. *Biomass Bioenergy* 82: 3–12. https://doi.org/10.1016/j.biombioe.2015.04.025.

Amon, T., Mayr, H., Eder, M. et al. (2009). EU Agro Biogas Project. In: *Ikke angivet. ETA—Renewable Energies*, 1551–1555. Hamburg, Germany: European Biomass Conference and Exhibition.

Anastasios, A. (2013). Soft sensor development and process control of anaerobic digestion, PhD Thesis, University of Exeter.

Anonymous (2011). BBSRC support for bioenergy and industrial biotechnology: recommendations to encourage UK science and technology for the energy and chemicals industries. *Industrial Biotechnol.* 7 (1): 41–52. https://doi.org/10.1089/ind.2011.7.041.

Arunasri, K., Ravella, S.R., Hobbs, P.J. et al. (2010a). Microbial diversity of a methanogenic community present in a commercial biogas plant. Molecular biology of Archaea II, 16–18 August, Cambridge, UK.

Arunasri, K., Ravella, S.R., Retter, A. et al. (2010b). Isolation and phylogenetic analysis of extracellular polymeric substances (EPS) producing bacterial strains from a Iiogas plant. Biofilms 4 International Conference, 1–3 September, Winchester, UK. https://biofilms4.files.wordpress.com/2009/06/biofilms-handbook-final-1-pq.pdf

Arunasri, K., Shivaji, S., Hobbs, P.J. et al. (2015). Molecular characterization of Euryarchaeal community within an anaerobic digester. In: *Rumen Microbiology: From Evolution to Revolution* (ed. A.K. Puniya, R. Singh and D.N. Kamra), 293–311. Springer Nature.

Atasoy, M., Owusu-Agyeman, I., Plaza, E., and Cetecioglu, Z. (2018). Bio-based volatile fatty acid production and recovery from waste streams: current status and future challenges. *Bioresour. Technol.* 268: 773–786.

Awty-Carroll, D., Ravella, S.R., Clifton-Brown, J. et al. (2020). Using a Taguchi DOE to investigate factors and interactions affecting germination in *Miscanthus sinensis*. *Sci. Rep.* 10 (1): 1–11. https://doi.org/10.1038/s41598-020-58322-x.

Banks, C.J., Heaven, S., Zhang, Y., and Baier, U. (2018). Food waste digestion: anaerobic digestion of food waste for a circular economy. In: *IEA Bioenergy Task*, vol. *37* (ed. J.D. Murphy), 12. 978-1-910154-58-8 (eBook electronic edition), 978-1-910154-57-1 (printed paper edition).

Baruah, J., Nath, B.K., Sharma, R. et al. (2018). Recent trends in the pretreatment of lignocellulosic biomass for value-added products. *Front. Energy Res.* 6: 141. https://doi.org/10.3389/fenrg.2018.00141.

Beauchemin, K.A., Ungerfeld, E.M., Eckard, R.J. et al. (2020). Review: fifty years of research on rumen methanogenesis: lessons learned and future challenges for mitigation. *Animal* 14, S1, s2–s16 https://doi.org/10.1017/S1751731119003100.

Besi, M.D. and McCormick, K. (2015). Towards a bioeconomy in Europe: national, regional and industrial strategies. *Sustainability (Switzerland)* 7 (8): 10461–10478. https://doi.org/10.3390/su70810461.

Bhatia, S.K. and Yang, Y.H. (2017). Microbial production of volatile fatty acids: current status and future perspectives. *Rev. Environ. Sci. Biotechnol.* 16: 327–345. https://doi.org/10.1007/s11157-017-9431-4.

Bhatia, R., Winters, A., Bryant, D.N. et al. (2020). Pilot-scale production of xylo-oligosaccharides and fermentable sugars from Miscanthus using steam explosion pretreatment. *Bioresour. Technol.* 296: 122285. https://doi.org/10.1016/j.biortech.2019.122285.

BIO-TIC (2015). The Bioeconomy Enabled: A roadmap to a thriving industrial biotechnology sector in Europe. [Online]. Available at: www.industrialbiotech-europe.eu. Accessed 1 September 2015. https://cordis.europa.eu/project/id/312121/reporting

Bonk, F., Popp, D., Weinrich, S.L. et al. (2018). Ammonia inhibition of anaerobic volatile fatty acid degrading microbial communities. *Front. Microbiol.* 9 (November): 2921. https://doi.org/10.3389/fmicb.2018.02921.

Bryant, D.N., Morris, S.M., Leemans, D. et al. (2011). Modelling real-time simultaneous saccharification and fermentation of lignocellulosic biomass and organic acid accumulation using dielectric spectroscopy. *Bioresour. Technol.* 102 (20): 9675–9682. https://doi.org/10.1016/j.biortech.2011.07.084.

Charlton, A., Elias, R., Fish, S. et al. (2009). The biorefining opportunities in Wales: understanding the scope for building a sustainable, biorenewable economy using plant biomass. *Chem. Eng. Res. Des.* 87: 1147–1161.

Chen, B., Wan, J., Shu, L. et al. (2017). Smart factory of Industry 4.0: key technologies, application case, and challenges. *IEEE Access* 6: 6505–6519. https://doi.org/10.1109/ACCESS.2017.2783682.

Committee on Climate Change. (2018). Biomass in a low-carbon economy, November. www.theccc.org.uk.

De Buck, V., Polanska, M., and Van Impe, J. (2020). Modeling biowaste biorefineries: a review. *Front. Sustain. Food Syst.* 4: 11. http://doi.org/10.3389/fsufs.2020.0001.

Demirel, B., Ergun, S., Neumann, L. et al. (2009). Performance and behaviour of the microbial community of an anaerobic biogas digester using sugar beet silage as mono-substrate. *Biosyst. Eng.* 102 (4): 444–452. https://doi.org/10.1016/j.biosystemseng.2009.01.008.

El-Gendy, N.S., Madian, H.R., and Amr, S.S.A. (2013, 2013). Design and optimization of a process for sugarcane molasses fermentation by *Saccharomyces cerevisiae* using response surface methodology. *Int. J. Microbiol.* 815631, 1–9. doi: 10.1155/2013/815631.

European Commission (2011a). Bio-Based Economy for Europe: State of Play and Future Potential—Part 1; Report on the European Commission's Public On-Line Consultation;

European Commission: Brussels, Belgium. Available online: http://ec.europa.eu/research/consultations/bioeconomy/bio-based-economy-for-europe-part1.pdf

European Commission (2011b). Bio-Based Economy in Europe: State of Play and Future Potential—Part 2; Summary of the Position Papers Received in Response of the European Commission's Public On-Line Consultation; European Commission: Brussels, Belgium. Available online: http://ec.europa.eu/research/consultations/bioeconomy/bio-based-economy-for-europe-part2.pdf

European Commission (2012). Innovating for sustainable growth: a bioeconomy for Europe. In Directorate-General for Research and Innovation DE-B, Agriculture and Food. Publications Office of the European Union, Brussels, Belgium.

Ferry, J.G. (2020). *Methanosarcina acetivorans*: a model for mechanistic understanding of aceticlastic and reverse methanogenesis. *Front. Microbiol.* 11: 1806. https://doi.org/10.3389/fmicb.2020.0180.

Gagliano, M.C., Ismail, S.B., Stams, A.J.M. et al. (2017). Biofilm formation and granule properties in anaerobic digestion at high salinity. *Water Res.* 121: 61–71. https://doi.org/10.1016/j.watres.2017.05.016.

Gagliano, M.C., Sudmalis, D., Pei, R. et al. (2020). Microbial community drivers in anaerobic granulation at high salinity. *Front. Microbiol.* 11: –235. https://doi.org/10.3389/fmicb.2020.00235.

Glynn, J.M. (2017). Pfizer CentreOne. https://w.J.M.ww.lifescienceleader.com/doc/fermentation-optimization-using-comparative-statistics-to-enhance-large-scale-process-productivity-0001.

Goldrick, S., Mercer, E., Montague, G. et al. (2014). Control of an industrial scale bioreactor using a PAT analyser. In IFAC Proceedings Volumes (IFAC-PapersOnline), pp. 6222–6227. doi: https://doi.org/10.3182/20140824-6-za-1003.02589.

Goldrick, S., Ştefan, A., Lovett, D. et al. (2015). The development of an industrial-scale fed-batch fermentation simulation. *J. Biotechnol.* 193: 70–82. https://doi.org/10.1016/j.jbiotec.2014.10.029.

Goldrick, S., Duran-Villalobos, C.A., Jankauskas, K. et al. (2019). Modern day monitoring and control challenges outlined on an industrial-scale benchmark fermentation process. *Computers Chem. Eng.* 130: 106471, 1–15. https://doi.org/10.1016/j.compchemeng.2019.05.037.

Gregory, M.E. Thornhill, N.F. (1992). Use of a graphical programming environment for the rapid development of custom software for fermentation process control. In IFAC Symposia Series, 25, 2, 231–234. doi: https://doi.org/10.1016/s1474-6670(17)50359-6.

Harmsen, P., Huijgen, W., Bermudez, L., Bakker, R. (2010). Literature review of physical and chemical pretreatment processes for lignocellulosic. Report number ECN-E-10-013. Energy Research Centre of The Netherlands, Petten, The Netherlands.

Hendriks, A.T.W.M. and Zeeman, G. (2008). Pretreatments to enhance the digestibility of lignocellulosic biomass. *Bioresour. Technol.* 100 (1): 10–18.

Hobbs, P.J., Ward, A., Ravella, S.R. et al. (2008). Optimisation of biogas production from a microbial perspective. In Koutev, V. (ed.), Proceedings of the 13th RAMIRAN International

Conference Potential for Simple Technology Solutions in Organic Manure Management, 11–14 June, Albena, Bulgaria. ISBN 978-954-9067671-6-3.

Hobbs, P.J., Rao, R.S., Ward, A.J. et al. (2010). Benchmarking commercial biogas plants to improve performance. In Marques dos Santos Cordovil, C.S.C., Ferreira, L. (eds), RAMIRAN 2010: Proceedings of the 14th International Conference of the FAO ESCORENA Network on the Recycling of Agricultural, Municipal and Industrial Residues in Agriculture, Lisbon, Portugal, 13–15 September. ISBN972866947X, 9789728669478.

Hodgson, E.M., Nowakowski, D.J., Shield, I. et al. (2011). Variation in Miscanthus chemical composition and implications for conversion by pyrolysis and thermo-chemical bio-refining for fuels and chemicals. *Bioresour. Technol.* 102, 3, 3411–3418.

Hodgson, E., Lewys-James, A., Rao Ravella, S. et al. (2016a). Optimisation of slow-pyrolysis process conditions to maximise char yield and heavy metal adsorption of biochar produced from different feedstocks. *Bioresour. Technol.* 214: 574–581. https://doi.org/10.1016/j.biortech.2016.05.009.

Hodgson, E., Ruiz-Molina, M., Marazza, D. et al. (2016b). Horizon scanning the European bio-based economy: a novel approach to the identification of barriers and key policy interventions from stakeholders in multiple sectors and regions. *Biofuels Bioprod. Biorefin.* 10 (5): 508–522. https://doi.org/10.1002/bbb.1665.

Horiuchi, J. (2002). Fuzzy modelling and control of biological processes. *J. Biosci. Bioeng.* 94: 574–578.

Huang, H., Gu, T., and Moo-Young, M. (2005). Data acquisition and control of a 22 L B. Braun fermenter using LabVIEW. *Chem. Eng. Commun.* 192 (1–3): https://doi.org/10.1080/00986440590473290.

Huang, S.R., Chen, H.T., Chung, C.H. et al. (2012). Fermentative hydrogen production using a real-time fuzzy controller. *Int. J. Hydrogen Energy* 37 (20): 15581–15575. https://doi.org/10.1016/j.ijhydene.2012.02.078.

IB 2025 (2009). Maximising UK Opportunities from Industrial Biotechnology in a Low Carbon Economy, A report to government by the Industrial Biotechnology Innovation and Growth Team. London, Department for Business, Enterprise and Regulatory Reform, May.

IEA BIOENERGY Task42 BIOREFINING (2014). Sustainable and synergetic processing of biomass into marketable food and feed ingredients, chemicals, materials and energy (fuels, power, heat), August. Paris, International Energy Agency. www.ieabioenergy.com; www.IEA-Bioenergy.Task42-Biorefineries.com.

Jackson, B.E., Bhupathiraju, V.K., Tanner, R.S. et al. (1999). *Syntrophus aciditrophicus* sp. nov., a new anaerobic bacterium that degrades fatty acids and benzoate in syntrophic association with hydrogen-using microorganisms. *Arch. Microbiol.* 171 (2): 107–114. doi: 10.1007/s002030050685.

Jünemann, S., Kleinbölting, N., Jaenicke, S. et al. (2017). Bioinformatics for NGS-based metagenomics and the application to biogas research. *J. Biotechnol.* 261: 10–23.

Kaya, İ., Erdoğan, M., Karaşan, A. et al. (2020). Creating a road map for Industry 4.0 by using an integrated fuzzy multicriteria decision-making methodology. *Soft Comput.* 24 (23): 17931–17956. https://doi.org/10.1007/s00500-020-05041-0.

Keegan, K.P., Glass, E.M., and Meyer, F. (2016). MG-RAST, a metagenomics service for analysis of microbial community structure and function. In: *Microbial Environmental Genomics (MEG). Methods in Molecular Biology*, vol. 1399 (ed. F. Martin and S. Uroz), 207–233. New York, NY: Humana Press https://doi.org/10.1007/978-1-4939-3369-3_13.

Kuprijanov, A., Schaepe, S., Aehle, M. et al. (2012). Toolbox for automatic control of fermentation processes for the production of recombinant therapeutic protein. 7th International Conference on Electrical and Control Technologies, 3–4 May, Kaunas, Lithuania.

Lim, J.W., Park, T., Tong, Y.W., and Yub, Z. (2020). The microbiome driving anaerobic digestion and microbial analysis. *Adv. Bioenergy* 5: 1–61. https://doi.org/10.1016/bs.aibe.2020.04.001.

Lu, Y., Slater, F., Bello-Mendoza, R. et al. (2013). Shearing of biofilms enables selective layer based microbial sampling and analysis. *Biotechnol. Bioeng.* 110 (10): 2600–2605. https://doi.org/10.1002/bit.24947.

Manzoor, S. et al. (2016). Complete genome sequence of *Methanoculleus bourgensis* strain MAB1, the syntrophic partner of mesophilic acetate-oxidising bacteria (SAOB). *Stand Genomic Sci.* 11 (80): 1–9. https://doi.org/10.1186/s40793-016-0199-x.

Markard, J., Wirth, S., and Truffer, B. (2016). Institutional dynamics and technology legitimacy— a framework and a case study on biogas technology. *Res. Policy* 45 (1): 330–344. https://doi.org/10.1016/j.respol.2015.10.009.

Martínez-Pérez, N., Cherryman, S.J., Premier, G.C. et al. (2007). The potential for hydrogen-enriched biogas production from crops: scenarios in the UK. *Biomass Bioenergy* 31 (2–3): 95–104. https://doi.org/10.1016/j.biombioe.2006.07.003.

Meyer, R. (2017). Bioeconomy strategies: contexts, visions, guiding implementation principles and resulting debates. *Sustainability (Switzerland)* 9 (1031): 1–32. https://doi.org/10.3390/su9061031.

Meyer, A.K.P., Ehimen, E.A., and Holm-Nielsen, J.B. (2018a). Future European biogas: animal manure, straw and grass potentials for a sustainable European biogas production. *Biomass Bioenergy* 111: 154–164. doi: 10.1016/j.biombioe.2017.05.013.

Meyer, F., Bagchi, S., Chaterji, S. et al. (2018b). MG-RAST version 4—Lessons learned from a decade of low-budget ultra-high-throughput metagenome analysis. *Brief. Bioinform.* 20 (4): 1151–1159. https://doi.org/10.1093/bib/bbx105.

Mkondweni, N., Tzoneva, R., and Harisson, S. (2005). Integrating LabVIEW capabilities for monitoring and supervisory control with a B-Braun biotech Gmbh unit for fermentation processing and direct control. *Int. J. Eng. Educ.* 21 (1 Part 1): 63–74.

Moll, L., Wever, C., Völkering, G., and Pud, R. (2020). Increase of Miscanthus cultivation with new roles in materials production—a review agronomy. 10: 308, 1–321. https://doi.org/10.3390/agronomy10020308.

Morais, A.R.C. and Bogel-Lukasik, R. (2013). Green chemistry and the biorefinery concept. *Sustain. Chem. Process.* 1: 18, 1–3. https://doi.org/10.1186/2043-7129-1-18.

Mu, Y., Yu, H.Q., and Wang, Y. (2006). The role of pH in the fermentative H2 production from an acidogenic granule-based reactor. *Chemosphere* 64 (3): 350–358. https://doi.org/10.1016/j.chemosphere.2005.12.048.

National Research Council (2015). Industrialization of Biology: A Roadmap to Accelerate the Advanced Manufacturing of Chemicals. Washington, DC: National Academies Press. https://doi.org/10.17226/19001.

Nattrass, L., Biggs, C., Bauen, A. et al. (2016). The EU Bio-Based Industry: Results from a Survey; Joint Research Centre (JRC) Technical Report. Brussels, Belgium, Joint Research Centre, Volume EUR 27736 EN.

O'Donohue, M. (2014). BIOCORE Final publishable summary report. Toulouse. Retrieved from http://www.biocore-europe.org/file/BIOCOREfinal report (1).pdf.

OECD (2009). The Bioeconomy to 2030: Designing a Policy Agenda. Main findings and conclusions. in Development OfEC-oa. OECD, Paris.

OECD (2011). *Future Prospects for Industrial Biotechnology*. Paris: OECD Publishing https://doi.org/10.1787/9789264126633-en.

Pérez-Camacho, M.N. and Curry, R. (2018). Regional assessment of bioeconomy options using the anaerobic biorefinery concept. *Proc. Inst. Civil Eng. Waste Resource Manage.* 171 (4): 104–113. https://doi.org/10.1680/jwarm.17.00015.

Pitcairn, J., Warmington, J., Gandy, S. et al. (2017). Biorefining potential for Scotland: mapping bioresource arisings across Scotland. *Ind. Biotechnol.* 13 (6): 301–305. doi: 10.1089/ind.2017.29111.jpi.

Plugge, C.M., Henstra, A.M., Worm, P. et al. (2012). Complete genome sequence of *Syntrophobacter fumaroxidans* strain (MPOBT). *Stand Genomic Sci.* 7 (1): 91–106. https://doi.org/10.4056/sigs.2996379.

Prochnow, A., Heiermann, M., Plöchl, M. et al. (2009). Bioenergy from permanent grassland—a review: 1. *Biogas. Bioresour. Technol.* 100: 4931–4944. https://doi.org/10.1016/j.biortech.2009.05.070.

Purdy, S.J., Maddison, A.L., Nunn, C.P. et al. (2017). Could Miscanthus replace maize as the preferred substrate for anaerobic digestion in the United Kingdom? Future breeding strategies. *GCB Bioenergy* 9 (6): 1122–1139. https://doi.org/10.1111/gcbb.12419.

Rao, R.S., Kumar, C.G., Prakasham, R.S., and Hobbs, P.J. (2008). The Taguchi methodology as a statistical tool for biotechnological applications: a critical appraisal. *Biotechnol. J.* 3: 510–523. https://doi.org/10.1002/biot.200700201.

Ravella, S.R., James, S.A., Bond, C.J. et al. (2010a). *Cryptococcus shivajii sp. nov.*: a novel basidiomycetous yeast isolated from biogas reactor. *Curr. Microbiol.* 60 (1): 12–16. https://doi.org/10.1007/s00284-009-9493-9.

Ravella, S.R., Quiñones, T.S., Retter, A. et al. (2010b). Extracellular polysaccharide (EPS) production by a novel strain of yeast-like fungus *Aureobasidium pullulans*. *Carbohydr. Polym.* 82 (3): 728–732. https://doi.org/10.1016/j.carbpol.2010.05.039.

Ravella, R., Hobbs, P., Rogers, J. et al. (2011). Evolution of organisms associated with methanogenic communities during the initiation of bioenergy fermentation. In 1st International Conference on Biogas Microbiology, Helmholtz-Zentrum für Umweltforschung—UFZ, Deutsches BiomasseForschungsZentrum (DBFZ), Leipzig, 14– 16 September, Germany.

Ravella, S.R., Gallagher, J., Fish, S., and Prakasham, R.S. (2012). Overview on commercial production of xylitol, economic analysis and market trends. In: *D-xylitol* (ed. S.S. da Silva and A.K. Chandel), 291–306. Berlin, Heidelberg: Springer-Verlag.

Ravella, S.R., Bryant, D., Clifton-Brown, J. Gallagher, J. (2019). Pre-treatment of Miscanthus using steam explosion for biogas production. In Conference Proceedings—6th International Conference on Renewable Energy Gas Technology (REGATEC), Malmö, Sweden

Rettenmaier, N., Harter, R., Himmler, H. et al. (2014). Integrated sustainability assessment of the BIOCORE biorefinery concept, IFEU—Institute for Energy and Environmental Research Heidelberg, Heidelberg, Germany. http://www.biocore-europe.org/file/BIOCORE_D7_5_Environmental%20assessment_2014-05-15.pdf.

Rocha, I., Ferreira, E.C. (2001). An integrated system for advanced monitoring and control of fed-batch fermentations of recombinant *E. coli*. IFAC Computer Applications in Biotechnology, IFAC Proceedings Volumes (IFAC-PapersOnline), vol. 34 (5). Québec City, Canada. https://doi.org/10.1016/S1474-6670(17)34244-1.

Rodriguez, C., Alaswad, A., Benyounis, K.Y. et al. (2017). Pre-treatment techniques used in biogas production from grass. *Renew. Sust. Energy Rev.* 68 (2): 1193–1204. http://doi.org/10.1016/j.rser.2016.02.022.

Rødsrud, G., Lersch, M., and Sjöde, A. (2012). History and future of world's most advanced biorefinery in operation. *Biomass Bioenergy* 46: 46–59. https://doi.org/10.1016/j.biombioe.2012.03.028.

Scarlat, N., Dallemand, J.F., Monforti-Ferrario, F. et al. (2015). The role of biomass and bioenergy in a future bioeconomy: policies and facts. *Environ. Dev.* 15: 3–34. https://doi.org/10.1016/j.envdev.2015.03.006.

Schlüter, A., Bekel, T., Diaz, N.N. et al. (2008). The metagenome of a biogas-producing microbial community of a production-scale biogas plant fermenter analysed by the 454-pyrosequencing technology. *J. Biotechnol.* 136 (1–2): 77–90. https://doi.org/10.1016/j.jbiotec.2008.05.008.

Schröder, T., Lauven, L.-P., Sowlati, T. et al. (2019). Strategic planning of a multi-product wood-biorefinery production system. *J. Clean. Prod.* 211: 1502–1516. https://doi.org/10.1016/j.jclepro.2018.12.004.

Shafiei, M., Kabir, M.M., Zilouei, H. et al. (2013). Techno-economical study of biogas production improved by steam explosion pretreatment. *Bioresour. Technol.* 148: 53–60. https://doi.org/10.1016/j.biortech.2013.08.111.

Sierocinski, P., Bayer, F., Yvon-Durocher, G. et al. (2018). Biodiversity–function relationships in methanogenic communities. *Mol. Ecol.* 27: 4641–4651.

Sierra, A.R., Zika, E., Lange, L. et al. (2021). The bio-based industries joint undertaking: a high impact initiative that is transforming the bio-based industries in Europe. *N. Biotechnol.*, 60 105–112. https://doi.org/10.1016/j.nbt.2020.09.003.

Simon, J., Trojanova, M., Zbihlej, J. et al. (2018). Mass customization model in food industry using Industry 4.0 standard with fuzzy-based multi-criteria decision making methodology. *Adv. Mech. Eng.* 10 (3): 1–10. https://doi.org/10.1177/1687814018766776.

Star-COLIBRI (2011). Joint European Biorefinery Vision for 2030. The Seventh Framework Programme of the European Union. EC, Brussels, Belgium.

Theuerl, S., Herrmann, C., Heiermann, M. et al. (2019a). The future agricultural biogas plant in Germany: a vision. *Energies* 12: 396, 1–32.

Theuerl, S., Klang, J., and Prochnow, A. (2019b). Process disturbances in agricultural biogas production—causes, mechanisms and effects on the biogas microbiome: a review. *Energies* 12: 365. 1–20.

UK Bioeconomy Strategy (2018). https://www.gov.uk/government/publications/bioeconomy-strategy-2018-to-2030.

USDA (2011). US Billion-Ton Update: Biomass Supply for a Bioenergy and Bioproducts Industry. R.D. Perlack and B.J. Stokes (Leads), ORNL/TM-2011/224. Oak Ridge National Laboratory, Oak Ridge, TN. 227 pp.

Vasickaninova, A., Bakosova, M., Meszaros, A. (2017). Control of a biochemical process using fuzzy approach. In Proceedings of the 2017 21st International Conference on Process Control, PC 2017. doi: https://doi.org/10.1109/PC.2017.7976209.

Wainaina, S., Lukitawesa, K., Awasthi, M. et al. (2019). Bioengineering of anaerobic digestion for volatile fatty acids, hydrogen or methane production: a critical review. *Bioengineered* 10: 437–458.

Walker, D.J., Gallagher, J., Winters, A. et al. (2018). Process optimisation of steam explosion parameters on multiple lignocellulosic biomass using Taguchi method: a critical appraisal. *Front. Energy Res.* 6: 46, 1–13. https://doi.org/10.3389/fenrg.2018.00046.

Wang, B., Wang, Z., Chen, T., and Zhao, X. (2020). Development of novel bioreactor control systems based on smart sensors and actuators. *Front. Bioeng. Biotechnol.* 8: 7, 1–15. https://doi.org/10.3389/fbioe.2020.00007.

Ward, A.J. (2016). Near-infrared spectroscopy for determination of the biochemical methane potential: state of the art. *Chem. Eng. Technol.* 39: 611–619.

Ward, A.J., Hobbs, P.J., Holliman, P.J., and Jones, D.L. (2008a). Optimisation of the anaerobic digestion of agricultural resources. *Bioresour. Technol.* 99: 7928–7940.

Ward, A., Hobbs, P.J., Holliman, P. et al. (2008b). Software sensor monitoring and expert control of biogas production. Proceedings of the 13th RAMIRAN International Conference, 11–14 June, Albena, Bulgaria.

Ward, A.J., Hobbs, P.J., Holliman, P.J., and Jones, D.L. (2011a). Evaluation of near infrared spectroscopy and software sensor methods for determination of total alkalinity in anaerobic digesters. *Bioresour. Technol.* 102: 4083–4090.

Ward, A.J., Bruni, E., Lykkegaard, M.K. et al. (2011b). Real time monitoring of a biogas digester with gas chromatography, near-infrared spectroscopy, and membrane-inlet mass spectrometry. *Bioresour. Technol.* 102: 4098–4103.

Williams, L., Gallagher, J., Bryant, D., and Ravella, S. (2017). Anaerobic digestion and the use of pre-treatments on lignocellulosic feedstocks to improve biogas production and process economics. In: *Advances in Biofeedstocks and Biofuels: Biofeedstocks and their Processing*, vol. 1 (ed. L.K. Singh and G. Chaudhary), 121–147. Beverly, MA: Wiley.

Winkler, M.K.H., Kleerebezem, R., Kuenen, J.G. et al. (2011). Segregation of biomass in cyclic anaerobic/aerobic granular sludge allows the enrichment of anaerobic ammonium oxidizing bacteria at low temperatures. *Environ. Sci. Tech.* 45 (17): 7330–7337. https://doi.org/10.1021/es201388t.

Wolf, M.J., Ninov, V., Babel, H. et al. (2012). A laboratory experiment for teaching bioprocess control part 1: Hardware setting. In IFAC Proceedings Volumes (IFAC-PapersOnline). doi: https://doi.org/10.3182/20120619-3-RU-2024.00033.

Zaimes, G.G., Vora, N., Chopra, S.S. et al. (2015). Design of sustainable biofuel processes and supply chains: challenges and opportunities. *Processes* 3: 634–663. https://doi.org/10.3390/pr3030634.

Zetterholm, J., Bryngemark, E., Ahlström, J. et al. (2020). Economic evaluation of large-scale biorefinery deployment: a framework integrating dynamic biomass market and techno-economic models. *Sustainability (Switzerland)* 12 (17): 7126, 1–27. https://doi.org/10.3390/su12177126.

Zhang, L., Loh, K.C., Lim, J.W., and Zhang, J.X. (2019). Bioinformatics analysis of metagenomics data of biogas-producing microbial communities in anaerobic digesters: a review. *Renew. Sust. Energy Rev.* 100: 110–126. https://doi.org/10.1016/j.rser.2018.10.021.

Index

Lignocellulose Bioconversion Through White Biotechnology, First Edition. Edited by Anuj Kumar Chandel.

Printed and bound by CPI Group (UK) Ltd, Croydon, CR0 4YY
26/08/2022
03144062-0001